Fluid Mech
Turbomachines

Madan Mohan Das

Formerly Professor
Civil Engineering Department
Assam Engineering College
and
Emeritus Fellow of AICTE
Director of Technical Education
Government of Assam

PHI Learning Private Limited
Delhi-110092
2014

₹ 450.00

FLUID MECHANICS AND TURBOMACHNIES
Madan Mohan Das

ISBN-978-81-203-3523-3

Sixth Printing ... ... September, 2014

Published by Asoke K. Ghosh, PHI Learning Private Limited, Rimjhim House, 111, Patparganj Industrial Estate, Delhi-110092 and Printed by Rajkamal Electric Press, Plot No. 2, Phase IV, HSIDC, Kundli-131028, Sonepat, Haryana.

To

My Parents
Madhab Das and Jayanti Das

Contents

Preface

This book on Fluid Mechanics and Turbomachines is written as a textbook for the beginners in this field and more specifically for 3rd or 4th semester students of Civil, Mechanical, Chemical, Electrical and Aeronautical Engineering. It is written with an aim to provide fundamental knowledge to students on buoyancy, floatation, kinematics of fluid flow, flow through pipes, orifices, mouthpieces, weir and open channel. Details of laminar flow and its application, turbulent flow, boundary layer concept, dimensional analysis and model investigations, compressible flow, flow around submerged objects are presented in different chapters. Second part of the book starts with impact of jets to present a clear picture of turbomachines like turbines and different pumps.

Attempts are made to present the chapters in a simple and easy-to-follow language. Solved examples are also presented to reinforce the understanding of the theory even by an average student. Deduction of formulae, worked out examples and problems (with answers) to be solved by students have been presented in S.I. units.

The author has been teaching hydraulic, fluid mechanics, advanced hydraulic engineering, hydrology, flow through porous media and water power engineering for the last 37 years in both B.E. and M.E. levels. Along with teaching the above subjects, dissertation works of M.E. students and Ph.D. theses works on the above fields have been continued till today in the department of civil engineering in Assam Engg. College under Guwahati University.

I have been motivated to write such a textbook from my students and family members. My appreciation and gratefulness have been conveyed to Professor Emeritus, D.I.H. Barr (D.Sc.), Retired Professor of University of Strathelyde, Glasgow, who has been encouraging me in writing books, publising papers, guiding Ph.D. Scholars since I met him in 1975 as my Ph.D. guide.

MADAN MOHAN DAS

Chapter 1

Fluid Properties

1.1 INTRODUCTION

This chapter consists of the definition of fluid and fluid mechanics, historical background of the development of the subject, different properties such as density, specific weight, specific gravity, viscosity, surface tension, compressibility, etc. and their units of measurement in SI units. A few numerical examples are solved and some problems are given for readers to solve. A list of references is also provided at the end of the chapter.

1.2 FLUID MECHANICS AND ITS HISTORICAL BACKGROUND

Fluid is a substance which is capable of flowing. It has no definite shape of its own but conforms to the shape of the container. Fluid continues to deform when subjected to shear force. When fluid is at rest, no shearing forces can act on it and thus all forces are normal to the planes on which they act.

A liquid is a fluid which varies slightly with temperature and pressure. Under ordinary conditions, liquids are difficult to compress and therefore, they are treated as being mostly incompressible except in some situations such as water hammer, etc. It forms a free surface if exposed to atmosphere.

A gas is also a fluid which is compressible and possesses no definite volume but goes on expanding until its volume is equal to that of the container.

A vapour is a gas whose temperature and pressure are such that it is very close to the liquid state. Steam may be considered as vapour because its state is not far from that of water.

Fluid mechanics is the study of these fluids at rest or in motion. This study deals with the static, kinematic and dynamic aspects of fluid. The study also takes into account the conservation of mass, momentum, Newton's law of motions, and laws of thermodynamics. Thus, the study of fluid mechanics is inter-disciplinary. It is studied by engineers, scientists, physicists, chemists, mathematicians, geologists, physiologists, meteorologists, and geophysicists because the flow of fluid covers almost all the fields of science and technology.

1.2.1 Historical Background

The application of fluid mechanics began with the motion of stones, arrows and spears. Irrigation aqueducts have been found in prehistoric ruins in the Indus and the Nile rivers, the Tigris-

Euphrates river valley, and both in Egypt and in Mesopotamia. The Archimedes principles in the third century B.C. formulated the well-known laws of floating bodies. Aristotle in the fourth century B.C. studied the motion of bodies in thin media and voids. Roman aqueducts were built in the fourth century B.C. Leonardo Da Vinci (1452–1519) correctly described many flow phenomena. He designed the first canal lock. However, up to the time of Da Vinci, the concept of fluid motion had been considered to be more of art than science. Galileo (1564–1642) also contributed a lot to the science of mechanics.

The ideas concerning the steady flow continuity equation in rivers, flow from a container, the barometer and some qualitative concepts of flow resistance in rivers came from the Italian School of Hydraulics which included Castelli (1577–1644), Torricelli (1608–1647) and Ganglielmini (1655–1710). Newton (1642–1727), in addition to his well-known laws of motions, proposed that fluid resistance is proportional to the velocity gradient which is known as Newton's law of viscosity. He also made experiments on the drag force on spheres.

Eighteenth century mathematicians such as Daniel Bernoulli, Leonard Euler (Swiss) and Clairaut and d' Alembart (French) developed the mathematical science of fluid mechanics, i.e. Hydrodynamic. The surface waves concept was proposed by *Lagrange* (1736–1818), Laplace (1749–1827) and Gestner (1756–1832).

The eighteenth century experimentalists included Poleni for weir flow, de Pitot for measuring the velocity of water in river by a tube (now known as Pilot tube), and Antonie Chezy who developed the equation of average velocity in 'open' channel flow known till today as Chezy's formula. Borda worked extensively on orifices, Du Buot who was the pioneer of the French school Hydraulics and Venturi who devised a metre of changing cross-section to measure discharge known as Venturimeter.

In the nineteenth century, Frenchman Coulomb (1736–1806), and Prony (1755–1839) concluded tests and arrived at conclusions relating to resistance to flow. The development and analysis of hydraulic turbines were undertaken by French engineer Burdin (1790–1873), Fourneyman (1802–1867), Coriolis (1792–1843) and American engineer Francis (1815–1892), Englishman Smith (1808–1874) and Swede Ericsson (1803–1889). Extensive works on pipe flow were done by the German Hagen (1797–1889), Frenchman Poiseuille (1799–1869), Weisbach (1806–1871).

Frenchman Saint-Venant (1797–1886) analysed sonic orifice and contributed a lot on unsteady gradually varied flow equations which are still today the basic governing equations of open channel hydraulics. Frenchman Dupuit (1804–1866), Bresse (1822–1883), and Bazin (1829–1917) and Irishman Manning (1816–1897) contributed immensely to hydraulics of open channel. The Frenchman Darcy (1803–1858) did a lot work on pipe friction and his equation on pipe friction is still used in pipe flow hydraulics. Englishman William Froude (1810–1879) and his son Robert Froude (1846–1924) carried out comprehensible testing on ship models. The Froude number is an important parameter in open channel flow.

Classical and applied hydrodynamics were advanced quite a bit by Navier (1785–1836), Cauchy (1842–1929), Poisson (1781–1840), and Saint–Venant and Bussinesq (1842–1922) in France, and in England by Airy (1801–1892), Stokes (1819–1903), Reynolds (1842–1912), Lord Kelvin (1824–1907), Lord Rayleigh (1842–1894), and Lamb (1849–1934), Helmholt (1821–1894) and Kirchhoff (1824–1887) in Germany and Joukowsky (1847–1921) in Russia.

Towards the end of the nineteenth century, theoretical hydrodynamics based on Euler's equation of motion for non-viscous (ideal) fluid had been developed to a high level. But by the end of this century, not enough data was available for comparison in the two fields of hydrodynamics and hydraulics. In 1904, it was Prandtl (1875–1953) in Germany who induced the concept of a boundary layer, a thin layer very close to the boundary in which viscous effects are concentrated. This proved to be the concept which unified all aspects of modern fluid mechanics, aerodynamics, hydraulics, gas dynamics and convective heat transfer. Prandtl is rightly considered to be the father of modern fluid mechanics. Details of history of hydraulics, fluid mechanics have been given by Rouse and Simon[1].

1.3 FLUID PROPERTIES

1.3.1 Density or Mass Density (ρ)

It is a measure of concentration of mass, i.e. mass per unit volume is called density.

$$\text{density} = \rho = \frac{\text{Mass of fluid}}{\text{Volume of fluid}}.$$

The density of liquid may be considered constant as liquids are mostly incompressible while the density of gases changes with pressure and temperature. In SI units, mass density is expressed in kg/m^3. For water, ρ is 1000 kg/m^3.

1.3.2 Specific Weight (w)

It is a force of gravity on the mass contained in a unit volume, i.e. specific weight w is the weight per unit volume

$$w = \frac{\text{Weight of fluid}}{\text{Volume of fluid}} = \frac{\text{Mass} \times \text{Acceleration}}{V} = \frac{\text{Mass of fluid}}{\text{Volume of fluid}} \times g$$

$$\therefore \quad w = \rho g \tag{1.1}$$

The specific weight of water is (9.81 × 1000) newton/m^3 in SI units.
It is equal to 9810 N/m^3.

1.3.3 Specific Volume (v)

It is the volume per unit mass of fluid.

$$\therefore \quad \text{specific volume} = \frac{1}{\rho}$$

It is reciprocal to density and expressed in m^3/kg. It is most commonly used in gases.

1.3.4 Specific Gravity (s)

It is a term used to compare the density of a substance with that of water and hence

$$s = \frac{\text{Density of fluid}}{\text{Density of water}}$$

$\therefore$ density of fluid = specific gravity (s) × density of water (ρ_w)

1.3.5 Viscosity

Viscosity is that property of real fluids which distinguishes them from ideal or non-viscous fluids. It is a measure of its resistance to flow. The resistance to flow is measured as a total shear force, with the shear stress being the shear force per unit area. Newton postulated that shear stress within the fluid is proportional to the rate of change of velocity normal to the flow. This rate of change of velocity in called 'velocity gradient'.

From Figure 1.1, velocity gradient at depth y from the boundary is:

$$\frac{du}{dy} = \lim_{\Delta y \to 0}\left(\frac{\Delta u}{\Delta y}\right).$$

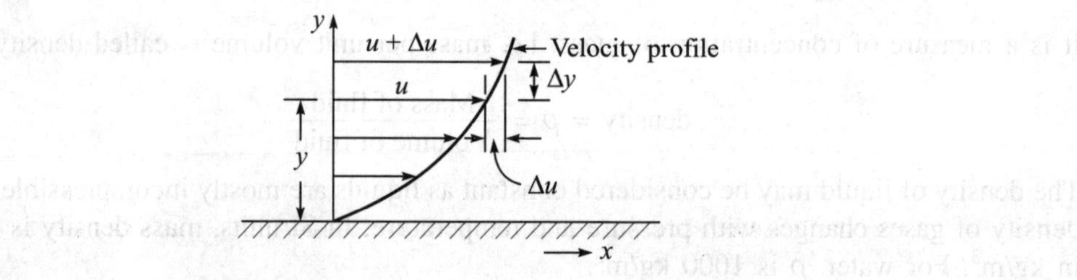

Figure 1.1 Velocity profile and velocity gradient.

Thus, Newton's law of viscosity may be written as:

$$\lambda \propto \frac{du}{dy}, \quad \text{where } \lambda \text{ is the shear stress.}$$

$$\therefore \qquad \lambda = \mu\frac{du}{dy} \tag{1.2}$$

where $\mu(mu)$ is the constant of proportionality and is called the dynamic viscosity.

$$\therefore \qquad \mu = \frac{\lambda}{\frac{du}{dy}}. \tag{1.3}$$

The unit of μ may be obtained from Equation (1.3) as

$$\mu = \frac{\text{Shear stress}}{\text{Shear strain}} = \frac{\text{Force/Area}}{\text{Length/Time} \times \frac{1}{\text{Length}}} = \frac{\text{Force} \times \text{Time}}{(\text{Length})^2}.$$

Thus unit of μ in SI units is N–s/m^2
MKS unit kg–sec/m^2
CGS unit dyne–sec/m^2 = poise

For two-dimensional (2–D) flow, $\lambda = \mu\left(\frac{du}{dy} + \frac{dv}{dx}\right).$ (1.4)

In turbulent flow,
$$\lambda = (\mu + \rho\epsilon)\frac{du}{dy} \tag{1.5}$$

where $\in$ is the eddy viscosity, which is not a fluid property and depends largely on flow. The value $\rho\in$ is 200 times larger than μ have been measured. Kinematic viscosity $\upsilon(nu)$ is defined as the ratio of μ and ρ, i.e.

$$\upsilon = \frac{\mu}{\rho}. \tag{1.6}$$

The unit of υ is $$\frac{N-s}{m^2 \times \left(\frac{\text{Mass}}{L^3}\right)} = \frac{(\text{Mass} \times \text{Acceleration}) \times \text{Time}}{(\text{Length})^2 \times \frac{\text{Mass}}{L^3}}$$

$$\upsilon = \frac{\text{Mass} \times \frac{L}{T^2} \times T}{L^2 \times \frac{\text{Mass}}{L^3}} = \frac{L^2}{T} \text{ i.e. Area/Time.}$$

In SI units, it is written as m^2/s.
In MKS units it is as $\text{metre}^2/\text{sec}$
CGS unit is as cm^2/sec = 1 stoke.

It is seen that the unit of kinematic viscosity is L^2/T and is independent of the force term, and, therefore, the name kinematic is given. Due to the simplicity of the unit, υ is commonly used in practice.

The viscosity of liquids decreases with an increase in temperature, while for gases, it increases with an increase in temperature.

1.3.6 Classification of Fluids

Common fluids such as water, air, kerosene, glycerine, etc. follow the Newton's law of viscosity, i.e. Equation (1.2). There are some other fluids which do not obey this law. Therefore, fluids are first classified as Newtonian and non-Newtonian fluids. Viscous behaviour of different fluids has been shown in Figure 1.2.

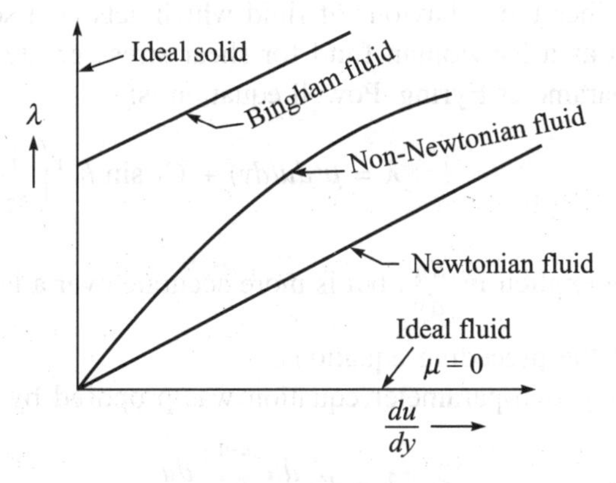

Figure 1.2 Viscous behaviour of fluids.

Although, these non-Newtonian fluids are not uncommon, their viscous behaviour is not yet completely understood. It was Wilkinson[2] who first classified fluids in the above two groups. The non-Newtonian group has the following three sub-groups:

(i) Fluids for which the λ depends only on the shear rate and independent of time of application of shear stress λ.
(ii) Fluids for which λ depends not only on the shear rate but also on time the fluid has been sheared.
(iii) Visco-elastic fluids which exhibit the characteristics of elastic solids and viscous fluids.

Metzner[3,4] classifies fluid into four general catagories. These are discussed below.

(i) ***Purely viscous fluids:*** These include both Newtonian and non-Newtonian fluids for which shear stress depends only on the shear rate and is time-independent. Water and air are Newtonian fluids. Gases and low-molecular weight liquids are almost always Newtonian. A number of equations have been used to describe the viscous behaviour of non-Newtonian fluids.

(a) The power law equation is:

$$\lambda = K\,(du/dy)^n \tag{1.7}$$

where K is the consistency index and n is the flow behaviour index.
For Newtonian fluid $K = \mu$ and $n = 1$.

(b) The Ellis equation is:

$$du/dy = \lambda_1/\mu_0 + \left(\frac{\lambda}{K}\right)^{\frac{1}{n}}. \tag{1.8}$$

This equation facilitates correction for the inaccuracy of the power law equation at low shear rates.

(c) The Bingham equation is:

$$\lambda = \lambda_1 + \mu_B(du/dy) \tag{1.9}$$

which describes the behaviour of fluid which acts as a solid for shear stress less than λ_1 and as a Newtonian fluid for shear stress greater than λ_1.

(d) The three-parameter Eyring–Powell equation is:

$$\lambda = \mu(du/dy) + C_1 \sin h^{-1}\left(\frac{1}{e_2}\frac{du}{dy}\right) \tag{1.10}$$

which is not explicit in $\frac{du}{dy}$, but is more accurate over a larger-range of shear rates than that of the preceding equations.

(e) The following two-parameter equation was proposed by Ostwald–de Wacle

$$\lambda = K\left|\frac{du}{dy}\right|^{n-1}\cdot\frac{du}{dy}. \tag{1.11}$$

when $n = 1$, it represents Newton Equation (1.2) when $K = \mu$.

(ii) ***Time-dependent fluids:*** One of these is a thixotropic fluid whose viscosity decreases with time under constant shear rate, while the other is a rheopectic fluid in which viscosity increases with time.

(iii) ***Visco-elastic fluids:*** Materials such as pitch, flour dough, molted polymers and a few solids exhibit the characterstics of both elastic solids and viscous fluids.

(iv) ***Most complex rheological systems:*** Some fluids exhibit the characterstics of all the three categories mentioned above.

Works on non-Newtonian fluids like thick gooey substances such as paste, printers, ink and slurries are investigated mostly by chemical engineers and by people other fields of science and engineering. Bugliarello, *et al*[5] have presented an informative discussion on civil engineering aspects of non-Newtonian fluids. Engineers in the field in many areas are increasingly being confronted with the flow of non-Newtonian fluids in both pipe as well as the design and selection of pumps for the fluids. Bugliarello[5,6] concludes that experiments have shown that the frictional effects for some non-Newtonian fluids are less than those for a Newtonian fluid under equivalent turbulent conditions resulting in a lower pressure drop in pipes and reduced drag on bodies submerged in liquids. The later situations of interest in naval hydrodynamics. Fredrickson[7] presented the principles and applications of rheology which is the study of non-Newtonian fluids.

1.3.7 Surface Tension (σ) and Capillarity

Liquids possess certain properties such as cohesion and adhesion due to molecular attraction. Cohesion refers to the intermolecular attraction between molecules of the same liquid. Adhesion implies the attraction between the molecules of a liquid and those of a solid boundary surface in contact with the liquid. Cohesion enables a liquid to resist tensile force whereas adhesion enables it to stick to another body. Surface tension is caused by cohesion between liquid particles at the surface and its effects occur at liquid–gas or liquid–liquid interfaces. On the other hand, capillarity, i.e. the rise or fall of a liquid surface in a small tube immersed in liquid is due to both cohesion and adhesion.

Surface tension is usually denoted by σ (sigma) and is measured in force/unit length, i.e. N/m. The value of the surface tension for water at room temperature is 0.0725 N/m while for mercury it is equal to 0.051 N/m.

Surface tension (σ) in a liquid droplet:

If the droplet is cut into two equal halves, the surface tension force acting on the circumforce $= \sigma \times$ length of circumforce $= \pi d \cdot \sigma$ where d is the diameter. If p is the intensity of pressure inside the droplet, the pressure force acting on the area $\left(\frac{\pi}{4} d^2\right)$ is equal to $p \cdot \frac{\pi}{4} d^2$. For equilibrium condition, these two forces are equal, i.e. $\sigma \, \pi \, d = \frac{\pi}{4} p d^2$.

$$\therefore \qquad \sigma = \frac{pd}{4}. \tag{1.12}$$

If tubes of a small diameter open at both ends, are dipped in water and mercury, the level of liquid in the tube will rise in the case of water and fall in the case of mercury as shown in

Figure 1.3. The rise of the liquid is called capillary rise and its fall is called capillary fall. The surface of the liquid in the small tube is called 'meniscus'. If the liquid wets the tube (i.e. due to adhesion), the meniscus in concave upward and the liquid level rises in the tube (i.e. in the case in water). If the tube liquid does not wet the tube due to cohesion, meniscus is convex upward and the liquid level falls in the tube and is seen in the case of mercury.

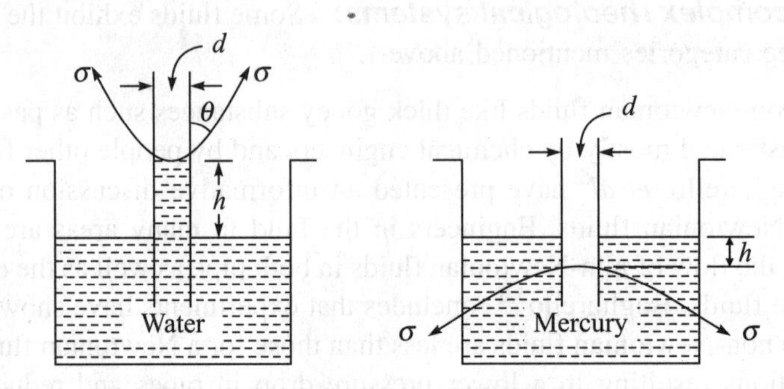

Figure 1.3 Capillary rise and fall.

Let h be the rise of level in the tube,

d be the diameter of tube,

σ be the surface tension, and

θ be the angle of contact between the liquid and the glass tube.

The weight of the column of water in a height h = vertical component of surface tension force, i.e.

$$w(\pi/4d^2)h = (\sigma \pi d)\cos\theta$$

$$h = \frac{4\sigma\cos\theta}{wd} \tag{1.13}$$

θ for water is equal to 25° 32′ and for mercury it is 128° 52′.

1.3.8 Vapour Pressure (p_v)

All liquids have the tendency to vaporise, i.e. to change from a liquid state to a gaseous state. When the liquid is enclosed and confined in a closed container, then ejected vapour molecules get accumulated in the space between the liquid surface and the top surface of the container. This accumulated vapour of liquid exerts a partial pressure on the liquid surface which is known as vapour pressure of the liquid.

The vapour pressure of liquids is significant in barometer, pump-piping system from an elementary point of view and in the formation of cavitations in zones where pressure falls below vapour pressure. Once the pressure at any point of flow system, e.g. pump-piping system becomes equal to or less than the vapour pressure, vaporisation of liquid takes place resulting in a pocket of dissolved gases and vapours. The bubbles of vapour thus formed are carried away by the flowing liquid into a zone of high pressure where they collapse, giving rise to high impact pressure. The pressure developed by collapsing bubbles is so high that the material in the adjoining boundaries gets eroded and cavities are formed. This phenomenon is called 'cavitation'. Therefore, in the practical field, care must always been taken to ensure that vapour pressure is not formed in any portion of the pipe-pumping system, penestock-turbine system, etc. Mercury has the least vapour pressure, hence it is an excellent fluid to use in barometer, thermometers and other pressure measuring devices.

1.3.9 Compressibility or Elasticity (*K*)

Fluids may be deformed by viscous shear and compressed by external forces applied to a volume of fluid. Liquids are less compressible and are usually assumed to be incompressible in most of the situations except in some situations such as in a field where the valve in a long pipeline is suddenly closed, thereby producing water hammer pressure. In such situations, water is assumed to be compressible. Other fluids such as gases can be compressed by external forces and gases are taken to be compressible.

Thus, the compressibility of fluid is defined in terms of the average bulk modulus of elasticity, i.e.

$$K = \frac{p_2 - p_1}{(\text{Volume}_2 - \text{Volume}_1)/\text{Volume}_1} \tag{1.14}$$

where Volume_2 and Volume_1 are volumes at pressure p_2 and p_1, respectively. Thus, Equating

(1.14) becomes $K = -\dfrac{dp}{\dfrac{d\forall}{\forall}}$. (1.15)

The negative sign indicates a situation wherein pressure increases and volume decreases.

EXAMPLE 1.1 One litre of crude oil weighs 9.6 N. Calculate its specific weight, density and specific gravity.

Solution: Volume = $(1/1000)\text{m}^3$, weight $W = 9.6$ N

$$\therefore \quad \text{Specific weight } w = \frac{W}{\text{Volume}} = \left(\frac{9.6}{1/1000}\right) = 9600 \text{ N/m}^3 \qquad \textbf{Ans.}$$

$$\therefore \quad \text{Density } \rho = \frac{w}{g} = \frac{9600}{9.81} = 978.593 \text{ kg/m}^3 \qquad \textbf{Ans.}$$

$$\therefore \quad \text{Specific gravity} = s = \frac{\text{Density of liquid}}{\text{Density of water}} = \frac{978.593}{1000} = 0.978593 \qquad \textbf{Ans.}$$

EXAMPLE 1.2 The velocity distribution for a flow over a flat plate is given by $u = \frac{3}{2}y - y^{3/2}$, where u is the point velocity in metre/sec at a distance y metre above the plate. Determine the shear stress at $y = 9$ cm. Assume dynamic viscosity to be 8 Poise.

Solution: $u = \dfrac{3}{2}y - y^{3/2}$

$$\frac{du}{dy} = \frac{3}{2} - \frac{3}{2}y^{1/2}$$

$$\left.\frac{du}{dy}\right|_{\text{at } y=0} = 3/2 \text{ and } \left.\frac{du}{dy}\right|_{\text{at } y=0.09} = 3/2 - 3/2\,(.09)^{1/2} = 1.05$$

The value of $\mu = 8$ Poise $= \frac{8}{10}$ N-s/m = 0.8 N-s/m (in SI units)

$\therefore$ Shear stress at $y = 9$ cm, $\lambda\ 0.9 = 0.8 \times 1.05 = 0.84$ N/m^2 **Ans.**

EXAMPLE 1.3 A plate 0.025 mm distant from a fixed plate, moves at a speed of 50 cm/sec and requires a force of 1.471 N/m^2 to maintain this speed. Determine the fluid viscosity between the plates in Poise.

Solution: Distance between plates = 0.025 mm = 0.025×10^{-3}m

Velocity of moving plate = 50 cm/sec = 0.5 m/sec

Force on the upper plate to move 1.471 N/m^2 = λ

$$\therefore \quad \lambda = \mu\frac{du}{dy},\ du = u_2 - u_1 = 0.5 - 0 = 0.5 \text{ m/sec}$$

$$1.471 = \mu\frac{0.5}{.025 \times 10^{-3}},\ \text{therefore, } \mu = 7.355 \times 10^{-5} \text{ Ns/m}^2$$

$$= 7.355 \times 10^{-4} \text{ Poise}$$ **Ans.**

EXAMPLE 1.4 The surface tension of water in contact with air at 20°C is given as 0.716 N/m. The pressure inside the droplet of water is to be 0.147 N/cm^2 greater than outside pressure. Calculate the diameter of the droplet.

Solution: $\sigma = 0.0716$ N/m

$p = 0.147$ N/cm^2 = 0.147×10^4 N/m^2

By formula $p = \frac{4\sigma}{d}$ or $d = \frac{4\sigma}{p} = \left(\frac{0.0716 \times 4}{0.147 \times 10^4}\right)$ m

$d = 0.1948$ cm

$d = 1.948$ mm **Ans.**

EXAMPLE 1.5 In a stream of glycerine in motion, at a certain point the velocity gradient is 0.25 m/sec/metre. The mass density of the fluid is 1268.4 kg/m^3 and kinematic viscosity is 6.30×10^{-4} m^2/sec. Calculate the shear stress at the point.

Solution: We have the equation

$$\lambda = \mu\frac{du}{dy},\ \frac{du}{dy} = 0.25 \text{ m/sec/metre}$$

$$\upsilon = \frac{M}{\rho} = 6.30 \times 10^{-4}$$

$$\therefore \quad \mu = \rho \times 6.30 \times 10^{-4}$$

$\therefore$ $\lambda = (1268.4 \times 6.30 \times 10^{-4}) \times 0.25\ \mu = (1268.4 \times 6.30 \times 10^{-4})$

$\lambda = 0.19973\ \text{N/m}^2 = 0.2\ \text{N/m}^2$ **Ans.**

EXAMPLE 1.6 Assuming the bulk modulus of elasticity of water is 2.07×10^6 KN/m^2 at standard atmospheric conditions, determine the increase of pressure necessary to produce 1% reduction in volume at the same temperature.

Solution: $K = 2.07 \times 10^6\ \text{KN/m}^2,\ \dfrac{d(\text{volume})}{\text{volume}} = 0.01\ (\because\ 1\%)$

$$\therefore \quad K = -\frac{dp}{\frac{d\forall}{\forall}} \qquad (1.15)$$

$$\therefore \quad dp = K \times \left(-\frac{d\forall}{\forall}\right) = 2.07 \times 10^6 \times 0.01 = 2.07 \times 10^4\ \text{KN/m}^2$$ **Ans.**

EXAMPLE 1.7 Find the capillary rise of water in a tube of 0.03 cm diameter if the surface tension of water is 0.0735 N/m

Take contact angle $\theta = 0°$

Solution: We have the formula:

$$h = \frac{4\pi \cos\theta}{\omega d} = \left(\frac{4 \times 0.0735 \times 1}{(1000 \times 9.81) \times \dfrac{0.3}{100}}\right) \text{m}$$

$$h = 0.099898 \text{ m}$$

$$h = 9.9898 \text{ cm}$$

$$h^2 = 9.99 \text{ cm.}$$ **Ans.**

EXAMPLE 1.8 Find the kinematic viscosity of an oil having density 981 kg/m^2. The shear stress at a point is 0.2452 N/m^2 and velocity gradient is 0.2 per second.

Solution: Using equation $\lambda = \mu \dfrac{du}{dy}$

$$0.2452 = \mu\ (0.2)$$

$$\therefore \quad \mu = 1.226\ \text{Ns/m}^2$$

$$\therefore \quad \upsilon = \frac{\mu}{\rho} = \frac{1.226}{981} = 1.2497 \times 10^{-3}\ \text{m}^2\text{/sec}$$

$$= 12.497 \times 10^{-2}\ \text{m}^2\text{/sec}$$

$$= 12.497 \times 10^{-2} \times 10^4\ \text{cm}^2\text{/sec}$$

$$= 0.12497 \times 10^4\ \text{cm}^2\text{/sec}$$

$$= 12.497\ \text{cm}^2\text{/sec}$$

$$= 12.497\ \text{STOKE.}\ (\because\ \text{cm}^2\text{/sec} = \text{STOKE})$$ **Ans.**

EXAMPLE 1.9 At a certain point in castor oil, the shear stress is 0.2158 N/m^2 and velocity gradient is 0.218/sec. If the density of castor oil is 959.5 kg/m^3, find the kinematic viscosity.

Solution: We have the equation $\lambda = \mu\left(\frac{du}{dy}\right)$

i.e. $$0.2158 = \mu \times 0.218$$

$\therefore$ $$\mu = 0.99 \text{ N-s/m}^2$$

$\therefore$ $$\upsilon = \frac{\mu}{\rho} = \frac{0.99}{959.5} = 1.03 \text{ m}^2\text{/sec} = 1.03 \times 10^4 \text{ cm}^2\text{/sec}$$

$$= 10.3 \text{ stokes}$$ **Ans.**

1.4 CONCLUSION

A general introduction has been presented. Definition of fluid, fluid mechanics, historical background of the development of the subject by different investigators since ancient times are also presented. Different fluid properties, their definitions, properties, units of measurement are given. Few numerical examples are solved. Some numerical problems with answers and a list of references are included at the end.

PROBLEMS

1.1 Find the kinematic viscosity of an oil having density 980 kg/m^3 when at a certain point in the oil, the shear stress is 0.25 N/m^2 and velocity gradient is 0.3/sec.
(**Ans.** 8.49 Stokes)

1.2 The velocity distribution of a fluid over a plate is given by $u = (3/4)\, y - y^2$, where u is e velocity in m/sec at a distance of y metres above the plate, determine the shear stress t $y = 0.15$ m. Take dynamic viscosity of the fluid to be 8.5×10^{-5} kg–sec/m^2.
(**Ans.** 3.825×10^{-5} kg/m^2)

1.3 Find the surface tension in a soap bubble of 30 mm density when inside pressure is 1.962 N/m^2 above atmosphere. (**Ans.** 0.00735 N/m)

1.4 Determine the capillary rise and fall in a glass tube of 3.0 mm diameter when it is immersed vertically in: (a) water (b) mercury. Take the surface tension of mercury and water as 0.52 N/m and 0.0725 N/m respectively, in contact with air. Specific gravity of mercury is 13.6. (**Ans.** (0.966 cm, –03275 cm))

1.5 One litre of crude oil weighs 9.6 N. Calculate its specific weight, density and specific gravity. (**Ans.** 9600 N/m^2, 979.6 kg/m^3, 0.9786)

1.6 In a stream of glycerine in motion, at a certain point the velocity gradient is 0.25/sec. The density of the fluid is 1268.43 kg/m^3 and kinematic viscosity is 6.30×10^{-4} m^2/sec. Calculate shear stress at the point. (**Ans.** 0.19973 N/m^2)

1.7 A capillary tube of diameter 1.5 mm is dipped in: (a) water and (b) mercury. Find the capillary rise for each case. Surface tension for water is taken as 0.075 N/m and for mercury as 0.52 N/m. The contact angle θ may be taken for water as 25° and for mercury as 130°. (**Ans.** h_w = 1.85 cm, h_m = –0.67 cm)

1.8 What should be the minimum size of a glass tube to be selected to measure the water level if it is desired that the capillary rise be limited to 0.25 mm? Surface tension of water may be taken as 0.075 N/m. (**Ans.** d = 12.2 cm)

REFERENCES

1. Hunter Rouse and Ince Simon, *History of Hydraulics*, Iowa City, Iowa Institute of Hydraulic Research, State University of Iowa, USA, 1958.
2. Wilkinson, W.L., *Non-Newtonian Fluids*, New York, Pergamon Press, 1960.
3. Metzner, A.B., *Flow of Non-Newtonian Fluids*, Section 7 of Handbook Fluid, V.L. Stricter, Dynamics (Ed.) McGraw-Hill, Inc., New York, 1961.
4. Metzner, A.B., *Heat Transfer in Non-Newtonian Fluids*, *Advances in Heat Transfer*, Vol. 2, Academic Press, p. 397, New York, 1965.
5. Bugliarello, V.C., C.E. Behin, E.M. Carver, J.F. Krokosky and R.L. Schiftman, *Non-Newtonian Flow*, Civil Engg., Vol. 35, pp. 68–70, 1965.
6. Bugliarello, V.C., *Some Considerations on the Analysis and Design of Hydraulic Machinery for Non-Newtonian Fluid*, Proc., Tenth Congress, International Assoc. for Hy. Research, Vol. A, Paper No. 4.1., 1963.
7. Fredrickson, A.G., *Principles of Application of Rheology*, Prentice-Hall, Inc., Englewood Cliff, New Jersey, 1964.

Chapter 2

Fluid Pressure and Its Measurement

2.1 INTRODUCTION

Pressure of fluid and its measurements under different situations by different manometers and other pressure gauges are presented in this chapter. Fluids considered here for discussion are both compressible and incompressible. Pascal's law, pressure of gas at any point, isothermal and adiabatic processes, entropy, enthalpy, types of manometers, aerostatics, few numerical examples relating to above are presented in this chapter. Attempts will also be made to provide few problems for readers to solve and some references for further study of the ambitious students at the end.

2.2 FLUID PRESSURE AT A POINT

Liquids have a free surface and occupy a definite volume. Gases do not possess a free surface, fill the container completely and are also compressible. When a certain mass of fluid is held in static equilibrium by being confined in solid boundaries, it exerts pressure on the solid boundary perpendicular to the surface. Fluid at rest cannot sustain shear force and hence there is no tangential force component. This normal force exerted by the static fluid on the surface is called 'fluid pressure'. Since liquids and gases have different properties, separate analysis is required to find the pressure at a point for liquids and gases.

2.2.1 Pressure at a Point in a Liquid

Figure 2.1 shows a tank with a horizontal bottom filled with liquid upto a depth h. A small area dA of the tank at the bottom is considered. Let w be the specific weight of the liquid. The pressure force exerted on the area dA = weight of water column with area dA and height $h = w(hdA)$.

$\therefore$ Intensity of pressure at depth h

$$p = \frac{\text{Force}}{\text{area}} = \frac{whdA}{dA}$$

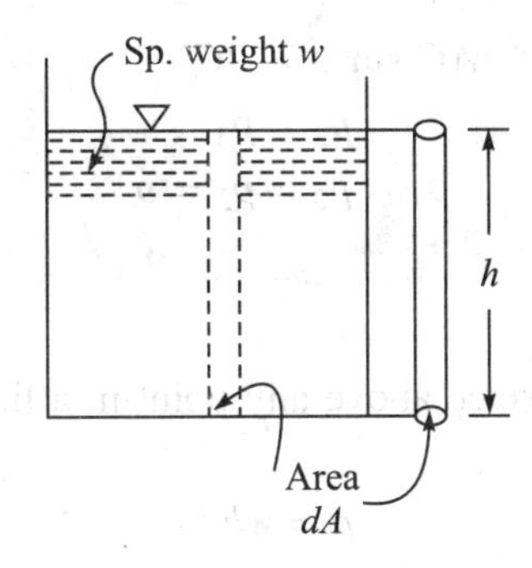

Figure 2.1 Pressure at point in liquid.

$$\therefore \qquad p = wh = \rho gh \qquad (2.1)$$

i.e. pressure p is directly proportional to the depth.

2.2.2 Pascal's Law of Fluid Pressure at a Point

Pascal states that the intensity of pressure at a point in a fluid at rest is the same in all directions.

Within the fluid mass, an elemental prism ABC of fluid is considered. Let ds be the width of the prism. Let p_1, p_2 and p_3 be the pressure intensities on faces AB, BC and AC. Let face AB be horizontal, BC be vertical and face AC be at θ with the horizontal. (See Figure 2.2.).

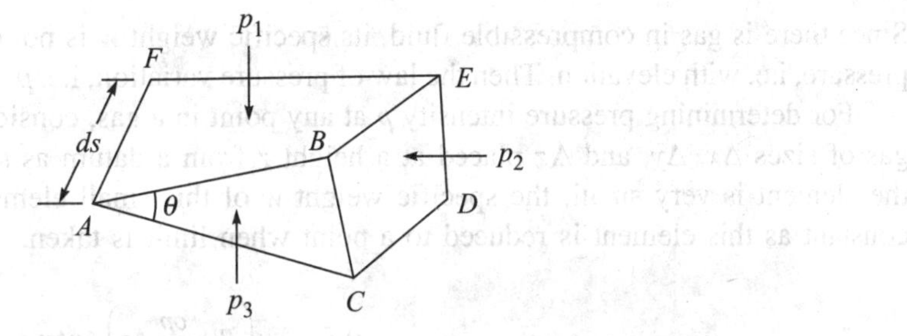

Figure 2.2 An elemental prism *ABC* to derive Pascal's law.

Now Force on the face $AB = p_1\ (AB \cdot ds)$

Force on the face $BC = p_2\ (BC \cdot ds)$

Force on the face $AC = p_3\ (AC \cdot ds)$

The above forces act at right angles to faces AB, BC and AC, respectively. Since the elemental prism is very small, neglecting this very small weight, the three forces must form a system of equilibrium.

Resolving vertically $\qquad p_1\ ABds = p_3\ ACds \cos\theta$

But $\qquad AC \cos\theta = AB$

$\therefore \qquad p_1 = p_3$

Resolving horizontally

$$p_2\ BCds = p_3\ ACds \sin\theta$$

But $$AC \sin \theta = BC$$

$\therefore$ $$p_2 = p_3$$

$\therefore$ $$p_1 = p_2 = p_3 = p \tag{2.2}$$

2.2.3 Pressure Head

The vertical height of the free surface above any point in a liquid at rest is known as pressure head.

From Equation (2.1), $$p = wh$$

$\therefore$ $$h = \frac{p}{w} \tag{2.3}$$

This pressure h is expressed as the pressure in terms of the m. or cm. of the liquid.

It must be remembered that pressure given by Equation (2.1) is the total pressure. The atmospheric pressure p_a is acting on the free surface, then the total pressure below the liquid surface is:

$$p = p_a + wh \tag{2.1a}$$

However, the surface pressure p_a is disregarded in many of the analyses.

2.2.4 Pressure at Any Point in a Gas

Since there is gas in compressible fluid, its specific weight w is not constant, and changes with pressure, i.e. with elevation. Then the law of pressure variation, i.e. $p = wh$, is not applicable here.

For determining pressure intensity p at any point in a gas, consider a very small element of gas of sizes Δx, Δy, and Δz placed at a height z from a datum as shown in Figure 2.3. Since the element is very small, the specific weight w of this small element may be assumed to be constant as this element is reduced to a point when limit is taken.

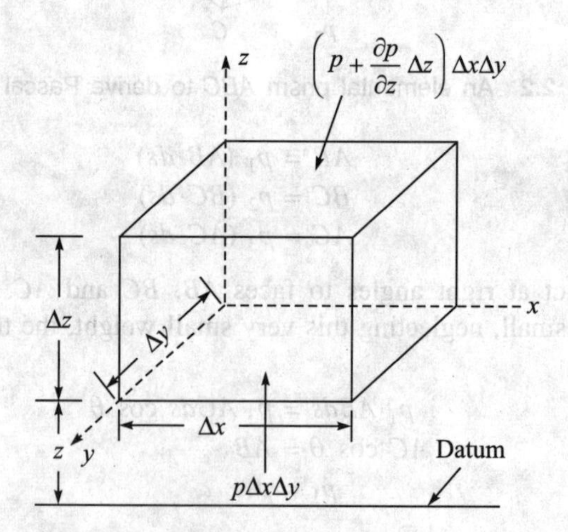

Figure 2.3 Elemental gas prism to calculate pressure in compressible fluid.

If p is the intensity of pressure at the bottom, then pressure intensity at the top of the element is $\left(p+\frac{\partial p}{\partial z}\Delta z\right)$. If w is the specific weight of the element, then the weight of the element is $w(\Delta x \cdot \Delta y \cdot \Delta z)$.

Since the element is in equilibrium,

$$p\Delta x \cdot \Delta y - \left(p+\frac{\partial p}{\partial z}\Delta z\right)\Delta x \cdot \Delta y - w(\Delta x \cdot \Delta y \cdot \Delta z) = 0$$

Simplifying
$$-\frac{\partial p}{\partial z} - w = 0$$

or
$$\partial p = -w\partial z$$

or
$$dp = -wdz$$

$\therefore$
$$dp = -\rho g\, dz \tag{2.4}$$

The minus sign in Equation (2.4) indicates that pressure decreases as the altitude or elevation increases. Equation (2.4) cannot be integrated as the specific weight w is variable unless some known algebraic relation between w and p is established. If dz is zero, dp is zero, i.e. the pressure remains constant over any horizontal plane in a fluid. Equation (2.4), if integrated for incompressible fluid (i.e. w should be constant) and finding the constant of integration, the same Equation (2.1a) or (2.1) is obtained.

2.3 ISOTHERMAL AND ADIABATIC PROCESSES

When the physical properties of gas such as pressure, density and temperature are changed due to a change in the compression or expansion of the gas, it is said to undergo a process. A gas may be compressed or expanded under any one of two processes, i.e. isothermal process or adiabatic process.

Isothermal process is that process in which the temperature is held constant, and it is governed by Boyle's law

i.e.
$$pv = p_1 v_1 = p_2 v_2 = \text{Constant}$$

or
$$\frac{p}{\rho} = \frac{p_1}{\rho_1} = \frac{p_2}{\rho_2} = \text{Constant} \quad \because \text{ specific volume } v = \frac{1}{\rho} \tag{2.5}$$

On the other hand, during the course of a process, if the gas neither absorbs heat from nor does it give out heat to its surroundings, then the process is called adiabatic. Adiabatic process is governed by the following law:

$$pv^k = p_1 v_1^k = p_2 v_2^k = \text{Constant}$$

or
$$\left(\frac{p}{\rho^k}\right) = \left(\frac{p_1}{\rho_1^k}\right) = \left(\frac{p_2}{\rho_2^k}\right) = \text{Constant} \tag{2.6}$$

where k = ratio of two specific heats = $\left(\frac{c_p}{c_v}\right)$

i.e. c_p is the specific heat at constant pressure and c_v is the specific heat at constant volume. Normally the value k of air and other diatomic gases is in the usual range of temperature $k = 1.4$.

2.4 EQUATION OF STATE

Equation of state is defined as the equation which gives the relation between the pressure, temperature and specific volume of a gas. For perfect gas, equation of the state is:

$$p \cdot v = RT$$

or
$$p = \frac{1}{v} RT \text{ i.e. } p = \rho RT \tag{2.7}$$

where p is the absolute pressure in N/m^2 or kgf/m^2

v is the specific volume $= \frac{1}{\rho}$

T is the absolute temperature = (273 + $t°$) centigrade

R is a gas constant in kgf-m/kg K or (J/kgk)

R = 29.2 kgf/kg°k or 287 J/kgk for air

i.e. p and ρ is taken as kgf/m^2 and kg/m^2, R = 29.2 kgf – m/kg°k and if p and ρ is in N/m^2 and kg/m^2, then R = 287J/kgk.

The other details of continuity, energy and momentum equation of compressible fluid like gases will be discussed later in Chapter 15 on Compressible Flow. Only the following two basic definitions of entropy and enthalpy of gas are defined.

2.4.1 Entropy

Entropy of gas may be defined as the measure of maximum heat energy available for conversion into work. It is a property of the gas and varies with absolute temperature and is static. If ΔH is the heat transferred per unit weight of gas in a small interval of time, and if T is the absolute temperature of the gas at that instant, then the change in entropy $\Delta\phi$ is defined by the relation

$$\Delta\phi = \frac{\Delta H}{T}$$

or
$$(\theta_2 - \theta_1) = \int_{T_1}^{T_2} \frac{\Delta H}{T} \tag{2.8}$$

For isothermal process, since there is no change in the heat constant

$$\theta_2 - \theta_1 = 0 \quad \text{or} \quad \theta_2 = \theta_1 = \theta = \text{Constant} \tag{2.9}$$

A process in which entropy does not change is termed as isoentropic.

2.4.2 Enthalpy

The sum of the internal energy and pressure volume product (converted into heat unit) is termed as enthalpy. It is purely a mathematical quantity.

2.5 ATMOSPHERIC, ABSOLUTE, GAUGE AND VACUUM PRESSURE

The pressure exerted by atmosphere upon all surfaces with which it is in contact is called 'atmospheric pressure'. It varies with altitude and can be measured by barometer. As such, it is also called parametric pressure. At sea level, this atmospheric pressure in SI unit at 15°C is 101.3 KN/m^2. It is also expressed in head form which is 10.3 m of water or 760 mm of mercury.

Fluid pressure is measured from two arbitrary datum levels i.e. (1) from absolute zero pressure and (2) from local atmospheric pressure. (See Figure 2.4).

If the pressure is measured from absolute zero (or from complete vacuum) datum, it is called absolute pressure. When measured with respect to atmospheric pressure as datum, it is called gauge pressure.

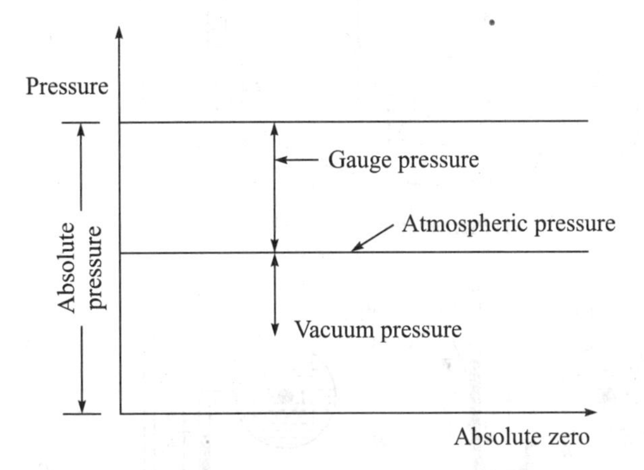

Figure 2.4 Relationship of different pressures.

2.6 MEASUREMENT OF PRESSURE

The pressure of fluid is measured by: (1) manometers, and (2) mechanical gauges.

2.6.1 Manometers

Manometers are classified as:

(a) Simple manometers
(b) Differential manometers

By using a simple manometer, the pressure at a point of the vessel or pipe can be measured, and differential manometer is used to measure the difference of pressure between two points of the pipe or vessel.

Simple Manometer

This is a glass tube having one end connected to the point where pressure has to be measured and the other open end is exposed to atmosphere. The common type of simple manometers are:

(i) Piezometer
(ii) U–tube manometer
(iii) Single column manometer

(i) **Piezometer:** This is the simplest form of manometer as shown in Figure 2.5. The liquid will rise to a height h in the glass tube for the point A.

∴ The pressure p at A is

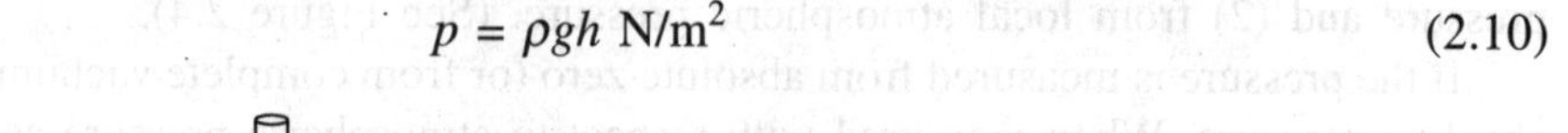

$$p = \rho g h \text{ N/m}^2 \tag{2.10}$$

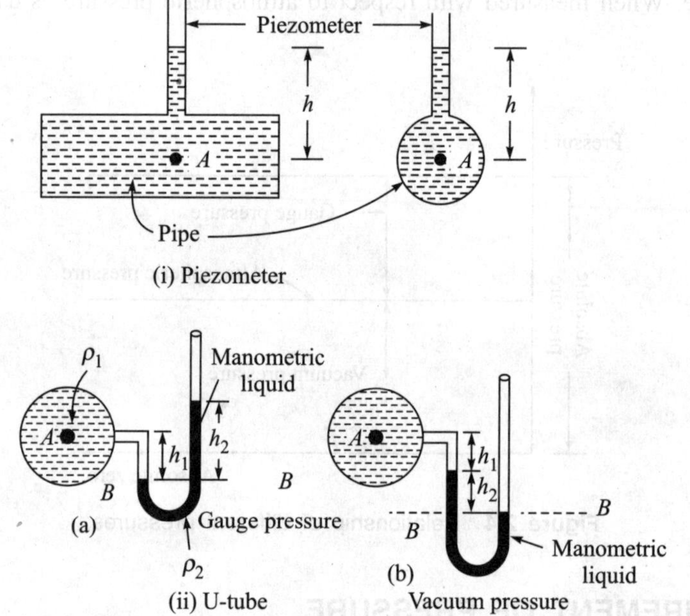

Figure 2.5 Piezometer and U-tube manometer.

(ii) **U–tube manometer:** It is a glass tube bent in the form of U, one end of which is connected to the point where pressure has to be measured and other end is connected after formation of U shape open to atmosphere. Manometric fluid, which is normally mercury, is filled in the U–tube. Manometer fluid may be a different liquid, heavier than the liquid whose pressure has to be measured.

(a) For gauge pressure. Let p be the pressure at A. Let AA be the datum line. Let specific gravity of the liquid whose pressure has to be measured be S_1 and specific gravity of the manometer liquid be S_2.

$$\therefore \qquad \rho_1 = 1000\ S_1 \quad \text{and} \quad \rho_2 = 1000\ S_2$$

As pressure along a horizontal line is the same,
the pressure above BB in [Figure 2.5ii(b)] of figure in the left column $= p + \rho_1 g h_1$

The pressure above BB in the right column = $\rho_2 gh_2$
Equating the two pressures

$$p + \rho_1 gh_1 = \rho_2 gh_2$$

$$\therefore \quad p = (\rho_2 gh_2 - \rho_1 gh_1) \tag{2.11}$$

(b) For vacuum pressure [Figure 2.5ii(b)]

Pressure above $BB = \rho_2 gh_2 + \rho_1 gh_1 + p$

Pressure at BB in the right column = 0
Equating the two pressures $\rho_2 gh_2 + \rho_1 gh_1 + p = 0$

$$\therefore \quad p = -(\rho_2 gh_2 + \rho_1 gh_1) \tag{2.12}$$

(iii) **Single column manometer (vertical):** Figure 2.6 shows single column vertical manometer. It is a modified form of U–type manometer having a large reservoir with a cross-sectional area about 100 times more than the U–tube. Let p be the pressure at point A. Let a be the area of the small glass tube. Let A be the area of a large reservoir.

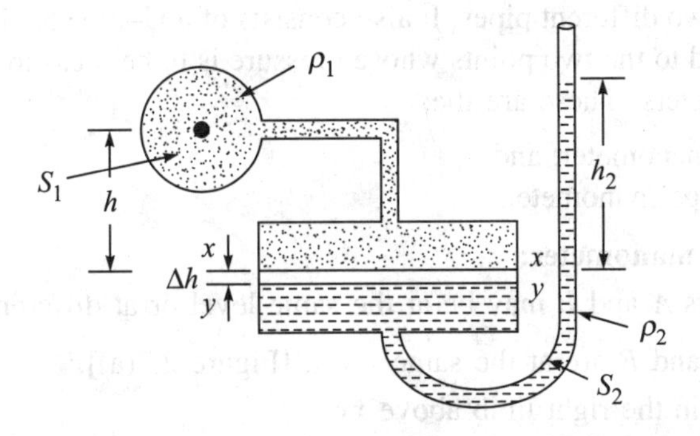

Figure 2.6 Vertical single column manometer.

Due to the rise of h_2 in the glass tube, the fall in reservoir is Δh

$$\therefore \quad A \cdot \Delta h = ah_2 \;\therefore\; \Delta h = \frac{a}{A} h_2$$

Taking yy as the datum,
pressure in the right limb = $\rho_2 \times g \times (\Delta h + h_2)$
pressure in the left limb above yy

$$= \rho_1 g(h_1 + \Delta h) + p$$

Equating the two pressures:

$$\rho_2 g(h_2 + \Delta h) = \rho_1 g(h_1 + \Delta h) + p$$

Putting the value $\Delta h = \dfrac{a}{A} h_2$

$$\rho_2\, g\left(h_2 + \frac{a}{A}h_2\right) = \rho_1 g\left(h_1 + \frac{a}{A}h_2\right) + p$$

$$p = \frac{a}{A}h_2\,(\rho_2 g - \rho_1 g) + (h_2\,\rho_2 g - h_1 \rho_1 g) \qquad (2.13)$$

As area A is very large as compared to a, $\frac{a}{A}$ may be neglected,

Then $$p = (\rho_2 g h_2 - \rho_1 h_1 g) \qquad (2.14)$$

If the right limb of the U–tube is inclined at θ to the horizontal and L is the length of the heavy liquid moved above xx, then

$$h_2 = L \sin\theta$$

$$\therefore \quad p = (L\sin\theta)\,\rho_2 g - h_1\,\rho_1 g \qquad (2.14a)$$

Differential Manometers

Differential manometers are devices used to measure the difference of pressure between two points in a pipe or in two different pipes. It also consists of a U–tube having heavy liquid whose two ends are connected to the two points whose pressure is to be measured. There are two types of differential manometers. These are the:

(i) U–tube type manometer, and
(ii) Inverted U–type manometer.

(i) **U–tube type manometer:**

The two points A and B may be at the same level or at different levels.

(a) When A and B are at the same level, [Figure 2.7(a)].

Pressure in the right limb above xx

$$= \rho_2 g h + \rho_1 x g + p_B$$

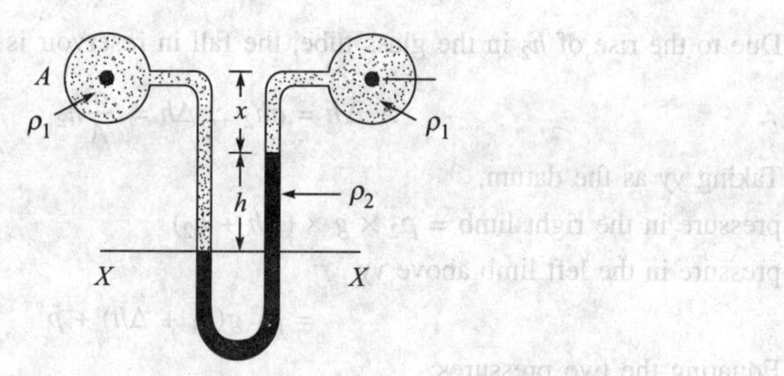

Figure 2.7(a) Differential manometer measuring points at same level.

Pressure in the left limb above xx

$$\rho_1(h + x)\,g + p_a$$

Equating the two pressures, $\rho_2 hg + \rho_1 gx + p_B$

$$= \rho_1 hg + \rho_1 xg + p_a$$

$$\therefore \quad p_A - p_B = \rho_2 hg - \rho_1(h + x)g + \rho_1 gx$$

$$p_A - p_B = \rho_2 hg - \rho_1 hg = gh(\rho_2 - \rho_1)$$

$$\frac{p_A}{\rho g} - p_B/\rho g = \frac{gh}{\rho g}(\rho_2 - \rho_1) \Rightarrow \frac{p_A}{w} - \frac{p_B}{w} = h(s_2 - s_1) \tag{2.15}$$

Consider the differential manometer shown in Figure 2.7(b) whose measuring points A and B are at different levels. The pressure above xx on the left

$$= \rho_1(h + x)\, g + p_A$$

Pressure in the right limb above xx

$$= \rho_1 hg + \rho_1 yg + p_B$$

Equating the pressure and simplifying, we get:

$$p_A - p_B = \rho_2 hg + \rho_1 yg - \rho_1(h + x) \tag{2.16}$$

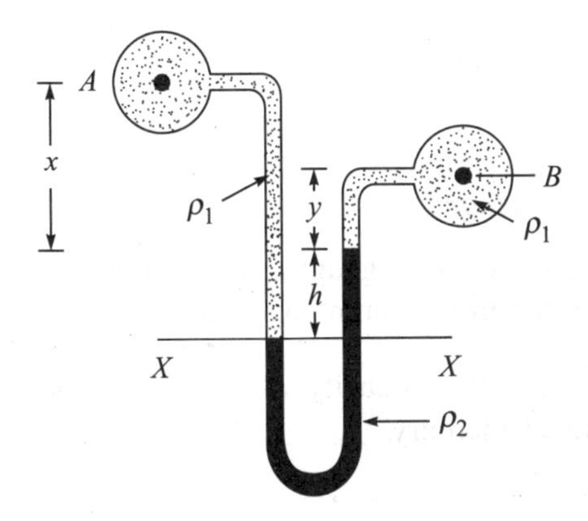

Figure 2.7(b) Differential manometer measuring points at different levels.

(ii) **Inverted U–type manometer**

Let A and B be at the same level but having different liquid of density ρ_3 and ρ_1 xx is the datum as shown in Figure 2.8.

Pressure in the left limb below xx

$$p_A - \rho_1 g h_1$$

Pressure in the right limb below xx

$$p_B - \rho_3 h_2 g - \rho_2 hg$$

Equating the two pressures and simplifying, we get:

$$p_A - p_B = \rho_1 h_1 g - \rho_3 h_2 g - \rho_2 hg \tag{2.17}$$

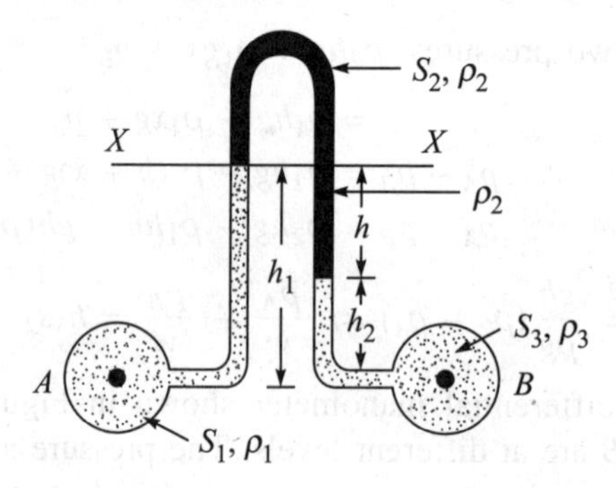

Figure 2.8 Inverted U–type manometer.

2.6.2 Mechanical Gauges

Mechanical gauges are pressure measuring devices which embody an elastic element that deflects under pressure. A pointer moves under pressure against a graduated circumferencial scale. These gauges are normally used to measure high pressure. There are different types of mechanical gauges. They are the:

(i) Bourdon type pressure gauge
(ii) Diaphragm pressure gauge
(iii) Bellows pressure gauge
(iv) Dead weight pressure gauge

Out of the above different types of gauges, the Bourdon type, which was developed by E. Bourdon (1808–1884), is the most commonly used.

EXAMPLE 2.1 Express pressure intensity of 73.575 N/cm^2 in all pressure units. Take barometer reading as 76 cm of mercury.

Solution: (a) Gauge units

$$p = 73.575 \text{ N/cm}^2 = 73.575 \times 10^4 \text{ N/m}^2$$

$$= \frac{73.575 \times 10^4}{\rho g} = \frac{73.575 \times 10^4}{9.81 \times 1000} = 75 \text{ m of water}$$

$$= \frac{73.575 \times 10^4}{13.6 \times (9.81 \times 1000)} = 5.5147 \text{ m of mercury.}$$

(b) Absolute units

Absolute pressure = Gauge pressure + Atmospheric pressure
Atmospheric pressure = 76 cm of mercury

$$= \frac{76 \times 13.6}{100} = 10.34 \text{ m of water}$$

$$= \left(\frac{76 \times 13.6}{100} \times \rho g\right)$$

$$= \left(\frac{76 \times 13.6}{100} \times 1000 \times 9.81\right) \text{N/m}^2$$

$$= (10.1396 \times 10^4)\text{N/m}^2$$

$$= 10.1396 \text{ N/cm}^2$$

$\therefore$ Absolute pressure $= (73.575 + 10.1396) = 83.7146 \text{ N/cm}^2$

$$= 83.7146 \times 10^4 \text{ N/m}^2$$

Absolute pressure head is $(75 + 10.34) = 85.34$ m of water

$$= (5.5147 + 0.76) = 6.2747 \text{ m of mercury.}$$

EXAMPLE 2.2 Find the depth of water in sea where the pressure intensity is 100.5525N/cm^2. Specific gravity of sea water is 1.025.

Solution: We have the equation $p = wh$. Here $w = 1.025 \times (1000 \times 9.81)$

$\therefore$ $100.5525 \times 10^4 = (1000 \times 9.81 \times 1.025)h$

$\therefore$ $$h = \frac{100.5525 \times 10^4}{9.81 \times 1000 \times 1.025} = 100 \text{ m}$$ **Ans.**

EXAMPLE 2.3 The pressure intensity at a point is given as 4.9 N/cm^2. Find the corresponding height of fluid when it is: (a) water, (b) oil of specific gravity 08.

Solution: We have the equation $p = wh \Rightarrow 4.9 \times 10^4 = (1000 \times 9.81)h$

$\therefore$ $h = 5$ m of water **Ans.**

For oil of specific gravity 0.8, $$h = \frac{4.9 \times 10^4}{(1000 \times 9.81 \times .8)} = 6.25 \text{ m of oil.}$$ **Ans.**

EXAMPLE 2.4 An open tank contains water upto a depth of 1.5 m and above it an oil of specific gravity 0.8 upto a depth of 2 m. Find the pressure intensity: (a) at the interface of two liquids, (b) at the bottom of the tank.

Solution: Pressure intensity at interface, i.e. up to a depth of 2 m from free surface with oil of specific gravity 0.8.

$$p_{\text{interface}} = (0.8 \times 1000 \times 9.81) \times 2 = 1.5696 \times 10^4 \text{ N/m}^2$$

$$= 1.5696 \text{ N/cm}^2$$ **Ans.**

Pressure at bottom, $p_{\text{bottom}} = [1.5696 + (1000 \times 9.81 \times 1.5)] \text{ N/m}^2$

$$= 3.0411 \times 10^4 \text{ N/m}^2$$

$$= 3.0411 \text{ N/cm}$$ **Ans.**

EXAMPLE 2.5 A simple manometer shown in Figure 2.9 is used to measure the pressure of oil of specific gravity 0.8, flowing in a pipeline. Its right limb is open to atmosphere and its left limbs connected to the pipe. The centre of the pipe is 9 cm below level of mercury (specific gravity 13.6) in the right limb. If the difference of mercury level in two limbs is 15 cm, determine the absolute pressure of the oil in the pipe in N/cm^2.

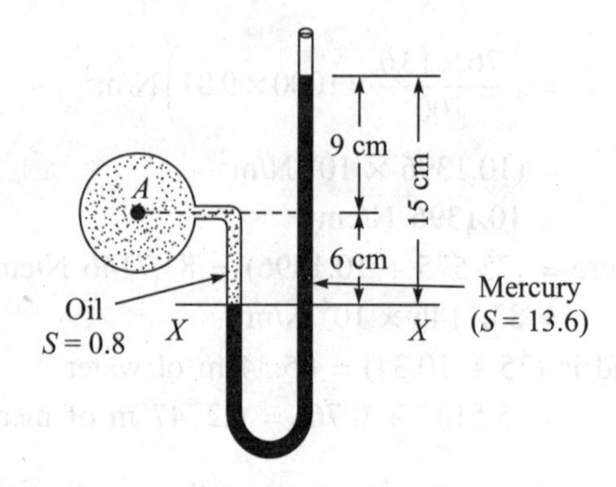

Figure 2.9 Visual of Example 2.5.

Solution: Taking *xx* as datum and pressure at centre of pipe *A* as p_A equating the pressure above *xx*,

$$p_A + (0.8 \times 1000 \times 9.81) \times .06 = (13.6 \times 1000 \times 9.81) \times 0.15$$

$$\therefore \quad p_A = 1.95415 \times 10^4 \text{ N/m}^2$$

$$= 1.95415 \text{ N/cm}^2$$

$\therefore$ Absolute pressure at *A*

$$= p_A + \text{atmospheric pressure}$$

$$= (1.95415 + 10.13) = 12.08415 \text{ N/cm}^2$$ **Ans.**

$\because$ Atmospheric pressure at sea level = 10.13 N/cm^2

EXAMPLE 2.6 A U–tube differential manometer shown in Figure 2.10 connects pipes *A* and *B*. Pipe *A* contains a liquid of specific gravity 1.594 under a pressure of 10.3 N/cm^2 and pipe *B* contains oil of specific gravity 0.8 under a pressure of 17.16 N/cm^2. Pipe *A* lies 2.5 m above pipe *B*. Find the difference of pressure measured by mercury as the fluid-filling U–tube if the mercury level in the left limb will remain 1.5 m below *B*.

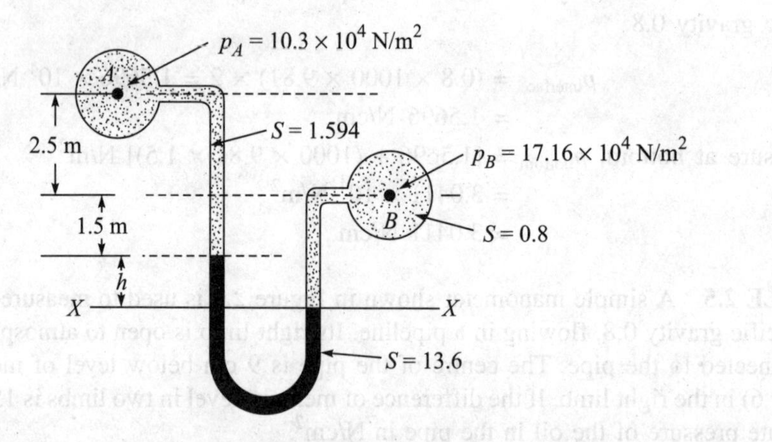

Figure 2.10 Visual of Example 2.6.

Solution: Let the difference of the mercury level in the manometer be $h(m)$. (Figure 2.10).

Taking xx as datum, pressure in the left limb above xx

$$= 10.8 \times 10^4 + (1.594 \times 9.81 \times 1000) \times 4 + (13.6 \times 9.81 \times 1000)h$$

Pressure in the right limb above xx

$$= (0.8 \times 9.81 \times 1000)\ (h + 1.5) + 17.16 \times 10^4$$

Equating the two forces:

$10.3 \times 10^4 + (1.594 \times 9.81 \times 1000) \times 4 + (13.6 \times 9.81 \times 1000)h = (.8 \times 9.81 \times 1000)h + 0.8 \times 9.81 \times 1000 \times 1.5 + 17.16 \times 10^4$

or $(13.6 \times 9.81 \times 1000)h - (0.8 \times 9.81 \times 1000)h = 0.8 \times 9.81 \times 1000 \times 1.5 + 17.16 \times 10^4 - 10.3 \times 10^4 - (1.594 \times 9.81 \times 1000 \times 4)$ $(12.5568 \times 10^4)h = 1.1772 \times 10^4 + 17.16 \times 10^4 - 10.3 \times 10^4 - 6.2548 \times 10^4$

or
$$(12.5568 \times 10^4)h = 1.7424 \times 10^4$$

$\therefore$
$$h = \frac{1.7424 \times 10^4}{12.5568 \times 10^4} = 0.13876 \text{ m of mercury}$$

$$h = 13.876 \text{ cm of mercury}$$ **Ans.**

EXAMPLE 2.7 The left limb of a U–tube mercury manometer shown in Figure 2.11 is connected to a pipeline conveying water, with the level of mercury in the left limb being 60 cm below the centre of the pipeline and right limb being open to atmosphere. The level of mercury in the right limb is 45 cm above that of the left limb containing a liquid of specific gravity 0.88 to a height of 30 cm. Find the pressure in the pipe.

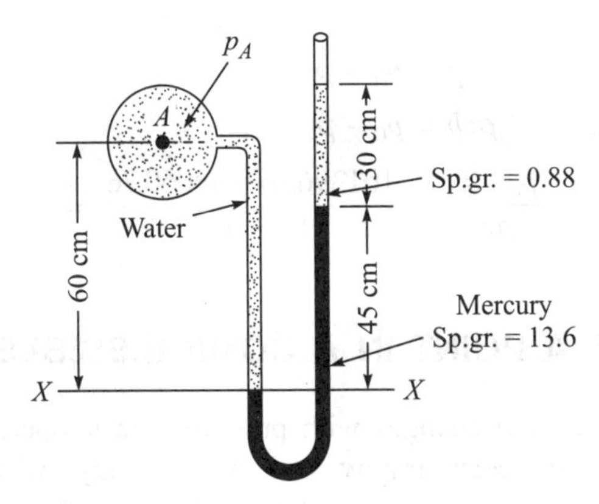

Figure 2.11 Visual of Example 2.7.

Let XX be the datum. Writing the pressure equation for both the left and right limbs above XX in pressure head form, we get: (Figure 2.11)

$$\frac{p_A}{\rho g} + 60 = 0.45 \times 13.6 + 0.30 \times 0.88$$

$$\frac{p_A}{\rho g} + 0.6 = 6.12 + 0.264 = 6.384$$

$$\therefore \quad \frac{p_A}{\rho g} = 6.384 - 0.6 = 5.784 \text{ m of water}$$

$$\therefore \quad p_A = (5.784 \times 1000 \times 9.81)\text{N/m}^2 = 5.6741 \times 10^4 \text{ N/m}^2$$ **Ans.**

EXAMPLE 2.8 The barometric pressure at sea level is 760 m of mercury while at the top of the mountain it is 735 mm, if density of air is assumed to be constant 1.22 kg/m^3, what is the elevation of the mountain?

Solution: Pressure at sea level p_0 = 760 mm of mercury

$$= \left(\frac{760}{1000} \times 13.6 \times 1000 \times 9.81\right) \text{N/m}^2$$

$$= 101396.16 \text{ N/m}^2$$

Pressure at mountain p = 735 mm of mercury

$$= \left(\frac{735}{1000} \times 13.6 \times 1000 \times 9.81\right) \text{N/m}^2$$

$$= 98060.76 \text{ N/m}^2$$

Density of air = 1.2 kg/m^2 (constant with elevation)

Let h be the height of mountain assumed from the sea level.

Pressure decreases as the elevation above sea level increases. Here, density of air is given constant at any level.

$$\therefore \quad p = p_0 - \rho g h$$

$$\rho g h = p_0 - p$$

$$\therefore \quad h = \frac{p_0 - p}{\rho g} = \frac{101396.16 - 98060.76}{1.2 \times 9.81} = 283.33 \text{ m}$$ **Ans.**

2.7 PRESSURE AT A POINT IN A COMPRESSIBLE FLUID

For compressible fluid density ρ changes with pressure and temperature. Such problems are encountered in aeronautics, oceanography and meteorology, which are connected with atmospheric air wherein the density, pressure, temperature changes with elevation. Thus fluid with variable density, Equation (2.4) cannot be integrated unless a relationship between pressure p and density ρ is established.

We know for gases, Equation (2.7)

$$\left.\begin{aligned} p &= \rho R T \\ \text{or} \quad \rho &= \frac{p}{RT} \end{aligned}\right\} \qquad (2.7)$$

Equation (2.4) now may be written as:

$$\frac{dp}{dz} = -\rho g = -\frac{p}{RT}g$$

$$\frac{dp}{p} = \frac{-g}{RT}dz \tag{2.18a}$$

If T is constant, it is an isothermal process, and it may be integrated as:

$$\int_{p_0}^{p} \frac{dp}{p} = -\frac{g}{RT} \int_{z_0}^{z} dz$$

or

$$\log\left(\frac{p}{p_0}\right) = -\frac{g}{RT}(z - z_0)$$

where p_0 is the pressure when $z = z_0$. If datum is taken at z_0,

$$z_0 = 0, \text{ then}$$

$$\log_e\left(\frac{p}{p_0}\right) = -\frac{g}{RT}z$$

∴

$$\frac{p}{p_0} = e^{-\frac{g}{RT}z}$$

or pressure p at height z is

$$p = p_0 e^{-\frac{g}{R_T}z} \tag{2.18b}$$

or

$$\frac{p}{p_0} = e^{-\frac{gz}{RT}} \quad \text{or} \quad \frac{\rho}{\rho_0} = e^{-\frac{gz}{RT}} \tag{2.18c}$$

If the pressure is not constant, the process follows adiabatic laws, i.e. Equation (2.6) is:

$$\frac{p}{\rho^k} = \text{Constant} = c \tag{2.6}$$

$$\rho^k = \frac{p}{c}$$

∴

$$\rho = \left(\frac{p}{c}\right)^{1/k}$$

Now Equation (2.4) may be written for the adiabatic process:

$$dp = -\rho g dz = -\left(\frac{p}{c}\right)^{1/k} g dz$$

$$\frac{dp}{\left(\frac{p}{c}\right)^{1/k}} = -g dz$$

or

$$c^{\frac{1}{k}} \frac{dp}{p^{1/k}} = -g dz$$

or

$$c^{1/k} p^{-1/k} dp = -g dz$$

$$\text{Integration of } \int_{p_0}^{p} c^{1/k}\, p^{-1/k}\, dp = -g \int_{z_0}^{z} dz$$

and further simplification of pressure p is obtained as:

$$p = p_0 \left[1 - \frac{k-1}{k} \frac{gz}{RT_0} \right]^{\frac{k}{k-1}} \tag{2.19}$$

for the adiabatic process.

Similarly temperature T at any point in compressible fluid can be worked out to be

$$T = T_0 \left[1 - \frac{k-1}{k} \cdot \frac{gz}{RT_0} \right] \tag{2.20}$$

This temperature changes with elevation i.e. $\frac{dT}{dz}$ is called temperature in lapse.

This temperature lapse L may be obtained as:

$$L = \frac{dT}{dz} = \frac{-g}{R} \left[\frac{k-1}{K} \right] \tag{2.21}$$

If $K = 1$, which means the isothermal process, $\frac{dT}{dz} = 0$, i.e. the temperature is constant with the height.

2.8 AEROSTATICS

Aerostatics is a subject of interest in meteorology and aeronautics. It deals with the determination of the static properties of the atmosphere. Considerable importance has been given to aerostatics by aeronautical engineers and meteorologists. It is already mentioned that air is compressible, and its density, in general, is a function of pressure and temperature. We have solved Equation (2.4), establishing a relation among altitude z, temperature T and pressure (Equation 2.18).

Experimental evidence of the variation of temperature with altitude in different layers of the atmosphere is shown in Figure 2.12.

From the stand-point of a single gas and continuum analysis, the layers of primary importance are the troposphere and stratosphere. Therefore, the atmospheric properties of these two layers by using aerostatic Equation are: (2.18a)

$$\text{i.e.} \quad \frac{dp}{p} = -\frac{g}{RT} dz \tag{2.18a}$$

2.8.1 The Troposphere

The troposphere is the first layer of the atmosphere that lies between the earth's surface and stratosphere, and in this layer the temperature–altitude variation may be written as:

$$T = T_0 - LZ \tag{2.22}$$

where $T_0 = 288°K$ is the absolute sea level temperature and L is the constant lapse rate. In the standard atmosphere adopted for aeronautical calculation, this lapse rate L is taken as 6.5°K/Km.

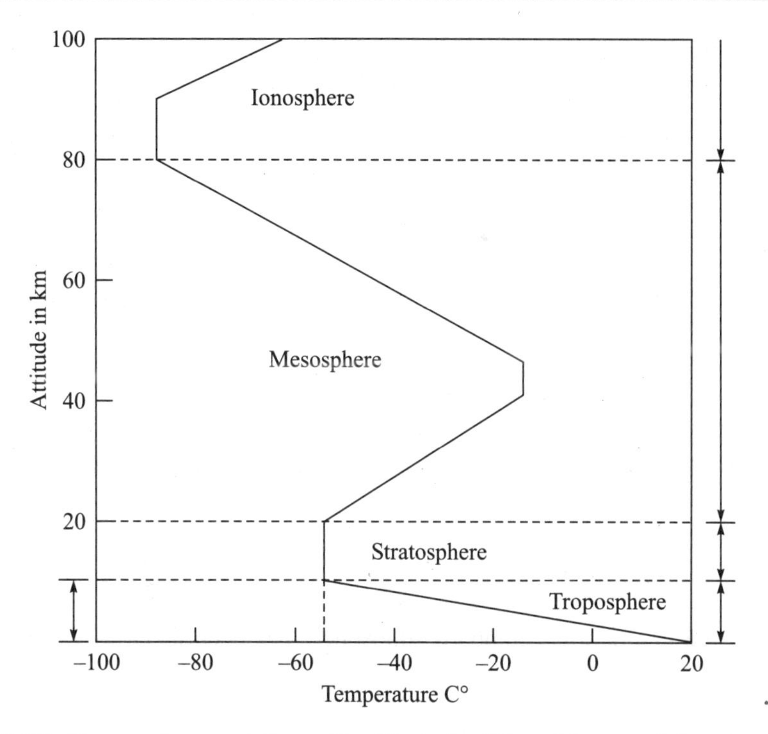

Figure 2.12 Temperature variation with altitude in different layers of the atmosphere.

Substituting the derivative of Equation (2.22) i.e. $dT = -Ldz$ i.e. $dz = -\dfrac{dT}{L}$

(2.18a), gives

$$\frac{dp}{p} = \frac{gdT}{RL} \tag{2.23}$$

Integrating Equation (2.23) $\log_e p = \dfrac{g}{RL} \log_e T + C$ (2.24a)

To obtain constant of integration, at sea level,

$$T = T_0,\ p = p_0$$

Equation (2.24a) becomes,

$$\log_e p_0 = \frac{g}{RL} \log_e T_0 + C \tag{2.24b}$$

$\therefore$

$$C = \log_e p_0 - \frac{g}{RL} \log_e T_0$$

Putting C in Equation (2.24a) gives

$$\log_e p = \frac{g}{RL} \log_e T + \log_e p_0 - \frac{g}{RL} \log_e T_0$$

or

$$\log_e \left(\frac{p}{p_0}\right) = \frac{g}{RL} \log_e \left(\frac{T}{T_0}\right) \tag{A}$$

Putting $T = T_0 - Lz$, i.e. Equation (2.22) in (A), we get:

$$\log_e\left(\frac{p}{p_0}\right) = \frac{g}{RL}\log_e\left(1-\frac{Lz}{T_0}\right) \quad (2.24c)$$

$$\therefore \quad \frac{p}{p_0} = \left(1-\frac{Lz}{T_0}\right)^{\frac{g}{RL}} \quad (2.24d)$$

The density–altitude relationship, which can be derived from the equation of state for perfect gas with the aid of Equation (2.24d), is:

$$\frac{\rho}{\rho_0} = \frac{w}{w_0} = \left(1-\frac{Lz}{T_0}\right)\left(\frac{g}{RL}-L\right) \quad (2.25)$$

If Eqs. (2.24d) and (2.25) are combined, the relation between pressure and density is found to be:

$$\left(\frac{\rho}{\rho_0}\right) = \left(\frac{p}{p_0}\right)^{(1-RL/g)} \quad (2.26)$$

At sea level, the international standard atmosphere has been taken as:

$$p_0 = 101.13 \text{ KN/m}^2$$
$$T_0 = 288.15°\text{K}$$
$$\rho_0 = 1.225 \text{ Kg/m}^3$$
$$R = 287.1 \text{ J/kg°k}$$
$$L = 6.5°\text{k/Km}$$

With the above values of R and L, Equation (2.22) and Equations. (2.24d) to (2.26) yield, respectively,

$$\frac{T}{T_0} = 1 - 2.26 \times 10^{-5}$$

$$\left.\begin{aligned}\frac{p}{p_0} &= (1-2.26\times10^{-5}z)^{5.26}\\ \frac{\rho}{\rho_0} &= (1-2.26\times10^{-5}z)^{4.26}\end{aligned}\right\} \quad (2.27)$$

$$\frac{\rho}{\rho_0} = \left(\frac{p}{p_0}\right)^{0.8097}$$

The minimum value of Z in Equation (2.27) is 11 km.

2.8.2 The Stratosphere

The layer of air above the troposphere is called stratosphere (shown in Figure 2.11). The temperature in this layer is more or less constant as shown by a vertical line (around –56°C) in this layer. The distance of this layer is between 11 km and 20 km above sea level. Now by letting subscript 1, indicate the condition at the intersection between the stratosphere and the troposphere. Equation (2.27) gives:

$$\left.\begin{aligned}\frac{T_1}{T_0} &= 0.751\\ \frac{p_1}{p_0} &= 0.223\\ \frac{\rho_1}{\rho_0} &= 0.298\end{aligned}\right\} \quad (2.28)$$

For an isothermal process (i.e. constant temperature), Equation (2.18a) can be written as:

$$\frac{dp}{p} = -\frac{g}{R}\cdot\frac{dz}{T_1} \quad (2.29a)$$

By integrating

$$\frac{p}{p_1} = e^{-(g/RT_1)\,(Z-Z_1)} \quad (2.29b)$$

The equation of state for a perfect gas in this case is reduced to:

$$\rho = \frac{p}{RT_1}$$

or

$$\frac{\rho}{\rho_1} = \frac{p}{p_1} = e^{-(g/RT_1)\,(Z-Z_1)} \quad (2.29c)$$

Equation (2.29c) represents the variation of pressure and density in a static compressible fluid with constant temperature. In the above discussion, g is assumed to be constant with altitude. The layer of air above the stratosphere is the mesosphere, which extends upto 80 km in which the temperature first increases, then remains constant for a few kms and then, falls between –80°C and –100°C. Above the mesosphere, the last layer is the ionosphere, wherein the temperature just above 80 km remains constant and then goes on increasing. Aeronautical engineers are mainly concerned with the first two layers, i.e. upto a height of 20 km.

EXAMPLE 2.9 What is the air pressure and temperature at 18.288 km?

Solution: The height 18.288 km is within the stratosphere.

For temperature, $T_s = T_0 - LZ$

$= 288.15° - 6.5 \times 18.288$

$= 169.278°$ k

$$\frac{p_s}{p_0} = \left(1 - \frac{6.5\times 18.288}{288.15}\right)^{\frac{9.81}{287.1\times 6.5}} = 0.9972$$

$\therefore$ $p_s = 0.9972 \quad p_0 = (0.9972 \times 101.18) = 100.897\ \text{KN/m}^2$ **Ans.**

EXAMPLE 2.10 Calculate pressure at a height of 8000 m above the mean sea level if the atmospheric pressure is 101.3 kN/m^2 and the temperature is 15°C at sea level assuming: (i) Air is incompressible, (ii) Pressure variation follows adiabatic laws, (iii) Pressure follows the isothermal law. Take the density of air at mean sea level to be equal to 1.285 kg/m^3. Neglect variation g with elevation.

Solution: Here z = 8000 m, p_0 = 101.3 kN/m^2 = 10.13 × 10^4 N/m^2

t_0 = 15°C, T_0 = 273 + 15 = 288°k, density of air $\rho = \rho_0$ = 1.285 kg/m^3

(i) When air is incompressible, $dp = -\rho g dz$

Integrating $\int_{p_0}^{p} dp = \int_{z_0}^{z} -\rho g/dz$

or $(p - p_0) = -\rho g\,[z - z_0]$

$\therefore\ p = p_0 - \rho g z \qquad \because\ z_0 = 0$ as datum

$$p = (10.13 \times 10^4 - 1.285 \times 9.81 \times 8000)\ \text{N/m}^2$$

$$p = 453.2\ \text{N/m}^2$$ **Ans.**

(ii) When pressure variation is in isothermal law:

Using Equation (2.18), $p = p_0\, e^{-gz/RT}$

$$\therefore \quad p = p_0\, e^{-gz\cdot\frac{1}{RT}} \qquad \left\{\begin{aligned} &\because p_0 = \rho_0 RT \\ &\therefore \frac{p_0}{\rho_0} = RT \\ &\therefore \frac{\rho_0}{p_0} = \frac{1}{RT} \end{aligned}\right.$$

$$p = p_0\, e^{-gz\frac{\rho_0}{p_0}}$$

$$p = 10.13 \times 10^{-4}\, e^{-9.81\times 8000\times\left(\frac{1.285}{10.13\times 10^4}\right)}$$

$$= 10.13 \times 10^4\, e^{-0.99552} = 10.13 \times 10^4 \times \frac{1}{e^{.99512}} = 37433.28\ \text{N/m}^2$$

$$= 37.433\ \text{KN/m}^2$$ **Ans.**

(iii) When pressure variation is in adiabatic law, using Equation (2.19):

$$p = p_0\left[1 - \frac{k-1}{K}\cdot\frac{gZ}{RT_0}\right]^{\frac{R}{R-1}} \quad \text{Ratio of specific heat } k = 1.4 \text{ (taken)} \because \frac{\rho_0}{p_0} = \frac{1}{RT_0}$$

$$p = 10.13 \times 10^4\left[1 - \frac{1.4-1}{1.4}\cdot\frac{9.81\times 8000\times 1.285}{10.13\times 104}\right]^{\frac{1.4}{1.4-1}}$$

$$p = 10.13 \times 10^4\ [0.715564]^{\frac{1.4}{.4}}$$

$$p = 10.13 \times 10^4 \times 0.309934 = 31396.348\ \text{N/m}^2$$

$$= 31.396\ \text{kN/m}^2$$ **Ans.**

EXAMPLE 2.11 An aeroplane is flying at an altitude of 4000 m. Calculate the pressure around the aeroplane, given the lapse rate in the atmosphere of 0.0065°k/m. Neglect variation of g with altitude. The pressure and temperature at ground level are 10.143 N/cm^2 and 15°C, respectively. The density of air is given as 1.285 kg/m^3.

Solution: Lapse rate $\frac{dT}{dz} = -0.0065°\text{k/m}, z = 4000$ m

$$p_0 = 10.143 \times 10^4 \text{N/m}^2,\ t_0 = 15°\text{C},\ T_0 = 273 + 15 = 288°\text{k}$$
$$\rho_0 = 1.285 \text{ kg/m}^3$$

∴ The temperature at 4000 m height $= T_0 + \frac{dT}{dz} = 288 + (-0.0065) \times 4000$

$$= 288 - 26 = 262°\text{k}$$

According to Equation 2.21:

$$L = \frac{dT}{dz} = -\frac{g}{R}\left[\frac{k-1}{k}\right]$$

$$-0.0065 = -\frac{9.81}{R}\left[\frac{k-1}{k}\right] \quad \text{(A)}$$

∴ $$p_0 = \rho_0 R\, T_0$$

∴ $$R = \frac{p_0}{\rho_0 T_0} = \frac{10.143 \times 10^4}{1.285 \times 288} = 274.076$$

(A) becomes $-0.0065 = -\frac{9.81}{274.076}\left(1 - \frac{1}{k}\right)$

$$1.781494 = 9.81\left(1 - \frac{1}{k}\right)$$

$$0.1816 = 1 - \frac{1}{k}$$

$$\frac{1}{k} = 1 - 0.1816$$

$$k = 1.2218$$

Pressure is given by Equation (2.19) as follows:

$$p = p_0\left[1 - \frac{k-1}{k} \cdot \frac{gz\,\rho_0}{p_0}\right]^{\frac{k}{k-1}}$$

$$p = 101430\left[1 - \left(\frac{1.2218 - 1}{1.2218}\right)\frac{9.81 \times 4000 \times 1.285}{101430}\right]^{\frac{1.2218}{.2218}}$$

$$\left\{\begin{array}{l} \because p_0 = \rho_0 R T_0 \\ \therefore R T_0 = \dfrac{p_0}{\rho_0} \\ \therefore \dfrac{1}{R T_0} = \dfrac{\rho_0}{p} \end{array}\right.$$

$$P = 60240.129 \text{ N/m}^2$$
$$P = 6.024 \text{ N/cm}^2 \quad \textbf{Ans.}$$

2.9 CONCLUSION

Equations of fluid pressure at a point for both incompressible liquid and compressible gases have been derived in this chapter. Isothermal and adiabatic process concept and corresponding equations are also deduced. Different pressure measurement devices like piezometer and manometers, and their equations for finding the pressure of liquid, have been presented systematically. The concepts of pressure head, gauge pressure, absolute pressure, absolute temperature, and gas content are given. A few examples are also solved to understand the theoretical equations derived in this chapter for both liquid and gases. A few problems with answers are enclosed for interested readers to solve. References are given at the end for further study of students and readers.

PROBLEMS

2.1 The petrol of specific gravity 0.8 flows upwards through a vertical pipe. *A* and *B* are two points on the pipe, *B* bends 30 cm higher than *A*. Connections are led from *A* and *B* to a U–tube manometer having mercury. If the difference of pressure between *A* and *B* is 1.7658 N/cm^2, find the length of difference of mercury level in the manometer.

(**Ans.** 12.2 cm.)

2.2 Find the depth of a point below water surface in sea where the pressure intensity is 100.5525 N/cm^2. The specific gravity of sea water is 1.025. (**Ans.** 100 m)

2.3 Convert the pressure of 100 m of water to:

(a) Kerosene of specific gravity 0.81

(b) Carbon tetrachloride of specific gravity 1.6

(**Ans.** (a) 123.4 m of kerosene (b) 62.5 of Carbon tetrachloride)

2.4 The inverted U–tube manometer is connected to pipes at *B* and *C* with water as shown in Figure 2.13. Manometric fluid is of specific gravity 0.8. Find the difference pressure at *C* and *B*.

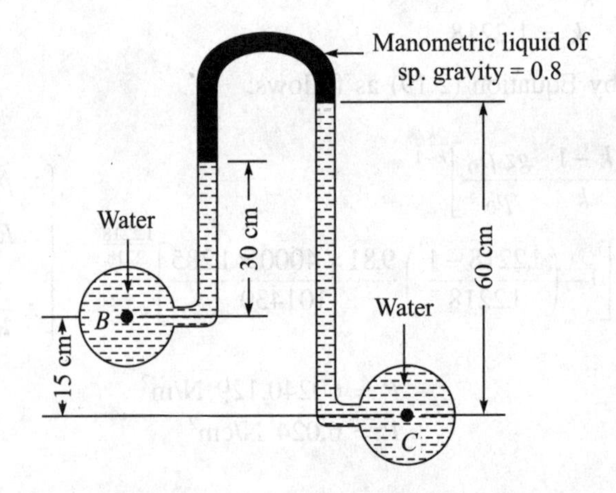

Figure 2.13 Visual for Problem 2.4

If the pressure at $C = 6.867$ N/cm^2 and atmosphere pressure 74.73 cm of mercury, find the pressure at B.

[**Ans.** $p_C - p_B = 0.17658$ N/cm^2 $p_B = 6.69$ N/cm^2 (gauge) $p_B = 16.65738$ N/cm^2(Abs)]

2.5 The atmospheric pressure at sea level is 101.3 KN/m^2 and temperature is 15°C. Calculate the pressure at 8000 m above sea level assuming: (i) Isothermal variation of pressure and density and, (ii) Adiabatic variation of pressure and density. Assume that the density of air at sea level is 1.285 kg/m^3. Neglect variation of g with elevation.

(**Ans.** (i) 37.45 kN/m^2 (ii) 31.5 kN/m^2)

2.6 A tank contains a liquid of specific gravity 0.8. Find the absolute and gauge pressure at a point 2 m below the free surface. The atmospheric pressure head is 760 mm of mercury. (**Ans.** 117092 N/m^2, 15696 N/m^2)

REFERENCES

1. Yuan, S.W., *Foundations of Fluid Mechanics*, Prentice-Hall of India, New Delhi, 1976.
2. Olson, R.M., *Engineering Fluid Mechanics*, International Text Book Company, Seranton, Pennsylvania, 1967.
3. Rouse, H., *Fluid Mechanics for Hydraulic Engineer*, McGraw-Hill Inc., 1938.
4. Rouse, H., *Elementary Mechanics of Fluid*, John Wiley and Sons, Inc., New York, 1946.
5. Harris, C.W., *Hydraulics*, Chapman & Hall, New York, 1936.
6. Lewitt, E.H., *Hydraulics and Mechanics of Fluid*, Pitman, 1955.
7. Modi, P.N. and S.M. Seth, *Hydraulics and Fluid Mechanics including Hydraulic Machines*, 3rd ed., New Delhi, 1977.
8. Bansal, R.K., *Fluid Mechanics and Hydraulic Machines*, Laxmi Publications, New Delhi, 1983.
9. Mott, R.L., *Applied Fluid Mechanics*, Prentice-Hall, Upper Saddle River, New Jersey 1999.

Chapter 3

Hydrostatic Forces on Surfaces

3.1 INTRODUCTION

This chapter deals with the forces of liquid at rest on surfaces which may be vertical, horizontal, inclined, or curved. The surfaces may be submerged partially or fully. The force exerted by the liquid on the surface without any shear stress $\left(\text{i.e. } \lambda = \mu \dfrac{du}{dy}\right)$ is called 'hydrostatic force'. In the design of some hydraulic structures like dams, gates, tanks, etc., the knowledge of this hydrostatic pressure or force is very important. The point at which this total hydrostatic pressure acts is called the centre of pressure.

3.2 HYDROSTATIC FORCE ON PLANE VERTICAL SURFACE

In Figure 3.1, a plane surface is shown immersed in a liquid of specific weight w. Let A be the total area of the surface on which this hydrostatic total pressure P acts. Let $\bar{y}$ be the depth of the CG of the area from the free surface. G is the CG of the surface and CP is the centre of pressure at a depth y_{cp} from the free surface.

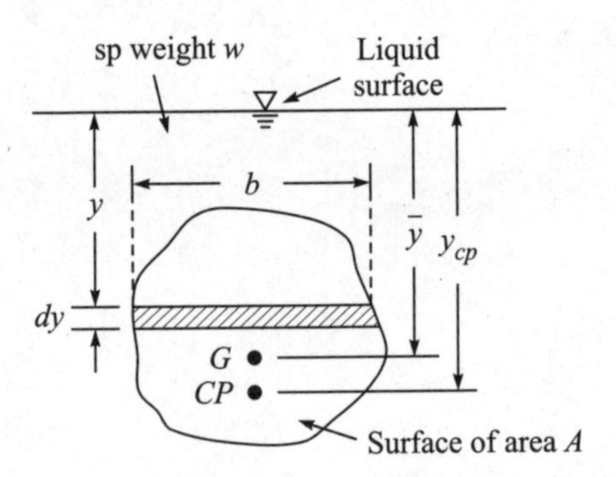

Figure 3.1 To derive equation of total hydrostatic force.

The total area A is divided into a number of small parallel strips. The force on any one of the strips is first calculated and the integrating total force on the whole area is then obtained.

Let a small strip of area dA be at depth y from the free surface pressure intensity on the strip, $p = \rho gy$ by Equation (3.1)

$$dA \text{ of the strip} = b\ dy$$

Total force on the strip $dp = p \cdot bdy = Pgy \cdot bdy$

$$dp = Pgb\ ydy.$$

Total pressure on the whole surface is obtained by integrating

$$P = \int dp = \int Pgbdy = Pg \int by\ dy$$

But

$$\int b \times y \times dy = \int ydA$$

$$= \text{moment of surface area about the free surface}$$

$$= A\bar{y}$$

$\therefore$

$$P = \rho g A \bar{y} \tag{3.1}$$

for water $\rho = 1000\ \text{kg/m}^3$

Depth of centre of pressure (y_{cp}):

Depth of centre of pressure y_{cp} is obtained by using the 'Principal of moment' which states that the moment of the resultant force about an axis is equal to the sum of moments of the components about the same axis. Hence force P about the free surface $= P \times y_{cp}$ (A)

The moment of dP, acting on a strip about free surface,

$$= dP \times y$$

$$= \rho gy\ bdy \cdot y$$

$\therefore$ Sum of the moments of all such forces about the force surface of liquid $= \int \rho gy\ bdy \cdot y$

$= \rho g \int y^2 dA$

But $\int y^2 dA = \int by^2\ dy$ = moment of inertia of the surface about the free surface

$$= I_0$$

$\therefore$ Sum of the moments $= \rho g I_0$ (B)

Now (A) = (B) gives

$$P \cdot y_{cp} = \rho g\ I_0$$

or

$$\rho g\ A\bar{y} = P$$

$\therefore$

$$\rho g\ A\bar{y} \cdot y_{cp} = \rho g\ I_0$$

$\therefore$

$$y_{cp} = \frac{I_0}{A\bar{y}} \tag{3.2}$$

Theorem of parallel axis gives

$$I_0 = I_G + A\bar{y}^2$$

Substituting in Equation (3.2)

$$y_{cp} = \frac{I_G + A\bar{y}^2}{A\bar{y}} = \frac{I_G}{A\bar{y}} + \bar{y}$$

$$y_{cp} = \bar{y} + \frac{I_G}{A\bar{y}} \tag{3.3}$$

3.3 FORCE ON PLANE HORIZONTAL SURFACE

Consider a plane surface immersed in a liquid of specific weight w and its surface is held horizontal at a depth y as shown in Figure 3.2. Let A be the area of the surface.

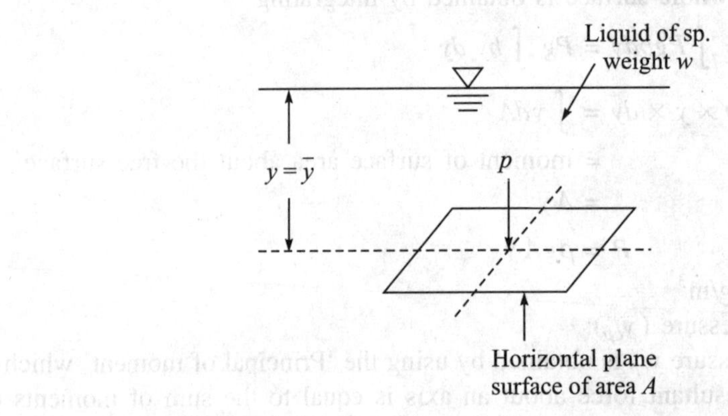

Figure 3.2 Total pressure on a horizontal plane surface.

If p is the intensity of pressure at the depth $y = \bar{y}$ at any point on the surface, then p, derived by Equation (3.1), is:

$$P = wy = w\bar{y}$$

$\therefore$ Total force P on area $A = pA$

$$\therefore \qquad P = wyA = w\bar{y}A = \rho g\bar{y}A \tag{3.1}$$

which is the same as Equation (3.1).

3.4 FORCE ON AN INCLINED SURFACE

A plane surface of area A is immersed in a liquid of specific weight w at an angle (inclined) θ to the water surface as shown in Figure 3.3. The extended line from the surface intersects the free surface at 0 and 00 is drawn perpendicular to this extended plane surface.
Let $\bar{y}$ be the depth of CG of the plane surface below free surface, and $\bar{x}$ be the inclined distance of CG from OO

A small strip of area dA lying at depth y from the free surface is considered, and this strip is at a distance x from OO.

For the small strip, the intensity of pressure p is assumed to be constant and equal to $p = wy$

$\therefore$ Total pressure on the strip $= dP = (wy)dA$

$$\because \qquad y = x \sin\theta$$

$$dP = w(x\sin\theta)dA$$

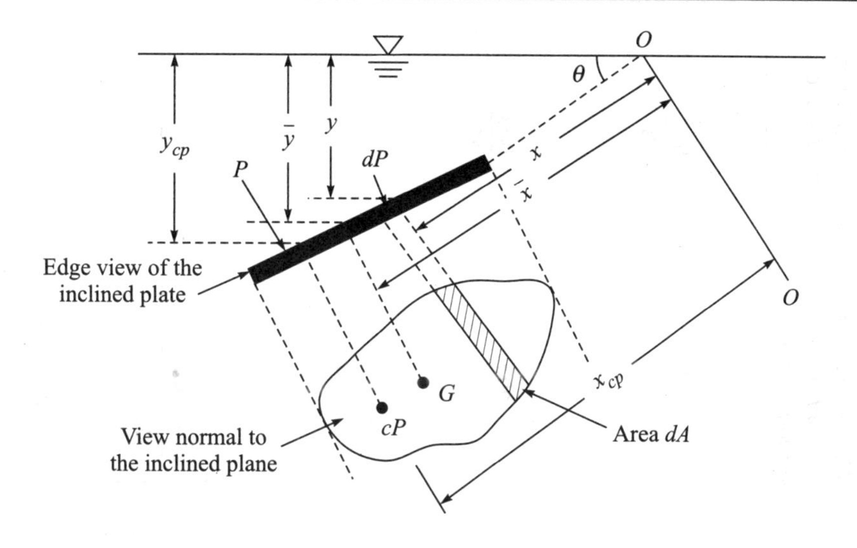

Figure 3.3 Total force on an inclined plane.

Integrating the above expression,

$$\int dP = \int w(x \sin \theta) dA$$

$$P = w \sin \theta \int x dA \quad \text{(A)}$$

$\int x dA$ represents the sum of the first moments of the area of the strips about OO, which is equal to the product of area A and inclined distance of the centroid of the surface area $\bar{x}$ from axis OO, i.e. $\int y dA = A\bar{x}$

$\therefore$

$$P = w \sin \theta \cdot A\bar{x}$$

or

$$P = wA\ (\bar{x} \sin \theta)$$

$\therefore$ $P = w A \bar{y}$ which is the same as Equation (3.1) $\because$ $\bar{x} \sin \theta = \bar{y}$

It shows that the total pressure of liquid on vertical, horizontal and inclined immersed surfaces represents the same Equation (3.1), i.e.

$$P = wA\bar{y} \quad (3.1)$$

Depth of centre of pressure (y_{cp}).

Let y_{cp} be the depth of the centre of pressure from the free surface and its distance from OO is x_{cp}.

Total pressure on the strip is $dP = w(x \sin \theta) dA$

and its moment about OO is

$$(dP \cdot y) = w(x \sin \theta)\ dA \cdot x = w \sin \theta\ x^2\ dA$$

TABLE 3.1 The Moments of Inertia and Other Geometric Properties of Some Important Plane Surfaces

	Plane surface	*C G from base*	*Area*	*Moment of inertia about an axis passing through CG*	*Moment of inertia about base*
1.	Rectangle $\bar{x} = d/2$	$\bar{x} = \frac{d}{2}$	bd	$\frac{bd3}{12}$	$\frac{bd^3}{3}$
2.	Triangle	$\bar{x} = \frac{h}{3}$	$\frac{1}{2}bh$	$\frac{bh3}{36}$	$\frac{bh3}{12}$
3.	Circle	$\bar{x} = \frac{d}{2} = r$	$\frac{\pi d^2}{4}$	$\frac{\pi d^4}{64}$	–
4.	Trapezium	$\bar{x} = \left(\frac{2a+b}{a+b}\right)\frac{h}{3}$	$\left(\frac{a+b}{2}\right)h$	$\left[\frac{a^2+4ab+b^2}{36(a+b)}\right]h^3$	–
5.	Parabola $\frac{3}{5}h$	$\bar{x} = \frac{2}{5}h$	$\frac{2}{5}hbh$	$\frac{8}{175}bh^3$	

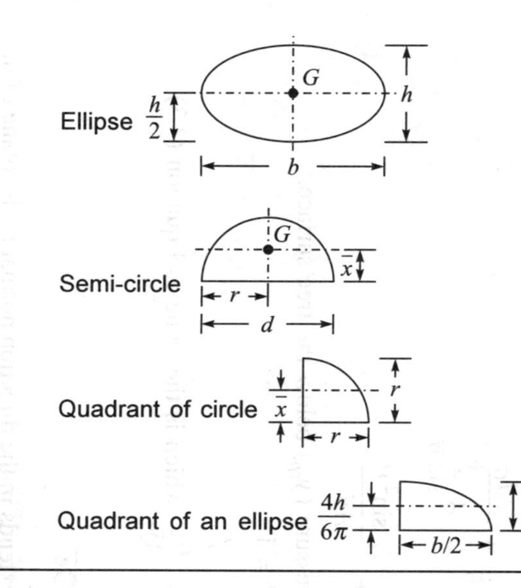

6.	Ellipse	$\bar{x} = \frac{h}{2}$	$\frac{\pi}{4} bh$	$\frac{\pi bh^3}{64}$	–
7.	Semi-circle	$\bar{x} = \frac{4r}{3\pi}$	$\frac{\pi d^2}{8}$	$\left(\frac{\pi}{8} - \frac{8}{9\pi}\right) r^4$	$\frac{\pi r^4}{8}$
8.	Quadrant of circle	$\bar{x} = \frac{4}{3}\frac{r}{\pi}$	$\frac{\pi r^2}{4}$	$\left(\frac{\pi}{4} - \frac{16}{9\pi}\right) r^4$	
9.	Quadrant of an ellipse	$\bar{x} = \frac{4h}{6\pi}$	$\frac{\pi}{16} bh$	$\left(\frac{\pi}{64} - \frac{1}{9\pi}\right) \frac{bh3}{4}$	–

Summing up the moments of total pressure on such small strips about OO and using the principle of moment, we get:

$$P\, x_{cp} = w \sin\theta \int x^2\, dA \tag{B}$$

From B, $\int x^2\, dA$ is the sum of the second moment of areas of the strips about OO, which is equal to the moment of inertia I_0 of the plane surface about OO, i.e.

$$I_0 = \int x^2 dA \tag{C}$$

Substituting (C) in (B), $P\, x_{cp} = w \sin\theta\, I_0$

$$\therefore \quad x_{cp} = \frac{w \sin\theta\, I_0}{P} = \frac{w \sin\theta\, I_0}{wA\,(\bar{x} \sin\theta)} \tag{D}$$

By the parallel axis theorem, the moment of inertia I_G

$$I_0 = I_G + A\bar{x}^2 \tag{E}$$

where I_G is the moment inertia of the area about an axis passing through the centroid G of the area parallel to OO. Moments of inertia about CO and base of different section's are presented in Table 3.1.

Introducing (E) in (D)

$$x_{cp} = \frac{w \sin\theta\,(I_G + A\bar{x}^2)}{wA\,\bar{x} \sin\theta}$$

$$x_{cp} = \bar{x} + \frac{I_G}{A\bar{x}}$$

Put $x_{cp} = y_{cp}/\sin\theta$, and $\bar{x} = \dfrac{\bar{y}}{\sin\theta}$

$$\frac{y_{cp}}{\sin\theta} = \frac{\bar{y}}{\sin\theta} + \frac{I_G}{A\dfrac{\bar{y}}{\sin\theta}}$$

$$y_{cp} = \bar{y} + \frac{I_G \sin^2\theta}{A\bar{y}} \tag{3.4}$$

which gives the vertical depth of centre of pressure (y_{cp}) below the free surface.

If $\theta = 90°$, the surface is vertical, $\sin 90° = 1$

$\therefore$ Equation (3.4) becomes:

$$y_{cp} = \bar{y} + \frac{I_G}{A\bar{y}}$$ which is the same as Equation (3.3)

3.5 FORCE ON CURVED SURFACE

ABC is the force of the curved surface which extends in the direction normal to the plane of the paper. Pressure always acts normal of the surface at any point. Consider a small elemental area

dA of the curve surface where force acts equal to dP normal to the surface as shown in Figure 3.4. The point of action dP is at a depth h from the liquid surface. Then $dP = (p \cdot dA) = whdA$ (A)

Integrating (A), we get: $$P = \int pdA \quad \text{(B)}$$

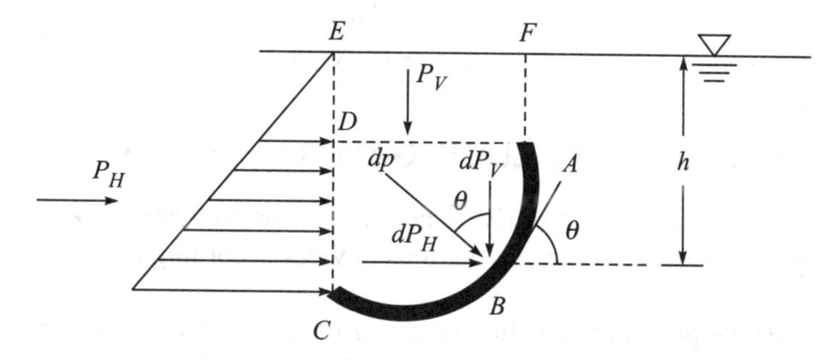

Figure 3.4 Total force on a curved surface.

As the direction of application of forces in the curved surface varies from point to point, direct integration of (B) is not possible.

Therefore, total pressure P may be resolved in P_H and P_V as the horizontal and vertical components, respectively.

Total press dP on the elementary area dA is resolved into two components as follows:

$$dP_H = dP \sin \theta = pdA \sin \theta = wh\ dA \sin \theta$$

and
$$dP_V = dp \cos \theta = pdA \cos \theta = wh\ dA \cos \theta$$

where θ is the inclination of elementary area with the horizontal as shown in Figure 3.4. Integrating dP_H and dP_V in the above two expressions, we get:

$$P_H = w \int h\ dA \sin \theta$$

and
$$P_V = w \int h\ dA \cos \theta$$

In the above two expressions, $dA \sin \theta$ is the vertical projection of dA and $dA \cos \theta$ is the horizontal component. Thus $(wh\ dA \sin \theta)$ represents the total pressure on the vertical projection of the elementary area dA and $\int wh\ dA \sin \theta$ represents the total pressure on the projected area of the curved surface on a vertical plane.

P_H = total pressure on projected area of curved surface on a vertical plane, the trace of which is represented by CD in Figure 3.4.

Further $(wh\ dA \cos \theta)$ represents the total force on horizontal projection of the dA of the curved surface and this is equal to the weight of volume of liquid vertically above dA.

Therefore, $(\int wh\ dA \cos \theta)$ represents the weight of the liquids vertically above the curved surface, i.e. weight of the volume of water in the portion ABCDEF shown in Figure 3.4 which is equal to P_V.

Then resultant of P_H and P_V is P and equal to

$$P = \sqrt{P_H^2 + P_V^2} \tag{3.5}$$

and the direction of resultant P is given by

$$\phi = \tan^{-1}\left(\frac{P_V}{P_H}\right) \tag{3.6}$$

when ϕ is the angle made by the resultant with horizontal.

3.6 TOTAL PRESSURE ON LOCK GATES

Lock gates are the pair of gates used in rivers and canals for navigation. The level of water in the lock gates may be lowered or raised by various devices according to the need of navigatory ships, boats etc.

Figure 3.5 shows the plan and elevation of a pair of lock gates. Let AB and BC be the two lock gates. Each gate is supported on two hinges at the top and bottom on A and C ends. In a closed position, the gates are at B.

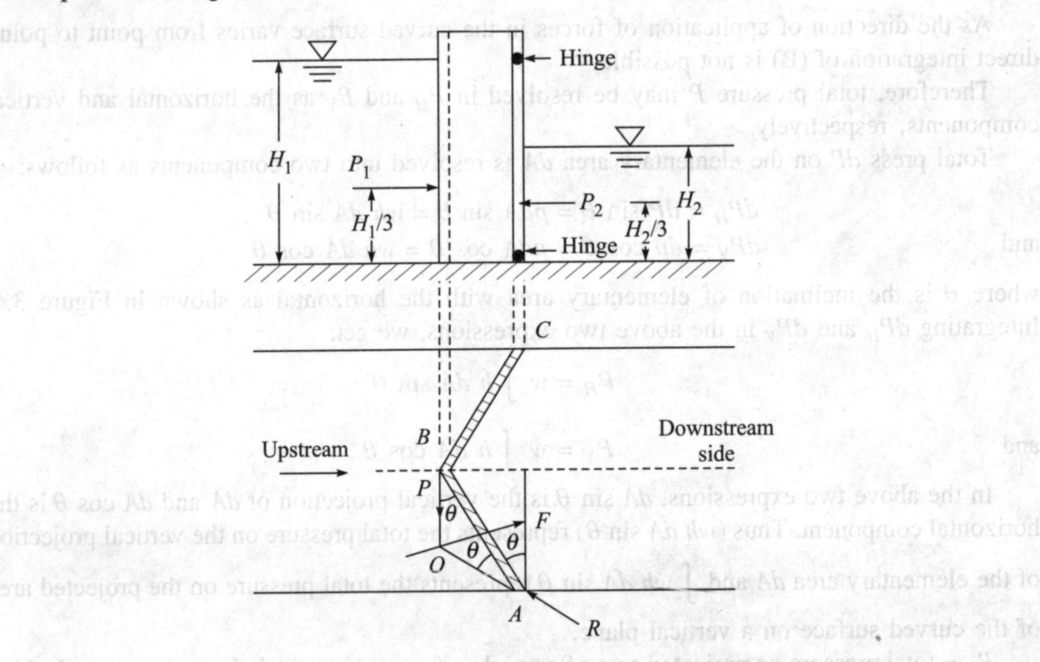

Figure 3.5 Plan and elevation of pair of lock gates.

Let F be the resultant force due to water on both sides of AB or BC acting perpendicular to the gate.

R is the reaction at the lower and upper hinges.

P is the reaction at the common contact surface of the two gates acting perpendicular to contact surface. Let P and F meet at O, then the reaction must pass through O for equilibrium. Let θ be the angle of inclination of the lock gate with the normal to the side of the lock. In the triangle ABO, $\angle OAB = \angle ABO = \theta$.

Resolving all the forces along the gate AB

$$R \cos \theta = P \cos \theta$$

$$\therefore \quad R = P$$

Resolving forces normal to the gate

$$R \sin \theta + P \sin \theta = F$$

or
$$R \sin \theta + P \sin \theta - F = 0$$

$$\because \quad R = P,\ F = 2P \sin \theta$$

$$\therefore \quad P = \frac{F}{2 \sin \theta} \tag{3.7}$$

If F and θ are known, P or R may be evaluated. θ is known from the angle made by the gates in closed position at B and the angle between the two gates is $(180° - 2\theta)$.

Let H_1 and H_2 be the depths upstream and downstream of the gates, respectively and P_1 and P_2 be the water pressure on the gate upstream and downstream, respectively.

Let L be the width of the gate.

$$\therefore \quad P_1 = \rho g A_1 \bar{h}_1 = \rho g (L H_1) \frac{H_1}{2} = \rho g L {H_1^2}/{2}$$

Similarly,
$$P_2 = \rho g L {H_2^2}/{2}$$

Resultant
$$F = P_1 - P_2 = \frac{\rho g L}{2} (H_1^2 - H_2^2)$$

If we substitute the known θ and F in Equation (3.7), P is known.

If R_t, R_b are reactions at the top and bottom hinges, respectively, $R = R_t + R_b$

The resultant water pressure F acts normal to the gate. Half of F is resisted by the hinges of one lock gate and other half is resisted by the hinges of the other gate. Also P_1 acts at ${H_1}/{3}$ and P_2 acts at ${H_2}/{3}$ from the bottom.

Taking the moment about the lower hinge:

$$R_t \sin \theta \cdot H = \frac{R}{2} \times {H_1}/{3} - \frac{P_2}{2} \times {H_2}/{3} \tag{A}$$

where H is the distance between the hinges.

Also resolving force horizontally:

$$R_t \sin \theta + R_b \sin \theta = \frac{P_1}{2} - \frac{P_2}{2} \tag{B}$$

From Equations (A) and (B), R_t and R_b can be computed.

3.7 PRACTICAL APPLICATIONS OF TOTAL PRESSURE

While designing dams, gates, tanks, etc. in practical field application, it is essential to assess the total pressure and centre of pressure. For example, in the design of a very high dam, the water force upstream is one of the most important forces to be considered in the design.

Consider a high concrete dam (Figure 3.6) constructed across a river so as to form a reservoir upstream of it for hydropower generation, flood control, irrigation and some other purposes. The upstream face is always in contact with static mass of water with a high depth H. Downstream of the dam, there is normally a dry bed or with a bed little depth of water.

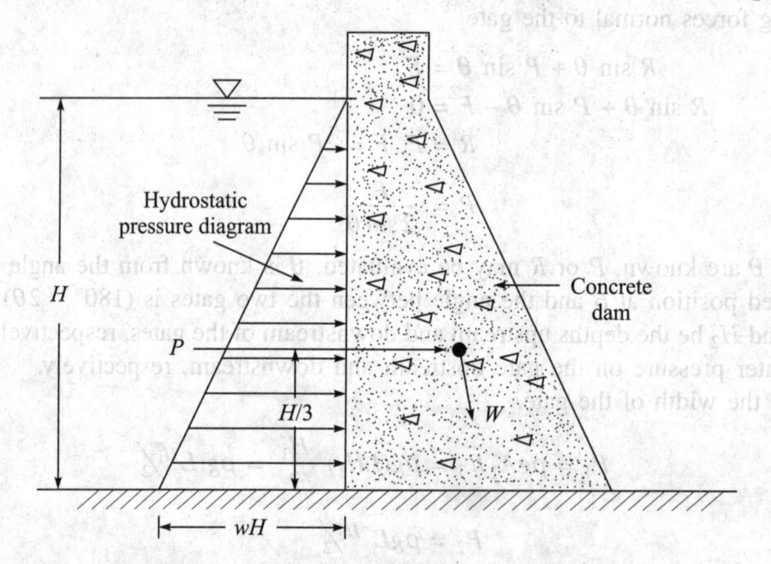

Figure 3.6 A concrete dam under hydrostatic force.

Consider a unit length of the dam. Total pressure per unit length of the dam due to depth H of water is computed by Equation (3.1) as follows:

$$P = wA\bar{y} = w(H \times 1) \times \frac{H}{2} = w\frac{H^2}{2} \qquad \text{(A)}$$

where w is the specific weight of water.

This force P may also be computed from the hydrostatic pressure diagram shown in Figure 3.6.

Equation (2.1) is $p = wh$ which gives the pressure intensity at a point, i.e. p increases linearly with depth as w is constant for the water liquid which is zero at $H = 0$ and equal to wH at depth H.

The area of the intensity or hydrostatic pressure diagram shown in Figure 3.6 is

$\frac{1}{2}(wH) \times H = \frac{1}{2}wH^2$, which is equal to the total pressure on the dam per unit length given in Equation (A). Thus, the area of pressure diagram is equal to the total water pressure acting on unit length or the dam. This P acts normal to the surface. The point where it acts is called the centre of pressure (y_{cp}). Applying Equation (3.3), we get:

$$y_{cp} = \bar{y} + \frac{I_G}{A\bar{y}} = \frac{H}{2} + \frac{\frac{1}{12}(1 \times H^3)}{(1 \times H)\cdot H/2} = \frac{2H}{3} \quad \left\{ \begin{array}{l} \because \bar{y} = H/2 \\ A = (1 \times H) \\ I_G = \frac{1}{12}(bd^3) \end{array} \right.$$

i.e. it acts at a depth $2H/3$ from the liquid surface or $H/3$ from the bottom.

This water pressure P exerted on the dam tends to overturn the structure which is resisted by the self-weight W of the dam acting vertically downwards through CG of the mass of the concrete. These two are the main forces acting on the dam. In addition to these main forces, other forces like uplift force acting vertically upwards by the seepage water, wave pressure, earthquake force, silt and ice pressure are also to be considered in design. Details of this design are provided in books on dam engineering.

Sluice gate is an opening used in hydraulic structure where the water pressure acts from both the upstream and downstream sides. Consider a sluice gate as shown in Figure 3.7. Sluice gates are provided in streams or in some hydraulic structures to control the flow or to dispose of extra storm or flood water. These gates may have water on both sides and accordingly, they are subjected to water pressure from both sides. It is essential to consider the maximum probable net water pressure that may act on the gate for its design.

Figure 3.7 shows the sluice gate of area A subjected to water pressure with a depth of water H_1 and H_2 from both sides as shown. Let h be the height of the sluice gate and P_1 and P_2 be the total water pressure acting on the gates from both sides.

Now $$P_1 = wA\,\bar{y}_1 = wA\left(H_1 - \frac{h}{2}\right)$$

and $$P_2 = wA\,\bar{y}_2 = wA\left(H_2 - \frac{h}{2}\right)$$

The resultant water pressure on the gate = $(P_1 - P_2)$

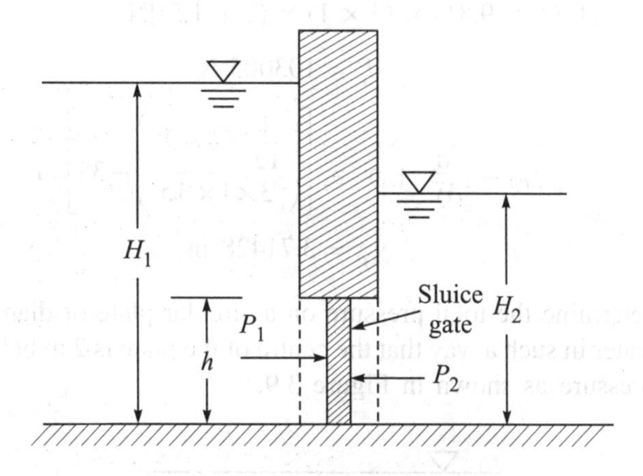

Figure 3.7 Pressure on sluice gate.

EXAMPLE 3.1 Determine the total pressure and depth of centre of pressure on a plane rectangular surface 1 m wide and 3 m deep when its upper edge is horizontal and: (a) Coincides with the water surface, (b) Is 2 m below the water surface as shown in Figure 3.8(a) and 3.8(b).

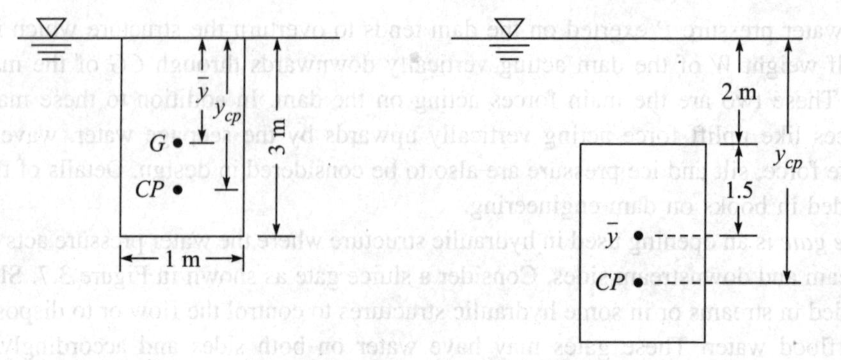

(a) Visual of the first part of Example 3.1 (b) Visual of the second part of Example 3.1

Figure 3.8

Solution: (a) $P = \rho g A \bar{y}$

$$\therefore \quad P = \left(1000 \times 9.81 \times 3 \times 1 \times \frac{3}{2}\right) \text{N}$$

$$y_{cp} = \frac{I_G}{A\bar{y}} + \bar{y} = \left[\frac{\frac{1}{12} \times 1 \times 3^3}{(3 \times 1) \times 3/2} + \frac{3}{2}\right] \text{m}$$

$$y_{cp} = 2 \text{ m}$$ **Ans.**

(b) $P = \rho g A \bar{y} = [(1000 \times 9.81) \times (3 \times 1) \times (2 + 1.5)]\text{N}$

$$P = 103005 \text{ N}$$ **Ans.**

$$y_{cp} = \frac{I_G}{A\bar{y}} + \bar{y} = \left[\left(\frac{\frac{1}{12} \times 1 \times 3^3}{3 \times 1 \times 3.5}\right) + 3.5\right] \text{m}$$

$$y_{cp} = 3.71428 \text{ m}$$ **Ans.**

EXAMPLE 3.2 Determine the total pressure on a circular plate of diameter 1.5 m, which is placed vertically in water in such a way that the centre of the plate is 2 m below the water surface. Find the centre of pressure as shown in Figure 3.9.

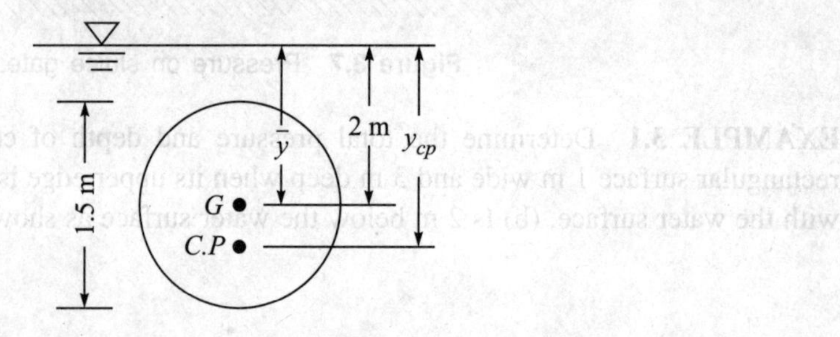

Figure 3.9 Visual of Example 3.2.

Solution: $P = \rho g A\bar{y} = [1000 \times 9.81 \times \pi/4(1.5)^2 \times 2]\text{N}$

$P = 34671.4$ N **Ans.**

$$y_{cp} = \frac{I_G}{A\bar{y}} + \bar{y} = \left[\frac{\frac{\pi \times 1.5^4}{64}}{\pi/4(1.5)^2 \times 2} + 2\right] \text{m}$$

$y_{cp} = 2.0703$ m **Ans.**

EXAMPLE 3.3 A circular plate 2.5 m in diameter shown in Figure 3.10 is immersed in water. Its greatest and least depth below the free surface are 3 m and 1 m, respectively. Find: (a) Total pressure on the face of the plate, (b) Depth of centre or pressure.

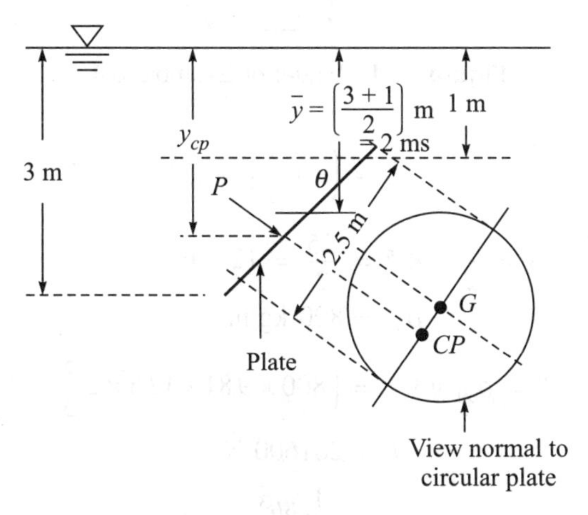

Figure 3.10 Visual of Example 3.3.

Solution: The visual of Example 3.3 is drawn in Figure 3.10.

$$P = \rho g A\bar{y} = \left[1000 \times 9.81 \times \frac{\pi}{4}(2.5)^2 \times \left(\frac{3+1}{2}\right)\right] \text{N}$$

$P = 96309.45$ N **Ans.**

$$y_{cp} = \frac{I_G \sin\theta}{A\bar{y}} + \bar{y}$$

$$\sin\theta = \frac{2}{2.5}$$

$\therefore$ $$\sin^2\theta = \frac{4}{6.25}$$

$\therefore$ $$y_{cp} = \left[\frac{\frac{-\pi}{64}(2.5)^4 \times \frac{4}{6.25}}{\frac{\pi}{4}(2.5)^2 \times 2}\right] + 2 = 2.125 \text{ m}$$

$\therefore$ $y_{cp} = 2.125$ m **Ans.**

EXAMPLE 3.4 Determine the total pressure and depth of centre of pressure on an isosceles triangular plate of base 5 m and altitude 5 m when it is immersed vertically in an oil with specific gravity of 0.8. The base of triangle is 1 m below the free surface of water as shown in Figure 3.11.

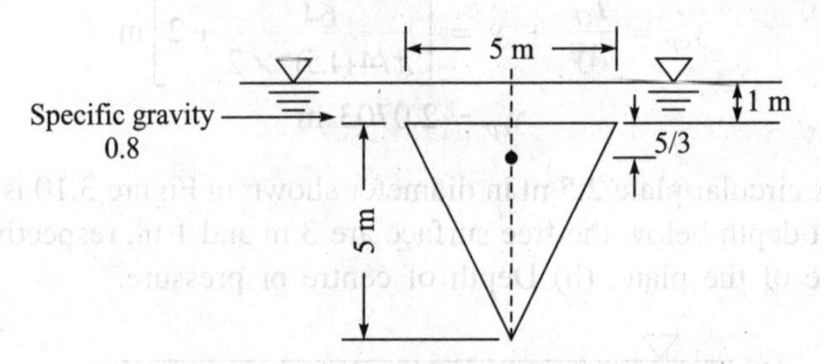

Figure 3.11 Visual of Example 3.4.

Solution:

$$\bar{y} = \left(1+\frac{5}{3}\right) = 2\frac{2}{3} \text{ m}$$

$$A = \frac{1}{2}5 \times 5 = \frac{25}{2} = 12.5 \text{ m}^2$$

$$\rho_{\text{oil}} = 800 \text{ kg/m}^3$$

$$\therefore \quad P = \rho_{\text{oil}}\, gA\,\bar{y} = \left(800 \times 9.81 \times 12.5 \times 2\frac{2}{3}\right) \text{ N}$$

$$P = 261600 \text{ N} \qquad \textbf{Ans.}$$

$$y_{cp} = \frac{I_G}{A\bar{y}} + \bar{y} = \frac{\frac{1}{36}bh^3}{A\bar{y}} + \bar{y}$$

$$y_{cp} = \left[\frac{\frac{1}{36}\times 5\times 5^3}{\frac{1}{2}5\times 5\times 2\frac{2}{3}} + 2\frac{2}{3}\right] \text{m}$$

$$y_{cp} = 3.1874 \text{ m} \qquad \textbf{Ans.}$$

EXAMPLE 3.5 A trapezoidal channel 2 m wide at the bottom and 1 m deep has a side slope 1 H: 1V. Determine the total pressure and depth of centre of pressure if the channel is closed by a vertical trapezoidal gate when it is full with water. (See Figure 3.12).

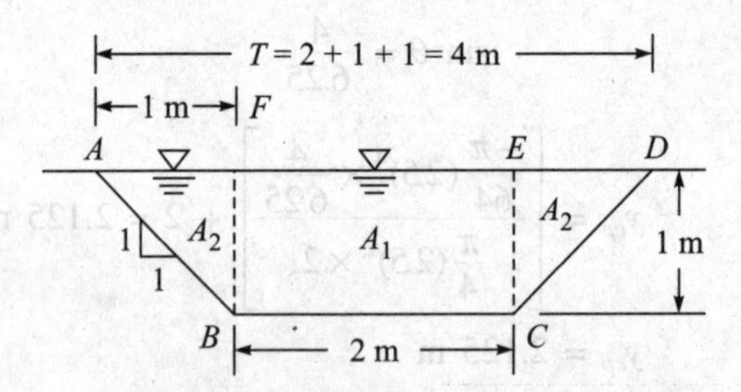

Figure 3.12 Visual of Example 3.5.

Solution: Top width T of the gate,

$$T = 2 + 1 + 1 = 4 \text{ m}$$

$$\text{Area of the gate ABCD} = \left(\frac{4+2}{2}\right) \times 1 = 3 \text{ m}^2$$

$$\bar{y} = \frac{A_1\bar{y}_1 + 2A_2\bar{y}_2}{(A_1 + 2A_2)} = \frac{2\times\frac{1}{2} + 2\times5\times\frac{1}{3}}{2+2\times0.5}$$

$$\bar{y} = 0.4444 \text{ m.}$$

$$P = \rho g A\bar{y} = [1000 \times 9.81 \times 3 \times 0.4444]\text{N}$$

$$P = 13079.996 \text{ N}$$ **Ans.**

$$I_G = \left(\frac{2^2 + 4\times2\times4 + 4^2}{36(2+4)}\right) \times 1^3 \text{ m}^4 \rightarrow \left[\text{Recall formula Table 3.1} \quad I_a = \left[\frac{a^2+4ab+b^2}{36(a+b)}\right]h^3\right]$$

$$I_G = 0.24074 \text{ m}^4$$

$$y_{cp} = \frac{I_G}{A\bar{y}} + \bar{y} = \frac{0.24074}{3\times0.4444} + 0.4444$$

$$y_{cp} = 0.62497 \text{ m}$$ **Ans.**

EXAMPLE 3.6 A cubical tank of 1.5 m side contains water at the bottom upto a depth of 0.6 m, its upper depth 0.9 m is filled with oil of specific gravity 0.9. Calculate the total pressure and depth of centre of pressure on one vertical side. (See Figure 3.13).

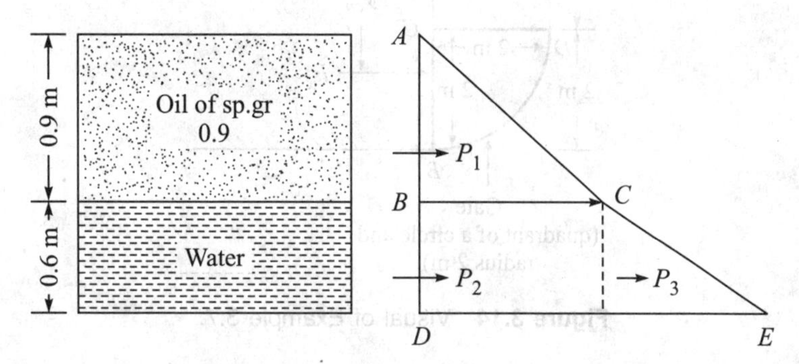

Figure 3.13 Visual of Example 3.6.

Solution: Intensity of pressure $BC = \rho_{0.9} \times g \times 0.9$

$$= 900 \times 9.81 \times .9 \text{ N/m}^2$$

$$= 7946.1 \text{ N/m}^2$$

Intensity of pressure $DE = 7946.1 + 1000 \times 9.81 \times 0.6 = 13832.1 \text{ N/m}^2$

The area of the pressure diagram ABDECA gives the pressure on the vertical wall per unit width.

$$\therefore \quad \text{Total pressure } P = \left[\frac{1}{2}\times 7946.1\times 0.9\times 1.5+\left(\frac{7946.1+13832.1}{2}\right)\times 0.6\times 1.5\right] \text{N}$$

$$P = 15163.8075 \text{ N}$$ **Ans.**

To find y_{cp}, take moment of forces about the top water surface

$$P\times y_{cp} = P_1\times \bar{y}_1 + P_2\bar{y}_2 + P_3\bar{y}_3$$

$$15163.8075\times y_{cp} = \left[\frac{1}{2}(7946.1)\times .9\times 1.5\times \frac{0.9\times 2}{3}\right] + \left[(7946.1\times .6\times 1.5)\times\left(0.9+\frac{0.6}{2}\right)\right]$$

$$+ \left[\frac{1}{2}(13832.1-7946.1)\times .6\times 1.5\times (.9+.6/3\times 2)\right]$$

$$y_{cp} = \frac{5363.6175\times 0.6+7151.49\times 1.2+2648.7\times 1.3}{15163.8075}$$

$$y_{cp} = 1.00524 \text{ m}$$ **Ans.**

EXAMPLE 3.7 Compute the horizontal and vertical components of the total force acting on a curved surface *AB* which is in the form of quadrant of a circle of radius 2 m as shown. Take the width of the gate as unity. (See Figure 3.14).

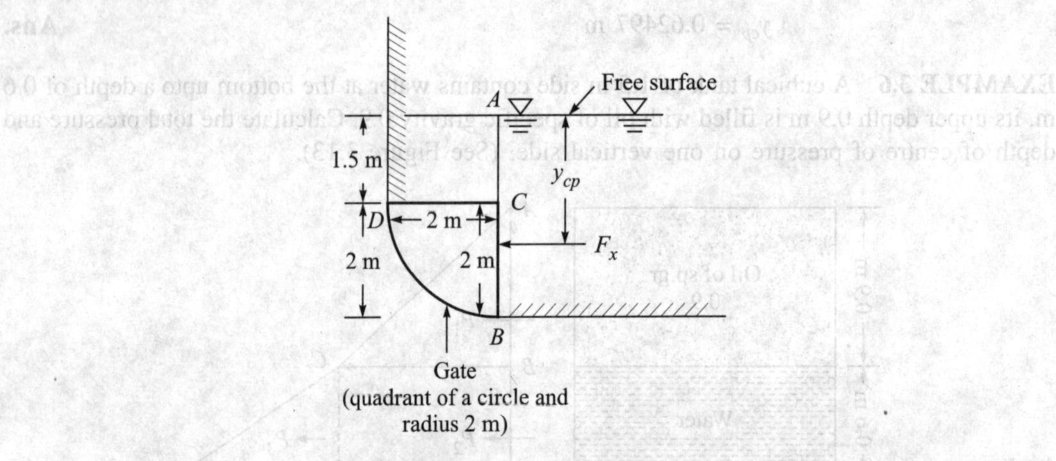

Figure 3.14 Visual of Example 3.7.

Solution: P_x = Total pressure in unit width of gate

= i.e. on projected area of curved surface *BD*

i.e. $P_x = PgA\bar{y} = 1000\times 9.81\times (BC\times 1)\times (1.5+1)$

$P_x = (1000\times 9.81\times 2\times 1\times 2.5)$ N

$P_x = 49050$ N.

Centre of pressure of P_x is y_{cp}

$$y_{cp} = \frac{I_G}{A\bar{y}} + \bar{y} = \frac{\frac{1}{2} \times 1 \times 2^3}{(2 \times 1) \times 2.5} + 2.5 = 2.6333 \text{ m}$$

$$P_y = \text{weight of water above } DB$$

$$= \left[(2 \times 1.5 \times 1) + \frac{1}{4}\pi 2^2 \times 1\right] \times 1000 \times 9.81 \text{ N}$$

$$= 60249.024 \text{ N}$$ **Ans.**

EXAMPLE 3.8 A gate having a shape of a quadrant of circle of 2 m radius has to resist water force as shown. Find the resultant water pressure and angle with horizontal the result acts. Take the unit width of gates. (See Figure 3.15)

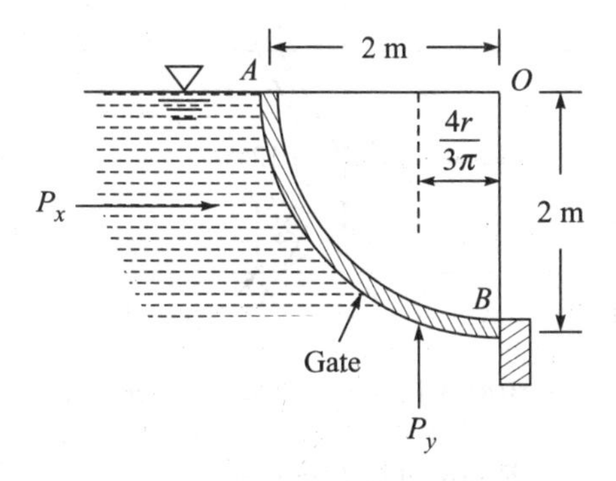

Figure 3.15 Visual of Example 3.8.

Solution:

$$P_x = \rho g(A)\bar{y} = \left[1000 \times 9.81 \times (2 \times 1) \times \frac{2}{2}\right] \text{N}$$

$$P_x = 19620 \text{ N}$$

$$P_y = \text{weight of water (imagined) supported } AB$$

$$= \rho g \text{ (volume in ABO/unit width)} \times \bar{y}$$

$$= 1000 \times 9.81 \times \left(\frac{\pi}{4}D^2\right) \times \frac{1}{4} \times 1 \times 1$$

$$= \left[9810 \times \frac{\pi}{4} \times \frac{4^2}{4} \times 1\right] \text{N}$$

$$= 30819.024 \text{ N.}$$

$\therefore$ $$R = \sqrt{P_x^2 + Py^2} = \sqrt{19620^2 + 30819.024^2} = 36534.321 \text{ N.}$$ **Ans.**

The angle made by R_1

$$\tan\theta = \frac{Py}{Px} = \frac{30819.024}{19620} = 1.57079$$

$\therefore$ $$\theta = 57.51°$$ **Ans.**

EXAMPLE 3.9 The lock gates are 6 m high and supported by two hinges placed at the top and bottom of the gate. They make an angle 120° in a closed position. The width of the gates is 5 m. The water level n/s and ds are 4 m and 2 m respectively. Determine the magnitudes of forces at top and bottom hinges. (See Figure 3.16).

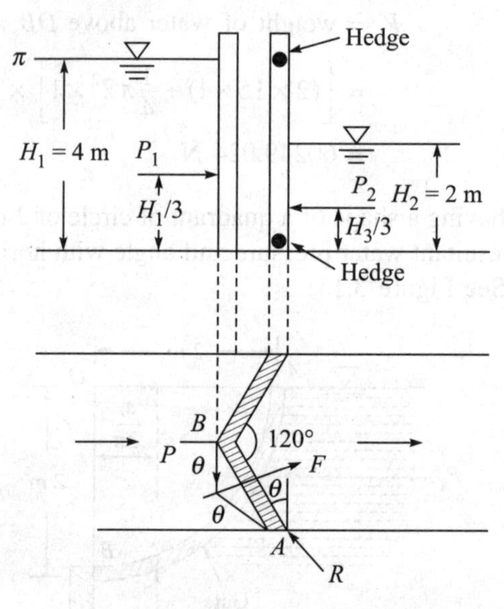

Figure 3.16 Visual of Example 3.9.

Solution:

$$R \cos \theta = P \cos \theta$$

$$\therefore \quad R = P$$

$R \sin \theta + P \sin \theta = F$ by resolving force perpendicular to gate

$$\because \quad R = P$$

$$\therefore \quad 2P \sin \theta = F$$

$$\therefore \quad P = \frac{F}{2 \sin \theta}$$

$$\text{Again } (180° - 20) = 120°$$

$$\therefore \quad 60° = 20°$$

$$\therefore \quad \theta = 30°$$

$$P_1 = \rho g \, A_1 \, \bar{y}_1 = (1000 \times 9.81 \times 5 \times 4 \times 2) \text{ N} = 392400 \text{ N}$$

$$P_2 = \rho g \, A_2 \, \bar{y}_2 = (1000 \times 9.81 \times 5 \times 2 \times 1) \text{ N} = 98100 \text{ N}$$

$$\therefore \quad F = F_1 - F_2 = 294300 \text{ N}$$

$$\therefore \quad P = \frac{F}{2 \sin \theta} = \frac{294300}{2 \sin 30°} = 294300 \text{ N}$$

$$R = P = 294300 \text{ N}$$

If R_t and R_b are the reactions at the top and bottom hinges, $R_t + R_b = R = 294300$ N
Taking moments of hinge reactions R_t, R_b and R about the bottom hinge

$$R_t \times 6 + R_b \times 0 = R \times x \qquad \text{(A)}$$

To find x, taking moment of P, P_1 and P_2 about the bottom hinges

$$P \times x = P_1 \times \frac{4}{3} - P_2 \times \frac{2}{3}$$

$$294300x = 392400 \times \frac{4}{3} - 98100 \times \frac{2}{3}$$

$$\therefore \quad x = 1.5555 \text{ m}$$

Putting x in (A)

$$R_t \times 6 = 294300 \times 1.5555$$

$$\therefore \quad R_t = 76297.275 \text{ N} \qquad \textbf{Ans.}$$

$$\therefore \quad R_b = R - R_t = 294300 - 76297.275$$

$$R_b = 218002.725 \text{ N} \qquad \textbf{Ans.}$$

EXAMPLE 3.10 A taintor gate shown in Figure 3.17 is 0.5 m in length. Find the: (a) Total horizontal pressure of water on the gate, and (b) Total vertical force of water against the gate. The resultant water pressure and its location.

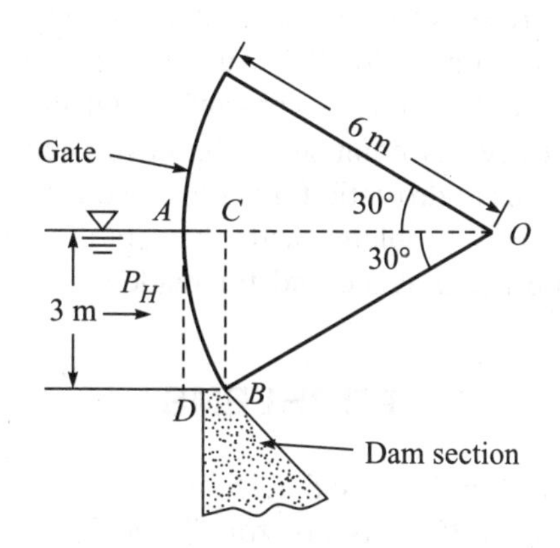

Figure 3.17 Visual of Example 3.10.

Solution: If P_H is the total horizontal water pressure,

$$P_H = \rho g \, (AD \times 0.5) \times \bar{y}$$

$$P_H = [1000 \times 9.81 \times (3 \times .5) \times 1.5]\text{N}$$

$$P_H = 22072.5 \text{ N} \qquad \textbf{Ans.}$$

$$Y_{cp} = \frac{I_G}{A\bar{y}} + \bar{y} = \frac{\frac{1}{12}(0.5) \times 3^3}{(.5 \times 3) \times 1.5} + 1.5$$

$Y_{cp} = 2$ m from top of liquid surface.

If P_V is the vertical water pressure

P_V is the weight of water pressure, it is the weight of the imaginary volume of ABC

$\therefore \quad P_V$ = volume in AB O A – volume in OBC

$$= \left[\left(\frac{\pi}{12} \times 6^2 \times 5\right) - \left(\frac{1}{2} 6 \cos 30° \times 3 \times 0.5\right)\right] + 9 \because \frac{30°}{2} = 15° = \frac{\pi}{12}, CO = 6 \cos 30°$$

$$= [(4.71238 - 3.8971)\ \rho g]\text{N}$$

$\therefore \quad P_V = 8056.7568$ N. **Ans.**

$\therefore$ Resultant $R = \sqrt{P_H^2 + P_V^2} = \sqrt{220728^2 + 8056.7568^2}$

$R = 23496.948$ N **Ans.**

Inclination of R with horizontal = θ

$$\tan\theta = \left(\frac{8056.756}{22072.5}\right) = 0.365$$

$\therefore \quad \theta = \tan^{-1}(20.05°)$ **Ans.**

3.8 CONCLUSION

Equation of hydrostatic pressure on vertical, horizontal, inclined and curved surface have been derived. Total hydrostatic force on hydraulic structures like dam, weir, barrage, lock gate, spillway crest gates, sluice gate, tank etc. is very important in the design of those structures. Equation of this hydrostatic force, its point of action have been derived in details. Numerical examples of determination of this hydrostatic force on sluice gate, spillway crest gate, lock gate, tank, curved gate are solved. Few problems with answers are enclosed for students to solve. References of some books are listed at the end for readers.

PROBLEMS

3.1 A rectangular plate of 2 m depth and 1.25 m width is immersed vertically in water so that the upper edge of 1.25 m is horizontal and is 1.6 m below the water surface. Determine the total pressure on the plate and depth of the centre of pressure.

(**Ans.** $P = 63765$ N, $y_{cp} = 2.7282$ m)

3.2 A circular plate 1.2 m in diameter is immersed vertically in water so that the centre of the plate is at 2 m below the water surface. Determine the total pressure and depth of centre of pressure.

(**Ans.** $P = 22192.22$ N, $y_{cp} = 2.04499$ m)

3.3 A circular plate of 2 m diameter is immersed in water so that the distance of its perimeter from the water surface varies from 1 m to 2.5 m. Find the water pressure acting on one side of the plate and the vertical distance of its centre of pressure from water surface.

(**Ans.** $P = 53927.53$ N $y_{cp} = 1.83$ m.)

3.4 A traitor gate shown in Figure 3.18 is retaining water. Calculate the horizontal water pressure and vertical water pressure assuming the unit metre width of the gate and the resultant of its line of action.

(**Ans.** P_H = 19620 N., P_V = 7109.3 N R = 20868.31 N $\theta = \tan^{-1}$ (19.91°))

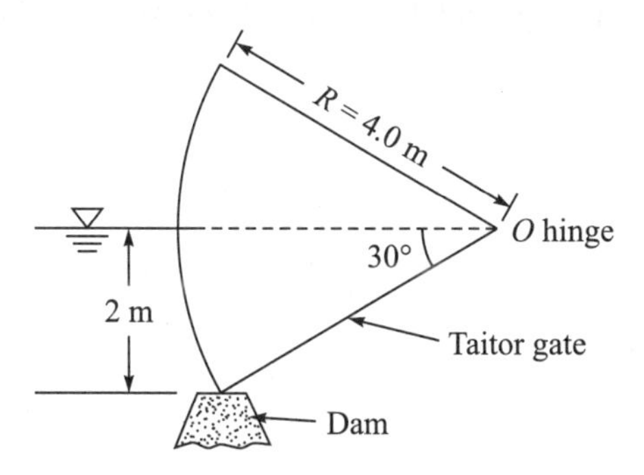

Figure 3.18 Visual of Problem 3.4.

3.5 A trapezoidal plate having its parallel sides $2a$ and a at a distance h apart is immersed in water with its $2a$ side towards the water surface vertically and at a distance h from the water surface. Find the water force and depth of centre of pressure in terms a and h.

$$\left(\textbf{Ans.}\ \ P = \frac{131\text{g}\, ah^2}{6}\ N,\ y_{cp} = \frac{3h}{2}\right)$$

3.6 A triangular gate, which has a base 1.5 m and an altitude of 2 m, lies in a vertical plane. The vertex of the gate is 1 m below the water surface in a tank of oil having specific gravity of 0.8. Find the pressure exerted by the oil of the gate and its depth or centre of pressure.

(**Ans.** P = 27468 N, y_{cp} = 2.43 m)

3.7 A rectangular sluice gate is situated on the vertical wall of a lock. The vertical side of the sluice is 6 m in length and the depth of centroid of area is 8 m below the water surface. Prove that the depth of the centre of pressure is given by 8.475 m.

3.8 A tank contains water upto a height of 1 m above the base. An immiscible liquid of specific gravity 0.8 is filled on the top of the water upto a height of 1.5 m. Calculate the total pressure on one side of the tank and the position of centre of pressure for one side of the tank which is 3 m wide.

(**Ans.** 76518 N, y_{cp} = 1.686 m)

REFERENCES

1. Rouse, H., *Elementary Mechanics of Fluid*, John Wiley & Sons, Inc., N.Y., 1946.
2. Streeter, V.L. and E.B. Waylie, *Fluid Mechanics,* McGraw-Hill, New York, 1983.
3. Bansal, R.K., *Fluid Mechanics and Hydraulic Machines,* Laxmi Publications, New Delhi, 1983.

4. Ramamruthum, S., *Hydraulics, Fluid Mechanics and Fluid Machines,* Dhanpat Rai and Sons, Delhi, 1980.
5. Modi, P.N. and S.M. Seth, *Hydraulics and Fluid Mechanics including Hydraulic Machines*, 3rd ed., Standard Book House, New Delhi, 1977.
6. Munson, B.R., D.F. Young, and T.H. Okiishi, *Fundamental of Fluid Mechanics*, Wiley Text Books, 2001.

Chapter 4

Buoyancy and Floatation

4.1 INTRODUCTION

An attempt is made in this chapter to present a detailed analysis of well known Archimedes principle of buoyancy and floatation. Centre of buoyancy, metacentre and its importance on floating bodies are analysed. Determination of metacentric height, both experimental and analytical, stability analyses of floating bodies are also presented.

4.2 BUOYANCY, BUOYANT FORCE AND CENTRE OF BUOYANCY

When a body is immersed in fluid either partially and fully, it is subjected to an upward force which tends to lift or buoy the body. This tendency of a body immersed or floating in fluid is known as buoyancy.

The force exerted by the fluid on the body to lift it against the force of gravity is called 'buoyant force'.

The point of application of this buoyant force on the body is called centre of buoyancy B.

The magnitude of this buoyant force is determined by the well-known Archimedes principle. In the third century B.C., Archimedes formulated the well-known laws of floating bodies. The Archimedes principle states that when a body is immersed wholly or partially, it is lifted up by a force and this buoyant force is equal to the weight of the volume of fluid displaced by the body, i.e. buoyant force is equal to the weight of the volume of fluid displaced by the body and acts upward opposite to the force of gravity.

Consider Figure 4.1 in which a body is immersed. An elemental cylinder of the body is at a depth y, height h, and has area da.

The downward force on the top of this small elemental cylinder is $p\,da$ where p is the intensity of pressure at depth y.

Upward force on this cylinder of height at the bottom h is equal to $(p + wh)\,da$.

Net upward force on the cylinder $= (p + wh)\,da - pda = whda$.

Since the body is imagined to be forced by such infinite number of elemental cylinders, net upward force on the body $= \Sigma wh\,da$

$$= w\Sigma hda$$
$$= w \times \text{volume of the body}$$
$$= w \times \text{volume of fluid displaced}$$
$$= \text{weight of the fluid displaced.}$$

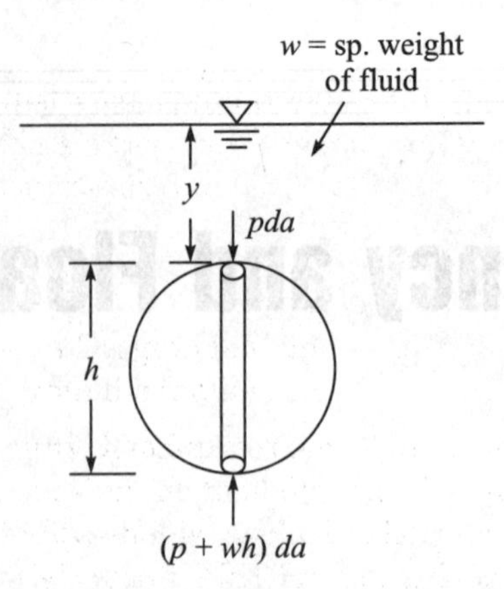

Figure 4.1 Buoyant force equal to the weight of volume of fluid displaced.

Thus upward force on the body = buoyant force F_B = weight of the fluid displaced. This buoyant force acts vertically upwards in the *CG* of the volume of water displaced and this *CG* is called the centre of buoyancy.

4.3 METACENTRE AND METACENTRIC HEIGHT

Metacentre is defined as the point about which the floating body starts oscillating when the body is given a small angular displacement.

Consider Figure 4.2(a) in which the body is floating in equilibrium. *G* is its centre of gravity and *B* is the centre of buoyancy. In the equilibrium condition, both *G* and *B* lie on the same vertical axis.

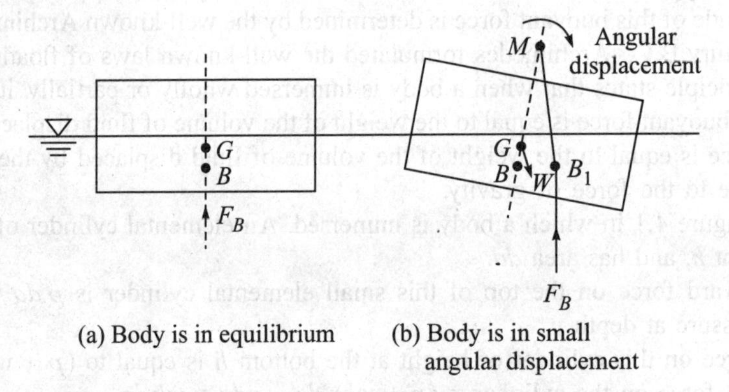

Figure 4.2 Metacentre *M* and Metacentric height M_G.

Let the body be given a small angular displacement clockwise as shown in Figure 4.2(b). The centre of buoyancy *B* in the tilted position will be shifted to a new position B_1 as the shape of volume displaced will be different. The vertical line through this new centre of buoyancy B_1 will intersect the normal axis of the body at *M*. This point *M* is called the metacentre. The body starts oscillating about this point M, as the force of buoyancy F_B tries to restore the body to its original position with the weight of the body W acting downwards. The distance MG is called metacentric height. Depending upon the position of M (if quite above G) after oscillating under the acting $F_B\uparrow$ and $W\downarrow$, the body will come to its original position. Metacentre or metacentric height is very important in the case of floating bodies like ships, boats, etc.

4.4 DETERMINATION OF METACENTRIC HEIGHT

Metacentric height may be determined by two different methods. These are:

(i) Experimental Method
(ii) Analytical Method

4.4.1 Experimental Method

The metacentric height of a floating vessel can be determined experimentally, provided that the centre of gravity G of the floating body is known. Consider Figures 4.3(a) and 4.3(b). Let m be a small known weight placed in the centre of the floating body. [Figure 4.3(a)].

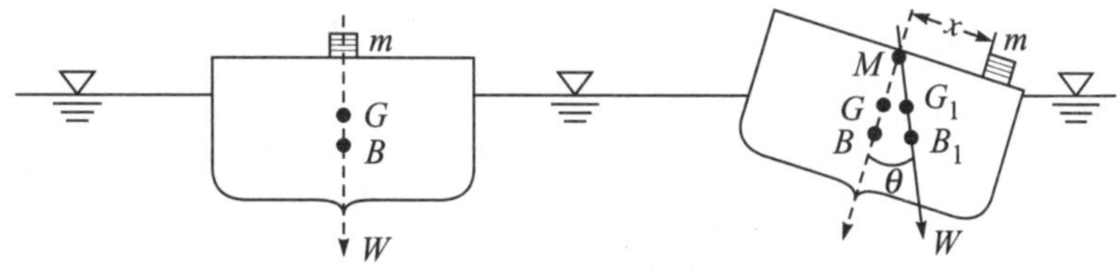

(a) Floating vessel is in equilibrium (b) Floating vessel is tilted by an angle θ by movement of weight m

Figure 4.3 Experimental determination of metacentric height.

Let W be the weight of the floating body (say a small ship), G be the centre of the gravity of the vessel, and B be the centre of buoyancy.

Let the small weight in Figure 4.3(a) move a distance x towards the right. The vessels will be tilted through an angle θ. This θ is measured by a plumbline and protector attached in the vessel. Due to the movement of the small weight m through distance x, the position of G and centre of buoyancy B are shifted to G_1 and B_1, respectively. The moment of the weight W of the vessel due to change G to $G_1 = W.\ MG \tan\theta$. The moment due to movement of weight m through a distance x is

$$= mx$$

For equilibrium in Figure 4.4(b),

$$W{\cdot}MG \tan\theta = mx.$$

$$\text{Metacentric height } MG = \frac{mx}{W\tan\theta} \tag{4.1}$$

4.4.2 Analytical Method

Figure 4.4(a) shows the floating body in equilibrium. G and B are the CG and centre of buoyancy of the body. A small angular displacement of θ is given to the body in the clockwise direction. After displacement, the new centre of buoyancy B_1 is shown. A vertical line through B_1 cuts the normal axis of the body at M. Hence M is the metacentre and MG is the metacentric height.

This angular displacement causes the wedge-shaped prism BOB' on the right [Figure 4.4(b)] to go inside the water while the identical prism AOA' on the left of the axis emerges out of the water. These two wedge-shaped prisms increase the buoyant force by dF_B on the light side and decrease buoyant force by the same magnitude of dF_B on the left side as shown in Figure 4.4(b), by their directions acting through the CG of the two wedges BOB' and AOA'. The centre of buoyancy is displaced from B to B_1, due to this angular displacement.

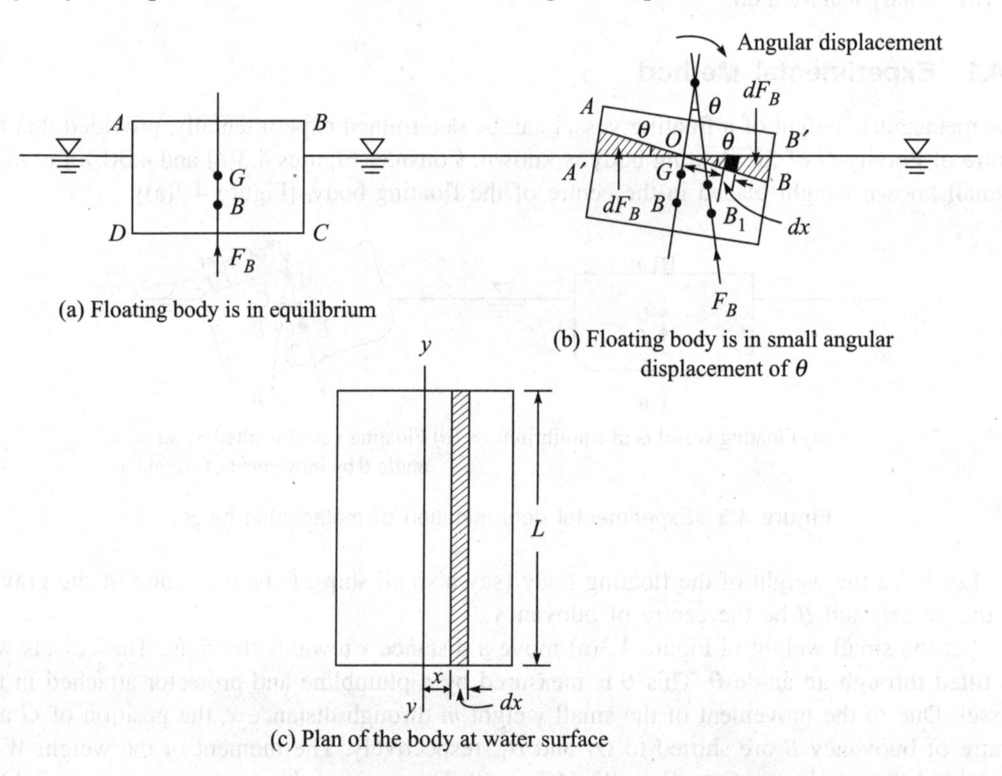

(a) Floating body is in equilibrium

(b) Floating body is in small angular displacement of θ

(c) Plan of the body at water surface

Figure 4.4 Analytical determination of metacentric height.

Figure 4.4(c) is the plan of the body at water line. Towards the right of the axis, a small strip dx at a distance x from centre is considered. Height of the strip $= x \times \angle BOB' = x\theta$

$$\text{Area of the strip} = x\theta dx$$

If L is the length of the floating body, volume of the strip $= Lx\theta dx$.

$$\text{Weight of the strip} = \rho g\ (Lx\theta dx)$$

Similarly weight of the strip at a distance x from the centre line towards the left side

$$= \rho g\ (Lx\theta dx)$$

The two weights act in opposite directions and thus they constitute a couple. The moment of this couple = weight of each strip × distance between them

$$= \rho g\ (Lx\theta dx)\ (x + x)$$
$$= 2\rho g x^2 \theta L dx$$

Moment of the couple for the wedge $= \int 2\rho g x^2 \theta L dx$ (A)

Moment of the couple due to shifting centre of buoyancy from B to B_1

$$= F_B \times BB_1$$
$$= F_B \times BM \times \theta\ (\because\ \theta \text{ is small})$$
$$= W \cdot BM\ \theta \quad \because\ W = F_B. \qquad \text{(B)}$$

Now equating (A) and (B), since these two couples are the same, we get

$$W \cdot BM \cdot \theta = \int 2\rho g x^2 \theta L dx.$$

$$W \cdot BM \cdot \theta = 2\rho g x \theta \int x^2 L dx.$$

$$W \cdot BM = 2\rho g \int x^2 L dx.$$

But $(L \cdot dx)$ from Figure 4.4(c) $= dA$

$$W \cdot BM = \rho g\ (2\int x^2 dA)$$

But Figure 4.4(c) shows that the term $(2\int x^2 dA)$ is second moment of area of the man of the body about yy at the water surface, i.e. moment of inertia I.

$$\therefore \qquad W \cdot BM = \rho g I$$

$$\therefore \qquad BM = \frac{\rho g I}{W}$$

But W = weight of the body

= weight of the fluid displaced by the body

= ρg × Volume of the water displaced

= ρg × Volume of the body submerged in liquid

= $\rho g \forall$ where $\forall$ = volume of the body submerged in liquid

$$\therefore \qquad BM = \frac{\rho g I}{\rho g \forall} = \frac{I}{\forall} \qquad (4.2)$$

$$MG = BM - BG$$

$$\therefore \qquad MG = \frac{I}{\forall} - BG \qquad (4.3)$$

where MG is the metacentric height.

4.5 CONDITIONS OF EQUILIBRIUM OF FLOATING BODY

The floating body has a tendency to return to its original positions or to a new position after it has been slightly displaced. Under the action of an external force, a floating or submerged body may be given a small linear displacement in the vertical or horizontal direction, or a slight angular displacement. Depending upon the relative position of centre of gravity and metacentre, a floating body has the following three conditions of stability or equilibrium:

(i) Stable Equilibrium
(ii) Unstable Equilibrium
(iii) Neutral Equilibrium

4.5.1 Stable Equilibrium

A floating body is said to be in a state of stable equilibrium if a small displacement produces a couple which tends to restore the body to its original position.

A floating body in Figure 4.5 has undergone a small angular displacement θ in the clockwise direction. The centre of buoyancy becomes B_1. The metacentre M is much above the centre of gravity G. Now buoyant FB and weight W produce a couple equal to ($W \cdot MG \tan \theta$) which is the restoring couple, tending to restore the body to its original position. Therefore, it is stated that if the metacentre M lies above the centre of gravity G of the body, i.e. $BM > BG$, then the body is in a stable equilibrium. For this condition to be achieved, G should always be much below the floating body as in the case of ships, boats, etc. for safety.

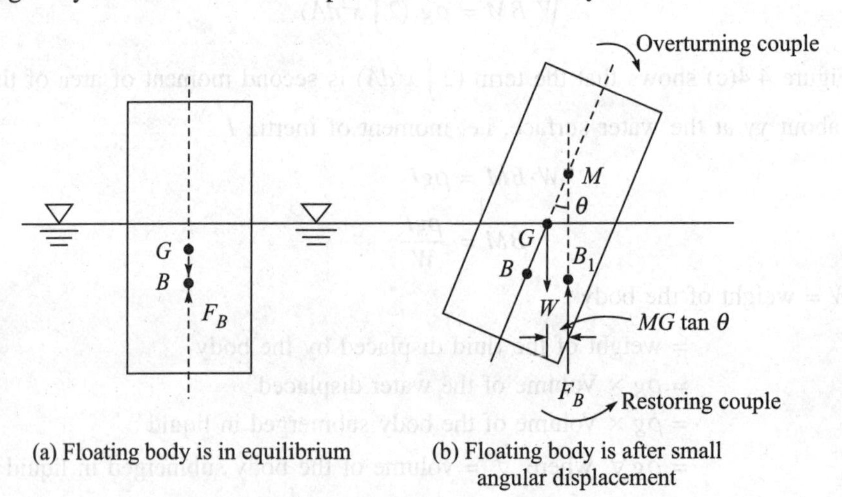

Figure 4.5 Floating body in stable equilibrium.

4.5.2 Unstable Equilibrium

As shown in Figure 4.6(a), when the floating body is slightly tilted in the clockwise direction, metacentre M comes below the centre of gravity G of the body. In Figure 4.6(b), both buoyant force F_B and weight W form a couple in the same clockwise direction, thus tending to increase the angular displacement of the body. Then the floating is considered to be in unstable equilibrium, i.e. when M lies below G or $BM < BG$.

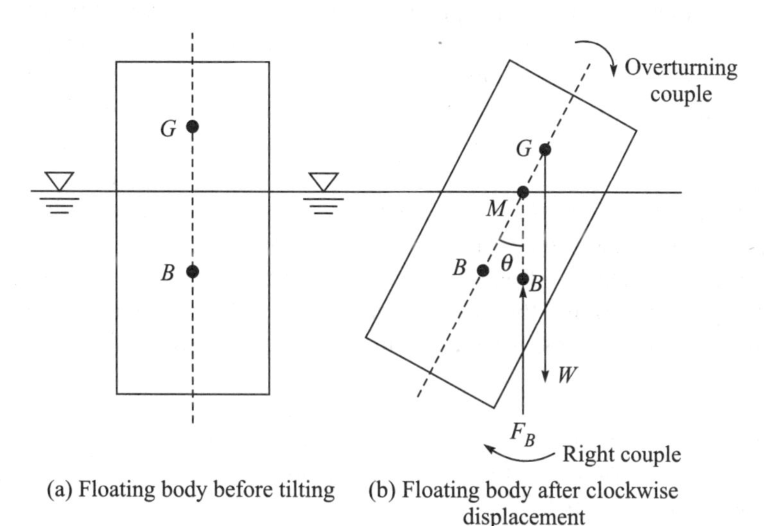

Figure 4.6 Floating body in unstable equilibrium.

4.5.3 Neutral Equilibrium

If the floating body in Figure 4.7(a) is tilted clockwise by a smaller displacement and it may so happen that after displacement, the vertical line passes through a new centre of buoyancy B_1, passes through the centre of gravity G of the body and then equal forces W and F_B act on the same line from the opposite direction and the floating body neither comes back to its original position nor heels further but remains in equilibrium in position shown in Figure 4.7(b). This condition is stated to be neutral equilibrium i.e. $BM = BG$.

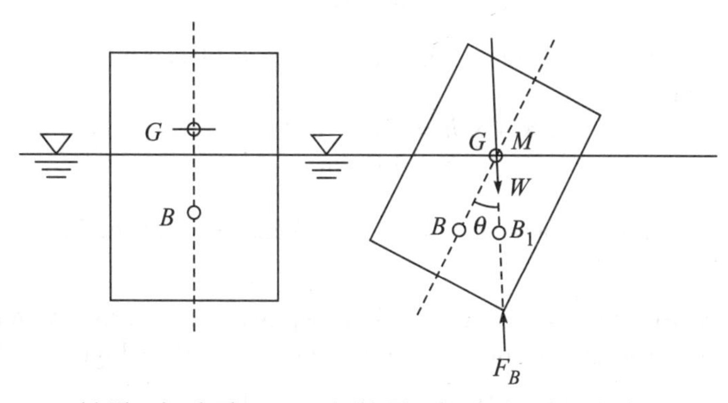

Figure 4.7 Floating body in neutral equilibrium.

4.6 NEED FOR EQUILIBRIUM STUDY

Floating bodies like boats, ships etc., should always be stable under any action of external force. The external forces that are usually encountered by these floating bodies are wave action, wind force, tidal and river currents, pressure due to manoeuvring of a boat or a ship in a curved path by a rudder or propeller action, anchor line pull, etc. These forces cause the floating body to heel. This heeling may also be caused due to the probable shifting of cargo or of passengers or any uneven distribution of load. Therefore, while designing such floating bodies, care must be taken to keep the metacentre always above the centre of gravity under any action of heeling by any of the forces. This can be achieved if the *CG* of the body is placed permanently at a much lower position by adding some heavy material near the bed known as ballast. It is a usual practice to check the metacentric height of a floating body like ship before putting it for operation i.e. $BM \gg BG$. In the practical field, the metacentric height of ocean-going ships varies from 0.3 m to 1.2 m while in warships, it varies from 1 m to 1.5 m. Some of river crafts may be as large as 3.6 m.

EXAMPLE 4.1 A wooden block of width 2 m, depth 1.5 m and length 4 m floats horizontally in water. Find the volume of water displaced and the position of centre buoyancy. The specific gravity of the wooden block is 0.7.

Solution: Volume of the wooden block = $2 \times 1.5 \times 4 = 12$ m^3

Since specific gravity of the wood is 0.7, density of the wood = 700 kg/m^3

Weight of the block = $\rho g \times$ volume = $700 \times 9.81 \times 12 = 82404$ N.

For equilibrium, the weight of water displaced = Weight of wooden box

= 82404 N

$$\therefore \quad \text{Volume of water displaced} = \frac{\text{Weight of water displaced}}{\text{Weight density of water}} \text{ m}^3$$

$$= \frac{82404}{9.81 \times 1000} \text{ m}^3$$

$$= 8.4 \text{ m}^3$$

Ans.

Volume of wooden block in water = Volume of water displaced if h is the depth up to which the block is immersed,

$$2 \times 4 \times h = 8.4$$

$$\therefore \quad h = 1.05 \text{ m}$$

$$\therefore \quad \text{Centre of buoyancy} = \frac{1.05}{2} = 0.525 \text{ m from the base}$$

Ans.

EXAMPLE 4.2 A wooden log of 0.8 m and 6 m long is floating in river water. Find the depth of the water when specific gravity of the wooden log is 0.7. (See Figure 4.8).

Solution: Weight of wooden log = Weight density $\times$ Volume

$$= (0.7 \times 1000) \times 9.81 \times \pi/4(0.8)^2 \times 6 \text{ N}$$

$$= 2111.150263 \times 9.81 \text{ N}$$

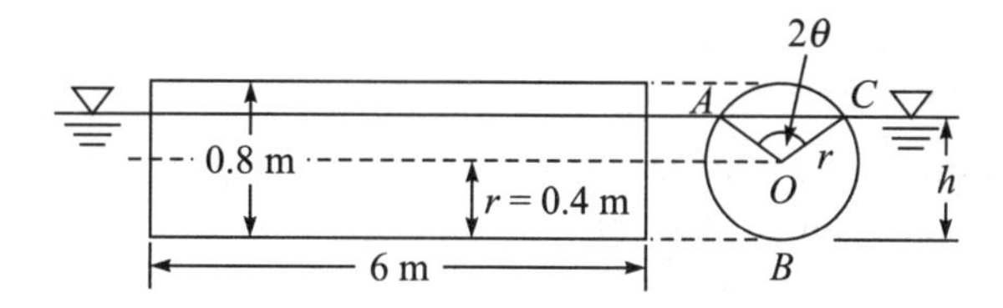

Figure 4.8 Visual of Example 4.2.

For equilibrium, weight of log = weight of water displaced

$$9.81 \times 2111.150263 = 1000 \times 9.81 \times \text{Volume of water displaced}$$

$\therefore$ Volume of water displaced = 2.11115 m^3

If h is the depth of immersion

$$\text{Area immersed} = \text{ABCOA} + \text{AOC}$$

$$= \pi r^2\left(\frac{360 - 2\theta}{360}\right) + \frac{1}{2} r \cos \theta \cdot 2r \sin \theta$$

$$= \pi r^2\left(1 - \frac{\theta}{180}\right) + r \cos \theta \cdot \sin \theta$$

$$\therefore \quad \frac{2.11115}{6} = \pi(0.4)^2\left(1 - \frac{\theta}{180}\right) + 0.4 \cos \theta \cdot \sin \theta$$

or $$0.35185 = 0.50265 - 0.002792\,\theta + 0.16 \cos \theta \sin \theta$$

or $$0.1508 = 0.002792\,\theta - 0.16 \cos \theta \sin \theta$$

$$\theta - 57.30659 \cos \theta \sin \theta = 54.01146$$

$$\therefore \quad 54.01146 = \theta - 57.3065 \cos \theta \cdot \sin \theta$$

by trial and error, $\theta = 71°$

$$\therefore \quad h = r + r \cos 71°$$

$$\therefore \quad h = 0.53 \text{ m}$$ **Ans.**

EXAMPLE 4.3 A stone weighs 490.5 N in air and 196.2 N in water. Determine the volume of the stone and its specific gravity.

Solution: Weight in air – weight in water = weight of the volume of water displaced

$\therefore$ Weight of volume water displaced = 490.5 – 196.2 = 294.3 N

$$\therefore \quad \text{Volume of water displaced} = \frac{294.3}{9.81 \times 1000}\ 0.03 \text{ m}^3$$ **Ans.**

$$\text{Density of the stone} = \text{Mass in air/volume}$$

$$= \frac{(490.5/9.81)}{0.03} = 1666\,\frac{2}{3}\ \text{kg/m}^3$$

$$\therefore \quad \text{Specific gravity of the stone} = \frac{\text{Density of stone}}{\text{Density of water}} = \frac{5000}{1000 \times 3} = \frac{5}{3}$$

$$= 1.66667$$ **Ans.**

EXAMPLE 4.4 A body of dimension is 2 m × 1 m × 3 m weighs 3924 N in water. Find its weight in air and specific gravity.

Solution: Volume of the body = (2 × 1 × 3) = 6 m^3

Weight in water = 3924 N

Volume of water displaced = volume of the body assuming full immersion

= 6 m^3

∴ Weight of water displaced = (9.81 × 1000 × 6) N = 58860 N

For equilibrium of the body,

Weight of the body in air – Weight of water displaced = Weight of the body

$$W_{air} - 588860 = 3924$$

$$\therefore \quad W_{air} = 588860 + 3924 = 62784 \text{ N}$$

$$\text{Mass of the body} = \frac{62704}{9.81} = 6400 \text{ kg/m}^3$$

$$\text{Density of the body} = \frac{6400}{6} = 1066.67$$

$$\therefore \quad \text{Specific gravity} = \frac{1066.67}{1000} = 1.06667$$ **Ans.**

EXAMPLE 4.5 Find the density of metallic body that floats in full submergence at the interface of mercury and water such that 40% of its volume is submerged in mercury and 60% in water.

Solution: Assume the volume of the body to be V m^3

Then 0.4 V m^3 is submerged in mercury and 0.6 V m^3 in water.

We know $F_B = W$

$$F_B = F_{B \text{ of mercury}} + F_{B \text{ of water}}$$

$F_{B \text{ of water}}$ = weight water displaced = (1000 × 9.81 × 0.6 V)N

$F_{B \text{ of mercury}}$ = weight of mercury displaced = (13.6 × 1000 × 9.81 × .4 V)N

weight of the body $W_B = \rho_B \times 9.81 \times V$

For equilibrium,

$$W_B = F_{B \text{ of water}} + F_{B \text{ of mercury}}$$

$$\therefore \quad \rho_B \times 9.81 \times V = 1000 \times 9.81 \times 0.6V + 13.6 \times 1000 \times 9.81 \times .4\,V$$

Dividing by 9.81 × V,

$$\rho_B = 1000 \times .6 + 13.6 \times 1000 \times .4$$

$$\rho_B = 600 + 5440$$

$$\rho_B = 6040 \text{ Kg/m}^3$$ **Ans.**

EXAMPLE 4.6 A uniform body 4 m long, 2 m wide, and 1 m deep floats in water. What is the weight of the body if the depth of immersion is 0.6 m. Determine its metacentric height.

Solution: Weight of the body = Weight of water displaced

= (1000 × 9.81 × 4 × 2 × 0.6)N

= 47088 N **Ans.**

To find metacentric height, use Equation (4.3), i.e. metacentric height $M_G = \frac{I}{\forall} - BG$

I is the moment of inertia about the axis yy in the plan of the body

$$\therefore \quad I = \frac{1}{12} \times 4 \times 2^3 = \frac{8}{3} \text{ m}^4$$

$\forall$ is the volume water displaced = $4 \times 2 \times .6 = 4.8 \text{ m}^3$
B is the centre of buoyancy, G is the CG of the body
If A is the bottom of the centre line in xx, i.e. vertical direction

$$AG = \frac{1}{2} = 0.5 \text{ m}$$

$$AB = \frac{0.6}{2} = 0.3 \text{ m}$$

$$\therefore \quad BG = AG - AB = 0.5 - 0.3 = 0.2 \text{ m}$$

$$\therefore \quad MG = \frac{8}{3 \times 4.8} - 0.2$$

Metacentric height MG is

$$\therefore \quad MG = 0.3555 \text{ m}$$ **Ans.**

EXAMPLE 4.7 A solid cylinder of diameter 3 m has a height 2 m. Find the metacentric height of the cylinder when it is floating with its axis vertical. The specific gravity of the cylinder is 0.7.

Solution: Depth of immersion of the cylinder = $(2 \times .7)$ m = 1.4 m
If A is at the base of vertical axis, G is the CG of the cylinder,
B is the centre of buoyancy,

$$AG = \frac{2}{2} = 1 \text{ m}, AB = \frac{1.4}{2} \text{ m} = 0.7 \text{ m} \therefore BG = AG - AB$$

$$\therefore \quad BG = 1 - 0.7 = 0.3 \text{ m}.$$

If MG is the metric height,

$$MG = \frac{I}{\forall} - BG, I = \frac{\pi}{64} D^4 = \frac{\pi \times 3^4}{64} = 3.976 \text{ m}^4$$

where I is the moment inertia of the cylindrical body (circular section in yy direction)

$$\forall = \text{volume of water displaced} = \left(\frac{\pi}{4} D^2 \times 1.4\right) \text{m}^3$$

$$= \left(\frac{\pi \times 3^2}{4} \times 1.4\right) \text{m}^3$$

$$\therefore \quad = 9.896 \text{ m}^3, \quad MG = \left(\frac{3.976}{9.896} - 0.3\right) \text{m}$$

$$MG = 0.1017778 \text{ m}$$ **Ans.**

EXAMPLE 4.8 A wooden cylinder of circular sections and uniform density, and specific gravity 0.6, is required to float in oil of specific gravity 0.8. If the diameter of the cylinder is d and its length is l, show that l cannot exceed $\sqrt{\frac{2}{3}}\, d$ for the cylinder to float with its longitudinal axis vertical. (See Figure 4.9).

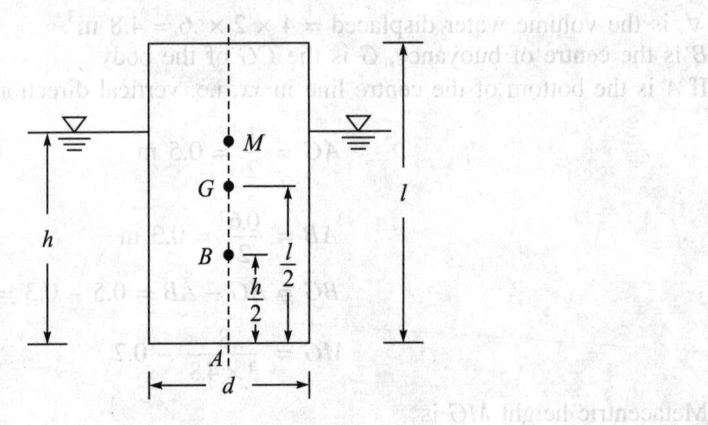

Figure 4.9 Visual of Example 4.8.

Solution: Let the depth of immersion = h

Then
$$\left(\frac{\pi d^2}{4} \times h\right) \times 0.8 = [(\pi/4)\, d^2 \times l] \times 0.6$$

$\therefore$
$$h = \frac{3}{4} l$$

$$AB = \frac{h}{2},\; AG = \frac{l}{2}$$

If M is the metacentre,
$$BM = \frac{I}{\text{Volume displaced}}$$

$$BM = \frac{\frac{\pi}{64} d^4}{\left(\frac{\pi}{4} d^2 \times \frac{3}{4} l\right)} = \frac{d^2}{12l}$$

$$BG = AG - AB = \frac{l}{2} - \frac{h}{2} = \frac{l}{2} - \frac{3l}{4 \times 2} = \frac{l}{8}$$

For stable equilibrium,
$$BM > BG$$

$$\frac{d^2}{12l} > \frac{l}{8}$$

or
$$\frac{d^2}{12l^2} > \frac{1}{8}$$

or
$$\frac{12l^2}{d^2} < 8$$

or
$$l^2 < \frac{8}{12} d^2$$

or
$$l < \sqrt{\frac{8}{12}}\, d$$

$\therefore$
$$l < \sqrt{\frac{2}{3}}\, d$$

Shown

EXAMPLE 4.9 A cone floating in water with its apex downwards has a diameter d and vertical height h (Figure 4.10). If the specific gravity of the cone is s, prove that for stable equilibrium.

$$h^2 < \frac{1}{4}\left(\frac{d^2 s^{1/3}}{1 - s^{1/3}}\right)$$

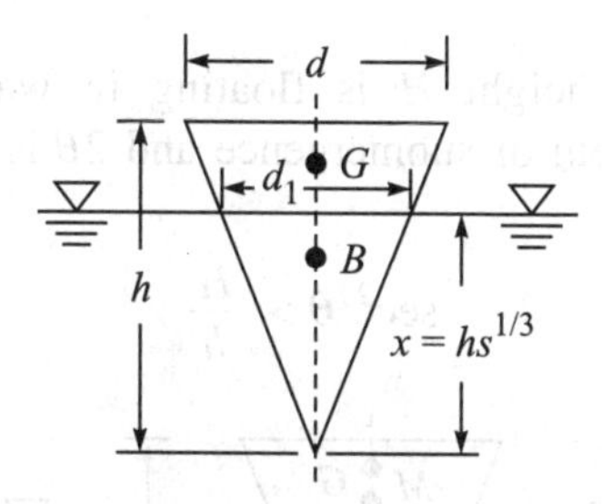

Figure 4.10 Visual of Example 4.9.

Solution: If the height of the portion of the cone under water is x,

then $\frac{1}{3} \cdot \frac{\pi d^2}{4} h \cdot ws = \frac{1\pi}{3} \frac{d_1^2}{4} \cdot xw$, where d_1 is the diameter of the cone at the water surface.

$$d^2 hs = d_1^2 x \tag{A}$$

further $\frac{d_1}{d} = \frac{x}{h}$ from the similar triangular

$\therefore$
$$d_1 = \frac{dx}{h} \tag{B}$$

Substituting d_1 from (B) in (A)

$$d^2 hs = \frac{d^2 x^2}{h^2} \cdot x$$

$$h^3 s = x^3$$

$$\therefore \qquad x = hs^{1/3} \qquad \text{(C)}$$

the CG of the cone is at a depth of one-fourth of the height

$$\therefore \qquad BG = \left(\frac{3}{4}h - \frac{3}{4}x\right) = \frac{3}{4}(h - hs^{1/3}) = \frac{3h}{4}(1 - s^{1/3})$$

If M is the metacentre, $BM = \dfrac{I}{\text{Volume of water displaced}} = \dfrac{\dfrac{\pi d_1^4}{64}}{\dfrac{1}{3}\dfrac{\pi d_1^2}{4}\cdot x}$

Substituting d_1 by Equation (B) and x by Equation (C) and simplifying

$$BM = \frac{3d^2 s^{1/3}}{16h}$$

For stable equilibrium $BM > BG$

$$\frac{3d^2 s^{1/3}}{16h} > \frac{3h}{4}(1 - s^{1/3})$$

or

$$h^2 < \frac{1}{4}\left(\frac{d^2 s^{1/3}}{1 - s^{1/3}}\right)$$

Proved.

EXAMPLE 4.10 A cone of height H is floating in water with its vertex downward (See Figure 4.11). If h is the depth of submergence and 2θ is the angle of the cone at vertex, show that for stable equilibrium,

$$\sec^2 \theta > \frac{H}{h}$$

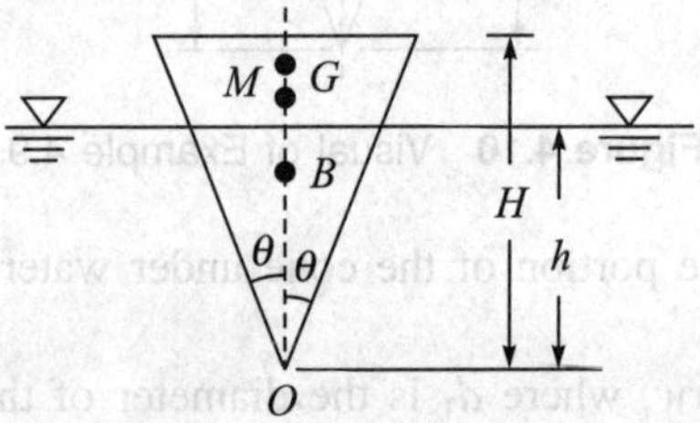

Figure 4.11 Visual of Example 4.10.

If G is the CG of the cone,

$$OG = \frac{3}{4}H$$

If B is the centre of buoyancy

$$OB = \frac{3}{4}h$$

If r is the radius of the cone at the water section,

$r = h \tan \theta$ where θ is the half the angle of cone of vertex

Volume of water displaced $= \frac{1}{3}\pi r^2 h$

Moment of inertia I at the water section $= \frac{\pi}{64}(2r)^4 = \frac{\pi r^4}{4}$

$$BM = \frac{I}{V} = \frac{\pi r^4}{4 \times \frac{1}{3}\pi r^2 h} = \frac{3}{4}\cdot\frac{r^2}{h} = \frac{3}{4}\cdot\frac{h^2 \tan^2\theta}{h} = \frac{3}{4} h \tan^2\theta$$

If M is the metacentre, then

$$OM = OB + BM = \frac{3}{4}h + \frac{3}{4}h \tan^2\theta = \frac{3}{4}h(1 + \tan^2\theta) = \frac{3}{4} h \sec^2\theta$$

For stable equilibrium $OM > OG$, metacentre should be above G.

$$\frac{3}{4} h \sec^2\theta > \frac{3}{4} H$$

$$\sec^2\theta > \frac{H}{h} \quad \text{shown.}$$

EXAMPLE 4.11 Show that a cylindrical buoy of 1 m diameter, 2 m high, weighing 7.848 KN will not float vertically in sea water of density 1030 kg/m^3.

Solution: If metacentre M lies below G, it cannot float vertically or it is unstable.

$$W = 7.848 \text{ KN} = 7848 \text{ N}$$

Weight of cylinder = Weight of water displaced

$$7848 = 1030 \times 9.81 \times \frac{\pi}{4} D^2 \times h \quad \text{where } h \text{ is the depth of immersion}$$

or
$$784 = 1030 \times 9.81 \times \frac{\pi}{4}(1)^2 \times h$$

$\therefore$
$$h = 0.989 \text{ m}$$

If A is the point at bottom of cylinder in the vertical axis

$$AB = \frac{0.989}{2} = 0.4945 \text{ m}$$

$$AG = \frac{2}{2} = 1 \text{ m}$$

$\therefore$
$$BG = AG - AB = (1.0 - 0.4945)\text{m} = 0.5055 \text{ m}$$

If M is the metacentre,

$$BM = \frac{I}{V} = \frac{\frac{\pi}{64}(1)^4}{\frac{\pi}{4}\times(1)^2\times 0.989} = 0.063 \text{ m}$$

$$BG - BM = (0.5055 - 0.063)\text{m} = 0.4425 \text{ m}$$

i.e. G is above metacentre M at a distance 0.4425 m

$\therefore$ The cylinder is unstable, i.e. it cannot float vertically.

4.7 CONCLUSION

It is seen that the contents of this chapter has the practical application in the field of navigation. Knowledge of metacentre, its height, importance, condition of equilibrium of floating bodies are important in inland water transports.

Solved examples are presented for better understanding of theoretical works presented for the students. Few problems are attached for students to solve. A list of references has been given at the end.

PROBLEMS

4.1 A wooden block of rectangular section 1.25 m wide, 2 m depth, 4 m long floats horizontally in sea water. The specific gravity of the wood is 0.64 and sea water density 1025 kg/m^3, find the volume of water displaced and position of centre of buoyancy.

(**Ans.** 6.25 m^3, 0.625 m above the base)

4.2 A wooden cylinder of diameter d and length $2d$ floats with its vertical. Check whether the cylinder in that position is stable or not. Locate its metacentre with reference of water surface. Specific gravity of wood is 0.6.

(**Ans.** Unstable, $0.584d$ below water surface)

4.3 A battle ship weighs 13,000 kn. On filling the ship one side with water, its weight is 60 kn its mean distance from the centre of the ship is 10 m, and the angle of displacement of the plumb line is 2°16′. Determine the metacentric height.

(**Ans.** 1.16 m)

4.4 An iceberg weighing a 15 kg/m^3 is in an ocean with a volume of 600 m^3 above the surface. Determine the total volume of the iceberg if specific weight of ocean water is 1025 kg/m^3.

(**Ans.** 5591 m^3)

4.5 A metallic body floats at the interface of mercury in such a way that 30% of its volume is submerged in mercury and 70% in water. Find the density of the metallic body.

(**Ans.** 4780 kg/m^3)

4.6 A body of dimension 0.5 m × 0.5 m × 1 m and of specific gravity 3.0 is immersed in water. Determine the least force required to lift the body.

(**Ans.** 2452.5 N)

4.7 A solid cylinder of diameter 5 m and height 5 m and floating with its vertical. State whether the cylinder is in stable or unstable equilibrium.

(**Ans.** Unstable)

4.8 A solid cone floats in water with its apex downwards. Determine the least apex angle of the cone for stable equilibrium. Specific gravity of the material of the cone is 0.7.

(**Ans.** 39°7′)

REFERENCES

1. Rouse, H. and J.W. Howe, *Basic Fluid Mechanics*, John Wiley & Sons. Inc., New York, 1953.
2. Prandtl, L., *Essential of Fluid Dynamics*, Hafner Publishing Company, New York, 1952.
3. Streeter, V.L., *Fluid Dynamics*, McGraw-Hill, New York.
4. Harris, C.W., *Hydraulics*, Chapman & Hall, New York, 1936.
5. Bansal, R.K., *Fluid Mechanics and Hydraulic Machanics*, Laxmi Publications, New Delhi, 1983.
6. Shapiro, A.H., 'Basic Equations of Fluid Flow', Section 2 of *Handbook of Fluid Dynamic*, edited by V.L. Streeter, McGraw-Hill, 1961.
7. Modi, P.N. and S.M. Seth, *Hydraulics and Fluid Mechanics including Hydraulic Machines*, Standard Book House, New Delhi, 1977.

Chapter 5

Kinematics of Fluid Flow

5.1 INTRODUCTION

Fluid kinematics is that branch of fluid mechanics which deals with motion of fluid particles without consideration of the character of the particles or the influence of forces upon their motion. From the kinematic point of view, types of flow are translation, rate of deformation and rotation of fluid particles. Each type of motion may occur in any combination. The basic understanding of this branch of mechanics is desirable since these studies constitute the groundwork on which dynamic results are constructed.

5.2 TYPES OF FLUID FLOW

The types of fluid flow are classified as:

(i) Steady and unsteady flows
(ii) Uniform and non-uniform flows
(iii) Laminar, turbulent and transitional flows
(iv) Compressible and in-compressible flows
(v) Rotational and irrotational flows
(vi) One-, two- and three-dimensional flows
(vii) Critical, super-critical and sub-critical flows

5.2.1 Steady and Unsteady Flows

Steady flow is defined as that type of flow in which fluid characteristics like velocity, pressure, density, etc. at a point in space do not change with time. Mathematically,

$$\left(\frac{\partial V}{\partial t}\right)_{x_0, y_0, z_0} = 0, \left(\frac{\partial p}{\partial t}\right)_{x_0, y_0, z_0} = 0, \left(\frac{\partial \rho}{\partial t}\right)_{x_0, y_0, z_0} = 0$$

where (x_0, y_0, z_0) is a fixed point in the fluid field.

Unsteady flow is that type in which the above flow characteristics at a point in space change with respect to time. Mathematically,

$$\left(\frac{\partial V}{\partial t}\right)_{x_0, y_0, z_0} \neq 0, \left(\frac{\partial p}{\partial t}\right)_{x_0, y_0, z_0} \neq 0, \left(\frac{\partial \rho}{\partial t}\right)_{x_0, y_0, z_0} \neq 0$$

5.2.2 Uniform and Non-uniform Flows

When certain fluid characteristics, like velocity, density and pressure do not change with respect to space at a given time, the flow is uniform. Mathematically,

$$\left(\frac{\partial V}{\partial s}\right)_{t=\text{constant}} = 0 \text{ etc.}$$

s is the space in flow direction.

If the above flow characteristics or parameters change with space at a given time, the flow is non-uniform flow. Mathematically,

$$\left(\frac{\partial V}{\partial s}\right)_{t=\text{constant}} \neq 0$$

5.2.3 Laminar, Turbulent and Transitional Flows

When fluid particles move along well-defined paths or streamlines and the paths are parallel and straight, the flow is laminar. Thus the particles move in laminars or layers gliding smoothly over the adjacent layer. Laminar flow is also called viscous flow. The velocity of fluid particles is usually small in laminar flow. If the particles with higher velocity move in a zig-zag manner and eddies are formed, the flow is called turbulent. These flows are normally determined by the Reynolds number which is the ratio of inertia force to viscous force. Reynolds number (R_e) is written as $R_e = \dfrac{VD}{\nu}$, where D is diameter of pipe, ν is the kinematic viscosity. In case of open channel flow, the diameter of pipe D may be replaced by four times the hydraulic radius $R\left(R = \dfrac{A}{P}\right)$, where A is the area of flow P is the wetted perimeter and thus R_e in the open channel may be written as $R_e = \dfrac{V \cdot 4R}{\nu}$. Experimental works in pipe shows that if $R_e < 2000$, the flow remains to be laminar, and if $R_e > 4000$, flow is turbulent and $2000 < R_e < 4000$, state of flow remains in a transition from laminar to turbulent. Therefore, investigators in hydraulics claim that the flow between this range of R_e (i.e. 2000–4000) is the transitional flow.

5.2.4 Compressible and Incompressible Flows

When the density (ρ) of the fluid flow changes from point to point, the flow is compressible. Mathematically, $\rho \neq$ constant. Normally fluid-like gases are compressible. Although most of the liquids have a constant ρ, they may also behave as compressible under pressure and temperature.

Incompressible flow is that type of flow in which the density remains constant in the flow field. Mathematically,

$$\rho = \text{constant}.$$

5.2.5 Rotational and Irrotational Flows

Rotational flow is that type of flow in which fluid particles, while flowing along the streamlines, rotate about their own axes and if they do not rotate while flowing along streamline, it is known as irrotational flow.

Also while flowing, the fluid element may undergo four types of movement, namely, (i) pure translation, (ii) a linear deformation, (iii) rotation, and (iv) angular or shear deformations. In Figure 5.1, (a), (b), (c), (d), (e) and (f) show the rotational flow, irrotational flow motion, translation, linear deformations, pure rotation and angular deformation of fluid, respectively.

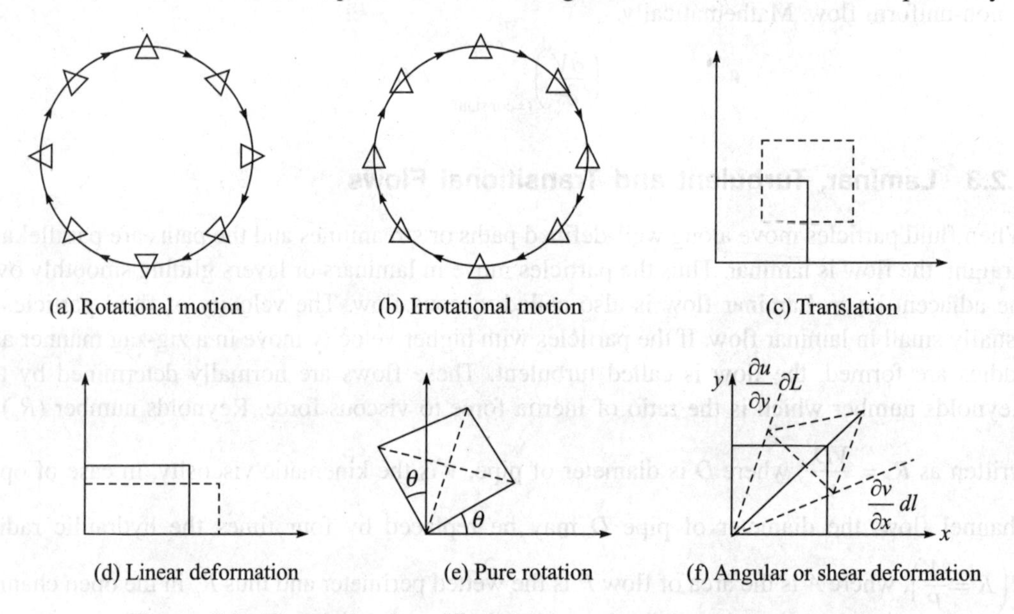

Figure 5.1 Figure showing different types of rotation and movement.

5.2.6 One-, Two- and Three-dimensional Flows

When velocity is a function of time and one space co-ordinates only, say x, it is called one-dimensional flow. If steady one-dimensional flow is a function of only one space co–ordinate; mathematically,

Steady 1-D: $u = f(x)$, $v = 0$, $w = 0$ unsteady 1-D, $u = f(x, t)$, $u = 0$, $w = 0$ where u, v, and w are velocity components in x, y and z directions. In steady two-dimensional flow, velocities are functions of two space co-ordinates, say x and y. Mathematically

Steady 2-D: $u = f_1(x, y)$, $v = f_2(x, y)$ and $w = 0$ unsteady 2-D: $u = f_1(x, y, t)$

$$v = f_2(x, y, t)$$
$$w = 0$$

In steady three-dimensional flow, fluid parameters are functions of three space co-ordinates (x, y and z) only. Mathematically,

Steady 3-D : $u = f_1(x, y, z)$, $v = f_2(x, y, z)$, $w = f_3(x, y, z)$

Unsteady 3-D : $u = f_1(x, y, z, t)$, $v = f_2(x, y, z, t)$, $w = f_3(x, y, z, t)$

5.2.7 Critical, Super-critical and Sub-critical Flows

These three types of flows normally occur in open channel flow where the gravity force is predominant. It is usually defined by Froude's number (F_r) which is the ratio of inertia force to gravity force. If $F_r = 1$, the flow is critical, if $F_r > 1$, the flow is super-critical flow, if $F_r < 1$, the flow is sub-critical.

5.3 RATE OF FLOW OR DISCHARGE (*Q*)

This is defined as the quantity of fluid flowing through a section of pipe or a channel (if it is a liquid). For an incompressible fluid like water, it is expressed as the volume of fluid flowing per second across the flow section. For compressible fluid-like gases, the rate of flow across the section is expressed in weight.

Then Q for liquid is expressed in m^3/sec or litres/sec and Q for gases is expressed in Kgf/sec or Newton/sec. Consider a liquid flow through a flow section of area A with average velocity V.

Then
$$Q = A \cdot V \tag{5.1}$$

5.4 CONTINUITY EQUATION IN 1-D

The equation based on the principle of conservation of mass is called continuity equation. In case of the fluid flowing through a pipe at all cross-sections, the quantity or mass of fluid flowing per second is constant. Consider cross-sections 1-1 and 2-2 of the pipe [(Figure 5.2a)]. If V_1, A_1, ρ_1 and V_2, A_2, ρ_2 are the velocities, cross-sectional areas of flow and densities at 1-1 and 2-2, respectively.

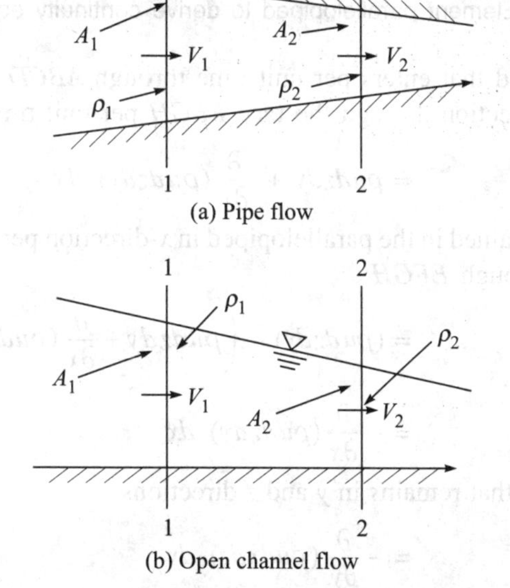

Figure 5.2 Flow of fluid.

Then mass rate of flow at 1-1 is $\rho_1 A_1 V_1$ and mass rate of flow at 2-2 is $\rho_2 A_2 V_2$. According to the conservation of mass

$$\rho_1 A_1 V_1 = \rho_2 A_2 V_2 = \text{Constant} \tag{5.2}$$

Equation (5.2) is applicable to compressible flow and is called continuity equation. If $\rho_1 = \rho_2 = \rho$ in case of liquid, Equation (5.2) in case of open channel flow for incompressible fluid in Figure (5.2b) becomes

$$A_1 V_1 = A_2 V_2 = AV = \text{Constant} \tag{5.3}$$

5.5 CONTINUITY EQUATION IN 3-D

Consider an elementary regular fluid parallelopiped with sides dx, dy and dz as shown in Figure 5.3. The velocity components in x, y and z directions are u, v, w respectively. Let ρ be the density of the fluid within the parallelopiped.

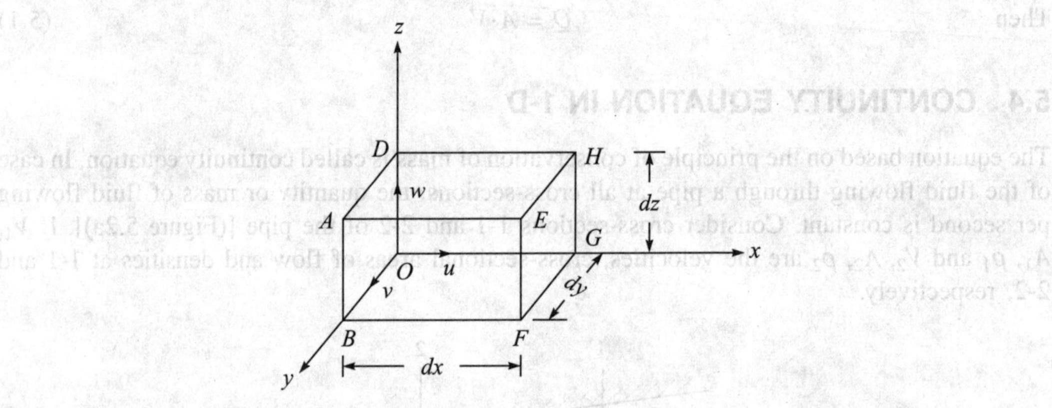

Figure 5.3 Element parallelopiped to derive continuity equation in 3-D.

Now the mass of fluid that enters per unit time through $ABCD$ is $(\rho u dy dz)$
Mass leaving in x direction through the face $EFGH$ per unit time

$$= \rho u dz dy + \frac{\partial}{\partial x}(\rho u dz dy)\, dx$$

$\therefore$ Mass of fluid that remained in the parallelopiped in x-direction per unit time = inflow through $ABCD$ – outflow through $EFGH$

$$= (\rho u dz dy) - \left(\rho u dz dy + \frac{\partial}{\partial x}(\rho u dz dy) dx\right)$$

$$= -\frac{\partial}{\partial x}(\rho u dz dy)\, dx$$

Similarly, mass fluid that remains in y and z directions

$$= -\frac{\partial}{\partial y}(\rho u dx dz)\, dy$$

and $$= -\frac{\partial}{\partial y}(\rho w dx dy) dz$$

The total mass that has remained in the parallelopiped per unit time

$$= -\left[\frac{\partial(\rho u)}{\partial x} + \frac{\partial(\rho u)}{\partial y} + \frac{\partial(\rho w)}{\partial z}\right] dx dy dz \quad \text{(A)}$$

Since mass of water within parallelopiped is neither created nor destroyed any increase of in the mass of water in this space per unit time is equal to the net total water mass is retained in the parallelopiped. Mass of water in the parallelopiped is $(\rho u \cdot dx \cdot dy \cdot dz)$ and rate of increase with time

$$\frac{\partial}{\partial t}(\rho dx dy dz) = \frac{\partial \rho}{\partial t}(dx dy dz) \quad \text{(B)}$$

By definition of conservation of energy (B) = (A)

$$\therefore \quad \frac{\partial p}{\partial t}(dx \cdot dy \cdot dz) - \left[\frac{\partial(\rho u)}{\partial x} + \frac{\partial(\rho v)}{\partial y} + \frac{\partial(\rho w)}{\partial z}\right] dx, dy, dz$$

Dividing by (dx, dy, dz) and taking the limit that the parallelopiped shrinks to a point (x, y, z), we have:

$$\frac{\partial p}{\partial t} + \frac{\partial(\rho u)}{\partial x} + \frac{\partial(\rho v)}{\partial y} + \frac{\partial(\rho w)}{\partial z} = 0 \quad (5.4)$$

which is the unsteady continuity equation in compressible flow in three-dimension.

If the flow is steady $\frac{\partial \rho}{\partial t} = 0$

Now Equation (5.4) becomes

$$\frac{\partial(\rho u)}{\partial x} + \frac{\partial(\rho v)}{\partial y} + \frac{\partial(\rho w)}{\partial z} = 0 \quad (5.5)$$

which is the steady continuity in 3-D for compressible flow. If the fluid is incompressible (i.e. say water), ρ = constant

$\therefore$ Equation (5.5) becomes

$$\frac{\partial u}{\partial x} + \frac{\partial v}{\partial y} + \frac{\partial w}{\partial z} = 0 \quad (5.6)$$

which is the continuity equation of steady flow in three–dimension. In vector notation

If $$V = \mathbf{i}u + \mathbf{j}v + \mathbf{k}w$$

$$\nabla = \frac{\partial}{\partial x}\mathbf{i} + \frac{\partial}{\partial y}\mathbf{j} + \frac{\partial}{\partial z}\mathbf{k}$$

then Equation (5.6) may be written as:

$$\text{div } \mathbf{V} = \nabla \cdot \mathbf{V} = \frac{\partial u}{\partial x} + \frac{\partial v}{\partial y} + \frac{\partial w}{\partial z} = 0 \quad (5.6a)$$

For two-dimensional flow $w = 0$
Equation (5.6) becomes

$$\frac{\partial u}{\partial x} + \frac{\partial v}{\partial y} = 0 \tag{5.7}$$

5.6 VELOCITY AND ACCELERATION IN 3-D

Let V be the resultant velocity at any point in fluid flow and u, v and w be its components in x, y and z directions. The velocity components are functions of space co-ordinates and time. Mathematically, they may be expressed as:

$$u = f_1(x, y, z, t)$$
$$v = f_2(x, y, z, t)$$
$$w = f_3(x, y, z, t)$$

and resultant
$$V = iu + iv + kw + \sqrt{u^2 + v^2 + w^2}$$

Let a_x, a_y and a_z be the total acceleration in x, y and z directions, respectively.
Total derivative of u,

$$du = \frac{\partial u}{\partial x}dx + \frac{\partial u}{\partial y}dy + \frac{\partial u}{\partial z}dz + \frac{\partial u}{\partial t}dt$$

Dividing by dt,

$$\frac{du}{dt} = \frac{\partial u}{\partial x}\cdot\frac{dx}{dt} + \frac{\partial u}{\partial y}\frac{dy}{dt} + \frac{\partial u}{\partial z}\cdot\frac{dz}{dt} + \frac{\partial u}{\partial t}$$

or
$$\frac{du}{dt} = u\frac{\partial u}{\partial x} + v\frac{\partial u}{\partial y} + w\frac{\partial u}{\partial z} + \frac{\partial u}{\partial t}$$

In the above expression, $\frac{du}{dt} = a_x$ (i.e. acceleration in the x-direction)

$\therefore$
$$a_x = \left(u\frac{\partial u}{\partial x} + v\frac{\partial u}{\partial y} + w\frac{\partial u}{\partial z}\right) + \frac{\partial u}{\partial t} \tag{5.8}$$

In above expression, the terms within brackets symbolise the convective acceleration and the term $\frac{\partial u}{\partial t}$ is the local acceleration.

Similarly

$$a_y = \left(u\frac{\partial v}{\partial x} + v\frac{\partial v}{\partial y} + w\frac{\partial v}{\partial z}\right) + \frac{\partial v}{\partial t} \tag{5.9}$$

$$a_z = \left(u\frac{\partial w}{\partial x} + v\frac{\partial w}{\partial y} + w\frac{\partial w}{\partial z}\right) + \frac{\partial w}{\partial t} \tag{5.10}$$

If the flow is one-dimensional, i.e. $V = \int (x, t)$

$$a = \frac{\partial V}{\partial t} + V\frac{\partial V}{\partial x} \qquad (5.11)$$

i.e. acceleration a = local acceleration + convective acceleration.

If the flow is steady, $\frac{\partial V}{\partial t} = 0$

If the flow is uniform, $V\frac{\partial V}{\partial x} = 0$

$\therefore$ $a = 0$ for steady uniform flow.

But if the flow is steady but non-uniform, $\frac{\partial V}{\partial t} = 0,\ V\frac{\partial V}{\partial x} \neq 0$

$$\therefore \qquad a = V\frac{\partial V}{\partial x} \qquad (5.11a)$$

Acceleration vector $\mathbf{A} = \mathbf{i}a_x + \mathbf{j}a_y + \mathbf{k}a_z$ (5.12)

$$= \sqrt{a_x^2 + a_y^2 + a_z^2}$$

EXAMPLE 5.1 A pipe through which water is flowing has diameters 40 cm and 20 cm at cross-sections 1 and 2, respectively. The velocity at section 1 is 5 m/sec. Find the velocity at section 2 and the rate of discharge.

Solution: Continuity equation gives $A_1V_1 = A_2V_2$

or $\pi/4(0.4)^2 \times 5 = \pi/4(0.2)^2\ V_2$

$\therefore$ $V_2 = 20$ m/sec **Ans.**

$\therefore$ $Q = A_2V_2 = \pi/4(.2)^2 \times 20 = 0.628318$ m^3/sec **Ans.**

EXAMPLE 5.2 A 40 cm diameter pipe converging in water, branches into two pipes of diameters 30 cm and 20 cm, respectively. If the average velocity in a 40 cm pipe is 3m/sec, find the discharge in this pipe. Also determine the velocity in a 20 cm pipe if the average velocity in 30 cm pipe is 2 m/sec.

Solution: Discharge in 40 cm pipe, $Q_{40} = AV = \pi/4(0.4)^2 \times 3 = 0.377$ m^3/sec

Discharge in 30 cm pipe, $Q_{30} = AV = \pi/4(0.3)^2 \times 2 = 0.14137$ m^3/sec

$\therefore$ Discharge in 20 cm pipe, $Q_{20} = Q_{40} - Q_{30} = (0.377 - 0.14137)$ m^3/sec

$= 0.23563$ m^3/sec

$\therefore$ $0.23563 = \pi/4(0.2)^2 \cdot V_{20}$

$\therefore$ $V_{20} = 7.5$ m/sec **Ans.**

EXAMPLE 5.3 The velocity vector in a fluid flow is given by

ie. $\mathbf{V} = 2x^3\mathbf{i} - 5xy\mathbf{j} + 4t\mathbf{k}$

Find the velocity and acceleration of a fluid particle at (1, 2, 3) at time $t = 1$

Solution: $V = 2x^3i - 5xyj + 4tk = ui + vj + kw$

$\therefore \quad u = 2x^3, v = -5xy, w = 4t$

at point (1, 2, 3), and $t = 1$

$$u = 2(1)^3 = 2, v = -5 \times 1 \times 2 = -10, w = 4 \times 1 = 4$$

$\therefore$ Velocity vector $\boldsymbol{V}$ at (1, 2, 3) = $2\boldsymbol{i} - 10\boldsymbol{j} + 4\boldsymbol{k}$

$\therefore$ Resultant velocity $= \sqrt{u^2 + v^2 + w^2}$

$$= \sqrt{2^2 + (-10)^2 + (4)^2}$$

$$= 10.95445 \text{ units}$$

Now $a_x = u\dfrac{\partial u}{\partial x} + v\dfrac{\partial u}{\partial y} + w\dfrac{\partial u}{\partial z} + \dfrac{\partial u}{\partial t}$

$$= 2x^3\,(6x^2) + (-5xy) \times 0 + 4t \times 0 + 0 = 2 \times 1^3 \times 6 \times 1^2 = 12 \text{ units}$$

$$a_y = u\frac{\partial v}{\partial x} + v\frac{\partial u}{\partial y} + w\frac{\partial v}{\partial z} + \frac{\partial v}{\partial t}$$

$$= 2x^3\,(-5y) + (-5xy)\,(-5x) + 4t \times 0 + 0$$

$$= 2(1)^3\,(-5 \times 2) + (-5 \times 1 \times 2)\,(-5 \times 1) = 20 + 50$$

$$= 30 \text{ units}$$

$$a_z = u\frac{\partial w}{\partial x} + v\frac{\partial w}{\partial y} + w\frac{\partial w}{\partial z} + \frac{\partial w}{\partial t}$$

$$= 2x^3(0) + (-5xy)\,(0) + 4t\,(0) + 4$$

$$= 0 + 0 + 0 + 4 = 4 \text{ units.}$$

Acceleration $A = a_x i + a_y j + a_z k = 12i + 30j + 4k$

$\therefore$ Resultant $A = \sqrt{12^2 + 30^2 + 4^2} = 32.557$ units

EXAMPLE 5.4 The velocity of flow is given by $V = i(xy^2) + j(-2yz^2) + k\left(\dfrac{2}{3}z^3 - zy^2\right)$ show that it satisfies the continuity equation in 3-D.

Solution: Velocity V in 3-D is $V = iu + jv + kw$

$$\therefore \quad = xy^2$$

$$\therefore \quad \frac{\partial u}{\partial x} = y^2$$

$$v = -2yz^2$$

$$\therefore \quad \frac{\partial v}{\partial y} = -2z^2$$

$$w = \frac{2}{3}z^3 - zy^2$$

$$\therefore \qquad \frac{\partial w}{\partial z} = \frac{2}{3} \cdot 3z^2 - y^2 = 2z^2 - y^2$$

$$\therefore \qquad \frac{\partial u}{\partial x} + \frac{\partial v}{\partial y} + \frac{\partial w}{\partial z} = y^2 - 2z^2 + 2z^2 - y^2 = 0$$

$\therefore$ The above equation satisfies the 3-D continuity equation in steady flow.

EXAMPLE 5.5 The following case represents the two velocity components, determine the third component of velocity such that it satisfies the continuity equation.

$$u = 4x^2 \quad \text{and} \quad v = 4xyz.$$

Solution: $\dfrac{\partial u}{\partial x} = 8x, \ \dfrac{\partial v}{\partial y} = 4xz$

Substituting $\dfrac{\partial u}{\partial x}$ and $\dfrac{\partial v}{\partial y}$ in the continuity equation,

$$8x + 4xz + \frac{\partial w}{\partial z} + 0$$

$$\frac{\partial w}{\partial z} = -8x - 4xz$$

or

$$\partial w = (-8x - 4xz)dz$$

$$\int dw = \int (-8x - 4xz)dz$$

$$w = -8xz - 4x\frac{z^2}{2} + \text{Constant of integration}$$

Constant of integration cannot be function of z, it is a $f(x, y)$

$$\therefore \qquad w = -8xz - 2xz^2 + f(x, y) \qquad \textbf{Ans.}$$

EXAMPLE 5.6 Calculate the unknown velocity component so that it satisfies the continuity equation. $u = 2x^2$, $v = 2xyz$

Solution: $\dfrac{\partial u}{\partial x} = 4x, \ \dfrac{\partial v}{\partial y} = 2xz$

The continuity equation gives:

$$4x + 2xz + \frac{\partial w}{\partial z} = 0$$

$$\therefore \qquad \frac{\partial w}{\partial z} = -4x - 2xz$$

$$\therefore \qquad \partial w = (-4x - 2xz)\partial z$$

Integrating

$$w = -4xz - \frac{2xz^2}{2} = -4xz - xz^2 \qquad \textbf{Ans.}$$

5.7 ROTATION AND VORTICITY

Fluid rotation about any axis is defined as the average rotation of any two mutually perpendicular segments in a plane normal to this axis. Vorticity is defined as twice the fluid rotation and is measured by the curl of velocity vector. If the fluid rotation or vorticity is zero, then the fluid is said to be irrotatational.

In vector rotation, the curl of velocity vector is equal to the fluid vorticity.

$$\text{Vorticity} = \text{curl } V = \nabla \times V = \begin{vmatrix} i & j & k \\ \dfrac{\partial}{\partial x} & \dfrac{\partial}{\partial y} & \dfrac{\partial}{\partial z} \\ u & v & w \end{vmatrix}$$

$$= \left(\frac{\partial w}{\partial y} - \frac{\partial v}{\partial z}\right) i + \left(\frac{\partial u}{\partial z} - \frac{\partial w}{\partial x}\right) j + \left(\frac{\partial v}{\partial x} - \frac{\partial u}{\partial y}\right) k \qquad (5.13)$$

For irrotational flow or motion the curl of the velocity vector is zero and coefficients of i, j, k in Equation (5.13) are each zero. Thus in case of irrotational flow in the x-y plane, $\dfrac{\partial u}{\partial y} = \dfrac{\partial v}{\partial x}$. For rotational flow, the curl of the velocity vector is non-zero and the coefficients i, j, k are also non-zero.

5.7.1 Rotation and Conditions for Irrotational Flow

In the in x-y plane, consider two mutually perpendicular line segments of length Δx and Δy as shown in Figure 5.4.

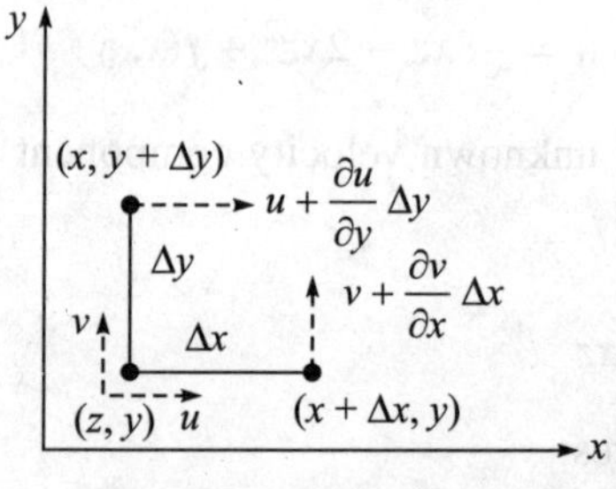

Figure 5.4 Rotation of two fluid-line segments.

The x component of velocity of the segment Δy at (x, y) is u, then at $(x, y, + \Delta y)$ is $u + \dfrac{\partial u}{\partial y}\Delta y$.

Similarly y component of segment Δx at (x, y) is v and at $(x + \Delta x, y)$ is $v + \dfrac{\partial v}{\partial x}\Delta x$. If $\dfrac{\partial u}{\partial x}$ and $\dfrac{\partial v}{\partial y}$ are both zero, both the line segments Δx and Δy can undergo only a translation, with no rotation.

Considering anti-clockwise rotation positive, the rotation of line segment Δx is the difference between the velocities at the ends of the segments divided by the length of the line segment.

This may be written as:

$$\frac{v + \left(\dfrac{\partial v}{\partial y}\right)\Delta x - v}{\Delta x} = \frac{\partial v}{\partial x}$$

and for line segment Δy, the rotation is

$$-\frac{u + \left(\dfrac{\partial u}{\partial y}\right)\Delta y - u}{\Delta y} = -\frac{\partial u}{\partial y}$$

The average of these is the rotation about the z-axis.

Similarly, ($\because$ angular velocity)

$$\left.\begin{aligned} \omega_z &= \frac{1}{2}\left(\frac{\partial v}{\partial x} - \frac{\partial u}{\partial y}\right) \\ \omega_y &= \frac{1}{2}\left(\frac{\partial u}{\partial z} - \frac{\partial w}{\partial x}\right) \end{aligned}\right\} \qquad (5.14)$$

$$\omega_x = \frac{1}{2}\left(\frac{\partial w}{\partial y} - \frac{\partial v}{\partial z}\right)$$

For no rotation $\omega_z = \omega_y = \omega_x = 0$

Thus for irrotational flow,

$$\left.\begin{aligned} \frac{\partial v}{\partial x} &= \frac{\partial u}{\partial y} \\ \frac{\partial u}{\partial z} &= \frac{\partial w}{\partial x} \\ \frac{\partial w}{\partial y} &= \frac{\partial u}{\partial z} \end{aligned}\right\} \qquad (5.15)$$

EXAMPLE 5.7 Show that following velocity components satisfy the continuity equation and that they represent irrotational flow:

$$u = (2x + y + z)t$$
$$v = (x - 2y + z)t$$
$$w = (x + y)t$$

Solution: $\dfrac{\partial u}{\partial x} = 2t, \dfrac{\partial v}{\partial y} = -2t$ and $\dfrac{\partial w}{\partial z} = 0$

$$\therefore \qquad \frac{\partial u}{\partial x} + \frac{\partial v}{\partial y} + \frac{\partial v}{\partial z} = 2t - 2t = 0$$

Hence they satisfy the continuity equation.

For irrotational flow, Equation (5.15) must be satisfied, i.e.

$$\text{i.e.} \qquad \frac{\partial v}{\partial x} = t, \frac{\partial u}{\partial y} = t \therefore \frac{\partial v}{\partial x} = \frac{\partial u}{\partial t}$$

$$\frac{\partial u}{\partial z} = t, \quad \frac{\partial w}{\partial x} = t$$

$$\therefore \qquad \frac{\partial u}{\partial t} = \frac{\partial w}{\partial x}$$

$$\frac{\partial w}{\partial y} = t, \quad \frac{\partial v}{\partial z} = t$$

$$\therefore \qquad \frac{\partial w}{\partial y} = \frac{\partial v}{\partial z}$$

Hence the flow is irrotational,

or if the curl of velocity vector is zero, the flow is irrotational, i.e.

$$\nabla \times V = \begin{vmatrix} i & j & k \\ \dfrac{\partial}{\partial x} & \dfrac{\partial}{\partial y} & \dfrac{\partial}{\partial z} \\ (2x+y+z)t & (x+2y+z)t & (x+y)t \end{vmatrix}$$

$$= (t - t)i + (t - t)j + k(t - t) = 0$$

Hence the flow is irrotational.

EXAMPLE 5.8 State if the flow represented by $u = 3x + 4y$ and $v = 2x - 3y$ is rotational or irrotational.

Solution: Rotation ω_z is given by $\omega_z = \frac{1}{2}\left(\frac{\partial v}{\partial x} - \frac{\partial u}{\partial y}\right)$

$$= \frac{1}{2}(2 - (+4)) = -1 \neq 0$$

Hence the flow is rotational. **Ans.**

5.8 POTENTIAL FLOW: VELOCITY POTENTIAL AND STREAM FUNCTION

Potential flow refers to the flow of an ideal, irrotational fluid flow. A study of such an ideal fluid in a real situation appears fruitless, but real fluid or viscous fluids in many situations exhibit the characteristics of ideal fluid.

A real or viscous fluid has a thin layer adjacent to the surface of the body in which the viscous effect is large and beyond this layer, the viscous effect is negligible. In this outer region, the flow of this viscous fluid is essentially the same as that of an ideal fluid. Therefore, the study and knowledge of ideal flow is useful in predicting or analysing the behaviour of real fluid.

5.8.1 Velocity Potential Function (ϕ)

This is defined as a scaler function of space and time such that its negative derivative with respect to any direction gives the fluid velocity in that direction. It is defined mathematically as $\phi = f(x, y, z)$ for steady such that

$$\left.\begin{aligned} u &= \frac{\partial \phi}{\partial x} \\ v &= -\frac{\partial \phi}{\partial y} \\ w &= -\frac{\partial \phi}{\partial z} \end{aligned}\right\} \tag{5.16}$$

The continuity equation for an incompressible steady flow is:

$$\frac{\partial u}{\partial x} + \frac{\partial v}{\partial y} + \frac{\partial w}{\partial z} = 0$$

Substituting u, v, w from Equation (5.16), we get:

$$-\frac{\partial}{\partial x}\left(\frac{\partial \phi}{\partial x}\right) + \frac{\partial}{\partial y}\left(-\frac{\partial \phi}{\partial y}\right) + \frac{\partial}{\partial z}\left(-\frac{\partial \phi}{\partial z}\right) = 0$$

$$\therefore \qquad \frac{\partial^2 \phi}{\partial x^2} + \frac{\partial^2 \phi}{\partial y^2} + \frac{\partial^2 \phi}{\partial z^2} = 0 \tag{5.17}$$

which is called the Laplace equation in ϕ.

For two-dimensional flow, Equation (5.17) becomes:

$$\frac{\partial^2 \phi}{\partial x^2} + \frac{\partial^2 \phi}{\partial y^2} = 0 \tag{5.18}$$

The essential properties of potential function are:

(i) Flow should be irrotational.
(ii) It represents the possible steady incompressible irrotational flow.

5.8.2 Stream Function (ψ)

This is defined as the scaler function of space and time such that its partial derivative with respect to any direction gives the velocity component at right angles to that direction. It is written or denoted by ψ and defined as only a two-dimensional flow. Mathematically, for steady flow it is defined as $\psi = f(x, y)$ such that

$$\left.\begin{aligned} \frac{\partial \psi}{\partial x} &= v \\ \frac{\partial \psi}{\partial y} &= -u \end{aligned}\right\} \tag{5.19}$$

The existence of ψ means a possible case of fluid flow. The flow may be rotational and irrotational.

The rotational component in the x-y plane is ω_z and given by Equation (5.14), i.e.

$$\omega_z = \frac{1}{2}\left(\frac{\partial v}{\partial x} - \frac{\partial u}{\partial y}\right)$$

Substituting the value v and u from Equation (5.19), we get:

$$\omega_z = \frac{1}{2}\left[\frac{\partial}{\partial x}\left(\frac{\partial \psi}{\partial x}\right) - \frac{\partial}{\partial y}\left(-\frac{\partial \psi}{\partial y}\right)\right]$$

$$\omega_z = \frac{1}{2}\left(\frac{\partial^2 \psi}{\partial x^2} + \frac{\partial^2 \psi}{\partial y^2}\right) \tag{5.20}$$

For irrotational flow, $\omega_z = 0$

$\therefore$ Equation (5.20) becomes $\dfrac{\partial^2 \psi}{\partial x^2} + \dfrac{\partial^2 \psi}{\partial y^2} = 0$ (5.21)

which is the Laplace equation in ψ.

The essential properties of ψ are:

(i) If y exists, it is a possible case of fluid flow which may be rotational or irrotational.
(ii) If ψ satisfies the Laplace equation, it is a possible case of irrotational flow.

5.8.3 Equipotential Lines

The lines of constant velocity potential ϕ are called equipotential lines.

$\therefore$ For equipotentional line, ϕ = constant

$$d\phi = 0$$

But $\phi = f(x, y)$ for steady flow

$$\therefore \quad d\phi = \frac{\partial \phi}{\partial x}dx + \frac{\partial \phi}{\partial y}dy$$

$$= -udx - vdy \qquad \because \frac{\partial \phi}{\partial x} = -u, \frac{\partial \phi}{\partial x} = -v$$

$$= -(udx + vdy)$$

But for equipotential line $d\theta = 0$

$$\therefore \quad -(udx + vdy) = 0$$

$$\therefore \quad \frac{dy}{dx} = -\frac{u}{v} \tag{5.22}$$

where $\dfrac{dy}{dx}$ is the slope of the equipotential line.

5.8.4 Streamlines and Flow Net

The lines of constant stream function are called streamlines.

$\therefore$ $$\psi = \text{constant}$$

$\therefore$ $$d\psi = 0$$

But $$\psi = f(x, y)$$

$\therefore$ $$d\psi = \frac{\partial \psi}{\partial x} dx + \frac{\partial \psi}{\partial y} d\psi$$

$$= vdx - udy \qquad \because \frac{\partial \psi}{\partial x} = v, \frac{\partial \psi}{\partial y} = -u$$

for a line of constant ψ,

$$d\psi = 0$$

$\therefore$ $$vdx - udy = 0$$

$\therefore$ $$vdx = udy$$

$\therefore$ $$\frac{\partial y}{\partial x} = \frac{v}{u} \tag{5.23}$$

$\frac{\partial y}{\partial x}$ is the slope of the streamline.

Equations (5.22) and (5.23), i.e. the slope of equipotential lines and streamlines, show that they are orthogonal since the product of two slope is equal to -1, i.e.

i.e. $$\frac{-u}{v} \times \frac{u}{v} = -1.$$

The two families of intersecting lines form a system of curvilinear squares called flow net, which has been explained below in Figures 5.5, 5.6 in a sluice gate and 5.7 in an earth dam.

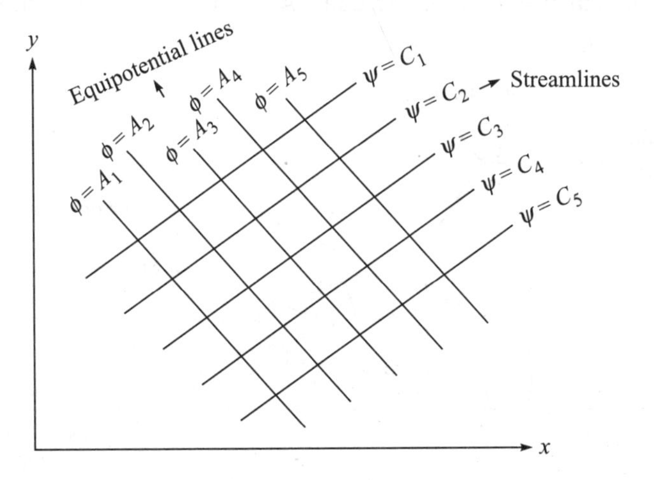

Figure 5.5 Elements of flow net in general.

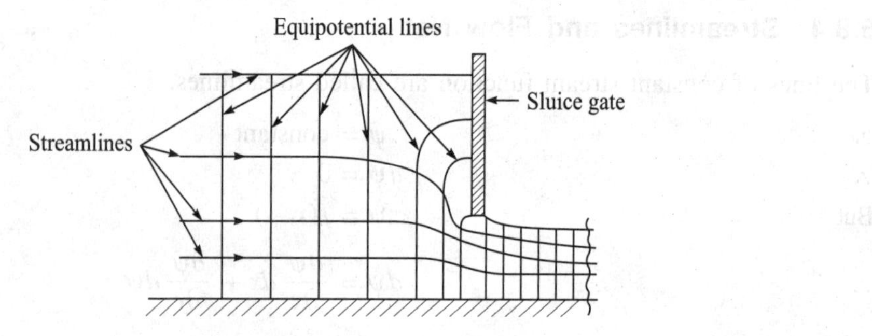

Figure 5.6 Flow net in a sluice gate.

The flow is along the direction tangential to the streamlines, the space between two adjacent streamlines may be considered as a flow channel and flow or discharge flowing through it is proportional to $(\psi_2 - \psi_1)$ where ψ_1 and ψ_2 are two adjacent streamlines.

The construction of flow net for the flows is restricted by certain conditions. These are:

(i) The flow should be steady.
(ii) The flow should be irrotational.
(iii) The flow is not governed by gravity force.

Different methods are used for drawing the flow nets. These are:

(i) The analytical method, which involves the solution of the Laplace equation in ϕ and ψ.
(ii) The graphical method, which is suitable for drawing the flow net of the fluid flow field of any shape. Here care must be taken to intersect the streamlines and boundary by equipotential lines perpendicularly.
(iii) The electrical analogy method.
(iv) The relaxation method.
(v) Helle Shaw or viscous flow analogy method.

The flow net can be used to determine the discharge. Consider an earth dam which is porous and water flows through the body of the dam. An earth dam with upstream of water H is considered (Figure 5.7).

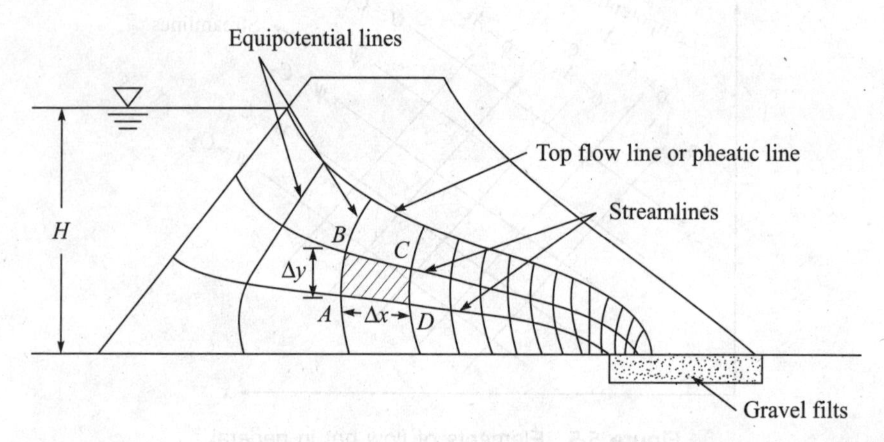

Figure 5.7 Flow net in earth dam and its use to determine the water flow through the body of the dam.

The flow net in the saturated zone of the dam is drawn by the graphical method. Two sets of lines, i.e. equipotential lines and streamlines, are drawn as curvilinear squares. Let the soil be isotropic. Assume the dam section of the unit width. Let f be the flow/unit width of the dam.

The flow through the square $ABCD$ (called flow held) by Darcy's law k is the hydraulic conductivity, ΔH is the energy or head drop between two equipotential lines, Δx and Δy are defined in Figure 5.7.

$$= \Delta q = K \cdot iA = k\left(\frac{\Delta H}{\Delta x}\right)\ (\Delta y \times 1)$$

$$\therefore \qquad \Delta q = k\left(\frac{\Delta H}{\Delta x}\right)\Delta y$$

But
$$\Delta H = \frac{\text{Total drop, i.e. total head causing the flow}}{\text{No. of investments into which total drop is equally divided}}$$

$$\Delta H = \frac{H}{N_d},\ N_d \text{ is the number drops in the complete net.}$$

$$\therefore \qquad \Delta q = K\frac{H}{N_d}\left(\frac{\Delta y}{\Delta H}\right)$$

$$= K\frac{H}{N_d}\ (\Delta y \simeq \Delta x \text{ taken while drawing the flow net})$$

q is the total flow through all the channels per unit width of the dam.

$$\therefore \qquad q = \Sigma\ \Delta\ q$$

$$\therefore \qquad q = k\frac{H}{N_d} \times \text{no. of flow channels}$$

$$q = k\frac{H}{N_d}N_f \tag{5.24}$$

q is the discharge passing through a flow net and is applicable to isotropic soil (i.e. soil of $kH = kv$).

5.8.5 Source, Sink, Circulation and Vortex

A line source is a line of unit length between two parallel planes, a unit distance apart from which fluid flows radially in all directions parallel to the planes [Figure. 5.8(a)].

In Figure 5.8(b), the radial velocity V_r is at a radial distance r from the origin in the x-y plane. The volumetric flow rate from this line source of unit length is $q = (2\pi r \cdot V_r)$

$$\therefore \qquad V_r = \frac{q}{2\pi r}$$

From the definition of velocity potential and stream function:

$$V_r = \frac{q}{2\pi r} = -\frac{\partial \phi}{\partial r} = -\frac{1}{r}\frac{\partial \psi}{\partial \theta}$$

$$\therefore \qquad \phi_{source} = -\frac{-q}{2\pi}\log_e r = -\frac{q}{4\pi}\log_e (x^2 + y^2) \qquad (5.25)$$

and

$$\psi_{source} = -\frac{q}{2\pi}\theta = -\frac{q}{2\pi}\tan^{-1}\left(\frac{y}{x}\right) \qquad (5.26)$$

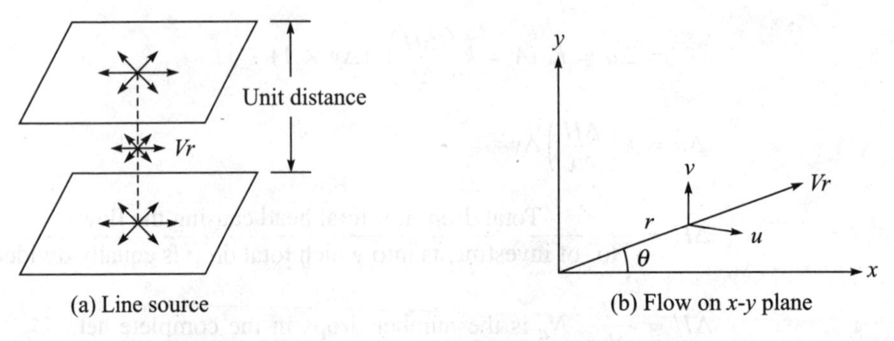

(a) Line source

(b) Flow on x-y plane

Figure 5.8 Flow from a two-dimensional source.

where q is equal to the strength of the source and $\frac{q}{2\pi}$ is the flow rate per unit of radian.

A sink is a negative source and velocity and stream functions for a sink are negative of those for a source

$$\therefore \qquad \phi_{sink} = \frac{q}{4\pi}\log_e (x^2 + y^2) = \frac{q}{2\pi}\log_e r \qquad (5.27)$$

$$\therefore \qquad \psi_{sink} = \frac{q}{2\pi}\tan^{-1}\left(\frac{y}{x}\right) = \frac{q}{2\pi}\theta \qquad (5.28)$$

The flow along a closed path is called circulation.

The mathematical concept of circulation is the line integral taken completely around a closed path.

Consider a closed path C in Figure 5.9(a) in the flow field with their streamlines. At any point of the curve, velocity V is the tangent of the streamline. The velocity component along the tangential direction of the closed path is $V\cos\alpha$ if α is the angle between the velocity direction and tangent of the closed path ds by definition, the circulation (Γ, gamma) around the closed path is:

$$\Gamma = \oint V\cos\alpha \, ds \qquad (5.29)$$

Next consider Figure 5.9(b) in which element path ds is the combination of the paths dx and dy. The line integral along the elemental path ds = line integral along dx + line integral along $dy = udx + vdy$.

Writing u and v in terms of velocity potentional function ϕ line integral along

$$ds = -\left(\frac{\partial \phi}{\partial x}dx + \frac{\partial \phi}{\partial y}dy\right)$$

i.e.

$$\Gamma = -d\phi = |d\phi| \qquad (5.30)$$

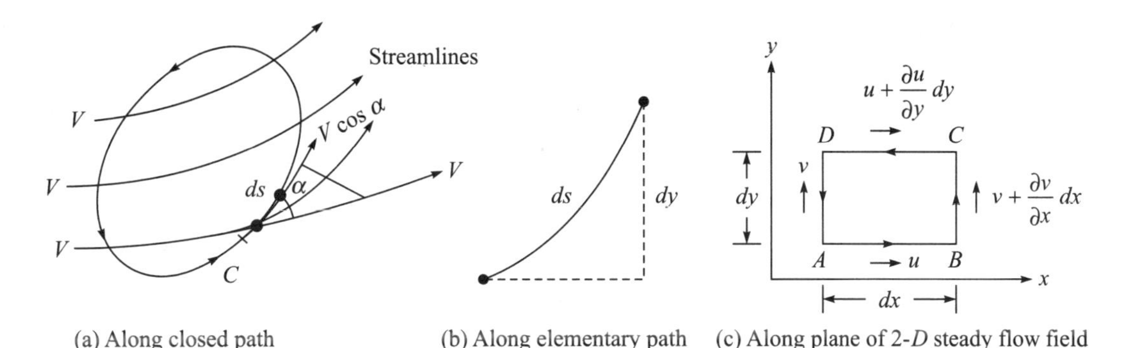

(a) Along closed path (b) Along elementary path (c) Along plane of 2-*D* steady flow field

Figure 5.9 Circulation along various paths.

i.e. the line integral from one point to another is equal to the change in velocity potential function. Now consider Figure 5.9(c) to find circulation around the sides of a rectangle *ABCD* of side *dx* and *dy*. The figure also shows the tangential velocities along the sides of the rectangle.

Integrating in anti-clockwise order *ABCDA*, the circulation

$$\Gamma = udx + \left(v + \frac{\partial v}{\partial x}dx\right)dy - \left(u + \frac{\partial u}{\partial y}dy\right)dx - vdy$$

or

$$\Gamma = dxdy\left[\frac{\partial v}{\partial x} - \frac{\partial u}{\partial y}\right] \tag{5.31}$$

Vorticity is the circulation per unity area

$$\therefore \qquad \text{Vorticity} = \frac{\Gamma}{dxdy} = \left(\frac{\partial v}{\partial x} - \frac{\partial u}{\partial y}\right) = 2w_z \tag{5.32}$$

where ω is the angular velocity about the z axis.

Equation (5.32) has already been shown in Section 5.7. This again means that if vorticity exists, the flow is rotational. When vorticity is zero,

$$\frac{\partial v}{\partial x} - \frac{\partial u}{\partial y} = 0, \text{ this condition is also already established.}$$

Vortex is a mass of rotating fluid and its motion is called vortex motion. It is of two types, i.e. free vortex motion and forced vortex motion.

A free vortex motion is that in which the fluid mass rotates without any external forces. Some of the common examples of free vortex are a whirlpool in a river, flow of liquid drained through an outlet provided at the bottom of a shallow vessel such as a wash handbasin, bathtub draining through outlets, etc.

A forced vortex is that mass of rotating fluid which is caused to rotate by external forces. A most common example of forced vortex motion is that of a vertical cylinder containing a liquid rotated about its central axis with a constant angular speed by an external force.

EXAMPLE 5.9 A stream function in two-dimensional flow is $\psi = 2xy$. Show that the flow is irrotational and determine the corresponding velocity potential function ϕ.

Solution: A stream function is irrotational if the Laplace equation is ratified, i.e.

$$\frac{\partial^2 \psi}{\partial x^2} + \frac{\partial^2 \psi}{\partial y^2} = 0$$

Substituting ψ,

$$\frac{\partial^2 (2xy)}{\partial x^2} + \frac{\partial^2 (2xy)}{\partial y^2} = 0$$

Hence, it represents irrotational flow.

$$\frac{\partial \phi}{\partial x} = -u = \frac{\partial \psi}{\partial y} = \frac{\partial (2xy)}{\partial y} = 2x \quad \text{(A)}$$

$$\frac{\partial \phi}{\partial y} = -v = \frac{\partial \psi}{\partial x} = \frac{-\partial (2xy)}{\partial x} = -2y \quad \text{(B)}$$

Integrating equation (A) $\phi = x^2 + f(y)$ (C)
differentiating with respect to y,

$$\frac{\partial \phi}{\partial y} = f'(y) \quad \text{(D)}$$

Equating (B) and (D)

$$f'(y) = -2y$$

Integrating $f(y) = -y^2 + C$ (E)

Where C is a constant of integration.
Substituting (E) in (C)

$$\phi = x^2 - y^2 + C$$ **Ans.**

EXAMPLE 5.10 A nozzle is so shaped that the velocity of flow along the centre line changes linearly from 2.5 m/sec to 16 m/sec in a distance of 3.75 m. Determine the magnitude of convective acceleration at the beginning and the end.

Solution: Rate of change V with respect to space S is:

$$\frac{\partial V}{\partial S} = \frac{16 - 2.5}{3.75} = 3.6 \text{ m/sec per metre.}$$

Convective acceleration is:

$$a_s = V \frac{\partial V}{\partial S}$$

$\therefore$ $a_{2.5} = (2.5 \times 3.6) = 9 \text{ m/sec}^2$ **Ans.**

and $a_{16} = (16 \times 3.6) = 57.6 \text{ m/sec}^2$ **Ans.**

EXAMPLE 5.11 The velocity components in two-dimensional flow are $u = ax$, $v = by$, show that $a = -b$, if the components satisfy the continuity equation.

Solution: The continuity equation is $\dfrac{\partial u}{\partial x} + \dfrac{\partial v}{\partial y} = 0$ (A)

$$u = ax \; v = by$$

$$\frac{\partial u}{\partial x} = a, \; \frac{\partial v}{\partial y} = b$$

Substituting in (A)

$\therefore$ $$a + b = 0$$

$\therefore$ $$a = -b \text{ shown.}$$

5.8.6 Pathline, Streakline, Streamline and Streamtube

A pathline [shown in Figure 5.10(a)] means a path or line described by a single fluid particle as it moves during a period of time. The pathline indicates the direction of velocity of the same fluid particle at a consecutive instant of time.

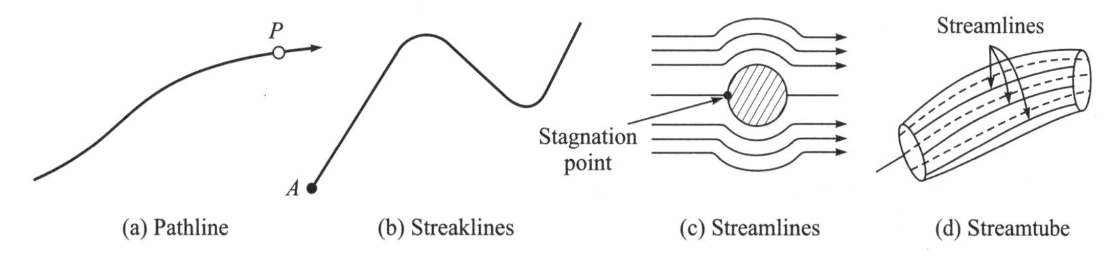

Figure 5.10 Description of flow patterns by fluid flow lines.

A streakline is the locus of the position fluid particles which have passed through a given point (say A) in space in succession. Consider a point $A(x_1, y_1, z_1)$ in space [Figure 5.10(b)]. The various fluid particles which have passed through A in succession lie in some position at a certain instant. The locus of these points constitutes a streakline.

Streamlines are lines which show the direction of velocity of the fluid at each point along the line. The tangent at any point is the direction of the velocity of the fluid particle at that point. Streamlines are drawn in Figure 5.10(c) around a submerged object.

A streamtube [Figure 5.10(a)] is an imaginary tubular space formed by a number of streamlines. It is the collection of streamlines which would form a small closed passage. There is no flow across the wall of a streamtube. This means that within the streamtube, a small but definite quantity of fluid will flow, i.e. the quantity of fluid entering at one end of a streamtube should be equal to the quantity leaving at the other end.

5.9 CONCLUSION

The concept of different types of fluid flow like steady, unsteady, compressible, incompressible, uniform, non-uniform, laminar, turbulent, transitional, critical, super-critical, rotational, irrotational, one-dimensional, two-dimensional, three-dimensional etc. are presented. Differential equations of velocity, acceleration, continuity in 1-D, 2-D and 3-D flows are derived. Rotation and vorticity, velocity potential and stream function, flow net along with their mathematical equations are also presented. Few worked out examples, problems with answer and a list of references are given at the end of the chapter.

Lamp[4], Robertson[5] gave a detailed analysis of hydrodynamics of fluid.

PROBLEMS

5.1 A 30 cm diameter pipe carries oil of specific gravity 0.8 at a velocity of 2 m/sec. At another section, the diameter is 20 cm. Find the velocity at this section and also the mass rate of flow of oil.

(**Ans.** 4.5 m/sec, 113 kg/sec)

5.2 A fluid is given by $V = ixy - j\,2yz^2 - k\left(zy^2 - \frac{2z^3}{3}\right)$

Prove that it is a case of possible steady incompressible fluid flow. Calculate the velocity and acceleration at the point (1, 2, 3).

(**Ans.** 36.7 units, 874.5 units)

5.3 Find the convective acceleration at the middle of a pipe which converges uniformly from 0.6 m diameter to 0.3 m diameter over a 3 m length. The rate of flow is 40 litres/sec. If the rate of flow changes uniformly from 40 litres to 80 litres in 40 seconds, find the total acceleration at the middle of the pipe at the 20th second.

(**Ans.** 0.0499 m/sec^2, 0.11874 m/sec^2)

5.4 For the velocity potential function $\phi = x^2 - y^2$, find the velocity components at the point (4, 5).

(**Ans.** $u = 8$, $v = -10$ units)

5.5 A fluid flow is given by $V = i(10x^3) - j(8x^3y)$ state whether flow is irrotational or rotational.

(**Ans.** Rotational)

5.6 If the stream function for steady flow is given by $\psi = y^2 - x^2$, determine whether the flow is rotational or irrotational. Then determine the velocity potential function ϕ.

(**Ans.** Irrotational, $\phi = -2xy + c$)

5.7 Determine which of the following pairs of velocity components u and v satisfy the continuity equation for two-dimensional incompressible flow:

(a) $u = c_x$; $v = -c_y$

(b) $u = (3x - y)$; $v = (2x + 3y)$

(c) $u = (x + y)$; $v = (x^2 - y)$

(**Ans.** a. Satisfy b. does not satisfy c. satisfy)

5.8 Calculate the unknown velocity component so that it satisfies the continuity equation. $u = 2x^2$, $v = 2xyz$, $w = ?$

(**Ans.** $w = -4xz - xz^2 + f(x, y)$)

5.9 Determine the corresponding stream function from the following velocity potential function, $\phi = x + y$.

(**Ans.** $\psi = y - x +$ constant)

5.10 Show that the vorticity for two-dimensional flow is equal to

$$\frac{\partial^2 \psi}{\partial x^2} + \frac{\partial^2 \psi}{\partial y^2}$$

REFERENCES

1. Shapiro, A.H., 'Basic Equations of Fluid Flow', Section 2 of *Handbook of Fluid Dynamics*, Edited by V.L. Streeter, McGraw-Hill, 1961.
2. Miline-Jhomson, L.M., *Theoretical Hydrodynamics*, Macmillan, New York, 1960.
3. Olson, R.M., *Engineering Fluid Mechanics*, International Textbook Company, Scranton, Pennsylvania, 1967.
4. Lamp, H., *Hydrodynamics*, Cambridge University Press, London, 1932.
5. Robertson, J.M., *Hydrodynamics in Theory and Application*, Prentice-Hall Inc., Englewood Cliffs, New Jersey, 1965.
6. Yuan, S.W., *Foundation of Fluid Mechanics*, Prentice-Hall of India, New Delhi, 1976.
7. Mott, R.L., *Applied Fluid Mechanics*, Prentice-Hall, Upper Saddle River, New Jersey, 1999.

Chapter 6

Dynamics of Fluid Flow

6.1 INTRODUCTION

The dynamics of fluid flow is the study of fluid motion caused by different forces. A variation of acceleration and energy takes place in the fluid motion due to these forces. Newton's second law of motion is applied to analyse this dynamics of fluid flow. Fluid is assumed to be incompressible. The various forces acting on fluid mass may be classified as body or gravity force (F_g), pressure force (F_p), viscous force (F_v), turbulent force (F_t), surface tension force (F_s), and compressible force (F_c).

6.2 EQUATIONS OF MOTION

According to Newton's second law of motion, the resultant force on any fluid element must be equal to the product of mass and acceleration of the element. The acceleration vector has the direction of the resultant force vector. Mathematically, it may be expressed as:

$$\Sigma F = M \cdot a \tag{6.1}$$

If force and acceleration are resolved along x, y and z directions, the corresponding equations may be written as:

$$\left.\begin{aligned} \Sigma F_x &= Ma_x \\ \Sigma F_y &= Ma_y \\ \Sigma F_z &= Ma_z \end{aligned}\right\} \tag{6.1a}$$

Considering all the forces only in the x direction, (say in one-dimensional), Equation (6.1a) may be written as:

$$M_{a_x} = F_{g_x} + F_{p_x} + F_{v_x} + F_{t_x} + F_{s_x} + F_{c_x} \tag{6.2}$$

In most of the problems, surface tension force Fs_x and force of compressibility Fc_x are not considered. Hence Equation (6.2) becomes:

$$M_{a_x} = F_{g_x} + F_{p_x} + F_{v_x} + F_{t_x} \tag{6.3}$$

Equation (6.3) is called the Reynolds equation which is useful in the analysis of turbulent flow.

Again if viscous force is predominant, F_{t_x} is negligible. Hence Equation (6.3) becomes:

$$F_{a_x} = F_{g_x} + F_{p_x} + F_{v_x} \tag{6.4}$$

Equation (6.4) is called the Navier-stokes equation.

If the viscous force is not significant or fluid is ideal, F_{v_x} is negligible, Equation (6.4) becomes,

$$M_{a_x} = F_{g_x} + F_{p_x} \tag{6.5}$$

Equation (6.5) is called Euler's equation of motion.

6.3 DERIVATION OF EULER'S EQUATION OF MOTION

As seen from Equation (6.5), only force of gravity and pressure are considered in Euler's equation.

To derive this Euler's equation, a streamline in direction of S is considered as shown in Figure 6.1. A cylindrical element of cross-section dA and length dS is considered. The pressure force in the direction of flow and the gravity force actions vertically downwards on this cylindrical element is:

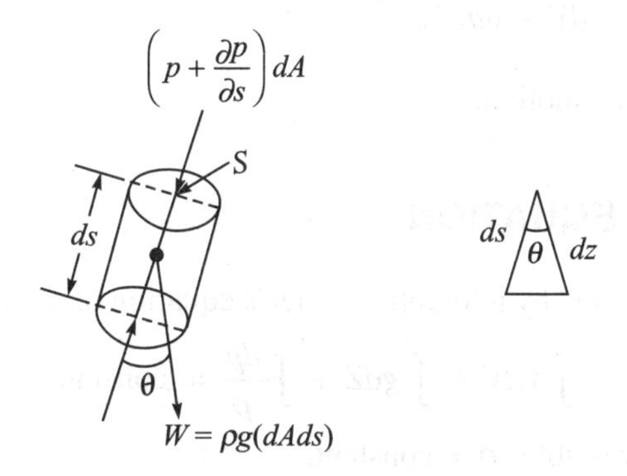

Figure 6.1 Pressure and gravity forces on fluid element.

(i) Pressure force pdA in the direction of S.

(ii) Pressure force $\left(p + \frac{dp}{ds}\right) dA$ opposite to the direction of flow.

(iii) Gravity force or weight of the cylindrical element acting vertically downwards equal to $\rho g(dAds)$.

Let θ be the angle between the direction of flow and line of action of gravity forces. The resultant pressure on the elemental cylindrical fluid is equal to the mass of the fluid multiplied by the acceleration a_s in the S-direction.

$$\therefore \qquad pdA - \left(p + \frac{\partial p}{\partial s} ds\right) dA - \rho g dAds \cos\theta = \rho(dAds)\, a_s \tag{A}$$

Total acceleration $a_s = \dfrac{\partial V}{\partial t} + V\dfrac{\partial V}{\partial s}$

If the flow is steady, local acceleration

$\dfrac{\partial V}{\partial t} = 0$ and convective acceleration $V\dfrac{\partial V}{\partial s}$ is not zero

$\therefore$ (A) becomes $-\dfrac{\partial p}{\partial s} ds\, dA - \rho g dA ds \cos\theta = \rho\, dA ds\, V\dfrac{\partial V}{\partial s}$

or
$$V\frac{\partial V}{\partial s} + g\cos\theta + \frac{1}{\rho}\frac{\partial p}{\partial s} = 0 \tag{B}$$

But
$$\cos\theta = \frac{dz}{ds}$$

(B) becomes $V\dfrac{\partial V}{\partial s} + g\dfrac{\partial z}{\partial s} + \dfrac{1}{\rho}\dfrac{\partial p}{\partial s} = 0$

Multiplying by ∂s, we get:

$$VdV + gdZ + \frac{dp}{\rho} = 0 \tag{6.6}$$

which is Euler's equation of motion.

6.4 BERNOULLI'S EQUATION

Bernoulli's equation is obtained by integrating Euler's equation of motion, i.e. Equation (6.6), i.e.

$$\int VdV + \int gdZ + \int \frac{dp}{\rho} = \text{constant.}$$

If the flow is incompressible, ρ = constant.

$$\therefore \qquad \frac{p}{\rho} + gZ + \frac{V^2}{2} = \text{constant}$$

Dividing by g, we get:

$$\frac{p}{\rho g} + Z + \frac{V^2}{2g} = \text{constant}$$

or
$$\frac{p}{w} + Z + \frac{V^2}{2g} = \text{constant} \tag{6.7}$$

Equation (6.7) is Bernoulli's equation.

Here $\dfrac{p}{w}$ is the pressure head or pressure energy per unit weight.

$\dfrac{V^2}{2g}$ is the kinetic energy per unit weight or kinetic head.

Z is the potential energy per unit weight or potential head.

Thus, $\left(Z+\frac{p}{w}+\frac{V^2}{2g}\right)$ is the total energy or head of flowing liquid per unit weight, which is constant at any section. This is the Bernoulli's equation named in honour of the Swiss mathematician Daniel Bernoulli who derived it from Euler's equation.

This total energy for both pipe and open channel has been shown in Figure 6.2 above any datum line.

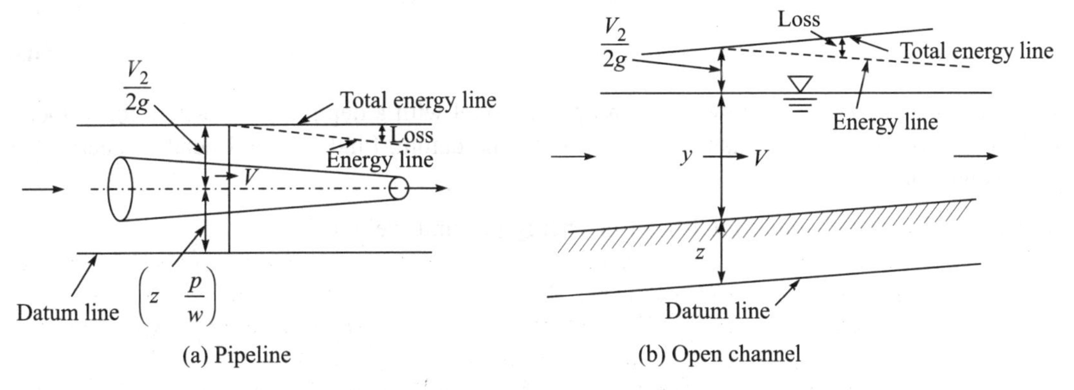

Figure 6.2 Total energy of flowing liquid in pipe and open channel.

In case of an open channel, the total energy head $H=\left(Z+y+\frac{V^2}{2g}\right)$ and in pipe $H=\left(Z+\frac{p}{w}+\frac{V^2}{2g}\right)$

Thus Bernoulli's equation gives for real fluid where loss occurs to resistance to flow,
Total energy at section 1 = Total energy at section 2 = Constant

$$\left(Z_1+\frac{p_1}{w}+\frac{V_1^2}{2g}\right) = \left(Z_2+\frac{p_2}{w}+\frac{V^2}{2g}\right) + \text{Losses} = \text{Constant for pipe}$$

$$\left(Z_1+y_1+\frac{V_1^2}{2g}\right) = \left(Z_2+y_2+\frac{V_2^2}{2g}\right) + \text{Losses} = \text{Constant for open channel.}$$

The assumptions made in Bernoulli's Equation (6.7) are: (i) The fluid is ideal, (ii) The flow is steady, (iii) The flow is incompressible, (iv) The flow is irrotational.

EXAMPLE 6.1 Water is flowing through a pipe of 100 mm diameter under a pressure of 19.62 N/cm^2 with mean velocity of flow 3 m/sec. Find the total head of water at the cross-section, which is 8 m above the datum line.

Solution: The total head above datum/unit weight according to Bernoulli's equation is:

$$H = Z + \frac{p}{w} + \frac{V^2}{2g} \quad \because \frac{p}{w} = \frac{p}{\rho g}$$

$$\frac{p}{w} = \frac{p}{\rho g} = \frac{19.62\times 10^4}{1000\times 9.81}\text{ m}$$

$$p = 19.6 \text{ N/cm}^2 = (19.6 \times 10^4) \text{ N/cm}^2$$

Given $Z = 8$ m and $V = 3$ m/sec

$$\text{Total head} = H = Z + \frac{p}{w} + \frac{V^2}{2g}$$

$$\therefore \quad H = \left(8 + \frac{19.62 \times 10^4}{1000 \times 9.81} + \frac{3^2}{2 \times 9.81}\right) \text{ m of water}$$

$$H = 28.457 \text{ m}$$ **Ans.**

EXAMPLE 6.2 Water is flowing in an open channel with a depth of 2 m where the velocity is 4 m/sec. The bed of the channel is 5 m above the datum. Find the total head or energy per unit weight of water.

Solution: For the channel, total head or energy per unit weight, is

$$H = Z + Y + \frac{V^2}{2g}$$

$$H = \left(5 + 2 + \frac{4^2}{2 \times 9.81}\right) \text{ m of water}$$

$$H = 7.8155 \text{ m}$$ **Ans.**

6.5 PRACTICAL USE OF BERNOULLI'S EQUATION

Bernoulli's equation of the total energy head concept is applied to various fluid flow problems. In this chapter, it is applied in the following flow measuring devices:

(i) Venturimeter
(ii) Orifice meter
(iii) Pitot tube
(iv) Nozzle meter

6.5.1 Venturimeter

Venturimeter is one of most widely used flow measuring devices which is based on Bernoulli's equation. The name 'venturimeter' is given in honour of the Italian physicist G.B. Venturi who first developed the principle in 1797, and which was developed in its present form by C. Herschel in 1887.

It consists of a short converging pipe, throat, diverging pipe and a piezometer or manometer to measure the difference of the pressure of flowing liquid between the pipe and throat section as seen in Figure 6.3.

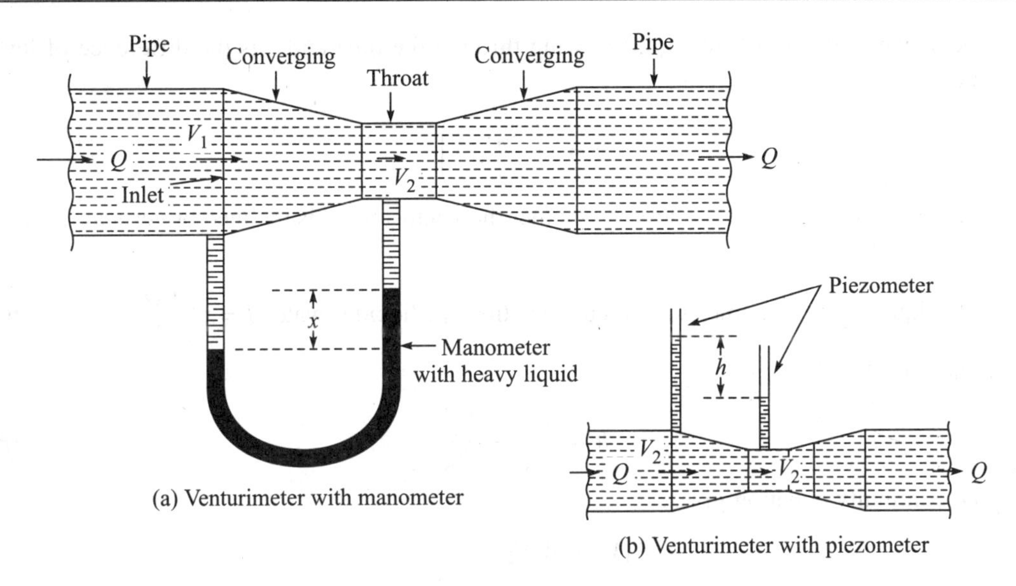

(a) Venturimeter with manometer

(b) Venturimeter with piezometer

Figure 6.3 Venturimeter.

Let Q be the discharge flowing through a pipe where the venturimeter is fitted.

Let d_1 = diameter of pipe at inlet.

p_1 = pressure intensity at inlet i.e. in the pipe.

V_1 = is the average velocity of pipe at inlet.

A_1 = area of the pipe at inlet = $\left(\frac{\pi}{4}d_1^2\right)$

Similarly d_2, p_2, V_2 and A_2 are diameter, pressure, velocity, and area at the throat.
Applying Bernoulli's equation at inlet and at the throat, we get:

$$Z_1 + \frac{p_1}{\rho g} + \frac{V_1^2}{2g} = Z_2 + \frac{p_2}{\rho g} + \frac{V_2^2}{2g}$$

If the pipe is horizontal, $Z_1 = Z_2$

$$\therefore \qquad \frac{p_1}{\rho g} + \frac{V_1^2}{2g} = \frac{p_2}{\rho g} + \frac{V_2^2}{2g}$$

or

$$\left(\frac{p_1}{\rho g} - \frac{p_2}{\rho g}\right) = \frac{V_2^2}{2g} + \frac{V_1^2}{2g} \qquad \text{(A)}$$

But $\left(\frac{p_1}{\rho g} - \frac{p_2}{\rho g}\right)$ is the difference of the pressure heads between the inlet and the throat and is equal to h metre measured by a manometer [Figure 6.3(a)] or by two piezometers [Figure 6.3(b)]. If measured by two piezometers, h is equal to the head of liquid flowing. If differential manometer is used, and if S_m is the specific gravity in the liquid in the manometer,

S_p is the specific gravity of the liquid flowing through the pipe and x is the difference of level in manometer, then

$$h = x\left(\frac{S_m}{S_p} - 1\right) \tag{6.8}$$

Equation (6.8) is valid regardless of whether the venturimeter is inclined or vertical from the manometric equation.

If the liquid in the manometer is mercury and the pipe liquid is water, $h = x\left(\frac{13.6}{1} - 1\right) = 12.6x$.

Equation (A) becomes:

$$h = \frac{V_2^2}{2g} - \frac{V_1^2}{2g} \tag{B}$$

The continuity equation gives:

$$A_1V_1 = A_2V_2$$

$$\therefore \quad V_1 = \frac{A_2}{A_1}V_2$$

Substituting V_1 in (B), we get:

$$h = \frac{V_2^2}{2g} - \left(\frac{A_2}{A_1}V_2\right)^2 \Big/ 2g$$

$$2gh = V_2^2 - \frac{A_2^2}{A_1^2}V_2^2 = V_2^2\left(\frac{A_1^2 - A_2^2}{A_1^2}\right)$$

$$\therefore \quad V_2 = \frac{A_1}{\sqrt{A_1^2 - A_2^2}}\sqrt{2gh}$$

$\therefore$ Discharge

$$Q = A_2V_2 = \frac{A_1A_2}{\sqrt{A_1^2 - A_2^2}}\sqrt{2gh} \tag{6.9}$$

Equation (6.9) gives the ideal discharge or theoretical discharge. Some loss occurs in contraction, expansion, friction etc. Therefore, the actual discharge is slightly less than that given by Equation (6.9). Introducing a coefficient of venturimeter, C_d ($\simeq$0.97 to .99). The actual discharge is:

$$Q_{\text{actual}} = C_d\frac{A_1A_2}{\sqrt{A_1^2 - A_2^2}}\sqrt{2gh} \tag{6.10}$$

EXAMPLE 6.3 A horizontal venturimeter with an inlet diameter 30 cm and throat diameter 15 m is used to measure the flow of water. The reading in the differential manometer connected to the inlet and throat is 10 cm of mercury. Determine the discharge if the coefficient of venturimeter is 0.98.

Solution: $A_1 = \frac{\pi}{4}(30)^2 = 706.858 \text{ cm}^2$

$$A_2 = \frac{\pi}{4}(15)^2 = 176.714 \text{ cm}^2$$

h = difference of mercury in differential manometer

$$= 10\left(\frac{13.6}{1} - 1\right) = 126 \text{ cm of water}$$

$$\therefore \quad Q = C_d \frac{A_1 A_2}{\sqrt{A_1^2 - A_2^2}} \sqrt{2gh}$$

$$= \left[0.98 \times \frac{706.858 \times 176.714}{\sqrt{706.858^2 - 176.714^2}} \sqrt{2 \times (9.81 \times 100) \times 126}\right] \text{cm}^3\text{/sec}$$

$= 88929.548 \text{ cm}^3\text{/sec}$

$= (88929.548/1000)$ litres/sec $\because$ 1 litre = 1000 cm^3

$= 88.929548$ litres/sec

$= 88.93$ lits/sec **Ans.**

EXAMPLE 6.4 A 30 cm × 15 cm venturimeter is inserted in a vertical pipe carrying water flowing in the upward direction. A differential mercury manometer connected to the inlet and throat gives a reading 30 cm. Find the discharge if $C_d = 0.98$.

Solution: $A_1 = \pi/4\ (30)^2 = 706.858 \text{ cm}^2$
$A_2 = \pi/4\ (15)^2 = 176.714 \text{ cm}^2$

$$h = \left(\frac{p_1}{\rho g} + z_1\right) - \left(\frac{p_2}{\rho g} + z_2\right)$$

$$= x\left(\frac{s_m}{s_p} - 1\right) = 30\left(\frac{13.6}{1} - 1\right) = 378 \text{ cm}$$

$$Q = \left(0.98 \times \frac{706.858 \times 176.714}{\sqrt{706.858^2 - 176.714^2}} \sqrt{2(9.81 \times 100) \times 378}\right) \text{cm}^3\text{/sec}$$

$= 154030.5062 \text{ cm}^3\text{/sec}$
$= 154.03$ lits/sec **Ans.**

EXAMPLE 6.5 If in Problem 6.4, instead of water, oil of specific gravity is flowing through the venturimeter, determine the rate of flow of oil in litre/sec.

Solution: Here $h = \left(\frac{p_1}{\rho g} + z_1\right) - \left(\frac{p_2}{\rho g} + z_2\right)$

$$= x\left(\frac{S_m}{S_{\text{oil}}} - 1\right) = 30\left(\frac{13.6}{0.8} - 1\right) = 480 \text{ cm}$$

$$Q = \left(0.98 \times \frac{706.858 \times 176.714}{\sqrt{706.858^2 - 176.714^2}}\sqrt{2 \times 9.81 \times 100 \times 480}\right) \text{cm}^3\text{/sec}$$

$$Q = 173572.7169 \text{ cm}^3\text{/sec}$$

$$Q = 173.572 \text{ litre/sec}$$

Ans.

EXAMPLE 6.6 A 30 × 15 cm venturimeter is provided in a vertical pipeline carrying oil of specific gravity 0.9, with the flow being upwards. The difference of elevation of the throat and entrance sections of the venturimeter is 30 cm. The differential mercury manometer shows a gauge deflection of 25 cm. Calculate the pressure difference at the inlet and the throat section. (See Figure 6.4).

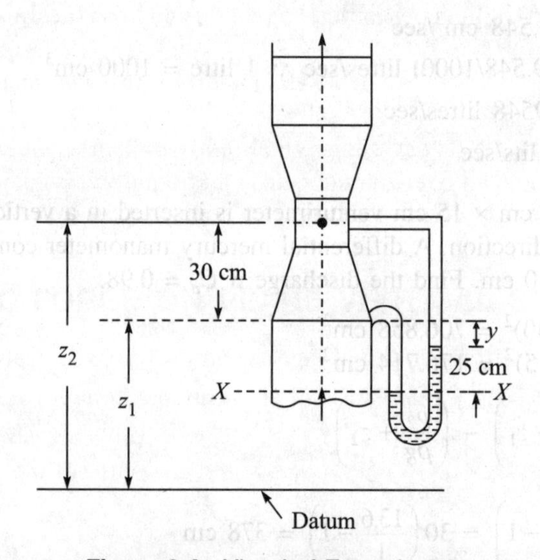

Figure 6.4 Visual of Example 6.6.

Solution: If p_1 is the pressure at inlet, p_2 at the throat writing manometric equation along x-x,

$$\frac{p_2}{\rho g} + (z_2 - z_1) + y + 25 = \frac{p_1}{\rho g} + y + 25$$

$$\frac{p_2}{\rho g} + (z_2 - z_1) + y + 25\left(\frac{13.6}{0.9}\right) = \frac{p_1}{\rho g} + y + 25$$

$$\left(\frac{p_1}{\rho g} - \frac{p_2}{\rho g}\right) + (z_1 - z_2) = 25\left(\frac{13.6}{0.9} - 1\right)$$

$$\frac{1}{\rho g}(p_1 - p_2) - 30 = 352.777$$

$\therefore \quad p_1 - p_2 = (352.777 + 30)\rho g$ But ρg for oil specific gravity 0.9

$p_1 - p_2 = (3.82777 \times 8829)$ N/m² $= (.9 \times 1000 \times 9.81) = 8829$

$= 33795.38$ N/m²

$$\therefore \quad p_1 - p_2 = \frac{33795.38 \text{ N}}{10000} \text{ N/cm}^2 = 3.379538 \text{ N/cm}^2$$ **Ans.**

6.5.2 Orifice Meter

This is also a device for measuring the discharge flowing through a pipe. It works in the same principle as that of the venturimeter with the application Bernoulli's and continuity equations. It is a flat circular plate which has a circular sharp-edged hole called 'orifice' which is concentric with the pipe. The diameter of the orifice varies from 0.4 to 0.8 times of the pipe diameter. A differential manometer or two piezometers are inserted at inlet which is about 1.5 to 2.0 times the distance upstream of the orifice and other one at about half the diameter of the orifice.

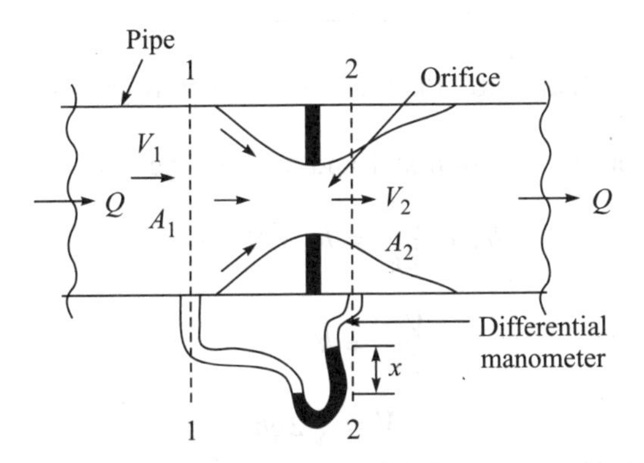

Figure 6.5 Orifice meter.

Applying Bernoulli's and continuity equation, the same equation of discharge like venturimeter is obtained, i.e.

$$Q = \frac{C_d \; A \cdot A_2 \sqrt{2gh}}{\sqrt{A_1^2 - A_2^2}}$$

Here C_d is much smaller than the venturimeter as some loss occurs due to contraction and expansion of flow. Here C_d varies from 0.60 to 0.65.

6.5.3 Pitot Tube

The pitot tube is another flow measuring device where Bernoulli's and continuity equations are used. Henri de Pitot, a French Engineer in 1732, adopted this tube for measuring velocity in the river Seine.

The basic principle used in this device is that if the velocity of flow at a particular point is reduced to zero, which is called the stagnation point, the pressure there increases due to conversion of kinetic energy. By measuring this pressure at this point, the velocity of flow may be determined. It consists of a glass tube open at both ends, bent at right angle, large enough in diameter to get rid of the capillary effect. The tube is dipped in water at about 0.6 of depth, of river.

With one open end facing the flowing stream as shown in Figure 6.6, the liquid enters the tube and rises to a height h above the level of water in the river. This is because, point A is the stagnation point where fluid is at rest, i.e kinetic energy becomes zero and is converted to pressure energy.

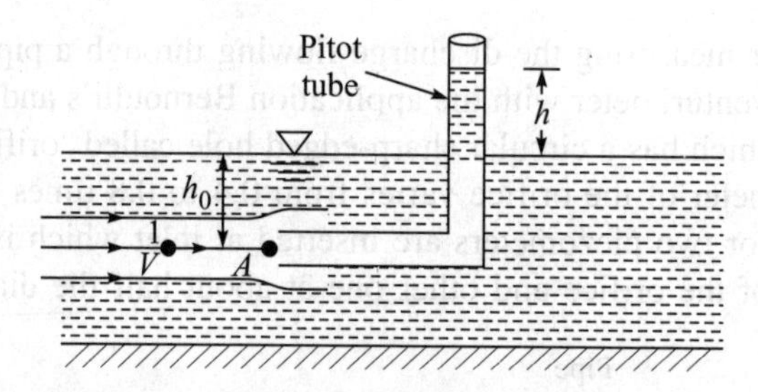

Figure 6.6 Pitot tube measurent in general.

Now applying Bernoulli's equation at 1 and at A, we get:

$$h_0 + \frac{V^2}{2g} = h_0 + 0 + h$$

$$\therefore \quad \frac{V^2}{2g} = h$$

$$\therefore \quad V = \sqrt{2gh} \tag{6.11}$$

Assuming some loss and error, we get:

$$V = C_v\sqrt{2gh} \tag{6.12}$$

This coefficient of the Pitot tube C_v varies from 0.97 to 0.98.

Once velocity is known, $Q = A \times V = AC_v\sqrt{2gh}$

$$\therefore \quad Q = C_v A \sqrt{2gh} \tag{6.13}$$

If it is used in the pipeline as shown in Figure 6.7.

(i) The difference of two piezometric heads $= h$

(ii) If differential manometer is used,

$$h = x\left(\frac{S_m}{S_w} - 1\right)$$

6.5.4 Nozzle Meter

It is also used to measure the discharge in a pipe. A convergent nozzle is used as shown in Figure 6.8. It is essentially a venturimeter where the throat and diverting portions are omitted.

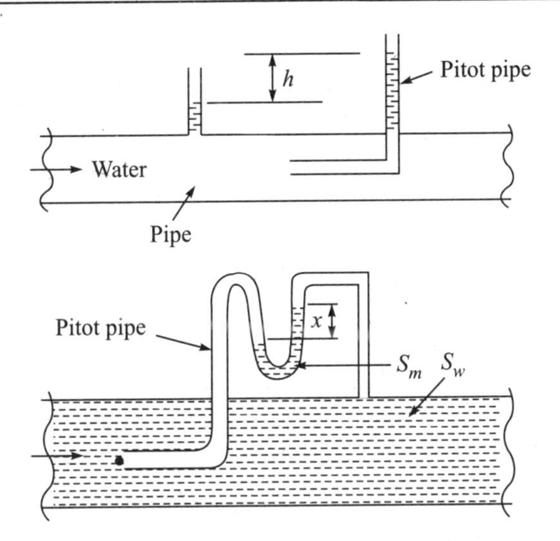

Figure 6.7 Pitot tubes with different pressure measurement arrangements/devices.

Two pressure gauges are at the upstream of the nozzle and at the nozzle point, respectively, which are inserted to measure the pressure difference. The same equation of discharge as in the case of the venturimeter is used. In the actual problem, the venturimeter is more widely used than the orifice meter and the nozzle meter.

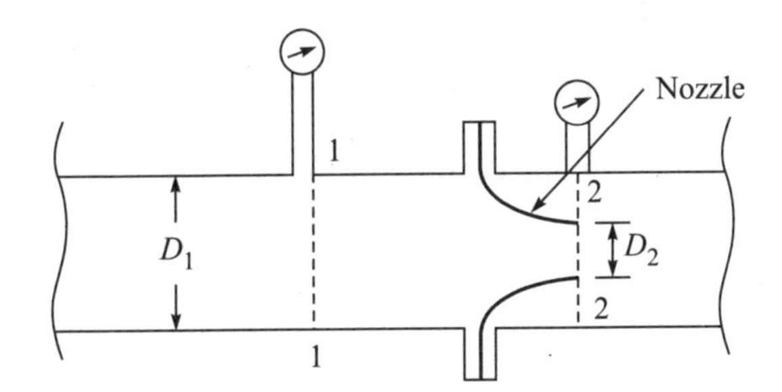

Figure 6.8 Nozzle meter.

EXAMPLE 6.7 An orifice meter with orifice diameter 15 cm is inserted in a pipe of 30 cm diameter. The pressure difference measured by a mercury oil differential manometer on the two sides of the orifice meter gives a reading of 50 cm of mercury. Find the rate of flow of oil of specific gravity 0.9 when the coefficient of discharge of the meter is 0.64.

Solution: $A_O = \frac{\pi}{4}(15)^2 = 176.714 \text{ cm}^2$ (A_o area at orifice)

$$A_P = D_A(30)^2 = 706.838 \text{ cm}^2 \quad (A_p \text{ area of the pipe})$$

$$h = \left[\frac{13.6}{0.9} - 1\right] \times 50 \text{ cm of oil}$$

$$= 705.555 \text{ cm of oil}$$

$$\therefore \qquad Q = C_d \frac{A_o A_P}{\sqrt{A_p^2 - A_p^2}} \sqrt{2gh}$$

$$Q = \left(0.64 \times \frac{176.714 \times 706.858}{\sqrt{706.858^2 - 176.74^2}} \times \sqrt{2 \times 9.81 \times 7.0555 \times 100} \right) \text{cm}^3/\text{sec}$$

$$Q = 137414.25 \text{ cm}^3/\text{sec}$$

$$Q = 137.41425 \text{ lits/sec}$$

Ans.

EXAMPLE 6.8 A Pitot tube place in the centre of a 30 cm pipe has one orifice pointing upstream and perpendicular to it. The mean velocity of the pipe is 0.8 times of central velocity. Find the discharge through the pipe if pressure difference between two orifices is 60 mm of water. Take the coefficient of Pitot tube to be 0.98.

Solution: Difference of pressure = 60 mm of water

= 0.06 m of water

Central velocity by Pitot tube = $C_v\sqrt{2gh} = 0.98\sqrt{2 \times 9.81 \times 0.06}$

= 1.063288 m/sec

Mean velocity $V = 0.8 \times 1.063288 = 0.850$ m/sec

$\therefore$ Discharge Q = area of pipe × V

$$Q = \left[\frac{\pi}{4}(0.3)^2 \times 0.8506 \right] \text{m}^3/\text{sec}$$

$$Q = 0.060127 \text{ m}^3/\text{sec}$$

Ans.

6.6 CONCLUSION

Equations of motion have been described with the application of Newton second law of motion. Euler's equation of motion considering gravity and pressure force is derived. Bernoulli's equation is obtained integrating this Euler's equation. Practical uses of Bernoulli's equation to develop flow measuring devices like venturimeter, orificemeter, Pitot tube etc. are also presented with numerical examples. Few problems are given for students to solve and a list of references is included at the end.

PROBLEMS

6.1 Water is flowing through a pipe having 30 cm and 20 cm at the bottom and upper end respectively. The intensity of pressure at the bottom end is 24.525 N/cm^2 and at the upper end is 9.81 N/cm^2. Determine the difference in datum head if the rate of flow through the pipe is 40 litres/sec. **(Ans.** 13.7 m)

6.2 A pipe through which water is flowing, has diameters 0.2 m and 0.1 m at cross-sections 1 and 2, respectively. The velocity of flow at 1 is 4 m/sec. Find the velocity head at sections 1 and 2 and the discharge. (**Ans.** 0.815 m, 83.047 m, $Q = 0.1256$ m^3/sec)

6.3 A pipe of diameter 0.4 m carries water at velocity of 25 m/sec pressure at section 1 and 2 are given 29.43 N/cm^2 and 22.563 N/cm^2, respectively while the datum at 1 and 2 are 28 m and 30 m. Find the loss of head between 1 and 2.

$$\left(\textit{Hints}: Z_1 + \frac{p_1}{\rho g} + \frac{V_1^2}{2g} = Z_2 + \frac{p_2}{\rho g} + \frac{V_2^2}{2g} + \text{Loss} \right)$$

(**Ans.** Loss = 56 m of water)

6.4 An oil of specific gravity 0.8 is flowing through a horizontal venturimeter having an inlet diameter of 20 cm and throat diameter 10 cm. The differential manometer with mercury shows a reading of 25 cm. Calculate the discharge if C_d of the venturimeter is 0.98. (**Ans.** 70.465 litres/sec)

6.5 A horizontal venturimeter with an inlet diameter of 20 cm and throat dia 10 cm is used to measure the flow of oil of specific gravity 0.8. The discharge of the oil is 60 litres/sec. Find the reading in the mercury manometer if C_d of the venturimeter is 0.98. (**Ans.** 18.12 cm)

6.6 A 20 cm × 10 cm venturimeter is inserted in a vertical pipe carrying oil of specific gravity 0.8, the flow of oil in upward direction. The difference of level between the throat and inlet section is 50 cm. The mercury differential manometer gives a reading of 30 cm of mercury. Find the discharge of the oil. Neglect losses.

(*Hints:* Neglect loss means $C_d = 1$) (**Ans.** 78.725 litres/sec)

6.7 An orifice meter with orifice diameter 10 cm is inserted in a pipe of 20 cm diameter. The pressure gauges fitted upstream and downstream of the orifice give readings of 19.62 N/cm^2 and 9.81 N/cm^2 respectively. If the coefficient of discharge is 0.6, find the discharge. (**Ans.** 68.21 litres/sec)

6.8 A Pitot tube is used to measure the velocity of water in pipe. The stagnation pressure head is 6 m and the static pressure head is 5 m. Calculate the velocity of flow taking the coefficient of Pitot tube to be 0.98. (**Ans.** 5.49 m/sec)

REFERENCES

1. Prandtl, *Essential of Fluid Dynamics*, Hafner Publishing Company, New York, 1952.
2. Robertson, J.M., *Hydrodynamics in Theory and Application*, Prentice-Hall Inc., Englewood Cliffs, 1965.
3. Lamp, H., *Hydrodynamics*, Cambridge University Press, London, 1932.
4. Liemann, H.W. and A. Roshko, *Elements of Gas Dynamics*, John Wiley & Sons. Inc., New York, 1956.
5. Miline–Thomson, L.M., *Theoretical Hydrodynamics*, Macmillan, New York, 1960.
6. Streeter, V., *Fluid Dynamics*, McGraw-Hill, New York, 1948.
7. Yuan, S.W., *Foundations of Fluid Mechanics*, Prentice-Hall of India, New Delhi, 1976.
8. Rouse, H., *Advanced Mechanics of Fluid*, John Wiley & Sons., Inc., London, 1965.
9. Munson, B.R., D.F. Young and Okiishi, *Fundamental of Fluid Mechanics*, Wiley Text Books, 2001.

Chapter 7

Flow Through Pipes

7.1 INTRODUCTION

A pipe is a closed conduit in which fluid flows under pressure. Pipes are commonly circular in cross-section. As the fluid flows under pressure, it always runs in full. This flowing fluid is always subjected to resistance due to shear forces between fluid particles, boundary walls of the pipe and between the fluid particles due to viscosity of the fluid. This resistance is known as *frictional resistance*. In order to overcome this resistance, certain energy of the flowing fluid is consumed. Thus a certain amount of fluid energy is lost due to this resistance. The flow of fluid may be laminar or turbulent, which will be dealt with in separate chapters. In this chapter, certain problems of pipe flow like different losses of energy—both major and minor—flow of pipe in series and parallel, flow from one reservoir to another by pipe, flow through siphon pipe, water hammer in brief, pipe networks, etc. will be discussed.

7.2 REYNOLDS EXPERIMENT: LAMINAR AND TURBULENT FLOW CONDITION

In 1883, with the help of a simple experiment conducted in a glass tube with coloured liquids, Osborne Reynolds determined the state of laminar and turbulent flow. He observed in the glass tube that at low velocity, coloured water moved in a straight path. With increasing velocity, he observed the flow of coloured water to be somewhat irregular. It is a critical or transition state. With a further increase in the velocity of flow, the fluctuation of the flowing filament of dye became more intense and eventually the dye diffused over the entire cross-section of the tube. This is the turbulent state of the flow. The velocity at which the flow begins to enter from the laminar to the turbulent state is called critical velocity.

Reynolds observed that the occurrence of these states of flow is governed by relative magnitudes of inertia and viscous force. He related the inertial force to viscous force by the dimensionless number R_e, i.e.

$$R_e = \frac{\text{Inertia force}}{\text{Viscous force}} = \frac{F_i}{F_v}$$

According to Newton's second law of motion, this inertial force F_1 is given by:

$$
\begin{aligned}
F_i &= \text{mass} \times \text{acceleration} \\
&= \rho(\text{volume}) \times \text{acceleration} \\
&= \rho L^3 \cdot \left(\frac{L}{T^2}\right) \\
&= \rho L^2 \left(\frac{L^2}{T^2}\right) \\
&= \rho L^2 V^2 \quad \because \frac{L}{T} = V
\end{aligned}
$$

Similarly viscous force F_v = shear stress × area

$$
\begin{aligned}
&= \mu \frac{\partial v}{\partial y} L^2 \\
&= \mu \left(\frac{V}{L}\right) L^2 \\
&= \mu V L
\end{aligned}
$$

$\therefore$

$$
\left.
\begin{aligned}
R_e = \frac{\rho L^2 V^2}{\mu V L} &= \frac{\rho V L}{\mu} \\
&= \frac{VL}{\frac{\mu}{\rho}} \\
&= \frac{VL}{\nu}
\end{aligned}
\right\} \tag{7.1}
$$

The dimensionless number is called 'Reynolds number' where ρ is the mass density, μ is the dynamic viscosity, $\frac{\mu}{\rho} = \nu$ is called kinetic viscosity, L is the characteristic length and in pipe flow, it is taken as diameter D.

Thus Reynolds number $= \frac{\rho VD}{\mu} = \frac{VD}{\nu}$ in pipe flow. With the help of further investigation, it was established that the flow is laminar upto approximately Reynolds number 2000.

Between 2000 and 4000, the transition legion or state exists. If Reynolds number is greater than 4000, the flow is turbulent. However, the critical value of Reynolds number is highly dependent on boundary geometry. As regards the flow between plates, flow in open channel, and flow around a sphere, this value varies to a great extent.

7.3 DARCY–WEISBACH EQUATION OF FRICTION LOSS

A Frenchman, H. Darcy, was the first to study the "Experimental researches on flow of water in pipes" in 1854, and then Julius Weisbach, a German, was associated with Darcy to derive the

Darcy–Weisbach equation of friction loss in pipe flow. In order to derive this equation, a horizontal pipe of cross-sectional area A carrying a fluid with an average velocity V is considered. Let 1 and 2 be two cross-sections of the pipe at a distance L apart. Let p_1 and p_2 be the pressure intensities at 1 and 2. Now applying Bernoulli's equation in the two sections:

$$\frac{p_1}{\rho g} + \frac{V_1^2}{2g} + Z_1 = \frac{p_2}{\rho g} + \frac{V_2^2}{2g} + Z_2 + \text{Loss of head } h_f \text{ due to friction}$$

Since the pipe is horizontal and the fluid is flowing with the same velocity, $Z_1 = Z_2$ and $V_1 = V_2 = V$

$$\therefore \qquad \frac{p_1}{\rho g} - \frac{p_2}{\rho g} = h_f \tag{A}$$

William Froude (1810–1879), an Englishman, performed an extensive experiment in a wooden box with a wooden board to study the resistance offered by different surfaces to the flowing water and gave the equation of frictional resistance force (F_R) equation as follows:

$$F_R = f'AV^n \tag{7.2}$$

where f' is the frictional resistance per unit area at unit velocity. A is the area equal (PL), where P is the wetted perimeter and n is the exponent, and for turbulent flow $n \approx 2$.

Now resolving the force horizontally,

$$p_1A = p_2A + F_R$$

$$(p_1 - p_2)A = f'(PL)V^2$$

$$(p_1 - p_2) = f'\left(\frac{P}{A}\right)LV^2$$

$$\text{Dividing by } \rho g, \quad \frac{p_1}{\rho g} - \frac{p_2}{\rho g} = \frac{f'}{\rho g}\left(\frac{P}{A}\right)LV^2 \tag{B}$$

The ratio of the cross-sectional area A to the wetted perimeter P is called hydraulic mean depth m

$$= \frac{A}{P} = \frac{\pi/4\, D^2}{\pi D} = \frac{D}{4}$$

$$\therefore \qquad \frac{P}{A} = \frac{4}{D}$$

Comparing Equation (A) and (B), we get:

$$h_f = \frac{4f'}{\rho g} \cdot \frac{LV^2}{D} = \frac{4f'}{w}\frac{LV^2}{D} \quad \because \rho g = w$$

Putting

$$\frac{4f'}{w} = \frac{f}{2g}$$

$$h_f = \frac{fLV^2}{2gD} \tag{7.3}$$

which is the Darcy–Weisbach equation of head loss due to friction, and f is called the friction factor which is a dimensionless parameter.

This friction factor f is not constant, as it varies with the roughness of the pipe surface and the Reynolds number of flow. In order to determine the loss of head due to friction correctly, it is important to assess the f correctly for the pipe surface and flow conditions.

When the flow is laminar, i.e. $R_e < 2000$

$$f = \frac{64}{R_e} \tag{7.4}$$

Blasius gave

$$f = \frac{0.316}{R_e^{1/4}} \tag{7.5}$$

when $R_e < 10^5$

Prandtl–Karman (1935), gave the following equation for smooth turbulent flow.

$$\frac{1}{\sqrt{f}} = 2\log_{10}\frac{R_e\sqrt{f}}{2.51} \tag{7.6}$$

and the following for rough turbulent flow:

$$\frac{1}{\sqrt{f}} = 2\log_{10}\left[3.71\frac{D}{k_S}\right] \tag{7.7}$$

from the experimental data of Nikuradse (1932).

Colebrook and White (1938) combined Equations (7.6) and (7.7) to present the following equation, known as the Colebrook–White equation, which is valid for any type of roughness (smooth to rough) in turbulent flow.

$$\frac{1}{\sqrt{f}} = -2\log_{10}\left(\frac{k_S}{3.71D} + \frac{2.51}{R_e\sqrt{f}}\right) \tag{7.8}$$

In Equations (7.7) and (7.8) k_S is the equivalent sand roughness size of Nikuradse.

In Equations (7.6) and (7.8), the solution for f is implicit. Therefore, the method of trial and error needs to be used for its evaluation. Direct approximate solutions for both equations are given by Barr (1973), which are expressed as:

$$f = 1/(1.785\log_{10} R_e - 1.424)^2 \tag{7.6a}$$

and

$$\frac{1}{\sqrt{f}} = -2\log_{10}\left[\left(\frac{K_s}{3.71D}\right) - \left(\frac{5.1286}{R_e^{0.89}}\right)\right] \tag{7.8a}$$

7.4 OTHER ENERGY LOSS IN PIPE FLOW

The various losses of energy in pipe flow are classified as:

(i) Major losses
(ii) Minor losses

Major loss occurs mainly due to friction which is already discussed in Section 7.3. It is very important for a long pipeline with higher velocity as h_f is directly proportional to length L and V^2 as seen from the Darcy–Weisbach equation.

The minor losses occur due to a change in the magnitude of velocity. For a long pipeline, minor losses are small as compared to loss due to friction, which may be neglected without serious error. In short pipelines, these minor losses are important. The following are the minor losses:

(I) Loss of energy due to sudden expansion of pipe, $h_L = \dfrac{(V_1 - V_2)^2}{2g}$

(II) Loss of energy due to sudden contraction, $h_L = 0.5\dfrac{V_2^2}{2g}$

(III) Loss of energy at entrance, $h_L = 0.5\dfrac{V^2}{2g}$

(IV) Loss of energy at exit, $h_L = \dfrac{V^2}{2g}$

(V) Loss of energy due to gradual enlargement and contraction, $h_L = K\dfrac{(V_1 - V_2)^2}{2g}$

(VI) Loss of energy at bends, $h_L = K\dfrac{V^2}{2g}$

(VII) Loss of energy in various pipe fittings, $h_L = K\dfrac{V^2}{2g}$

It can be seen that the minor losses are dependent on the velocity head. The above expressions of minor losses clearly show that they are quite small compared to head lost due to friction.

7.5 LOSS DUE TO SUDDEN ENLARGEMENT

In Figure 7.1, it can be seen that the pipe suddenly expanded from area A_1 to A_2. Eddies are formed at the two corners and these give a pressure opposing the flow direction equal to $p'(A_2 - A_1)$. If h_e is the loss of head due to sudden enlargement, we may write the Bernoulli's equation as follows:

i.e. Bernoulli's equation at 1-1 and at 2-2, we get:

$$\frac{p_1}{w} + \frac{V_1^2}{2g} + Z_1 = \frac{p_2}{w} + \frac{V_2^2}{2g} + Z_2 + h_e$$

$Z_1 = Z_2$, taking the pipe to be horizontal

$$\therefore \qquad h_e = \left(\frac{p_1}{w} - \frac{p_2}{w}\right) + \frac{1}{2g}(V_1^2 - V_2^2) \tag{A}$$

Change of momentum of control volume of liquid between 1-1 and 2-2

= Σ forces acting on this in the flow direction

$$[(\rho A_2 V_2)V_2 - (\rho A_1 V_1)V_1] = p_1 A_1 + p_1(A_2 - A_1) - p_2 A_2 = p_1 A_2 - p_2 A_2 = A_2(p_1 - p_2)$$

or

$$\rho A_2 V_2^2 - \rho A_1 V_1^2 = A_2(p_1 - p_2) \tag{B}$$

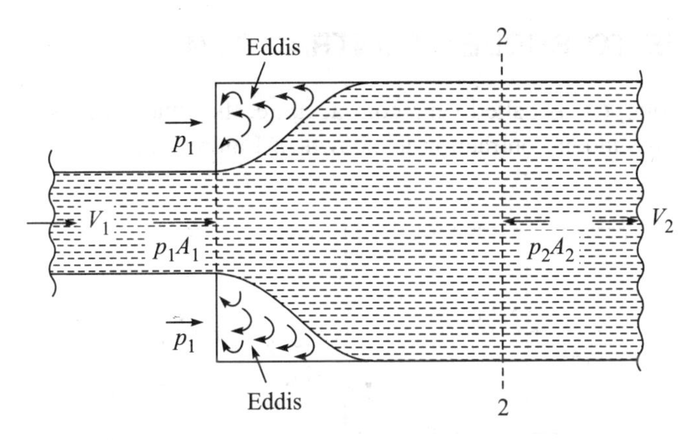

Figure 7.1 Sudden enlargement.

Continuity equation gives:

$$A_1V_1 = A_2V_2 \quad \therefore\ A_1 = \frac{V_2}{V_1}A_2$$

$\therefore$ (B) becomes $\rho A_2 V_2^2 - \rho\dfrac{V_2}{V_1}A_2V_1^2 = A_2(p_1 - p_2)$

$$\rho A_2(V_2^2 - V_1V_2) = A_2(p_1 - p_2)$$

$$V_2^2 - V_1V_2 = \frac{p_1}{\rho} - \frac{p_2}{\rho}$$

Dividing by g, we get:

$$\frac{V_2^2 - V_1V_2}{g} = \frac{p_1}{\rho g} - \frac{p_2}{\rho g} = \left(\frac{p_1}{w} - \frac{p_2}{w}\right)$$

$$\therefore \quad \left(\frac{p_1}{w} - \frac{p_2}{w}\right) = \frac{V_2^2 - V_1V_2}{g} \tag{C}$$

Substituting the value of $\left(\dfrac{p_1}{w} - \dfrac{p_2}{w}\right)$ from (C) in (A), we get:

$$h_e = \frac{V_2^2 - V_1V_2}{g} + \frac{V_1^2}{2g} - \frac{V_2^2}{2g} = \frac{2V_2^2 - 2V_1V_2 + V_1^2 - V_2^2}{2g}$$

$$h_e = \left(\frac{V_1^2 + V_2^2 - 2V_1V_2}{2g}\right)$$

$$h_e = \frac{(V_1 - V_2)^2}{2g} \tag{7.9}$$

which gives the loss of energy h_e due to sudden expansion. This loss of head equation was first obtained by J.C. Borda and L. Carnot and is therefore known as the Borda–Carnot equation of head loss.

7.6 LOSS DUE TO SUDDEN CONTRACTION

In the contraction zone, CC is called vena contracta, i.e. the smallest cross-section of flow. Let h_e be the loss due to sudden contraction as shown in Figure 7.2.

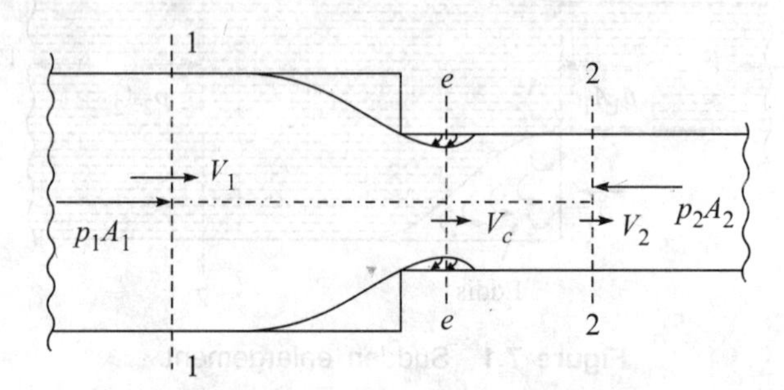

Figure 7.2 Sudden contraction.

Applying Equation (7.9) for sudden enlargement from CC to 22, we get:

$$h_c = \frac{(V_c - V_2)^2}{2g} = \frac{V_2^2}{2g}\left(\frac{V_c}{V_2} - 1\right)^2 \quad \text{(A)}$$

From continuity (applying in CC and 22),

$$A_c V_c = A_2 V_2$$

$$\frac{V_c}{V_2} = \frac{A_2}{A_c} = \frac{1}{\left(\frac{A_c}{A_2}\right)} = \frac{1}{C_c} \quad \text{(B)}$$

where C_c is the coefficient of contraction usually less than unity and is assumed to be 0.62.

Substituting $\frac{V_c}{V_2}$ in (A),

$$h_c = \frac{V_2^2}{2g}\left[\frac{1}{C_c} - 1\right]$$

$$h_c = K\frac{V_2^2}{2g} \qquad \text{where } K = \left(\frac{1}{C_c} - 1\right)^2 = 0.375$$

$$\therefore \quad h_c = 0.375\frac{V_2^2}{2g}$$

If C_c is not given, head loss due to contraction is taken as:

$$h_c = 0.5\frac{V_2^2}{2g} \quad (7.10)$$

EXAMPLE 7.1 The discharge through a pipe is 200 litres/sec. Find the loss of head when the pipe is suddenly enlarged from 150 mm to 300 mm diameter.

Solution: Continuity gives $A_1V_1 = A_2V_2 = 200$ litres/sec

$$\frac{\pi}{4}\left(\frac{150}{100}\right)^2 V_1 = \frac{\pi}{4}\left(\frac{300}{100}\right)^2 V_2 = 200 \text{ lits}$$

where V_1 and V_2 are in decimeter/sec $\quad\because$ 1 dm^3 = 1 litre

$\therefore$
$$V_1 = 113.1768 \text{ dm/sec}, \quad V_2 = 28.294 \text{ dm/sec.}$$

$\therefore$ Loss of head due to sudden enlargement is

$$h_c = \frac{(V_1 - V_2)^2}{2g} = \frac{(113.1768 - 28.294)^2}{2 \times 9.81 \times 10}$$

$$h_c = 36.723 \text{ dm}$$

$$h_c = 3.6723 \text{ m of water}$$ **Ans.**

EXAMPLE 7.2 A horizontal pipe of diameter 400 mm is suddenly contracted to 200 mm. The pressure intensities in the large and smaller ends of the pipe are given as 14.715 N/cm² and 12.753 N/cm², respectively. If $C_C = 0.62$, find the loss of head due to sudden contraction. Also determine the discharge.

Solution: $D_1 = 400 \text{ mm} = 0.4 \text{ m}$

$$A_1 = \pi/4(.4)^2 = 0.12566 \text{ m}^2$$
$$D_2 = 200 \text{ mm} = 0.2 \text{ m}$$
$$A_2 = \pi/4(.2)^2 = 0.03141 \text{ m}^2$$
$$p_1 = 14.715 \text{ N/cm}^2 = 14.715 \times 10^4 \text{ N/m}^2$$
$$p_2 = 12.753 \text{ N/m}^2 = 12.753 \times 10^4 \text{ N/m}^2$$

Applying Bernoulli's equation at 1-1 and 2-2, we get:

$$Z_1 + \frac{p_1}{w} + \frac{V_1^2}{2g} = Z_2 + \frac{p_2}{w} + \frac{V_2^2}{2g} + h_c$$

$\because$ horizontal, $Z_1 = Z_2$

$$\left(\frac{p_1}{w} - \frac{p_2}{w}\right) = \frac{1}{2g}(V_2^2 - V_1^2) + h_c$$

$$\frac{14.715 \times 10^4}{1000 \times 9.81} - \frac{12.753 \times 10^4}{1000 \times 9.81} = \frac{1}{2g}(V_2^2 - V_1^2) + h_c$$

$$(14.175 - 12.753) \times 10^4 = \frac{9810}{2g}(V_2^2 - V_1^2) + h_c \qquad \text{(A)}$$

Continuity gives $A_1V_1 = A_2V_2$

$\therefore$
$$V_2 = \frac{A_1}{A_2} V_1 = \frac{0.12566}{0.3141} V_1 = 4V_1 \quad \therefore V_1 = 0.25V_2$$

$$h_c = \frac{V_2^2}{2g}\left(\frac{1}{C_c} - 1\right) = 0.375\frac{V_2^2}{2g}$$

Equation (A) now becomes:

$$2g(15 - 13) = [V_2^2 - (0.25\ V_2)^2] + 0.375\ V_2^2$$

$$2 \times 9.81 \times 2 = (1 - 0.0625 + 0.375)\ V_2^2$$

$$39.24 = 1.3125\ V_2^2 \quad \therefore \quad V_2 = 5.4678 \text{ m/sec}$$

$$\therefore \quad Q = A_2V_2 = (.03141 \times 5.4678) \text{ m}^3\text{/sec}$$

$$Q = 0.17174 \text{ m}^3\text{/sec}$$ **Ans.**

$$h_c = \left(0.375 \frac{5.4678^2}{2 \times 9.81}\right) \text{ m of water}$$

$$\therefore \quad h_c = 0.5714 \text{ m of water}$$ **Ans.**

EXAMPLE 7.3 Calculate the discharge of water flowing through a pipe of diameter 300 mm when difference of pressure head between two ends of a pipe 400 m long is 5 m of water. The value of friction factor of the pipe is 0.00225.

Solution: $D = 300 \text{ mm} = 0.3 \text{ m}$

$A = \pi/4(.3)^2 = 0.07068 \text{ m}^2$

$L = 400 \text{ m}, f = 0.00225$

We know difference pressure head is equal to h_f

and
$$h_f = \frac{fLV^2}{2gD}$$

i.e.
$$5 = \frac{0.00225 \times 400 \times V^2}{2 \times 9.81 \times 0.3}$$

$$\therefore \quad V = 5.7184 \text{ m/sec}$$

$$\therefore \quad Q = A \cdot V = (0.07068 \times 5.7184) \text{ m}^3\text{/sec}$$

$$Q = 0.0404176 \text{ m}^3\text{/sec}$$ **Ans.**

EXAMPLE 7.4 An oil of specific gravity 0.8 is flowing through a pipe of diameter 300 mm at the rate of 500 lits/sec. Find the head lost due to friction and power required to maintain the flow for a length 1000 m. Take $\nu = 0.29$ stoke.

Solution: $D = 300 \text{ mm} = 0.3 \text{ m}, \therefore A = \frac{\pi}{4}(0.3)^2 = 0.07068 \text{ m}^2$

$$Q = 500 \text{ lits/sec} = 0.5 \text{ m}^3\text{/sec}$$

$$L = 1000 \text{ m}, V = \frac{Q}{A} = \frac{5}{0.07068} = 7.07413 \text{ m/sec}$$

$$\text{Reynolds number } R_e = \frac{V \times D}{\nu} = \frac{7.07413 \times 0.3}{0.29 \times 10^{-4}} = 7.318 \times 10^4 < 10^5$$

So we may use the Blasius equation to find the friction factor.

i.e.
$$f = \frac{0.316}{R_c^{1/4}} = 0.01921$$

$$\therefore \quad \text{Head lost due to friction } h_f = \frac{fLV^2}{2gD} = \frac{0.01921 \times 1000 \times 7.07413^2}{2 \times 9.81 \times 0.3}$$

$$h_f = 163.325 \text{ m}$$ **Ans.**

$$\text{Power required} = \frac{\rho g Q h_f}{1000} \text{ kw} = \frac{9.81 \times 0.8 \times 1000 \times .5 \times 163.325}{1000} \text{ kw}$$

$$= 640.318 \text{ kw.}$$ **Ans.**

EXAMPLE 7.5 Determine the discharge of water through a pipe of diameter 0.1 m and length of 60 m when one end of the pipe is connected to a tank and the other end of the pipe is open to atmosphere. The height of the water in the tank from the centre of the pipe is 4 m. The pipe is horizontal and the friction factor is 0.04. Consider minor losses. (See Figure 7.3).

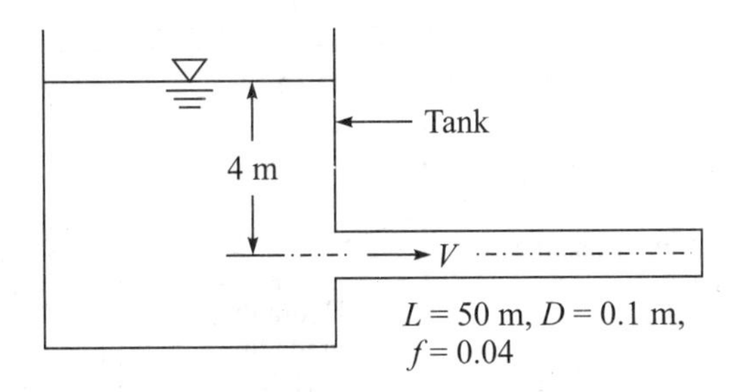

Figure 7.3 Visual of Example 7.5.

Solution: If V_2 is the velocity through the pipe, applying Bernoulli's equation at tank water surface and pipe outlet, we get:

$$\frac{p_1}{\rho g} + \frac{V_1^2}{2g} + Z_1 = \frac{p_2}{\rho g} + \frac{V_2^2}{2g} + Z_2 + \text{all losses}$$

Considering datum line through the centre line of the pipe,

$$0 + 0 + 4 = 0 + \frac{V_2^2}{2g} + 0 + (h_{\text{at entrance}} + h_f)$$

$\therefore$ p_1 and p_2 exposed to atmosphere, $p_1 = p_2 = 0$

and also $$Z_2 = 0,\ h_i = 0.5\frac{V_2^2}{2g},\ h_f = \frac{fLV_2^2}{2gD} = \frac{0.04 \times 50 \times V_2^2}{2 \times 9.81 \times 0.1}$$

$$\therefore \quad 4 = \frac{V_2^2}{2g} + 0.5\frac{V_2^2}{2g} + \frac{0.04 \times 50}{2 \times 9.81 \times 0.1} V_2^2 \quad \because \quad h_f = \frac{f L V_2^2}{2gD}$$

$$4 = \frac{V_2^2}{2g}[1 + 0.5 + 20]$$

$$\sqrt{\frac{4 \times 2 \times 9.81}{21.5}} = V_2$$

$$V_2 = 1.910558 \text{ m/sec}$$

$$\therefore \qquad Q = A \cdot V_2 = (\pi/4(0.1)^2 \times 1.910558) \text{ m}^3\text{/sec}$$

$$\therefore \qquad Q = 0.015005 \text{ m}^3\text{/sec} = 15.005 \text{ litres/sec} \qquad \textbf{Ans.}$$

7.7 HYDRAULIC GRADE LINE AND TOTAL ENERGY LINE

In the study of the flow of fluid in pipes (or water in open channel), the concepts of the Hydraulic Grade Line (HGL) and Total Energy Line (TEL) are very important.

These two lines may be understood clearly from the following discussion.

Consider a long pipe carrying liquid from a reservoir *A* to reservoir *B* (Figure 7.4).

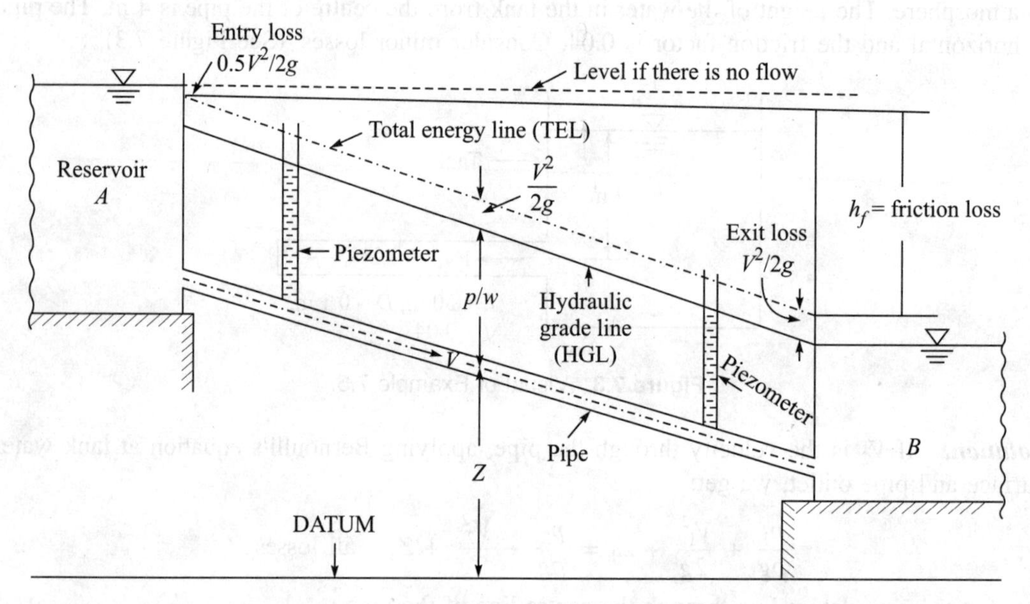

Figure 7.4 Hydraulic Grade Line (HGL) and Total Energy Line (TEL).

At some points, piezometers are installed. While the liquid flows under pressure, the liquid level in the piezometer will rise to a certain height, depending on the pressure at that point. If the levels of the piezometer are joined, the line obtained is known Hydraulic Grade Line (HGL). The height of the piezometer above the centre line of the pipe represents the pressure (p/w) head at that section. Loss of energy due to friction takes place along the length of the pipe and hence that HGL is gradually falling. Thus at any section, the vertical distance between the HGL and the centre line of the pipe is equal to the pressure head. If an arbitrary datum is considered, then $\left(z + \frac{p}{w}\right)$ represents the piezometric head. The Total Energy Line (TEL) as shown in the Figure 7.4 is at vertical distance of $V^2/2g$ above HGL. Entry loss, exit loss and friction loss, all are shown in the Figure.

Figures 7.5(a) and 7.5(b) show the HGL and TEL in two different situations.

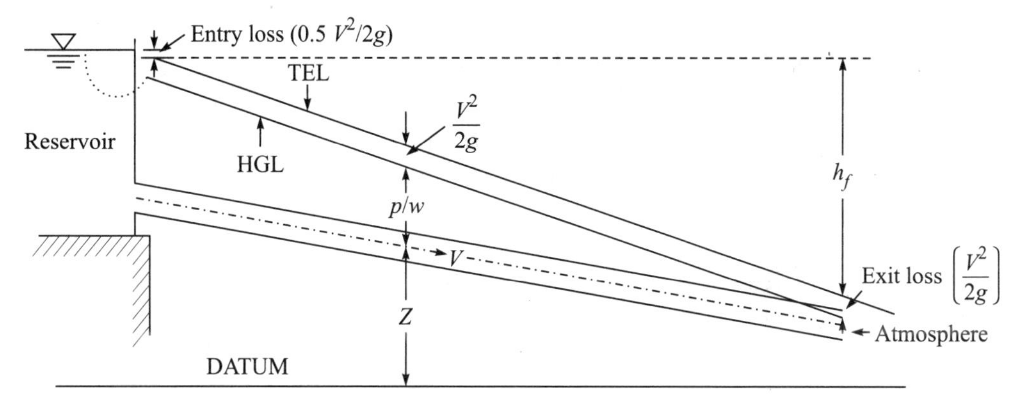

(a) HGL and TEL of pipe discharging to atmosphere

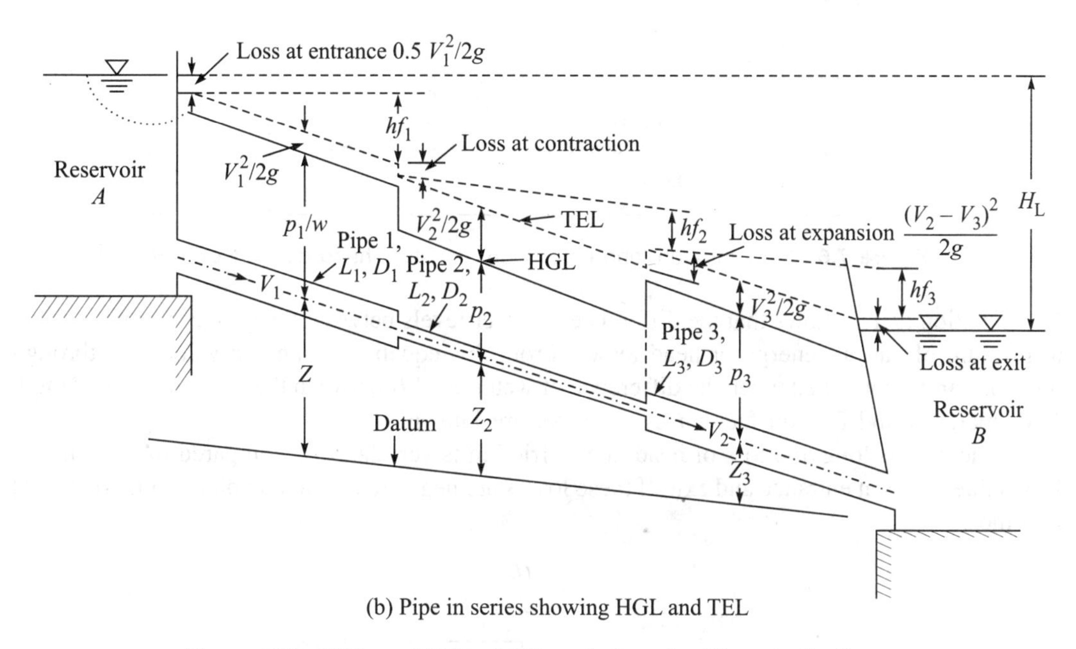

(b) Pipe in series showing HGL and TEL

Figure 7.5 HGL and TEL of different pipes in different situations.

7.8 FLOW THROUGH LONG PIPES UNDER CONSTANT HEAD *H*

Figure 7.6 shows a long pipeline of length L, and diameter D connecting two reservoirs A and B. Water flows through this long pipe from A to B under constant head $H = (H_A - H_B)$ with a velocity V.

Applying Bernoulli's equation at the water level in A and at the end section of the pipe at B,

$$Z_A + H_A = Z_B + H_B + \text{Losses}$$

$$Z_A + H_A = Z_B + H_B + \left(0.5\frac{V^2}{2g} + h_f + \frac{V^2}{2g}\right)$$

$$(Z_A + H_A) - (Z_B + H_B) = 0.5\frac{V^2}{2g} + \frac{SLV^2}{2gD} + \frac{V^2}{2g} = \left(1.5 + \frac{fL}{D}\right)\frac{V^2}{2g}$$

or

$$H = \left(1.5 + \frac{fL}{D}\right)\frac{V^2}{2g} \tag{7.11}$$

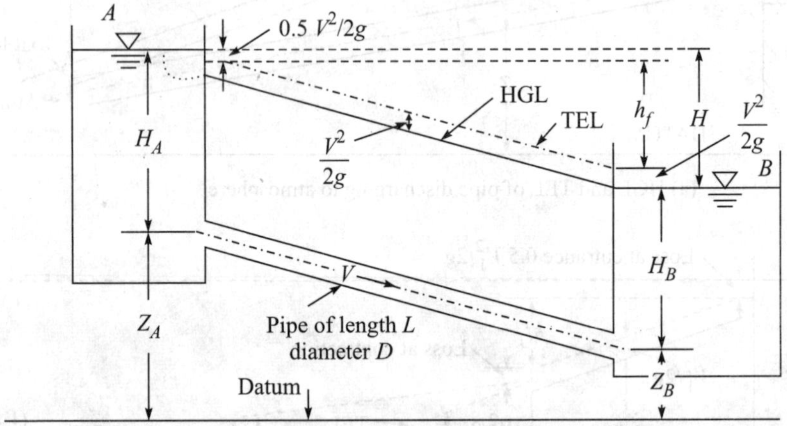

Figure 7.6 Flow through long pipe under constant head difference of *H*.

Equation (7.11) shows that the difference in water levels between the two reservoirs is equal to the sum of various energy or head losses. From this equation, unknown velocity V through the pipe can be computed when the difference of water level H between the two reservoirs, length L, diameter D and friction factor f of the pipe, are known.

If the pipe is long, the loss of head due to friction is very large as compared to other minor losses like losses at entrance and exit. If these losses are neglected for a long pipe, Equation (7.11) becomes:

$$H = \frac{fLV^2}{2gD}$$

$$\therefore \qquad V = \sqrt{\frac{2gDH}{fL}} \tag{7.12}$$

7.9 PIPE IN SERIES OR COMPOUND PIPE

Figure 7.5(b) in Section 7.7 is an example of a pipe in series or compound pipe. When a pipe connecting two reservoirs is made up of different diameters D_1, D_2, D_3, etc. and lengths L_1, L_2, L_3, etc. all are connected end to end (i.e. in series), it is called a compound pipe. In such cases, the discharge through each pipe is the same, i.e.

$$Q = \frac{\pi}{4}D_1^2V_1 = \frac{\pi}{4}D_2^2V_2 = \frac{\pi}{4}D_3^2V_3$$

and the difference of water level between the two reservoirs is equal to all head losses.

i.e. $$H_L = 0.5\frac{V_1^2}{2g} + \frac{f_1 L_1 V_1^2}{2gD_1} + \frac{0.5V_2^2}{2g} + \frac{f_2 L_2 V_2^2}{2gD_2} + \frac{(V_2 - V_3)^2}{2g} + \frac{f_3 L_3 V_3^2}{2gD_3} + \frac{V_3^2}{2g} \quad (7.13)$$

If the minor losses are neglected,

$$H_L = \frac{f_1 L_1 V_1^2}{2gD_1} + \frac{f_2 L_2 V_2^2}{2gD_2} + \frac{f_3 L_3 V_3^2}{2gD_3} \quad (7.14)$$

7.10 EQUIVALENT PIPE

A compound pipe of several diameters and lengths [Figure 7.5(b)] may be replaced by a single pipe of uniform diameter. This new pipe of uniform diameter is called 'equivalent pipe'. The uniform diameter of the pipe is also known as the equivalent diameter of the compound pipe. The size of equivalent pipe may be computed as follows. Neglecting minor losses, head lost (h_L) in the compound, the pipe of Figure 7.5(b) is:

$$h_L = \frac{f_1 L_1 V_1^2}{2gD_1} + \frac{f_2 L_2 V_2^2}{2gD_2} + \frac{f_3 L_3 V_3^2}{2gD_3}$$

Assuming $f_1 = f_2 = f_3$,

$$h_L = \frac{f}{2g}\left[\frac{L_1 V_1^2}{D_1} + \frac{L_2 V_2^2}{D_2} + \frac{L_3 V_3^2}{D_3}\right] \quad \text{(A)}$$

Continuity gives, $Q = \frac{\pi}{4} D_1^2 V_1 = \frac{\pi}{4} D_2^2 V_2 = \frac{\pi}{4} D_3^2 V_3$

$$\therefore \quad V_1^2 = \left(\frac{4Q}{\pi D_1^2}\right)^2, \; V_2 = \left(\frac{4Q}{\pi D_2^2}\right)^2, \; V_3 = \left(\frac{4Q}{\pi D_3^2}\right)^2$$

Substituting V_1^2, V_2^2 and V_3^2 in (A) and simplifying,

$$h_L = \frac{f}{2g} \cdot \frac{Q^2}{\left(\frac{\pi}{4}\right)^2}\left[\frac{L_1}{D_1^5} + \frac{L_2}{D_2^5} + \frac{L_3}{D_3^5}\right] \quad \text{(B)}$$

If V, D and L are the velocity, diameter and length of the equivalent pipe, respectively, and this pipe carries the same discharge and assuming the head lost in the equivalent pipe is the same as that of the compound pipe, the loss of the head in the equivalent pipe is:

$$h_L = \frac{fLV^2}{2gD} = \frac{f}{2g} \cdot \frac{Q^2 \cdot L}{\left(\frac{\pi}{4}\right)^2 D^5} \quad \text{(C)}$$

Equating the head loss h_L from (B) and (C), we get

$$\frac{L}{D^5} = \left[\frac{L_1}{D_1^5} + \frac{L_2}{D_2^5} + \frac{L_3}{D_3^5}\right] \quad (7.15)$$

Equation (7.15) is known as the Dupuits equation. If $L = L_1 + L_2 + L_3$, D can be computed. Thus this diameter D of the equivalent pipe may be used in the equivalent pipe.

7.11 PIPES IN PARALLEL

Figure 7.7 is an example of pipe in parallel connection. When the main pipe is divided into two or more pipes join again to continue as the main pipe, pipes are said to be parallel. As shown in Figure 7.7 head lost due to friction in both the pipes from A to B are the same.

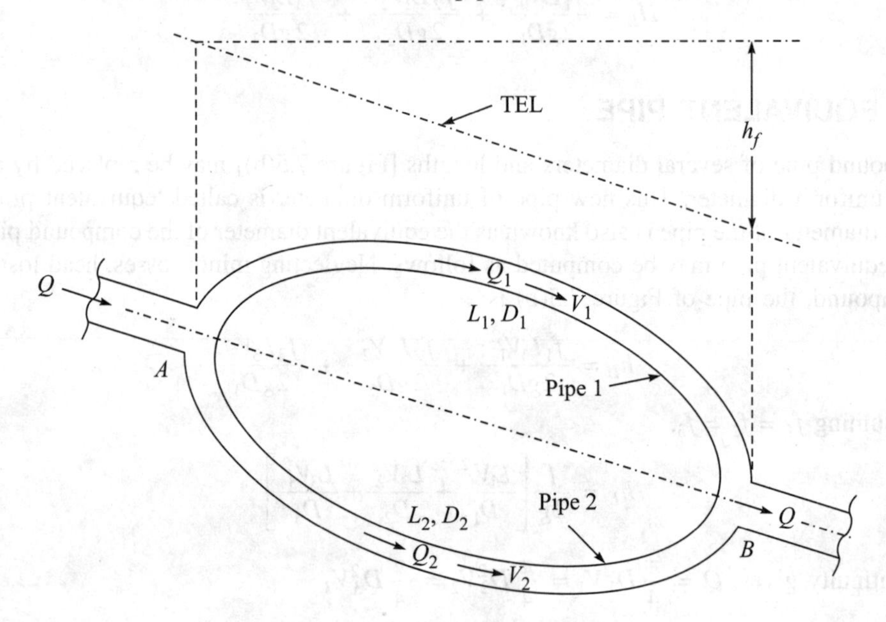

Figure 7.7 Pipes in parallel.

Here $Q = Q_1 + Q_2$ and $h_f = \dfrac{fL_1V_1^2}{2gD_1} = \dfrac{fL_2V_2^2}{2gD_2}$ if the values of f for both the pipes are the same.

$$\therefore \qquad h_f = \frac{f\,L_1V_1^2}{2gD_1} = \frac{f\,L_2V_2^2}{2gD_2} \tag{7.15a}$$

7.12 FLOW THROUGH A BYE-PASS PIPE

Sometimes in a city water supply mains, bye-pass pipes are necessary as shown in Figure 7.8. It is a diversion of main flow in a small pipe which later joins the main pipe at some downstream point.

If point (1) is the inlet of the pass and (2) is the outlet,

Difference of head between (1) and (2) is:

$$H_1 - H_2 = \frac{fLV^2}{2gD} = \frac{flv^2}{2gd} + K'\frac{v^2}{2g}$$

$$\therefore \qquad \frac{LV^2}{D} = \frac{lv^2}{d} + kv^2 \quad \text{where } k = \frac{k'}{f}$$

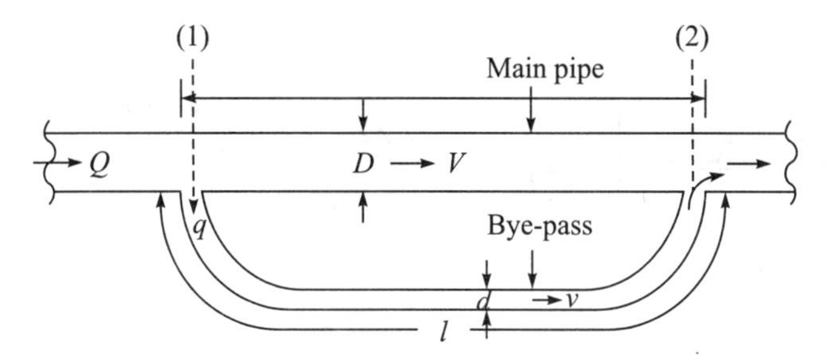

Figure 7.8 Flow through a bye-pass pipe.

$$\therefore \qquad V = \sqrt{\frac{D}{L}v^2\left(\frac{l}{d}+k\right)} = v\sqrt{\frac{D}{L}\left(\frac{l}{d}+k\right)}$$

$$Q = \frac{\pi}{4}D^2V, \quad q = \frac{\pi d^2}{4}v$$

$$\therefore \qquad \frac{Q}{q} = \frac{D^2V}{d^2v} = \frac{D^2}{d^2}\cdot\sqrt{\frac{D}{L}\left(\frac{l}{d}+k\right)}$$

$$\therefore \qquad \frac{Q}{q} = \frac{D^2}{d^2}\sqrt{\frac{D}{L}\cdot\frac{l}{d}+\frac{D}{L}k} = \frac{D^2}{d^2}\sqrt{\frac{D}{d}\cdot\frac{l}{L}+\frac{D}{L}k}$$

or

$$\frac{Q}{q} = \sqrt{\frac{D^5}{d^5}\cdot\frac{l}{L}+\frac{D^5}{d^4L}k}$$

or

$$\frac{Q}{q} = \sqrt{\left(\frac{D}{d}\right)^5\left[\frac{l}{L}+\frac{d}{L}k\right]}$$

or

$$\frac{Q}{q} = \sqrt{\left(\frac{D}{d}\right)^5\left(\frac{l+kd}{L}\right)}$$

or

$$\frac{q}{Q} = \frac{1}{\sqrt{\left(\frac{D}{d}\right)^5\left(\frac{l+kd}{L}\right)}}$$

$$\therefore \qquad q = \frac{Q}{\sqrt{\left(\frac{D}{d}\right)^5\left(\frac{l+kd}{L}\right)}} \qquad (7.16)$$

which is the relation between Q and q.

7.13 FLOW THROUGH SIPHON PIPE

A siphon is a long bent pipe used to carry water from a higher reservoir A to a lower reservoir B when the two reservoirs are separated by hill or higher level ground, as shown in Figure 7.9. Hydraulic grade line cuts the siphon at C and i.e. pressure at C and D is atmospheric.

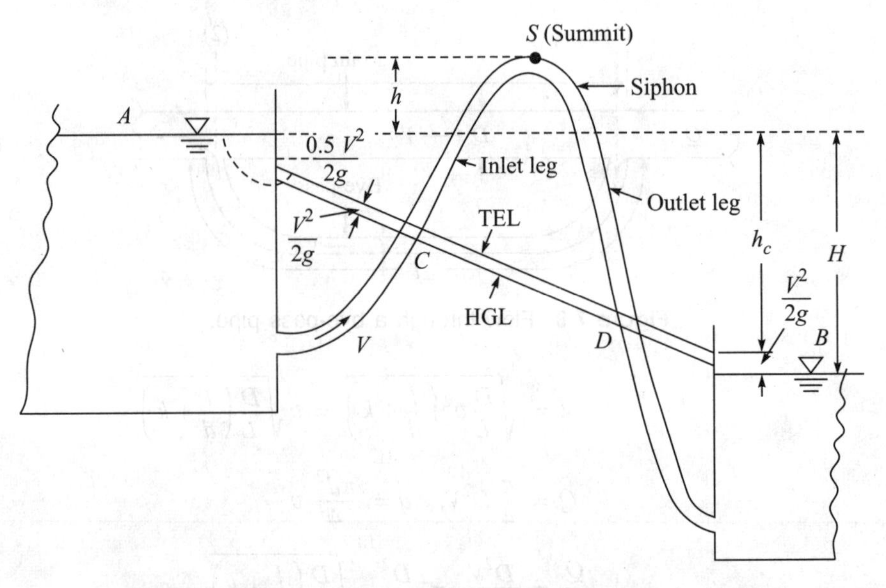

Figure 7.9 Siphon pipe carrying water from reservoir *A* to *B*.

The pressure in the siphon above *C* and *D* is below atmospheric or negative. At the highest point *S* i.e. in summit is the least. Theoretically, the pressure at *S* may be reduced to 10 m of water (atmospheric pressure is 10 m of water). This limit tends to a perfect vacuum and hence no flow will take place. In practice if the pressure is reduced to 2.4 m of water absolute or 7.6 m of water vacuum, dissolved gases come out of the water and collect at the summit *S* forming an air lock which may stop the flow. Therefore, the siphon should be laid in such a way that no portion of the pipe should be more than 7.6 m about the hydraulic gradient. Preferably, the summit *S* should always be below 7.6 m to ensure smooth flow in the siphon.

In order to allow the siphon to run the following conditions need to be satisfied.

Applying Bernoulli's equation at points *A* and *B*,

$$H = 0.5\frac{V^2}{2g} + \frac{V^2}{2g} + \frac{fLV^2}{2gD}$$

$$H = \frac{V^2}{2g}\left(1.5 + \frac{fL}{D}\right) \tag{7.17}$$

If *h* is the vertical distance between the upper reservoir level at *A* and submit *S*, apply Bernoulli's equation at points *A* and *S* to get

$$\frac{p_a}{w} = \frac{p_s}{w} + \frac{V^2}{2g} + 0.5\frac{V^2}{2g} + \frac{flV^2}{2gD} + h$$

where $\frac{p_a}{w}$ is the atmospheric pressure, $\frac{p_s}{w}$ is the absolute pressure at summit, *l* is the inlet leg (*CS*) of the siphon

or
$$h = \frac{p_a}{w} - \frac{p_s}{w} - \frac{V^2}{2g}\left(1.5 + \frac{fl}{D}\right) \tag{7.18}$$

For the siphon to carry the water safely, h should be lower than that given in Equation (7.18) or the inlet leg l may be adjusted so that the pressure at the submit is not reduced below the permissible limit.

EXAMPLE 7.6 Determine the difference of the water level between two reservoirs which are connected by a horizontal pipe of 30 cm, and length 400 m. The discharge through the pipe is 0.3 m^3/sec. Consider all losses and take f to be 0.032.

Solution: Considering all losses, if H is the difference of the water level

Then $$H = \frac{V^2}{2g}\left(1.5 + \frac{fl}{D}\right) \quad \text{(i.e. Equation 7.11)}$$

$$\text{Now} = f = 0.032,\ L = 400 \text{ m},\ D = 0.3 \text{ m},\ Q = 0.3 \text{ m}^3\text{/sec}$$

$$AV = Q$$

$$\therefore \quad \pi/4(0.3)^2 \times V = 0.3$$

$$\therefore \quad V = 4.2441 \text{ m/sec}$$

$$\therefore \quad H = \frac{V^2}{2g}\left(1.5 + \frac{fl}{D}\right) = \frac{4.2441^2}{2 \times 9.81}\left(1.5 + \frac{0.032 \times 400}{0.3}\right)$$

$$H = 40.547 \text{ m}$$ **Ans.**

EXAMPLE 7.7 A siphon of diameter 0.2 m connects two reservoirs A and B with a difference of elevation as 20 m. The length of the siphon is 500 m and the summit is 3 m above the water level of the upper reservoir. The length of the siphon from upper reservoirs to the summit is 100 m. Determine the discharge through the siphon and the pressure at the summit. Neglect minor losses. The friction factor of the pipe is 0.02.

Solution: Applying Bernoulli's theorem at the water surface at A and at B:

$$\frac{p_A}{w} + \frac{V_A^2}{2g} + Z_A = \frac{p_B}{w} + \frac{V_B^2}{2g} + Z_B + \frac{fLV^2}{2gD}$$

$$\frac{p_A}{w} = \frac{p_B}{w} = \text{atmospheric pressure},\ V_A = V_B = 0$$

$$\therefore \quad Z_A - Z_B = \frac{fLV^2}{2gD}$$

Let $$Z_A - Z_B = H = 20 \text{ m}$$

$$20 = H = \frac{0.02 \times 500 \times V^2}{2 \times 9.81 \times 0.2}$$

$$\therefore \quad V = 2.8012 \text{ m/sec}$$

$$\therefore \quad Q = A \times V = \pi/4(.2)^2 \times 2.8012 = 0.088 \text{ m}^3\text{/sec.}$$ **Ans.**

To find pressure at submit, i.e. $\frac{p_S}{w}$,

Apply Bernoulli's equation at A and submit S, taking the upper reservoirs level as datum

$$\frac{p_A}{w} + \frac{V_A^2}{2g} + Z_A = \frac{p_S}{w} + \frac{V_S^2}{2g} + Z_C + \frac{fLV_S^2}{2g}$$

$$0 + 0 + 0 - Z_C = \frac{p_S}{w} + \frac{V^2}{2g} + \frac{fLV}{2g} \quad \because V_S = V$$

$$-3 = \frac{p_S}{w} + \frac{2.8012^2}{2 \times 9.81} + \frac{0.02 \times 100 \times 2.8012^2}{2 \times 9.81 \times .2}$$

$$\therefore \quad \frac{p_S}{w} = -7.39928 \text{ m}$$ **Ans.**

EXAMPLE 7.8 Two reservoirs are connected by three pipes in series. The lengths of the pipes are 300 m, 170 m and 210 m, respectively and the corresponding diameters are 0.3 m, 0.2 m and 0.4 m. Determine the rate of flow if friction factors of the three pipes are 0.02, 0.0208, 0.0192, respectively. Consider only loss due to friction if the difference of water level is 12 m.

Solution: $$H = \frac{f_1 L_1 V_1^2}{2gD_1} + \frac{f_2 L_2 V_2^2}{2gD_2} + \frac{f_3 L_3 V_3^2}{2gD_3} \quad \text{(A)}$$

where $f_1 = 0.02$, $f_2 = 0.0208$, $f_3 = 0.0192$, $L_1 = 300$ m, $L_2 = 170$ m, $L_3 = 210$ m, $D_1 = 0.3$ m, $D_2 = 0.2$ m, $D_3 = 0.4$ m, $H = 12$ m.

Continuity gives $A_1V_1 = A_2V_2 = A_3V_3$

$$\therefore \quad V_2 = \frac{A_1}{A_2} V_1 = \frac{\frac{\pi}{4}(0.3)^2}{\frac{\pi}{4}(0.2)^2} V_1 = 2.25 V_1$$

$$V_3 = \frac{A_1}{A_3} V_1 = \frac{\frac{\pi}{4}(0.3)^2}{\frac{\pi}{4}(0.4)^2} V_1 = 0.5625 V_1$$

$$\therefore \quad \text{(A) becomes: } 12 = \frac{0.02 \times 300 V_1^2}{2 \times 9.81 \times 0.3} + \frac{0.0208 \times 170 \times (2.25)^2 V_1^2}{2 \times 9.81 \times 0.2} + \frac{0.0192 \times 210 \times (.5625)^2 \times V_1^2}{2 \times 9.81 \times 0.4}$$

$$12 = \frac{V_1^2}{2 \times 9.81}\left[\frac{0.02 \times 300}{0.3} + \frac{0.0208 \times 170 \times (2.25)^2}{0.2} + \frac{0.0192 \times 210 \times (0.5625)^2}{0.4}\right]$$

$$12 = (112.694376)\frac{V_1^2}{2 \times 9.81}$$

$$\therefore \quad V_1 = 1.4454 \text{ m/sec}$$

$\therefore$ $$Q = A_1V_1 = [\pi/4(0.3)^2 \times 1.4454] \text{ m}^3\text{/sec}$$

$\therefore$ $$Q = 0.102193 \text{ m}^3\text{/sec}$$

$$Q = 102.193 \text{ litre/sec}$$ **Ans.**

EXAMPLE 7.9 Three pipes of lengths 800 m, 400 m and 200 m and diameters of 0.6 m, 0.4 m and 0.2 m, respectively, are connected in series. These pipes are to be replaced by an equivalent pipe of length 1400 m. Assuming the friction factor of the compound pipe to be the same, determine the equivalent diameter of the equivalent pipe.

Solution: If D_1, D_2, and D_3 are the diameters of the compound pipes, and D is the diameter of the equivalent pipe, then applying the formula in Equation (7.15) of the equivalent pipe.

$$\frac{L}{D^5} = \frac{L_1}{D_1^5} + \frac{L_2}{D_2^5} + \frac{L_3}{D_3^5} \qquad \text{(A)}$$

where $L = 1400$ m, $L_1 = 800$ m, $L_2 = 400$ m, $L_3 = 200$ m,

$D_1 = 0.6$ m, $D_2 = 0.4$ m, $D_3 = 0.2$ m.

Substituting the above values in (A):

$$\frac{1400}{D^5} = \frac{800}{(0.6)^5} + \frac{400}{(0.4)^5} + \frac{200}{(0.2)^5}$$

$$\frac{1400}{D^5} = 10288.06584 + 39062.5 + 625000 = 974350.5658$$

$$D = 0.27 \text{ m}$$ **Ans.**

EXAMPLE 7.10 A main pipe is divided into two parallel pipes in series, which again forms one pipe. The length and diameter of the first parallel pipe is 1000 m and 0.5 m, while the length and diameter of the second pipe in series are 1000 m and 0.4 m, respectively. Find the discharges in each parallel pipe if the rate of flow in the main pipe is 1.5 m^3/sec. Assume the friction factors of both the parallel pipes to be equal.

Solution: If L_1 and D_1 are the length and diameter of the first pipe L_2 and D_2 are the length and diameter of the second parallel pipe, and f is the friction factor of the both the pipes.

Then the friction loss in both pipes is the same.

$$h_f = \frac{fL_1V_1^2}{2gD_1} = \frac{fL_2V_2^2}{2gD_2}$$

or

$$\frac{L_1V_1^2}{2gD_1} = \frac{L_2V_2^2}{2gD_2}$$

Here $L_1 = L_2 = 1000$ m, $D_1 = 0.5$ m, $D_2 = 0.4$ m

$\therefore$ $$\frac{1000 \times V_1^2}{2 \times 9.81 \times 0.5} = \frac{1000 \times V_2^2}{2 \times 9.81 \times 0.4}$$

$\therefore$ $$V_1 = \frac{V_2}{\sqrt{1.25}}$$

$$\therefore \quad Q_1 = \frac{\pi}{4}(0.5)^2 \times \frac{V_2}{\sqrt{1.25}} = 0.17562\ V_2$$

$$\therefore \quad Q_2 = \frac{\pi}{4}(0.4)^2 \times V_2 = 0.12566\ V_2$$

But $\quad Q_1 + Q_2 = 1.5$

$\therefore \quad 0.17562 V_2 + 0.12566 V_2 = 1.5$

$\therefore \quad V_2 = 4.9787$ m/sec

$\therefore \quad V_1 = 4.453$ m/sec

$\therefore \quad Q_1 = [\pi/4(.5)^2 \times 4.453]$ m^3/sec $= 0.8743$ m^3/sec **Ans.**

$\therefore \quad Q_2 = Q - Q_1 = (1.5 - 0.8743) = 0.6257$ m^3/sec **Ans.**

EXAMPLE 7.11 Three pipes of the same length L, same diameter D and the same friction factor f, are connected in series. Determine the diameter D of the pipe of the same length L and same friction factor f, which will carry the same discharge for the same head loss.

Solution: $h_f = h_{f1} = h_{f2} = h_{f3} = \dfrac{fLV^2}{2gD}$

Again $\quad Q_1 = Q_2 = Q_3$

$\therefore$ Total discharge to be carried by the new pipe $= Q = Q_1 + Q_2 + Q_3$

or $$Q = 3Q_1 \qquad \text{(A)}$$

If D_n is the new diameter of the single pipe of same length L which will carry the total discharge Q, i.e. $3Q_1$ with a new velocity V_n:

Then $$\frac{\pi}{4}(D_n)^2 \times V_n = 3(\pi/4)D^2 \cdot V$$

$$3 \times \frac{D^2}{D_n^2} = \frac{V_n}{V} \qquad \text{(B)}$$

Head loss in the single pipe is the same as the loss of three pipes.

$$\frac{fLV^2}{2gD} = \frac{fLV_n^2}{2gD_n}$$

$$\frac{D_n}{D} = \frac{V_n^2}{V^2} \quad \text{or} \quad \left(\frac{D_n}{D}\right)^{1/2} = \frac{V_n}{V}$$

Substituting $\dfrac{V_n}{V}$ in (B):

$$3\left(\frac{D^2}{D_n^2}\right) = \frac{D_n^{1/2}}{D^{1/2}}$$

$$3 = \frac{D_n^{5/2}}{D^{5/2}} \quad \therefore D_n^{5/2} = 3D^{5/2}$$

$$D_n = (3D^{5/2})^{2/5}$$

$$\therefore \quad D_n = 1.55184\ D \qquad \textbf{Ans.}$$

7.14 POWER TRANSMISSION THROUGH PIPES

Power may be transmitted by a flowing liquid through a pipe. It depends on the weight of the liquid that flows through the pipe, length of the pipeline, total head available at the end of the pipe (which is also dependent on the friction factor of the pipe), etc. Consider a *AB* of length *L* connected to a tank with a head of water *H* (See Figure 7.10). If *d* is the diameter of the pipe, *f* is the friction factor and *V* is the velocity of liquid in the pipe, then considering loss of energy due to friction only (assuming the pipe to be long)

$$h_f = \frac{fLV^2}{2gd}$$

Now head available at the end point *B* is $= H - h_f$

Weight *W* of water flowing through the pipe per sec,

$$W = (\rho g)\ (\text{volume of water/sec})$$

$$W = (\rho g)\ [(\pi/4d^2) \times V] = \frac{\pi}{4}\rho g d^2 V$$

The power *P* transmitted at the outlet *B* of the pipe:

P = (Weight of liquid/sec) × head available at *B*

or

$$P = \frac{\pi}{4}\rho g d^2 V \times (H - h_f) \text{ watts}$$

$$\therefore \quad P = \left[\frac{\pi}{4}\rho g d^2 V\ (H - h_f)\right]/1000 \text{ kw} \tag{7.19}$$

Efficiency (η) of power transmission:

$$\eta = \frac{\text{Power available at outlet of the pipe}}{\text{Power supply at the inlet of the pipe}}$$

$$\eta = \frac{W(H - h_f)}{W \times H}$$

$$\therefore \quad \eta = \left(\frac{H - h_f}{H} \times 100\right) p.c. \tag{7.20}$$

In order to obtain the condition for maximum efficiency, differentiate the power *P* (i.e. expression given by Equation 7.19) with respect to *V* and equate it to zero for maximum *P*.

$$\frac{d(P)}{dV} = \frac{\pi}{4} \cdot \frac{\rho g}{1000} d^2 \frac{d}{dV}\left(HV - \frac{fLV^3}{2gD}\right) = 0$$

$$\left(\frac{\pi}{4}\right)\left(\frac{\rho g}{1000}\right)d^2\left[H - \frac{fLV^2}{3 \times 290}\right] = 0$$

$$\therefore \qquad H - \frac{1}{3}\cdot\left(\frac{fLV^2}{2gd}\right) = 0$$

$$H - \frac{1}{3}h_f = 0 \quad \therefore\ H = \frac{h_f}{3} \qquad (7.21)$$

which is the condition for maximum efficiency.

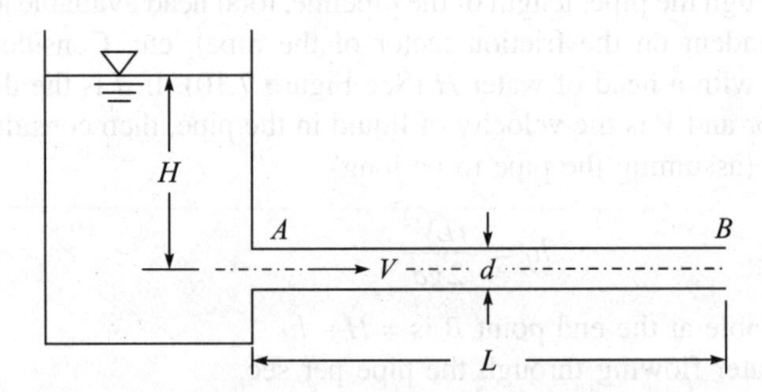

Figure 7.10 Power transmission through pipe.

EXAMPLE 7.12 A pipe of diameter 0.4 m, and length 400 m is used for the transmission of power under a head of 400 m at the inlet of the pipe. Find the maximum power available at the outlet of the pipe. Assume friction factor f of the pipe to be 0.025.

Solution: For power to be maximum, $H = 3h_f$

$$\therefore \qquad h_f = \frac{H}{3} = \frac{420}{3} = 140 \text{ m}$$

$$\text{Again} \quad h_f = \frac{fLV^2}{2gD} = \frac{0.25 \times 4000 \times V^2}{2 \times 9.81 \times 0.4} = 127.421 \text{ V}^2\text{m}$$

Equating the two h_f:

$$140 = 127.421\,V^2$$

$$\therefore \qquad V = 1.0482 \text{ m/sec}$$

$$\therefore \qquad Q = AV = \left[\frac{\pi}{4}(.4)^2 \times 1.0482\right] \text{ m}^3\text{/sec}$$

$$Q = 0.13172 \text{ m}^3\text{/sec}$$

Head available at outlet of the pipe is:

$$H - h_f, \text{ i.e.} = 420 - 140$$

$$= 280 \text{ m}$$

$$\therefore \qquad \text{Maximum power available} = \frac{\rho g Q \times 280}{1000} = \frac{1000 \times 9.81 \times 0.13172 \times 280}{1000}$$

$$= 361.8085 \text{ kw}$$

Ans.

7.15 WATER HAMMER IN PIPES

When water with high velocity flows in a long pipe or tunnel and when the valve at the end of the pipe is closed suddenly, the flow of water is stopped and momentum of the flowing water is destroyed which leads to a sudden high rise of pressure very close to the valve. The high rise of the pressure moves upstream with a speed equal to the velocity of the sound wave and downstream to the valve. This movement of pressure up and down continues until it is damped down by friction. The movement of high rise of pressure has a hammering action on the walls of the pipe and hence this is called water hammer. This water hammer situation occurs most commonly in hydroelectric projects, wherein a long pipeline is required to convey water from the reservoir to the powerhouse and commonly valves are required to close for no-load conditions. Since the pressure rise becomes too high, sometimes the pipe may burst. To ease out the high pressure, a high head hydropower plant is provided with a surge tank close to the valve to absorb a reasonable part of the pressure.

Pressure rise due to the water hammer depends on the velocity of the flow, the time of the valve closure, and elastic properties of the fluid and pipe material.

7.15.1 Pressure Intensity Due to Gradual Closure of Valve

Let L be the length, A be the area of the pipe.

V is the velocity before closing. p_i is the intensity of the pressure wave produced.

Mass of water in the pipe = ρAL

The valve is closed gradually in a time of T secs and water is brought to rest in t sections.

$$\text{Retardation} = \frac{\text{Change of velocity}}{\text{time}} = \frac{V - O}{T} = \frac{V}{T}$$

$$\text{Retarding force} = \text{mass} \times \text{retardation}$$

$$p_i\, A = \rho AL \frac{V}{T} = \frac{\rho ALV}{T}$$

$$\therefore \quad p_i = \frac{\rho AVL}{AT}$$

$$p_i = \frac{\rho VL}{T} \tag{7.22}$$

or

$$\frac{p_i}{\rho g} = \frac{\rho VL}{\rho g T}$$

$$\therefore \quad H = \frac{LV}{gT} \tag{7.23}$$

If closure of $T > \frac{2L}{C}$, it is said to be gradual,

and if $T < \frac{2L}{C}$, it is said to be sudden,

where C is the velocity of the pressure wave in the fluid media.

7.15.2 Sudden Closure: Pipe is Elastic and Water is Compressible

In sudden closure, the pressure rise is very high. Due to this very high rise of pressure, the pipe is assumed to be elastic and water is compressible.

Let K is be the bulk modulus of compression of water and E is the modulus of elasticity of pipe material, $\frac{1}{m}$ is the Poisson ratio for pipe material.

D is the diameter of the pipe.
t is the thickness of the pipe.
f_l and f_c are longitudinal and circumferential stress in pipe,

$$\therefore \qquad f_l = \frac{p_i D}{4t} \quad \text{and} \quad f_c = \frac{p_i D}{2t}$$

where p_i is the inertia or increase of pressure.

According to the rule of strength of materials, both f_c and f_l are the circumferential and longitudinal stress, and strain energy stored in pipe material per unit volume,

$$= \frac{1}{2E}\left(f_l^2 + f_c^2 - \frac{2 f_l f_c}{m}\right)$$

$$= \frac{1}{2E}\left[\left(\frac{p_i D}{4t}\right)^2 + \left(\frac{p_i D}{2t}\right)^2 - \frac{2\left(\frac{p_i D}{4t}\right)\left(\frac{p_i D}{2t}\right)}{m}\right]$$

$$= \frac{1}{2E}\left[\frac{p_i^2 D^2}{16t^2} + \frac{p_i^2 D^2}{4t^2} - \frac{p_i^2 D^2}{4mt}\right]$$

Taking $\frac{1}{m} = \frac{1}{4}$,

The strain energy stored in pipe material per unit volume

$$= \frac{1}{2E}\left[\frac{p_i^2 D^2}{16t^2} + \frac{p_i^2 D^2}{4t^2} - \frac{p_i^2 D^2}{16t^2}\right]$$

$$= \frac{1}{2E} \times \frac{p_i^2 D^2}{4t^2}$$

$$= \frac{p_i^2 D^2}{8Et^2}$$

Total strain energy stored in the pipe material $= \frac{p_i^2 D^2}{8Et^2} \times$ total volume of pipe material

$$= \frac{p_i^2 D^2}{8Et^2} \times (\pi D \times t) \times L$$

$$= \frac{\pi p_i^2 D^3 L}{8Et}$$

$$= \left(\frac{\pi D^2}{4}\right) \cdot \frac{p_i^2 DL}{2Et} \quad \because \frac{\pi}{4} D^2 = A$$

$$= \frac{p_i^2 ADL}{2Et} \qquad \text{(A)}$$

Loss of kinetic energy due to closure of valve

$$= \frac{1}{2} \text{mV}^2 = \frac{1}{2} (\rho AL) V^2 \qquad \text{(B)}$$

Gain of strain energy in water due to compression

$$= \frac{1}{2}\left(\frac{p_i^2}{K}\right) \times \text{volume of water}$$

$$= \frac{1}{2}\left(\frac{p_i^2}{K}\right) \times AL = \frac{1}{2} \cdot \frac{p_i^2 AL}{K} \qquad \text{(C)}$$

Now loss of kinetic energy = strain energy stored + Gain in strain energy stored

i.e. (B) = (A) + (C)

i.e.
$$\frac{1}{2} \rho AL\, V^2 = \frac{1}{2} \cdot \frac{p_2^2 ADL}{Et} + \frac{1}{2} \cdot \frac{p_i^2}{K} AL$$

Dividing by $\frac{1}{2} AL$, $\rho V^2 = p_i^2 \left(\frac{D}{Et} + \frac{1}{K}\right)$

$\therefore$
$$p_i = \frac{V}{\sqrt{\frac{1}{\rho}\left(\frac{1}{K} + \frac{D}{tE}\right)}} \qquad (7.24)$$

which gives the rise of pressure of sudden closure considering both the elasticity of pipe material and compressibility of water.

If the elasticity of pipe material is neglected, i.e. the pipe is rigid, the term $\frac{D}{tE}$ is zero as $E \to \propto$ for rigidity of the pipe material. Equation (7.24) becomes:

$$p_i = \frac{V}{\sqrt{\frac{1}{\rho} \cdot \frac{1}{K}}}$$

i.e.
$$p_i = V\sqrt{\rho K} = V\sqrt{\frac{K}{\rho} \cdot \rho^2}$$

But $\sqrt{\frac{K}{\rho}} = C$ velocity of pressure wave, which will be shown later in Chapter 15, i.e. (Compressible Flow)

$\therefore$
$$p_i = \rho VC \qquad (7.25)$$

which is the equation of pressure considering only water to be compressible.

If T is the time taken by the pressure wave to travel from the valve to the tank or reservoir and back to the valve again, and if L is the length of the pipe, the total distance moved in time T with velocity of pressure wave C, then

$$T = \frac{2L}{C} \tag{7.26}$$

As already stated if $T < \frac{2L}{C}$, it is sudden closure and if $T > \frac{2L}{C}$, it is gradual closure.

EXAMPLE 7.13 Water flows in a pipeline of length 2000 m, diameter 0.6 m with a velocity of 2 m/sec. The valve at the end is closed in 20 seconds. Find the rise of pressure if the velocity of pressure wave is 1420 m/sec.

Solution: L = 2000 m, d = 0.6 m, V = 2 m/sec, C = 1420 m/sec, T = 20 seconds

$$\therefore \quad \frac{2L}{C} = \frac{2 \times 2000}{1420} = 2.816 \text{ sec} \; \therefore \; T > 2.816 \text{ secs, closure is gradual.}$$

$$\therefore \quad p_i = \frac{\rho L V}{T} = \frac{1000 \times 2000 \times 2}{20} = 200000 \text{ N/m}^2$$

$$\therefore \quad p_i = \frac{200000}{10^4} = 20 \text{ N/cm}^2$$

$$p_i = 20 \text{ N/cm}^2$$

Ans.

EXAMPLE 7.14 In Example 7.13, water is compressible with bulk modular of compression of 19.62×10^4 N/cm^2. The pipe is rigid and the valve is closed in 1.5 seconds. Find the rise of pressure.

Solution: Here $K = 19.62 \times 10^4 \text{ N/cm}^2 = (19.62 \times 10^4 \times 10^4)\text{N/m}^2$

and $$\frac{2L}{C} = \frac{2 \times 2000}{1420} = 2.816 \text{ sec}$$

$$\therefore \quad T = 1.5 \text{ secs} \quad \therefore \; T < \frac{2L}{C}, \text{ it is sudden closure}$$

Here $$C = \sqrt{\frac{K}{\rho}} = \sqrt{\frac{19.62 \times 10^4 \times 10^4}{1000}} = 1400.714 \text{ m/sec}$$

$$\therefore \quad p_i = \rho V C = (1000 \times 2 \times 1400714) \text{ N/m}^2 = 280.1428 \times 10^4 \text{ N/m}^2$$

$$\therefore \quad p_i = 280.1428 \text{ N/cm}^2$$

Ans.

EXAMPLE 7.15 In Example 7.13, the thickness of the pipe is 1 cm and the valve is closed suddenly. Considering the pipe to be elastic and water to be compressible, find the rise of pressure if $K = 19.62 \times 10^4$ N/cm^2 and E (of pipe) = 19.62×10^6 N/cm^2. Calculate also circumferential and longitudinal stress developed in the pipe.

Solution: $E = 19.62 \times 10^6$ N/cm², $K = 19.62 \times 10^4$ N/cm²

$= 19.62 \times 10^{10}$ N/m² $= 19.62 \times 10^8$ N/m²

$V = 2$ m/sec, $t = 1$ cm $= 0.01$ m, $D = 0.6$ m

$$\therefore \quad p_i = \frac{V}{\sqrt{\frac{1}{\rho}\left(\frac{1}{K}+\frac{D}{tE}\right)}} = \frac{2}{\sqrt{\frac{1}{1000}\left(\frac{1}{19.62\times 10^8}+\frac{0.6}{0.01\times 19.62\times 10^{10}}\right)}}$$

$p_i = 221.47 \times 10^4$ N/m² $= 221.47$ N/cm² **Ans.**

$$f_c = \frac{p_i D}{2t} = \frac{221.47\times .6}{2\times 0.01} = 6644.1 \text{ N/m}^2$$ **Ans.**

$$f_l = \frac{p_i D}{4t} = \frac{221.7\times 0.6}{4\times 0.01} = 3322.05 \text{ N/m}^2$$ **Ans.**

7.16 UNSTEADY MOMENTUM EQUATION OF WATER HAMMER

Consider a small length dx of the pipe just after the valve closure in Figure 7.11. The pressure wave moves upstream, and the flow will be retarded. If α is the angle made by the centre line of the pipe with horizontal component of gravity force on dx is $W \sin \alpha$ in the direction force of resistance F_f.

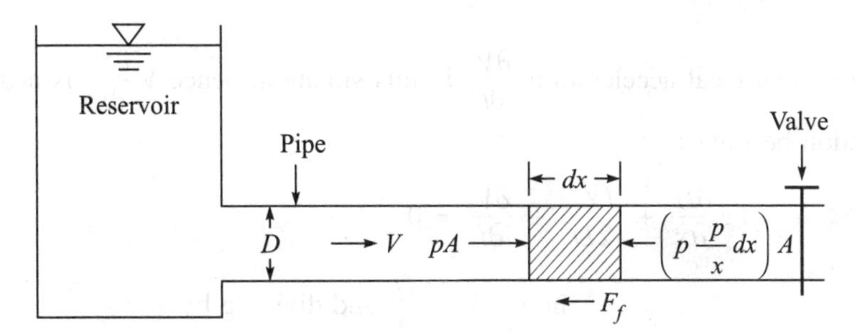

Figure 7.11 Water hammer situation after valve closure.

Writing Newton's equation:

$$\left(p+\frac{\partial p}{\partial x}dx\right)A - pA + F_f + W\sin\alpha = \rho(Adx)\left[-\left(\frac{\partial V}{\partial t}+V\frac{\partial V}{\partial x}\right)\right]$$

Negative sign of total acceleration is retardation:

$$\text{or} \quad \left(p+\frac{\partial p}{\partial x}dx\right)\frac{\pi}{4}D^2 - p\cdot\frac{\pi}{4}D^2 + f'(\text{area on which friction acts})\times V^2 + \rho g\left(\frac{\pi}{4}D^2\,dx\right)\sin\alpha$$

$$= \rho\left(\frac{\pi}{4}D^2\,dx\right)\left[-\left(\frac{\partial V}{\partial t}+V\frac{\partial V}{\partial x}\right)\right]$$

or $$\left(\frac{\partial p}{\partial x}dx\right)\frac{\pi}{4}D^2 + f'(\pi D\,dx)V^2 + \rho g(\pi/4)D^2\,dx\sin\alpha + \rho\frac{\pi}{4}D^2\,dx\frac{\partial V}{\partial t} + \rho\frac{\pi}{4}D^2\,dx\cdot V\frac{\partial V}{\partial x} = 0$$

Dividing by $\left(\frac{\pi}{4}D^2\right)dx$:

$$\frac{\partial p}{\partial x} + \frac{4f'}{D}V^2 + \rho g\sin\alpha + \rho\frac{\partial V}{\partial t} + \rho V\frac{\partial V}{\partial x} = 0$$

Again dividing by ρ:

$$\frac{1}{\rho}\frac{\partial p}{\partial x} + \left(\frac{8f'}{\rho}\right)\frac{V^2}{2D} + \frac{\partial V}{\partial t} + V\frac{\partial V}{\partial x} + g\sin\alpha = 0$$

or $$\frac{1}{\rho}\frac{\partial p}{\partial x} + \frac{fV^2}{2D} + \frac{\partial V}{\partial t} + V\frac{\partial V}{\partial x} + g\sin\alpha = 0 \quad (7.27)$$

Write the momentum equation in general form.

Further writing: $p = wH = \rho gH$,

$$g\frac{dH}{dx} + \frac{fV^2}{2D} + \frac{\partial V}{\partial t} + V\frac{\partial V}{\partial x} + g\sin\alpha = 0$$

Assuming $\alpha = 0$, i.e. the pipe to be horizontal, $\sin\alpha = 0$ again connective acceleration $V\frac{\partial V}{\partial x}$ is much smaller than local acceleration $\frac{\partial V}{\partial t}$ in this situation, hence $V\frac{\partial V}{\partial x}$ is neglected.

$\therefore$ The equation becomes:

$$g\frac{\partial H}{\partial x} + \frac{fV^2}{2D} + \frac{\partial V}{\partial t} = 0$$

now $V = \frac{Q}{A}$ and dividing by g

$$\therefore \quad \frac{\partial H}{\partial x} + \frac{fQ|Q|}{2gDA^2} + \frac{1}{gA}\cdot\frac{\partial Q}{\partial t} = 0 \quad (7.27a)$$

which is the simplified form of the momentum equation.

Q^2 is written as $Q|Q|$ as the discharge becomes positive and negative due to a change of velocity towards the reservoir and back to the valve.

7.17 UNSTEADY CONTINUITY EQUATION IN WATER HAMMER

A small length dx of the pipe as shown in Figure 7.12 is considered.

Volume of water entering dx in time dt:

$$d\forall = (\text{difference of velocity}) \times \text{area} \times dt$$

or $$d\forall = \left[V - \left(V - \frac{\partial V}{\partial x}dx\right)\right]\frac{\pi}{4}D^2 dt$$

or $$d\forall = \left(\frac{\partial V}{\partial x}dx\right)\frac{\pi}{4}D^2 dt \qquad \text{(A)}$$

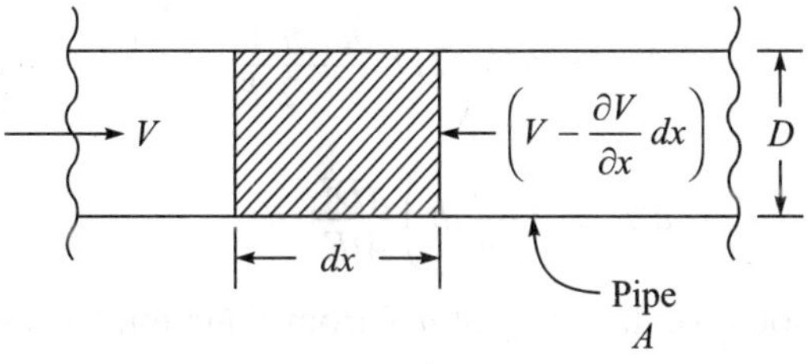

Figure 7.12 A small length *dx* of the pipe under water hammer pressure.

This $d\forall$ can enter only if the water is compressed and the pipe is expanded due to high water hammer pressure.

If $\partial\forall_w$ is obtained due to compressibility of water

and $\partial\forall_c$ is obtained due to elasticity of pipe material,

Then $$\partial\forall = \partial\forall_w + \partial\forall_c \qquad \text{(B)}$$

Now K = Bulk modulus of compression of water

$$K = \frac{\text{Rise of pressure in time } \partial t}{\dfrac{\text{increase in volume due to compressibility of water}}{\text{Original volume in } dx}}$$

$$K = \frac{p_i}{\dfrac{d\forall_w}{(\pi/4D^2)dx}} = \frac{\dfrac{\partial p}{\partial t}dt}{\dfrac{d\forall_w}{(\pi/4D^2)dx}} = \frac{\dfrac{\partial p}{\partial t}dt\cdot\dfrac{\pi}{4}D^2 dx}{d\forall_w}$$

$\therefore$ $$\partial\forall_w = \frac{\left(\dfrac{\partial p}{\partial t}dt\right)\left(\dfrac{\pi}{4}D^2 dx\right)}{K} \qquad \text{(C)}$$

Due to this increase of pressure in time dt, the pipe diameter expands from D to $(D + dD)$

$$E = \text{Modulus of Elasticity} = \frac{\text{hoop stress}}{\text{Stain}} = \frac{p_i D}{2t \times \dfrac{dD}{D}} = \frac{\left(\dfrac{\partial p}{\partial t}dt\right)\times D}{2t\,dD}$$

$\therefore$ $$dD = \frac{\left(\dfrac{\partial p}{\partial t}\right)dt \times D^2}{2tE}$$

Corresponding change in volume = $d\forall_p = \pi/4(D + dD)^2\, dx - \pi/4D^2 dx$

$$d\forall_p = \pi/4(2D\, dD)dx + (\pi/4)dD^2 dx$$

The term $(\pi/4)dD^2 \cdot dx$ is very small, hence neglected

$\therefore$
$$d\forall_p = (\pi/4)(2D\, dD)dx$$

$$d\forall_p = \pi/4\left(2D \cdot \frac{\partial p/\partial t \cdot dt\, D^2}{2tE}\right)dx$$

$\therefore$
$$d\forall_p = \left(\frac{\partial p}{\partial t}dt\right)\frac{\pi D^3}{4tE}\, dx \qquad \text{(D)}$$

Substituting in Equation (B), the value of $d\forall$ from Equation (A) and $\partial\forall_w$ and $\partial\forall_p$ from (C) and (D) respectively,

$$\left(\frac{\partial V}{\partial x}dx\right)\frac{\pi}{4}D^2 dt = \frac{\left(\frac{\partial p}{\partial t}dt\right)\frac{\pi}{4}D^2\, dx}{K} + \left(\frac{\partial p}{\partial t} \cdot dt\right)\frac{\pi D^3\, dx}{4tE}$$

Dividing by $(dx \cdot dt \cdot \pi/4D^2)$, gives:

$$\frac{\partial V}{\partial x} = \left(\frac{1}{K} + \frac{D}{tE}\right)\frac{\partial p}{\partial t} \qquad \text{(E)}$$

But velocity of the wave, i.e. Celerity = $a = \sqrt{\dfrac{K_e}{\rho}}$

where
$$K_c = \frac{1}{\left(\frac{1}{K} + \frac{D}{tE}\right)}$$

$$a^2 = \frac{K_e}{\rho} = \frac{1}{\left(\frac{1}{K} + \frac{D}{tE}\right)}$$

from (e)
$$\frac{\partial p}{\partial t} = \frac{1}{\left(\frac{1}{K} + \frac{D}{tE}\right)}\frac{\partial V}{\partial x}$$

$\therefore$
$$\frac{\partial p}{\partial t} = \rho a^2 \frac{\partial V}{\partial x}$$

putting
$$p = \rho g H:$$

$$\rho g \frac{\partial H}{\partial t} = \rho a^2 \frac{\partial V}{\partial x}$$

$\therefore$
$$\frac{\partial H}{\partial t} = \frac{a^2}{g}\frac{\partial V}{\partial x} \qquad \text{(F)}$$

But
$$V = \frac{Q}{A}, \quad \frac{\partial V}{\partial x} = \frac{1}{A}\frac{\partial Q}{\partial x}$$

∴ Equation (F) becomes:

$$\frac{\partial H}{\partial t} = \frac{a^2}{gA}\frac{\partial Q}{\partial x} \tag{7.28}$$

which is the continuity equation of unsteady flow in water hammer.

7.7.1 Methods of Solution of Momentum and Continuity Equations

These two Equations (7.27a) and (7.28) are non-linear partial differential equations for which an analytical solution is not possible. Therefore, they are usually solved by different numerical solutions like the method of characteristics, Lax method, predictor–corrector method, etc. with proper initial and boundary conditions.

As an example, the finite difference explicit Lax diffusive method has been discussed to solve the above equation.

In the *x*-*t* plane, along *x* the length *L* of the pipe is divided into number Δx and along *t* directions, it is divided into number Δt. (See Figure 7.13).

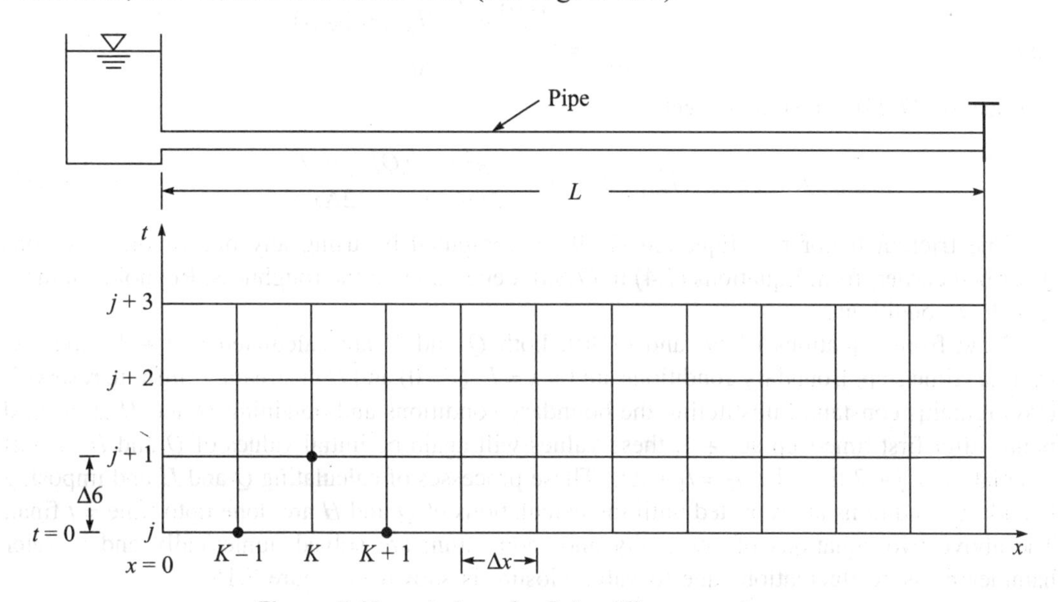

Figure 7.13 *x-t* plane for finite difference scheme.

Finite difference equations in this *x*-*t* plane may be written as follows:

$$\left.\begin{aligned}\frac{\partial H}{\partial x} &= \frac{H_{K+1}^{j} - H_{K-1}^{j}}{2\Delta x}\\ \frac{\partial Q}{\partial t} &= \frac{Q_K^{j+1} - \frac{1}{2}(Q_{K-1}^{j} + Q_{K+1}^{j})}{\Delta t}\end{aligned}\right\} \tag{7.29}$$

Substituting these finite difference equations in Equation (7.27a):

$$\frac{H_{K+1}^{j} - H_{K-1}^{j}}{2\Delta x} + \frac{1}{gA} \cdot \frac{Q_K^{j+1} - \frac{1}{2}(Q_{K-1}^{j} + Q_{K+1}^{j})}{\Delta t} + \frac{f}{2gDA^2} \left| Q_k^j \right| Q_k^j = 0$$

$$\therefore \quad Q_k^{j+1} = \frac{1}{2}(Q_{k-1}^{j} - Q_{k+1}^{j}) - \frac{gA\Delta t}{2\Delta x}(H_{k-1}^{j} - H_{k-1}^{j}) - \left(\frac{f\Delta t}{2DA}\right) \times \frac{1}{2}$$

$$\left| Q_{k-1}^{j} + Q_{k+1}^{j} \right| \cdot \frac{1}{2}(Q_{k_1}^{j} + Q_{k+1}^{j}) \tag{7.30}$$

In Equation (7.30), the values of Q and H at time j and grid position $K - 1$, K, $K + 1$ are known as initial conditions, so the values of Q in the next time step at $j + 1$, i.e. $t_1 = t_0 + \Delta t$ can be calculated.

Substituting:
$$\frac{\partial Q}{\partial x} = \frac{Q_{K+1}^{j} - Q_{K+1}^{j}}{2\Delta x}$$

and
$$\frac{\partial H}{\partial t} = \frac{H_K^{j+1} - \frac{1}{2}(H_{K-1}^{j} + Q_{K+1}^{j})}{\Delta t}$$

in Equation (7.28), we similarly get:

$$H_k^{j+1} = \frac{1}{2}(H_{k-1}^{j} + H_{k+1}^{j}) - \frac{a^2\Delta t}{2gA\Delta x}\frac{(Q_{K+1}^{j} - Q_{K-1}^{j})}{2\Delta x} \tag{7.31}$$

The friction factor f in Equation (7.29) is computed by using any one of the equations described earlier, from Equations (7.4) to (7.8a), depending on the roughness, Reynolds number and flow conditions.

Now from Equations (7.29) and (7.30), both Q and H are calculated at $j + 1$ time, i.e. $(t_0 + \Delta t)$ time step. Boundary conditions are (at $x = L$, $Q = 0$) and H upstream, if infinite reservoir (say) remains constant. Substituting the boundary conditions and obtaining Q and H at all grid points after first time step at $j + 1$, these values will again be initial values of Q and H to start calculation at $j + 2$ time, i.e. $t_2 = t_1 + \Delta t_1$. These processes of calculating Q and H and imposing boundary conditions are repeated until the calculations of Q and H are done upto time = t final. The above two equations of continuity and momentum are solved numerically and a water hammer pressure fluctuations due to valve closure is shown in Figure 7.15.

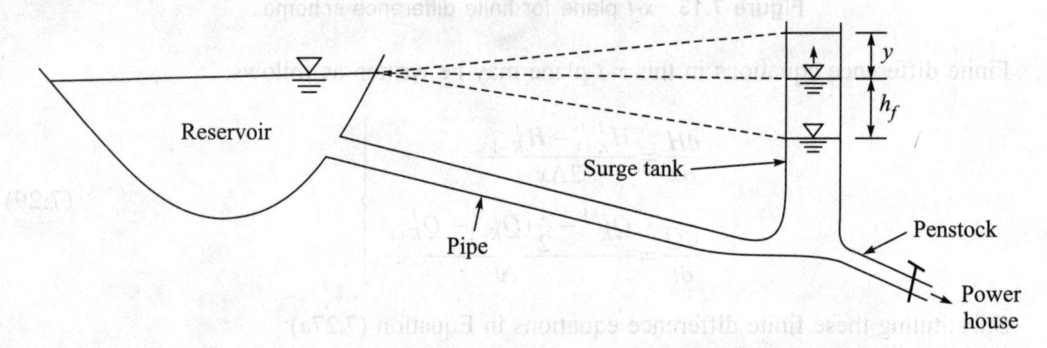

Figure 7.14 Surge tank in high head hydro project with a long pipe.

7.18 UNSTEADY CONTINUITY EQUATION IN SURGE TANK

In order to relieve the high rise pressure in the long pipe of a hydroelectric project with a high head, a surge tank as shown in Figure 7.14 is provided.

Thus, the surge tank provides the function of the initial flow to the turbine when the reservoir is at a long distance with long pipe as it takes some time to cause the flow to the turbine from the reservoir as soon as the turbine or penstock valve is obtained. Also, it relieves the long pipeline from extra high pressure rise when the valve is to be closed due to no-load condition. Thus, the rise of pressure due to the closing of the turbine valve created an unsteady flow in the surge

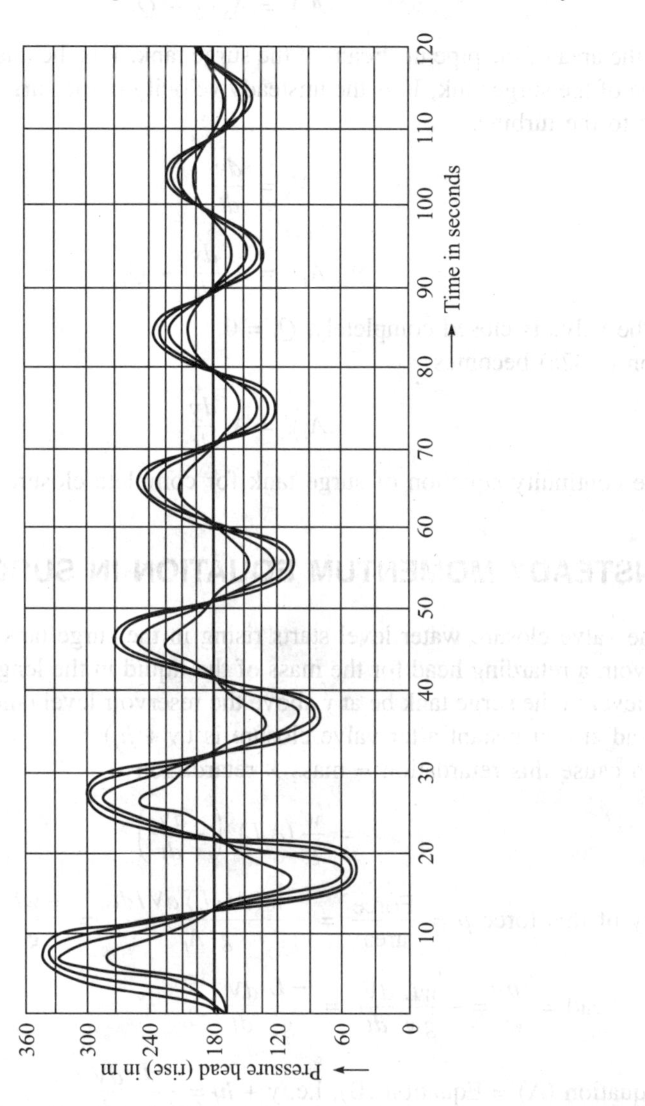

Figure 7.15 Solution of Equations (7.27a) and (7.28) for water hammer pressure head (H) fluctuations in four cross-sections of a pipe by numerical method. (Steady state pressure head at upstream end is 180 m.)

tank. The water level in the surge tank rises and falls again. Thus the rise and fall of the water level with time continues for some time until it damps down to the reservoir level.

When the turbines receive a constant discharge Q through the penstock of the surge tank, a steady state is maintained with a loss of head h_f as shown in Figure 7.14. As soon as the valve is closed, an unsteady situation is created and water in the surge first with unsteady velocity V_s rises up to a height y and falls with unsteady velocity, it again rises and thus the water level goes on fluctuating with unsteadiness until with time it is damped down.

For partial closure, continuity gives:

$$A_t V = A_S V_S + Q_t \tag{7.32}$$

where A_t is the area of the pipe upstream of the surge tank, V is the unsteady velocity in the pipe, A_s is the area of the surge tank, V_s is the unsteady velocity in the surge tank, and Q_t is the partial water going to the turbine.

Thus
$$V_s = \frac{dy}{dt}$$

and
$$A_t V = A_s \frac{dy}{dt} + Q_t \tag{7.32a}$$

When the valve is closed completely, $Q_t = 0$,
Equation (7.32a) becomes:

$$A_t V = A_s \frac{dy}{dt} \tag{7.32b}$$

which is the continuity equation of surge tank for complete closure.

7.19 UNSTEADY MOMENTUM EQUATION IN SURGE TANK

Just after the valve closure water level starts rising in the surge tank even more than the level of the reservoir, a retarding head for the mass of the liquid in the length L is created. Let at any instant, the level in the surge tank be at y above the reservoir level (shown in Figure 7.14). Total retarding head at that instant after valve closure is $(y + h_f)$. (A)

Force to cause this retardation = mass × retardation

$$= \frac{w}{g}(A_t L)\left(-\frac{dV}{dt}\right)$$

Intensity of this force $p = \dfrac{\text{Force}}{\text{area}} = \dfrac{-w(At \cdot L)\, dV/dt}{g \cdot A_t} = \dfrac{-wL}{g}\dfrac{dV}{dt}$

$$\therefore \quad \text{Pressure head} = \frac{p}{w} = -\frac{wL}{gw}\frac{dV}{dt} = \frac{-L}{g}\frac{dV}{dt} \tag{B}$$

Now Equation (A) = Equation (B), i.e. $y + h_f = \dfrac{-L}{g}\dfrac{dV}{dt}$

$$\therefore \qquad \frac{L}{g}\frac{dV}{dt} + y + h_f = 0 \tag{7.33}$$

which is the momentum equation in the surge tank.

From Equation (7.32b), $V = \dfrac{A_s}{A_t}\dfrac{dy}{dt}$ and $h_f = \dfrac{fLV^2}{2gD}$

Substituting the value of V in Equation (7.33):

$$\frac{A_s}{A_t}\frac{L}{g}\cdot\left(\frac{d^2y}{dt^2}\right) + \frac{fL}{2gD}\left(\frac{A_s}{A_t}\frac{dy}{dt}\right)^2 + y = 0$$

or

$$\frac{d^2y}{dt^2} + \frac{A_s^2 \cdot At}{A_t^2\, A_s L}\cdot\frac{fL}{2gD}\left(\frac{dy}{dt}\right)^2 + \left(\frac{A_t g}{A_s L}\right)y = 0$$

or

$$\frac{d^2y}{dt^2} + \left(\frac{A_s\, g f}{At \cdot 2g\, D}\right)\left(\frac{dy}{dt}\right)^2 + \left(\frac{A_t g}{A_s L}\right)y = 0 \tag{7.34}$$

or

$$\frac{d^2y}{dt^2} + \alpha\left(\frac{dy}{dt}\right)^2 + \beta y = 0 \tag{7.34a}$$

where

$$\alpha = \left(\frac{fA_s}{2DA_t}\right),\ \beta = \left(\frac{At \cdot g}{A_s \cdot L}\right)$$

7.20 NUMERICAL SOLUTIONS OF THE MOMENTUM EQUATION

Equation (7.34) is non-linear, and cannot be solved analytically. Therefore, numerical solutions are necessary. The different numerical solutions used to solve the above equations are:

(i) Conventional explicit finite difference method
(ii) Pressel's method of successive trials
(iii) Simple arithmetic mean method
(iv) Escande's method
(v) Jakobsen's method
(vi) Modified Jakobsen's method

A lot of work has been undertaken on the solution of the unsteady surge tank equations of continuity and momentum by various investigators. Laboratory works have also been undertaken to generate data of rise and fall of surge heights in the surge tanks. Such data and solutions are available with the author of this book. Figure 7.16 shows the plot of solution by Modified Jakobsen's method and the experimental results obtained by the author on this problem. Numerical solutions by finite difference, arithmetic mean and Pressel methods are also computed. Solutions have agreed quite well with the author's experimental data generated in the laboratory of Hydraulics, Assam Engineering College, Guwahati, Assam.

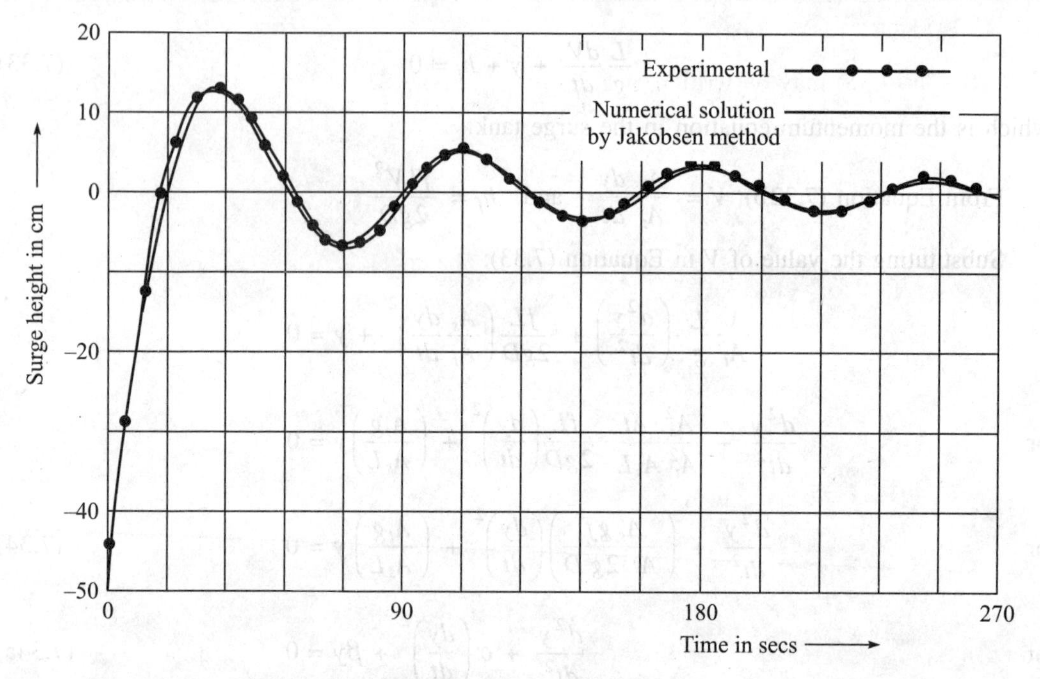

Figure 7.16 Theoretical and experimental plots of surge tank height fluctuation.

Similarly Equations (7.27a) and (7.28) are solved by the Lax diffusive explicit method already described as finite difference forms in Equation (7.30) for discharge and Equation (7.31) for pressure head are used. Plots of the solutions are given in Figures (7.15) and (7.16), respectively. The friction factor equation used in this solution was Barr's explicit solution of Equation (7.8a).

7.21 CLASSICAL SOLUTION OF SURGE TANK EQUATIONS

However, analytical solution of continuity and momentum equations of surge tank (neglecting friction) i.e. classical solution is possible. Although these classical solutions are not in use in the practical field, they give an idea of the type and trend of solution that may be available in the real field with the consideration of friction.

When friction is neglected, $h_f = 0$

Then momentum in Equation (7.32) becomes:

$$\frac{L}{g}\frac{dV}{dt} + y = 0 \tag{A}$$

from continuity, (i.e. Equation (7.32b)), $V = \dfrac{A_s}{A_t} \cdot \dfrac{dy}{dt}$

$$\therefore \text{ (A) becomes } \frac{L}{g}\frac{A_s}{A_t}\frac{d^2y}{dt^2} + y = 0 \tag{7.35}$$

which is a simple ordinary differential equation that can be solved analytically.

Equation (7.35) may be written as $\dfrac{d^2y}{dt^2} + \left(\dfrac{gA_L}{LA_s}\right)y = 0$ (7.35a)

Let $$D = \frac{d}{dt}$$

Equation (7.35a) becomes $D^2(y) + \left(\dfrac{gA_t}{LA_s}\right)y = 0$

$\therefore$ $$y\left(D^2 + \frac{gA_t}{LA_s}\right) = 0$$

$\therefore$ $$y \neq 0,\ D^2 + \frac{gA_t}{LA_s} = 0$$

$\therefore$ $$D = \sqrt{(-1)\frac{gA_t}{LA_s}}$$

or $$D_1 = i\sqrt{\frac{gA_t}{LA_s}}$$

and $$D_2 = (-i)\sqrt{\frac{gA_t}{LA_s}}$$

$\therefore$ Solution of y can be written as:

$$y = C_1 e^{i\sqrt{\frac{gA_t}{LA_s}}t} + C_2^{-i\sqrt{\frac{gA_t}{LA_s}}t}$$

$$y = C_1\left[\cos\sqrt{\frac{gA_t}{LA_s}}\,t + i\sin\sqrt{\frac{gA_t}{LA_s}}\,t\right] + C_2\left[\cos\sqrt{\frac{gA_t}{LA_s}}\,t - i\sin\sqrt{\frac{gA_t}{LA_s}}\,t\right]$$

To evaluate the constants C_1 and C_2
put Boundary conditions, at $t = 0$, $y = 0$

$\therefore$ $$O = C_1[1 + 0] + C_2[1 - 0]$$

$\therefore$ $$C_2 = -C_1$$

Putting $C_2 = -C_1$

$$y = C_1 \cos\sqrt{\frac{gA_t}{LA_s}}\,t + i\sin\sqrt{\frac{gA_t}{LA_s}}\,t - C_1\cos\sqrt{\frac{gA_t}{LA_s}}\,t + iC_1\sin\sqrt{\frac{gA_t}{LA_s}}\,t$$

$$y = (2Ci)\sin\sqrt{\frac{gA_t}{LA_s}}\,t = C\sin\sqrt{\frac{gA_t}{LA_s}}\,t \tag{A}$$

which shows that y is a sine function in t.

To evaluate C:

$$\frac{dy}{dt} = C \cos \sqrt{\frac{gA_t}{LA_s}}\, t \times \sqrt{\frac{gA_t}{LA_s}}$$

$$\frac{dy}{dt} = C \sqrt{\frac{gA_t}{LA_s}} \cos \sqrt{\frac{gA_t}{LA_s}}\, t \quad \text{(B)}$$

From continuity Equation (7.32b):

$$\frac{dy}{dt} = \frac{At}{As} V \quad \text{(C)}$$

$\therefore$ Equating (B) and (C):

$$\frac{A_t}{A_s} V = C \sqrt{\frac{gA_t}{LA_s}} \cos \sqrt{\frac{gA_t}{LA_s}}\, t$$

$$\therefore \quad V = \left(\frac{A_t}{A_s}\right) C \sqrt{\frac{gA_t}{LA_s}} \cos \sqrt{\frac{gA_t}{LA_s}}\, t$$

$$V = C \sqrt{\frac{gA_s}{LA_t}} \cos \sqrt{\frac{gA_t}{LA_s}}\, t$$

Put Boundary Conditions, at $t = 0$, $V = V_0$ (Steady State Velocity).

$$\therefore \quad V_0 = C \sqrt{\frac{gA_s}{LA_t}} \times 1$$

$$\therefore \quad C = V_0 \sqrt{\frac{LA_t}{gA_s}}$$

Substituting C in Equation (A):

$$y = V_0 \sqrt{\frac{LA_t}{gA_s}} \sin \sqrt{\frac{gA_t}{LA_s}}\, t \quad (7.36)$$

which is the solution of surge height y.

For maximum, $\sin \sqrt{\frac{gA_t}{LA_s}}\, t = 1 = \sin (\pi/2)$

$$\therefore \quad (\pi/2) = \sqrt{\frac{gA_t}{LA_s}}\, t \quad \therefore t = \frac{\pi}{2} \sqrt{\frac{LA_s}{gA_t}} \quad (7.37)$$

and

$$y_{max} = V_0 \sqrt{\frac{LA_t}{gA_s}} \quad (7.38)$$

Classical solution of surge height i.e. equation (7.36) is plotted in Figure 7.17.

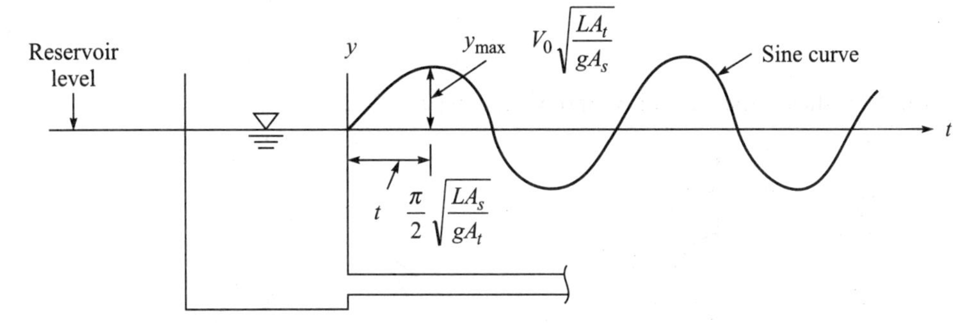

Figure 7.17 Classical solution of surge height showing maximum surge height and time of occurrence.

To find expressions for V and V_S, start with Equation (7.36)

i.e.

$$y = V_0\sqrt{\frac{LA_t}{gA_s}}\ \sin\sqrt{\frac{gA_t}{LA_s}}\ t$$

Differentiating with respect to t:

$$\frac{dy}{dt} = V_0\sqrt{\frac{LA_t}{gA_s}}\sqrt{\frac{gA_t}{LA_s}}\ \cos\sqrt{\frac{gA_t}{LA_s}}\ t$$

$$\frac{dy}{dt} = V_0\frac{A_t}{A_s}\ \cos\sqrt{\frac{gA_t}{LA_s}}\ t$$

But

$$\frac{dy}{dx} = \frac{A_t}{A_s}V$$

$\therefore$

$$\frac{A_t}{A_s}V = V_0\ \frac{A_t}{A_s}\ \cos\sqrt{\frac{gA_t}{LA_s}}\ t$$

$\therefore$

$$V = V_0\ \cos\sqrt{\frac{gA_t}{LA_s}}\ t \tag{7.39}$$

which gives the solution of V and gives the cosine curve.

For V to be maximum, $\cos\sqrt{\frac{gA_t}{LA_s}}\ t = 1$

$\therefore$

$$V_{max} = V_0 \tag{7.40}$$

Again

$$\frac{dy}{dt} = V_s = \frac{A_t}{A_s}V = \frac{A_t}{A_s}V_0\ \cos\sqrt{\frac{gA_t}{LA_s}}\ t$$

$$V_s = \frac{A_t}{A_s}\ V_0\ \cos\sqrt{\frac{gA_t}{LA_s}}\ t \tag{7.41}$$

$$V_{s\max} = \frac{A_t}{A_s} V_0 \quad (7.42)$$

Figure 7.18 shows the plot of V and V_s against t

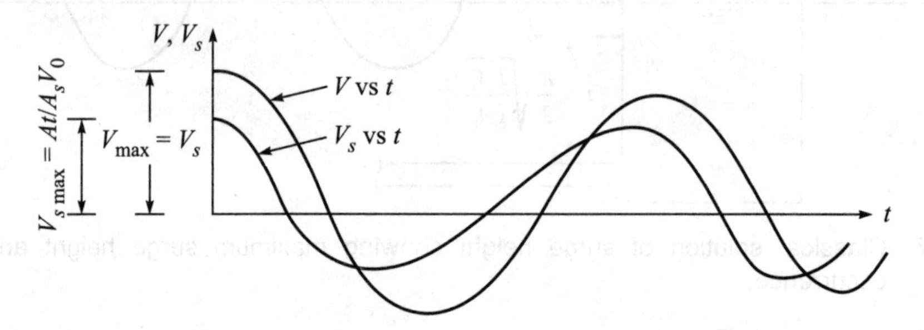

Figure 7.18 Plot of V and V_s against time t.

EXAMPLE 7.16 A cylindrical surge tank 4 m in diameter is connected to a 1 m diameter conduit, which is 150 m long. The initial steady state flow is 2 m³/sec. The penstock flow is abruptly closed. Neglecting friction, find the maximum rise of water level in the surge tank and its time of occurrence. Also find the maximum velocity in the surge tank.

Solution: $A_s = \frac{\pi}{4}(4)^2 = 12.5663 \text{ m}^2$

$$A_t = \frac{\pi}{4}(1)^2 = 0.7854 \text{ m}^2$$

$$L = 150 \text{ m}$$

$$Q = 2 \text{ m}^3/\text{sec}$$

$$\therefore \quad V_0 = \frac{Q}{A_t} = \frac{2}{0.7854} = 2.54647 \text{ m/sec}$$

Maximum rise of water level in the surge tank (neglecting friction) is:

$$y_{\max} = V_0\sqrt{\frac{LA_t}{gA_s}} = \left(2.54647 \times \sqrt{\frac{150 \times 0.7854}{9.81 \times 12.5663}}\right) \text{ m}$$

$$y_{\max} = 2.4893 \text{ m}$$ **Ans.**

$$\text{Time of occurrence of this } y_{\max} = \frac{\pi}{2}\sqrt{\frac{LA_s}{gA_t}} = \left(\frac{\pi}{2}\sqrt{\frac{150 \times 12.5663}{9.81 \times 0.7854}}\right) \text{ sec}$$

$$= 24.57 \text{ secs}$$ **Ans.**

$$\text{Maximum velocity in the surge tank} = V_0 \times \frac{A_t}{A_s} = \left(2.54647 \times \frac{0.7854}{12.5663}\right)$$

$$= 0.159155 \text{ m/sec}$$ **Ans.**

7.22 PIPE NETWORKS

In city's municipal water supply, a group of interconnected pipes forming several loops or circuit as shown in Figure 7.19 is called pipe networks.

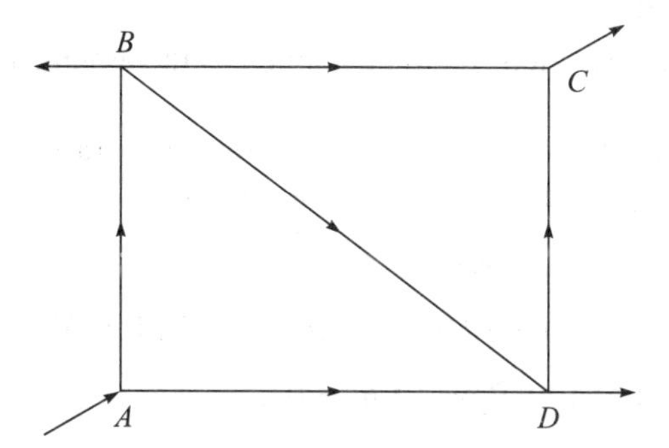

Figure 7.19 Pipe networks.

The main problem of pipe networks is in the determination of correct distribution of discharges through the various pipes with different diameters and different loops in a network. In the determination of discharges, flow conditions like continuity, friction losses, etc. are to be satisfied. Thus different conditions that need to be satisfied are:

(I) According to continuity, the flow coming to each junction must be equal to the flow going out of the junction.

(II) In each loop, the loss of head due to the flow in the clockwise direction must be equal to the loss of head due to the flow in the anti-clockwise direction. For example, for loop *ABD*, the sum of head loss in *AB* and *BD* must be equal to head loss in *AD*.

(III) The Darcy–Weisback equation must be satisfied for the flow in each pipe.

Head lost due to friction, $h_f = \dfrac{fLV^2}{2gD} = \dfrac{fL}{2gD}\dfrac{Q^2}{A^2} = \dfrac{fLQ^2}{2gD\left(\dfrac{\pi}{4}D^2\right)^2}$

$$h_f = \frac{fL}{2g\,(\pi/4)^2}\frac{Q^2}{D^5} = rQ^2 \tag{7.43}$$

where

$$r = \frac{fL}{2g\,(\pi/4)^2\,D^5} = \frac{fL}{12.1026\,D^5} \tag{7.44}$$

The Hardy Cross Method of Solution

The pipe network problem is complicated and cannot be solved analytically. Thus the method of successive approximation is used. The commonly used method of solution is the 'Hard Cross Method' named after original investigator. The method requires the following steps:

(1) Assume a suitable distribution of flow that satisfies the continuity in each junction.

(2) With the assumed Q values, compute the head lost for each pipe using Equation (7.43). Friction factor f, length L and diameter must be known, i.e. r must be known.

(3) Considering different loops, compute the net head lost around the loop taking the clockwise head loss as positive and the anti-clockwise head lost as negative. If the assumed flow in the pipe is correct, Σh_f should be zero so that the loop will be balanced. If it is not zero, the assumed flows are corrected introducing a correction ΔQ for the flows, till $\Sigma h_f \simeq 0$.

(4) The values of the correction ΔQ to be applied to the assumed flow of the loops may be obtained as follows:
If Q_A is the assumed flow and Q is the correct flow rate,
then $Q = (Q_A + \Delta Q)$ and head lost for the pipe is

$$h_f = r(Q_A + \Delta Q)^2$$

$$\Sigma h_f = \Sigma r(Q_A + \Delta Q)^2 = \Sigma r(Q_A^2 + \Delta Q^2 + 2Q_A\, \Delta Q)$$

For correct distribution of flows, $\Sigma h_f = 0$

$$\therefore \qquad \Sigma r(Q_A^2 + \Delta Q^2 + 2Q_A\, \Delta Q) = 0$$

Assuming ΔQ^2 to be small, it may be neglected

$$\therefore \qquad \Delta Q = -\frac{\Sigma r Q_A^2}{|2r\, Q_A|} \tag{7.45}$$

In Equation (7.45), the denominator $2rQ_A$ is the sum of the absolute terms hence it has no sign. If the head lost due to the clockwise movement is more than that due to the anti-clockwise, ΔQ will be negative. On the other hand, if the clock is less than anti-clockwise ΔQ is positive.

(5) Applying correction ΔQ in the loop, the second trial calculation is made for all loops and the process is repeated till ΔQ becomes negligible.

EXAMPLE 7.17 For a pipe network shown in Figure 7.20, determine the discharges in each pipe.

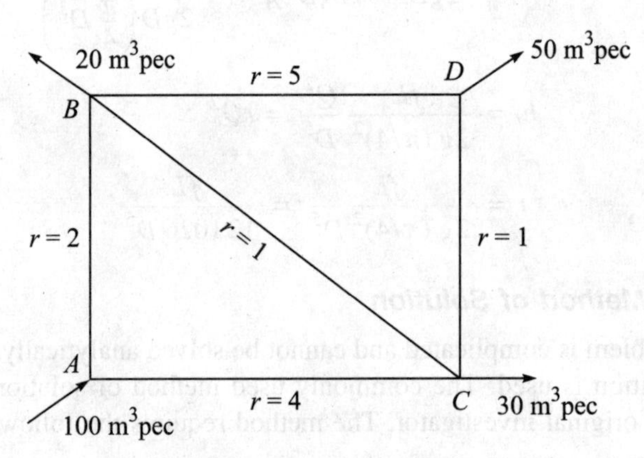

Figure 7.20 Visual for Example 7.17.

Solution: $r = \dfrac{fL}{12.0126\, D^5}$ and r values for all the pipes are given.

We have the formula $h_f = rQ^2$

Consider the first loop ABC. Assume the discharges as shown in Figure 7.21. Take rQ^2 + *Ve* in the clockwise and anti-clockwise direction as negative.

(a) Loop ABC

Pipe	rQ^2	$2rQ$
AB	9800	280
BC	1225	70
AC	−3600	240
	$\sum rQ^2 = 7425$	$\sum 2rQ = 590$

$$\therefore \quad \Delta Q = -\frac{\Sigma rQ^2}{|2rQ|} = \frac{(+7425)}{-590} = 12.584 \cong -13$$

(b) Loop BDC

BC	−1225	70
CD	−1225	70
BD	1125	150
	$\sum rQ^2 = -1325$	$\sum 2rQ = 290$

$$\therefore \quad \Delta Q = -\frac{(-1325)}{290} = +4.569 \simeq 5$$

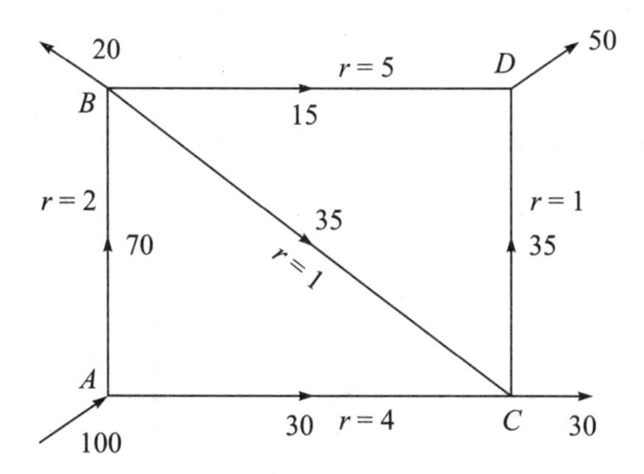

Figure 7.21 Visual for Example 7.17 after distribution discharges in pipes of the two loops.

Now apply the first correction in two loops and obtain Figure 7.22.

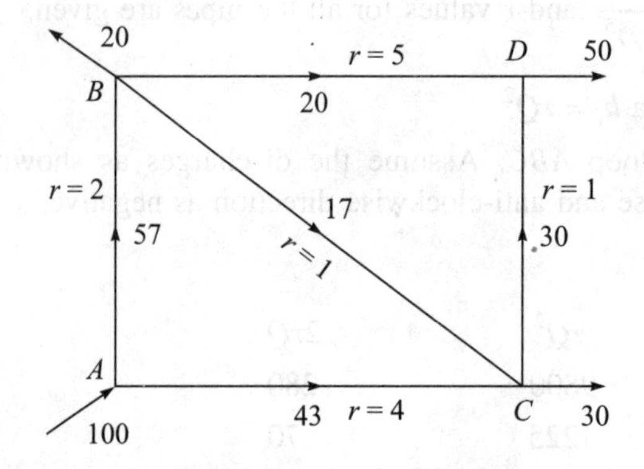

Figure 7.22 Visual for Example 7.17 after first correction.

Loop *ABC* again

AB	$2 \times 57^2 = 6498$	$2 \times 2 \times 57 = 228$
BC	$2 \times 17^2 = 289$	$2 \times 1 \times 17 = 34$
AC	$-4 \times 43^2 = -7396$	$2 \times 4 \times 43 = 344$
	$\sum rQ^2 = -609$	$\sum 2rQ = 606$

$$\therefore \quad \Delta Q = -\frac{(-609)}{606} \simeq 1$$

(c) Loop *BDC*

BC	$-1 \times 17^2 = -289$	$2 \times 1 \times 17 = 34$
CD	$-1 \times 30^2 = -900$	$2 \times 1 \times 30 = 60$
BD	$5 \times 20^2 = 2000$	$2 \times 5 \times 20 = 200$
	$\sum rQ^2 = 811$	$\sum 2rQ = 294$

$$\therefore \quad \Delta Q = \frac{-811}{294} = -2.758 \simeq -3$$

Applying again the second correction and obtain Figure 7.23.

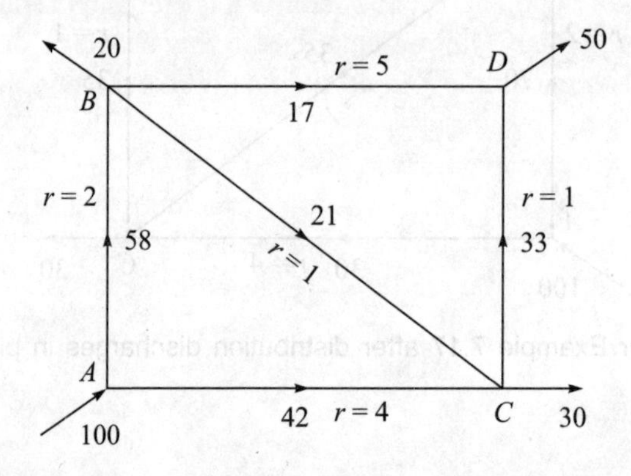

Figure 7.23 Visual for Example 17.17 after second correction.

Loop *ABC* (For another trial)

AB	$2 \times 58^2 = 6728$	$2 \times 2 \times 58 = 232$
AC	$-4 \times 42^2 = -7056$	$2 \times 4 \times 42 = 326$
CB	$1 \times 21^2 = 441$	$2 \times 1 \times 21 = 42$
	$\sum rQ^2 = 113$	$\sum 2rQ = 610$

$$\Delta Q = -\frac{113}{610} \simeq 0.1832 \simeq 0 \text{ (almost negligible)}$$

Loop *BDC*

BD	$5 \times 17^2 = 1445$	$2 \times 5 \times 17 = 170$
BC	$-1 \times 21^2 = -441$	$2 \times 1 \times 21 = 42$
CD	$-1 \times 33^2 = -1089$	$2 \times 1 \times 33 = 66$
	$\sum rQ^2 = -85$	$\sum 2rQ = 278$

$$\Delta Q = \frac{-85}{278} = -0.305 \qquad 0 \text{ (almost negligible)}$$

Since the discharges in Figure 7.23 equation produce negligible ΔQ so the discharges in Figure 7.23 are the corrected discharges or Answer of the example.

Thus

$$\left.\begin{aligned} AB &= 58 \text{ m}^3/\text{sec} \\ AC &= 42 \text{ m}^3/\text{sec} \\ BC &= 21 \text{ m}^3/\text{sec} \\ BD &= 17 \text{ m}^3/\text{sec} \\ CD &= 33 \text{ m}^3/\text{sec} \end{aligned}\right\} \quad \textbf{Ans.}$$

7.23 CONCLUSION

Flow of fluid through pipe is very important in different practical fields like water supply, irrigation, water power development, oil and natural gas etc. and therefore, knowledge of flow of fluid through pipe is essential. The presentations in this chapter from Section 7.2 to Section 7.14 are although quite common almost in all the text books of fluid mechanics, attempts have been made in this chapter to present these contents in a systematic and easy-to-follow method for students with numerically solved examples. Analysis of unsteady flow in pipe in water hammer situations with equations of continuity and momentum in pipe and surge tank are usually not available in many of the books on this subject. Finite difference solutions of these equations, verification of solutions with experimental data of the author, classical solutions of the nonlinear equations etc. are some extra important presentations of this chapter. Investigators have been made to present different different methods of solution of this unsteady pipe flow problems. The list of references includes research paper till to the present time. This may help the ambitious students to pursue further works in this field of pipe flow.

In early days, extensive leading works were presented by Blasius[1], Von Karman[2], Prandtl[3] and Nikurdse[4] and in water hammer situation by Alleevi[13], Angus[14]. In the subsequent works on pipe resistance in laminar to turbulent and transitional state Colebrook and White[5], Barr[15] and others. Barr's[15] transition control function is extensively used by Das[16] to study the resistance effect of different roughness pattern from laminar to turbulent in free surface profiles. It was shown by Borthakur[20], Choudhury[19], Das[21], Das and Sarma[22,23], Das–Saikia et at[22] that friction factor changes with state of flow condition which was originally proposed by Barr[6].

PROBLEMS

7.1 Find the head lost due to friction in a pipe 0.3 m diameter and length 100 m. The discharge flowing through the pipe is 0.28475 m^3/sec. Assume Darcy's friction factor to be 0.02. (**Ans.** 5.437m)

7.2 The rate of flow of water through a horizontal pipe is 0.3 m^3/sec. The diameter of the pipe is suddenly enlarged from 0.25 m to 0.5 m. The pressure intensity in the smaller pipe is 13.734 N/cm^2. Determine: (i) Loss of head due to sudden expression, (ii) Pressure intensity in the layer pipe, and (iii) Power lost due to enlargement.

(**Ans.** (i) 1.07 m (ii) 14.43 N/m^2 (iii) 3.15 kw)

7.3 Find the difference of elevations between the water surface of two reservoirs which are connected by a horizontal pipe of diameter 0.4 m and length 500 m. The rate of flow through the pipe is 200 litres/sec. Consider all losses and $f = 0.036$ (**Ans.** 11.79 m)

7.4 A siphon of diameter 0.15 m connects two reservoirs having a difference of level of 15 m. The length of the siphon is 400 m and the summit is 4 m above the water level of the upper reservoir. The length of the pipe from the upper reservoir to the summit is 80 m. Determine the discharge through the siphon and also the pressure at the summit. Neglect minor losses. Darcy's friction factor is 0.02.

(**Ans.** 41.52 litres, –7.281 m of water).

7.5 Three pipes of lengths 800 m, 600 m, and 300 m, and of diameters 0.4 m, 0.3 m and 0.2 m, respectively are connected in series. The ends of the compound pipe are connected to two reservoirs whose water surface levels are maintained at a difference of 15 m. Determine the discharge through the pipes if $f = 0.02$. What will be the diameter of the equivalent pipe of 1700 m if f remains to the same as that of the compound pipe.

(**Ans.** 0.0848 m^3/sec 0.2665 m)

7.6 Two pipes of length 2500 m each and diameter 0.8 m and 0.6 m, respectively are connected in parallel. If $f = 0.024$ for each pipe, and the total flow is equal to 250 litres/sec, find the discharges in each pipe. (**Ans.** 0.1683 m^3/sec, 0.0817 m^3/sec)

7.7 A pipe of diameter 0.3 m and length 3000 m is used to transmit power by water. The total head at the inlet is 400 m. Find the maximum power available at the outlet of the pipe. Take $f = 0.02$. (**Ans.** 667.07 kw)

7.8 A pipeline of length 2100 m is used for transmitting power up to 103 kw. The pressure at the inlet of the pipe is 392.4 N/cm^2. If the efficiency of transmission is 80%, find the diameter of the pipe. Take $f = 0.02$. (**Ans.** 4.15 cm)

7.9 A pipeline 2500 m long carries water with a velocity 1.5 m/sec. The diameter of the pipe is 0.5 m. A valve at the end of the pipe is closed in 25 seconds. Taking velocity of pressure to be 1460 m/sec, find the rise of pressure. (**Ans.** 15 N/cm^2)

7.10 If in *Problem 7.9*, the valve is closed in 2 secs, find the rise of pressure close to the valve. Assume pipe to be rigid, and water to be compressible with the bulk modulus of compression of water to be 19.62×10^4 N/cm^2. (**Ans.** 210.1 N/cm^2)

7.11 If in *Problem 7.9*, the thickness of the pipe is 0.01 m and the valve is closed suddenly, find the rise of pressure, considering the pipe to be elastic and the water to be compressible. Also find the circumferential and longitudinal stress if the bulk modulus of compression of water K and modulus of elasticity E of the pipe are 19.62×10^4 N/cm^2 and 19.62×10^6 N/cm^2, respectively.

(**Ans.** 17155.1 N/m^2, 4286.9 N/m^2, 2243.45 N/m^2)

7.12 A cylindrical surge tank of 5 m diameter is connected to a 1 m diameter conduit 200 m long. The initial steady state flow is 3 m^3/sec. The penstock valve is abruptly closed. Neglecting friction, find the maximum rise of water level in the surge tank and its time of occurrence. (**Ans.** 3.45 m, 35.46 secs)

7.13 For the pipe network given in Figure 7.24, determine the discharges in each pipe given that

$$r = \frac{fL}{12.0126\, D^5}.$$

(**Ans.** Students are advised to follow Example 7.17 to find the discharges in different pipes.)

Hint: Discharge coming to a joint must be equal to discharge going out of it.

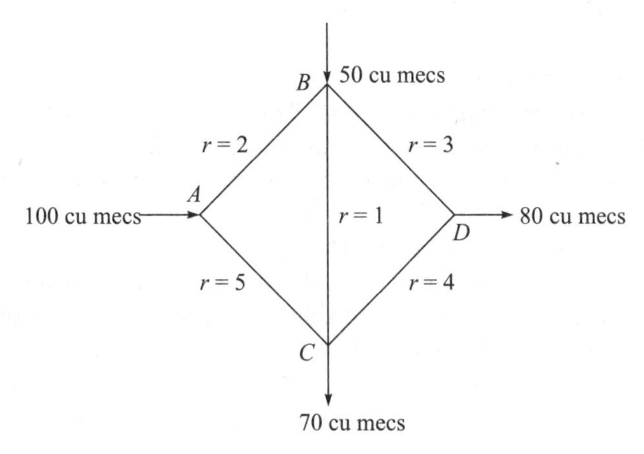

Figure 7.24 Visual of Problem 7.13.

REFERENCES

1. Blasius, H., *The Law of Similitude for Friction in Fluids*, For schungsheft des Vereins deutscher Ingenieure, No. 131, Berlin, 1913.

2. Von Karman, T. "Mechanical similitude and turbulence" *Proc. 3rd. International Conference for Applied Mechanics*, Stockholm, Vol. 1, pp. 85–93, 1930.
3. Prandtl, L., "Mechanics of viscous fluid", in W.F. Durand Editor in Chief, *Aerodynamic theory*, Springer-Verlag, Vol. III, div. G., Berlin, 1935.
4. Nikuradse, J. "Laws of turbulent flow in smooth pipe" For schungsheft des Vereins deutscher Ingeniewu, No. 354, Berun, 1932.
5. Colebrook, C.F. and C.M. White, "Experiments with fluid friction in roughened pipes", *Proc. Royal Society of London*, source A, Vol. 161, pp. 367–381, 1938.
6. Barr, D.I.H., Two additional methods of direct solution of Colebrook-White function", *Proc. ICE*, Partz, 1976.
7. Dryden, H.L., *Transition from Laminar to Turbulent Flow*, Princeton University Press, Section A of Turbulent Flows and Heat Transfer, edited by C.C. Lew, 1959.
8. Streeter, V.L., *Water Hammer Analysis*, ASCE, Vol. 95, Hydro. 6, Nov., 1969.
9. Das, M.M., "Computer Aided Solution of Surge Height and Comparison with Laboratory Data", *National Conference of Hydro-Power Development*, CBIP, April, New Delhi, 1997.
10. Chatterjee, P.N., *Fluid Mechanics for Engineers: Digital Computer Application*, Macmillan (India), 1995.
11. Pick Ford, J., *Analysis of Surges*, Macmillan, 1969.
12. Bergeron, L., *Water Hammer in Hydraulics and Wave Surges in Electricity*, ASME, Trans. 1961.
13. Allievi, L., *The Theory of Water Hammer*, Translated by ASME, 1929.
14. Angus, R.W., "Water Hammer in Pipes including those supplied by Centrifugal Pumps: Graphical Treatment" *Proc. Mech. Engrs*., Vol. 136, 1937.
15. Barr, D.I.H. "New forms of equations for co-relation of pipe resistance data", *Proc. ICE*, London, No. 53, pp. 383–390, September, 1973.
16. Das, M.M., "The effect of resistance in steady and unsteady gradually varied free surface profiles", *Ph.D. Thesis*, University of Stathclyde, Glasgow, 1978.
17. Johnson, R.D., *Differential Surge Tank*, Trans., ASCE, 78, 1915.
18. Jakobsen, B.F., *Surge Tanks*, Trans., ASCE, 85, 1922.
19. Choudhury, H. "Simulation of theoretical solution and experimental data on the theory of unsteady flow in surge tank", *M.E. Thesis*, Gauhati University, 1996.
20. Borthakur, K.C. "Experimental and Theoritical Studies of Water Hammer Pressure in surge Tank and Pressure Conduit", *Ph.D. Thesis*, Gauhati University, 1997.
21. Das, Mimi "Different computer aided solutions of unsteady equation of surge tank and their verfication against laboratory data.", *M.E. Thesis*, Gauhati University, 1999.
22. Das, Mimi, Sarma, A.K. and M.M. Das, "Effect of resistance parameter in the numerical solution of nonlinear unsteady equations of surge tank", *IHS Jour. of Hy. Engg*., Vol. 11, No. 2, pp. 101–110, 2005.
23. Das–Saikia, Mimi and Sarma, A.K. "Numerical modelling of water hammer with variable friction factor", *Proc. 2nd. International Conf. of Computational Mechanics and Simulations*, IIT Guwahati, 8-10 Dec., 2006.
24. Das–Saikia, Mimi and Sarma, A.K., "Simulation of water hammer flows with unsteady friction factor", ARPN, *Jour. of Engg*., ISSN, 1819–6608, Dec. 2006.

Chapter 8

Flow Through Orifices and Mouthpieces

8.1 INTRODUCTION

An orifice is a small opening of any shape such as circular, rectangular or triangular, provided on the side or bottom of a tank for the purpose of discharging liquid from the tank or the vessel on which it is provided. This orifice may discharge the liquids into the atmosphere or from one tank to another.

A mouthpiece is a short tube of length equal to two or three times the diameter of the tube, fitted to an orifice of the same diameter, provided in the tank or vessel containing the liquid. It is essentially an extension of the orifice through which also the fluid may be discharged.

8.2 CLASSIFICATION OF ORIFICES

(i) The orifices are classified as circular, rectangular, square or triangular on the basis of the shape of the cross-sectional area.

(ii) The orifices are classified as small or large orifices on the basis of the size of orifice and the head of water above the orifices. When the head of the fluid from the centre line of the orifice is more than five times its depth, it is called a small orifice and if the head is less than five times the depth, it is called a large orifice.

(iii) Orifices are also classified as sharp-edged or bell-mouthed depending upon the shape of the upstream edge of the orifices.

(iv) Further, they are classified as submerged or drowned, free discharging orifices. Submerged orifices may be fully submerged or partially submerged.

8.3 VELOCITY OF AN ORIFICE UNDER A HEAD *H*

A circular orifice is provided on the side of a tank. The flow takes place through this orifice under a head H as shown in Figure 8.1. The liquid flowing out of the orifice forms a jet of liquid whose area of flow is less than the area of the orifice. At section *CC*, the area is the minimum. At *CC*, the streamlines are straight and parallel to each other and perpendicular to the plane of the orifice. The section *CC* is at a distance of approximately half the diameter of the orifice. This section *CC* is called 'Vena Contracta'. Beyond the section *CC*, the jet diverges and is inclined downwards due to gravity.

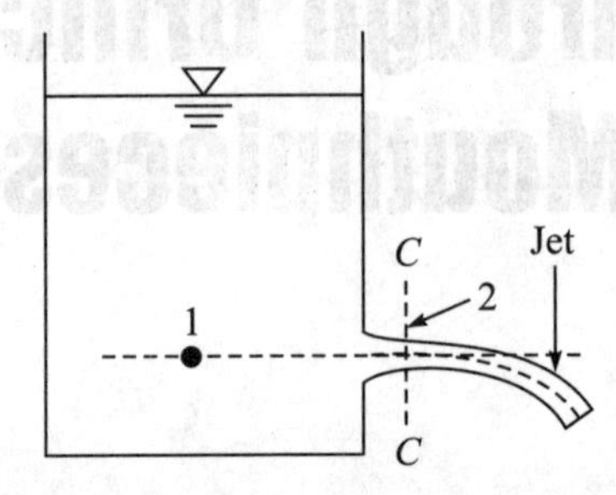

Figure 8.1 Flow through an orifice under a head *H*.

Considering two points 1 and 2, and applying Bernoulli's equation at 1 and 2 considering the centre line of the orifice as the datum i.e. 1 and 2 are on same datum:

$$\frac{p_1}{w} + \frac{V_1^2}{2g} + Z_1 = \frac{p_2}{w} + \frac{V_2^2}{2g} + Z_2$$

$Z_1 = Z_2$ V_1 is very small as compared to V_2, hence $\frac{V_1^2}{2g}$ is neglected, p_2 is atmospheric, as *CC* or 2 exposed to the atmosphere.

$$\therefore \quad \frac{p_1}{w} + 0 = 0 + \frac{V_2^2}{2g}$$

$$\therefore \quad p_1 = wH$$

$$\therefore \quad \frac{p}{w} = H$$

$$\therefore \quad H = \frac{V_2^2}{2g} \quad \therefore V_2 = \sqrt{2gH} \tag{8.1}$$

Equation (8.1) gives the theoretical velocity of the jet.

The actual velocity will be less than that given by Equation (8.1). This equation is known Torricelli's formula in honour of Evangelisa Torricelli who deduced this equation in 1643.

8.4 COEFFICIENT OF VELOCITY (C_v), CONTRACTION (C_c) AND DISCHARGE (C_d)

Coefficient of velocity (C_v) is the ratio of the jet, a Vena Contracta to the theoretical velocity. If *V* is the velocity of the jet (i.e. at vena contracta),

$$C_V = \frac{V}{\sqrt{2gH}}$$

$\therefore$
$$V = C_V\sqrt{2gH} \tag{8.2}$$

The values of C_V vary from 0.95 to 0.99 for different types of orifices.

Generally C_V is taken as 0.98.

Coefficient of contraction (C_C) is the ratio of area of the jet at Vena Contracta to the area of the orifice. If a is the area of the orifice and a_c is the are at C_c:

$\therefore$
$$C_c = \frac{a_c}{a} \tag{8.3}$$

or
$$a_c = C_c a$$

The values of C_c vary from 0.61 to 0.69. In general C_c is taken as 0.64.

Coefficient of discharge (C_d) is the ratio of actual discharge through the orifice to the theoretical discharge of the orifice. If Q is the actual discharge and Qth is the theoretical discharge, then:

$$C_d = \frac{Q}{Q\text{th}} = \frac{\text{Actual velocity} \times \text{Actual area}}{\text{Theoretical velocity} \times \text{Theoretical area}}$$

$$C_d = \left(\frac{C_V\sqrt{2gH}}{\sqrt{2gH}}\right) \times \left(\frac{a_c}{a}\right)$$

$$C_d = C_v \cdot C_c \tag{8.4}$$

EXAMPLE 8.1 The head of water over an orifice of diameter 5 cm is 12 m. Find the actual discharge and actual velocity of the jet at Vena Contracta. Take $C_d = 0.6$, $C_v = 0.98$.

Solution: Area of the orifice $a = \frac{\pi}{4}(0.05) = 0.0019635 \text{ m}^2$

$\therefore$
$$C_d = 0.6,\ C_v = 0.98,\ H = 12 \text{ m.}$$

$$\begin{aligned}
\text{Actual discharge} &= C_d \times \text{Theoretical discharge} \\
&= C_d \times \text{area of the orifice} \times \text{theoretical velocity} \\
&= 0.6 \times (0.0019635 \times \sqrt{2 \times g \cdot h}\,) \\
&= (0.6) \times (0.0019635 \times \sqrt{2 \times 9.81 \times 12}\,) \text{ m}^3\text{/sec} \\
&= 0.0180768 \text{ m}^3\text{/sec}
\end{aligned}$$
Ans.

$$\text{Actual velocity} = V = C_v\sqrt{2gH} = (0.98\sqrt{2 \times 9.81 \times 12}\,) \text{ m/sec}$$

$$V = 15.037 \text{ m/sec}$$
Ans.

EXAMPLE 8.2 The head of water over the centre of an orifice of diameter of 3 cm is 1.5 m. The actual discharge through the orifice is 2.35 lits/sec. Find the coefficient of discharge C_d.

Solution: $a = \frac{\pi}{4}(0.03)^2 = 0.000706858 \text{ m}^2$

$$Q = 2.35 \text{ lits/sec} = 0.00235 \text{ m}^3\text{/sec}$$

$$C_d = \frac{\text{actual discharge}}{\text{theoretical discharge}} = \frac{0.00235}{\sqrt{2 \times 9.81 \times 1.5} \times 0.000706858}$$

$$C_d = 0.61283 \simeq 0.613$$ **Ans.**

8.5 EXPERIMENTAL DETERMINATION OF C_d, C_V AND C_C

Figure 8.2 shows the experimental set-up for the determination of coefficients. Water is allowed to flow through an orifice under a constant head *H*. Constant head *H* is maintained by an inflow supplied by a pipe, i.e. the water entering through a supply inflow pipe is equal to the outflow through the orifice. After attaining the steady head *H* in the tank, water is collected in a measuring tank for a known time of *t* secs.

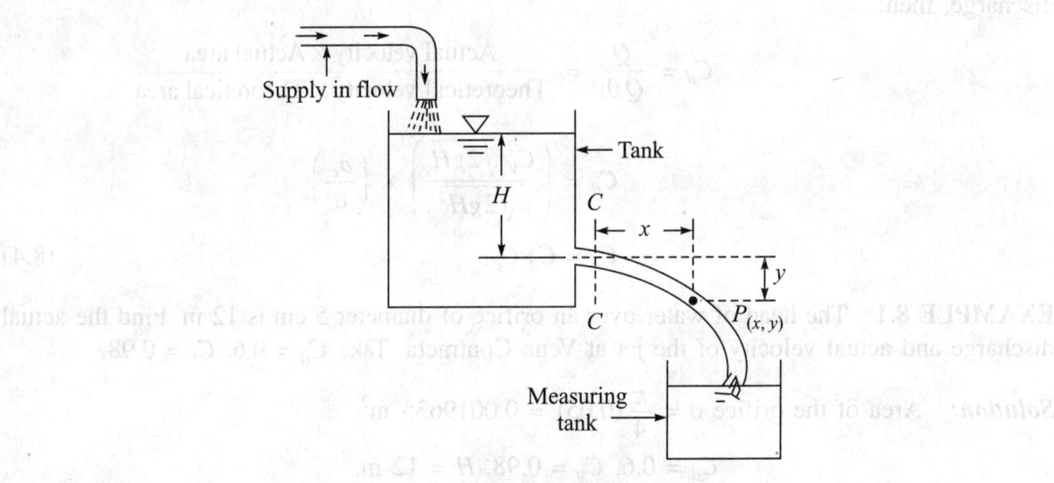

Figure 8.2 Experimental set-up to determine the coefficient.

Then actual $$Q = \frac{\text{Area of measuring tank} \times \text{height of water}}{t \text{ secs}}$$

If *a* is area of the orifice:

$$Q_{\text{theoretical}} = a\sqrt{2gH}$$

$$\therefore \quad C_d = \frac{\text{Actual } Q}{Q \text{ theoretical}} = \frac{Q}{a\sqrt{2gH}} \tag{8.5}$$

Let C_c be the Vena Contracta of the jet of water that comes out from the tank under this constant head *H*.

Select point P in the jet. A scale to measure the distance x from the Vena Contracta and vertical distance of P, i.e. y below the Vena Contracta is normally fitted in the experimental set-up.

Let x be the horizontal distance of P measured by the scale and y be the vertical distance of P from the centre of the Vena Contracta.

V is the actual velocity at the Vena Contracta.

Then the horizontal distance x in time t with velocity V,

$$x = Vt \tag{A}$$

and

$$y = \frac{1}{2} g t^2 \tag{B}$$

from (A) $t = \dfrac{x}{V}$

Substituting this t in (B):

$$y = \frac{1}{2} g (x/V)^2 = \frac{g x^2}{2V^2}$$

$$\therefore \quad V = \sqrt{\frac{g x^2}{2y}}$$

But

$$V_{\text{theoretical}} = \sqrt{2gH}$$

$$\therefore \quad \text{Coefficient of velocity} = C_v = \frac{V}{V_{\text{theoretical}}}$$

$$C_v = \frac{\sqrt{\dfrac{g x^2}{2y}}}{\sqrt{2gH}}$$

$$C_v = \frac{x}{\sqrt{4yH}} \tag{8.6}$$

Simply measuring x, y and H, coefficient of velocity C_v is obtained.

Equation (8.4) gives, $C_d = C_v \cdot C_c$

From known C_d and C_v, C_c could be obtained as:

$$C_c = \frac{C_d}{C_v} \tag{8.7}$$

EXAMPLE 8.3 A jet of water issuing from a sharp-crested vertical orifice under a constant head of 0.6 m has the horizontal (x) coordinate and vertical coordinate (y) measured from the Vena Contracta at a certain point as 0.1 m and 0.0045 m respectively. Find C_v and C_c if $C_d = 0.6$

Solution: $H = 0.6$ m, $x = 0.1$ m, $y = 0.0045$ m, $C_d = 0.6$.

C_v is given by Equation (8.6), i.e.

$$C_v = \frac{x}{\sqrt{4yH}} = \frac{0.1}{\sqrt{4 \times 0.0045 \times 0.6}}$$

$\therefore$ $$C_v = 0.96225$$ **Ans.**

Again Equation (8.7) gives $C_c = \frac{C_d}{C_v} = \frac{0.6}{0.96223}$

$\therefore$ $$C_c = 0.6235$$ **Ans.**

EXAMPLE 8.4 The head of water over an orifice of diameter 10 cm is 500 cm. Water coming out of the orifice is collected in a circular tank of diameter 2 m. The rise of water in the circular tank is 0.45 cm in 30 seconds. Also the coordinates of a certain point on the jet measured from the Vena Contracta are 100 cm and 5.2 cm vertical. Find C_v, C_d, C_c.

Solution: H = 500 cm = 5 m, d = 0.1 m, area of collecting $A_c = \pi/4(2)^2$
t = 30 secs. Rise in collecting tank = 0.45 in 30 secs. A_c = 3.14159 m^2
x = 1 m, y = 0.052 m

$$\text{Actual } Q = \frac{\text{Area of collecting tank} \times \text{rise}}{\text{time}} = \left(\frac{3.14159 \times 0.45}{30}\right) \text{m}^3/\text{sec}$$

$$Q_{act} = 0.047124 \text{ m}^3/\text{sec.}$$

$$Q_{th} = \text{area of orifice} \times \sqrt{2gH} = \left(\frac{\pi}{4}(0.1)^2 \times \sqrt{2 \times 9.81 \times 5}\right) \text{m}^3/\text{sec}$$

$$Q_{th} = 0.7779 \text{ m}^3/\text{sec.}$$

$\therefore$ $$C_d = \frac{Q_{act}}{Q_{th}} = \frac{0.047124}{0.07779}$$

$$C_d = 0.60578$$ **Ans.**

$$\text{Again } C_v = \frac{x}{\sqrt{4yH}} = \frac{1}{\sqrt{4 \times 0.052 \times 5}}$$

$$C_v = 0.9805$$

$\therefore$ $$C_c = \frac{C_d}{C_v} = 0.6177$$ **Ans.**

EXAMPLE 8.5 A tank has two identical orifices in one of its sides. The upper orifices is 4 m below the water surface and the lower one is 6 m below the water surface. If the C_v for both the orifices are same at 0.98, find the point of intersection of the jets.

Solution: C_v (for both) = 0.98
H_1 (head of first orifice) = 4 m
H_2 (head of second orifice) = 6 m

If P be the point of intersection of both the jets,

then x = horizontal distance for both orifice

y_1 = for first orifice

y_2 = for second orifice

$$y_1 = y_2 + (6 - 4) = (y_2 + 2)\text{m} \qquad \text{(A)}$$

Now
$$C_{v_1} = \frac{x}{\sqrt{4y_1 H_1}} = \frac{x}{\sqrt{4y_1 \times 4}}$$

$$C_{v_2} = \frac{x}{\sqrt{4y_2 H_2}} = \frac{x}{\sqrt{4y_2 \times 6}}$$

According to the problem, $C_{v_1} + C_{v_2}$'

$$\frac{x}{\sqrt{4y_1 \times 4}} = \frac{x}{\sqrt{4y_2 \times 6}}$$

or
$$4y_1 = 6y_2$$

Substituting y_1 from (A):

$$4(y_2 + 2) = 6y_2$$

or
$$4y_2 + 8 = 6y_2 \qquad \therefore \quad y_2 = 4 \text{ m}$$

$$C_{v_2} = \frac{x}{\sqrt{4y_2 \times H2}}$$

$$0.98 = \frac{x}{\sqrt{4 \times 4 \times 6}}$$

$\therefore$
$$x = 9.602 \text{ m.}$$

The point of intersection of the two jets at distance is 9.602 m from the orifices and 4 m below the second orifice. **Ans.**

8.6 DISCHARGE THROUGH LARGE ORIFICE

When an orifice is large (i.e. its head is less than five times its depth, of orifice), the discharge equation $Q = C_d\, a \sqrt{2gH}$ is not applicable as the top and bottom of the orifice are under two different heads H_1 and H_2, and hence velocity of the jet in the entire cross-section is not constant. Therefore, a different discharge equation is to be used.

8.6.1 When Orifice is Large Rectangular

In Figure 8.3, a large rectangular orifice of breadth b, and depth $d = (H_2 - H_1)$ is considered, discharging under a constant head H. Consider an elementary strip of thickness dh at a depth h from the reservoir or the tank water level.

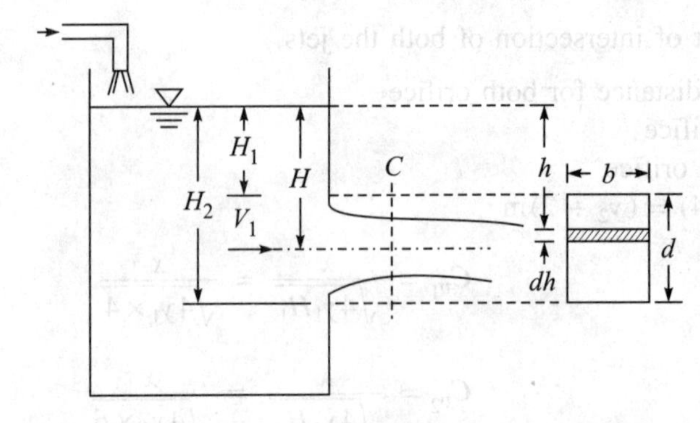

Figure 8.3 Determination discharge in large rectangular orifice.

Discharge through the strip = $dQ = Cd \times (bdh)\sqrt{2gH}$

$$dQ = Cd\sqrt{2gH}\ bh^{1/2}dh$$

Integrating dQ between limit h from H_1 to H_2, the discharge through the whole orifice is obtained:

$$\therefore \qquad \int dQ = \int_{H_1}^{H_2} Cd\sqrt{2g}\ bh^{1/2}dh$$

$$\therefore \qquad Q = C_d\sqrt{2g}\ b[H_2^{3/2} - H_1^{3/2}] \tag{8.8}$$

which is the discharge equation neglecting the velocity of approach. If V_1 is the velocity of approach, (which is usually small):

Equation (8.8) becomes

$$Q = C_d\sqrt{2g}\ b\left[\left(H_2 + \frac{V_1^2}{2g}\right)^{3/2} - \left(H_1 + \frac{V_1^2}{2g}\right)^{3/2}\right] \tag{8.8a}$$

8.6.2 When Orifice is Large Circular

Figure 8.4 shows a large circular orifice of diameter d discharging under a constant head H.

Consider a small strip dx at a distance x from the centre of the orifice.

$$\text{Area of the strip} = \sqrt{\left(\frac{d}{2}\right)^2 - x^2} \times 2 = 2\sqrt{\left(\frac{d}{2}\right)^2 - x^2}$$

and velocity through the strip = $C_v\sqrt{2g(H-x)}$

$$\therefore \quad \text{Flow through the strip} = dQ = C_v\sqrt{2g}\,(H-x)^{1/2}\cdot 2\sqrt{\left(\frac{d}{2}\right)^2 - x^2}$$

For integration, it is necessary to expand $(H - x)^{1/2}$ by using the binomial theorem

i.e. $$(H - x)^{1/2} = H^{1/2} - \frac{H^{-1/2}x}{2} - \frac{H^{-3/2}x^2}{8} - \frac{H^{-5/2}x^3}{16} \cdots$$

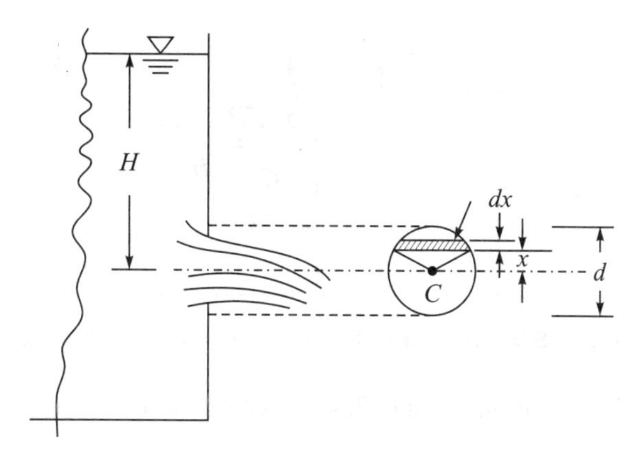

Figure 8.4 Discharge through large circular orifice.

$$\therefore\ dQ = 2C_v\sqrt{2g}\left[\left(\frac{d}{2}\right)^2 - x^2\right]^{1/2} H^{1/2} - \frac{\left\{\left(\frac{d}{2}\right)^2 - x^2\right\}^{1/2} x}{2H^{1/2}} - \frac{\left\{\left(\frac{d}{2}\right)^2 - x^2\right\} x^2}{8H^{3/2}} - \frac{\left\{\left(\frac{d}{2}\right)^2 - x^2\right\} x^3}{16H^{5/2}}$$

Integrating the above expression, we obtain:

$$Q = C_v\left(\frac{\pi d^2}{4}\right)\sqrt{2gH}\left[1 - \frac{d^2}{128\,\mathrm{H}^2} - \frac{5d^4}{1638\,\mathrm{H}^4} - \dots\right]$$

Introducing a coefficient of contraction C_c $\because$ Vena Contracta not considered.

$$Q = C_v \cdot C_c\left(\frac{\pi d^2}{4}\right)\sqrt{2gH}\left[1 - \frac{d^2}{128\,\mathrm{H}^2} - \frac{5d^4}{1638\,\mathrm{H}^4} - \dots\right]$$

$$\therefore \quad Q = C_d\left(\frac{\pi d^2}{4}\right)\sqrt{2gH}\left[1 - \frac{d^2}{128\,\mathrm{H}^2} - \frac{5d^4}{1638\,\mathrm{H}^4} - \dots\right] \tag{8.9}$$

which is the exact equation of discharge through a large circular orifice. Equation (8.9) shows that the term within the big bracket is smaller than unity. Hence the discharge through a large circular orifice is always less than the expression $Q = C_d\left(\frac{\pi}{4}d^2\right)\sqrt{2gH}$.

The exact value of C_d for the large orifice is not yet available, but a value of 0.6 may be used for approximate calculation for both large rectangular and circular orifice.

8.7 DISCHARGE THROUGH A FULLY SUBMERGED ORIFICE

Figure 8.5 shows a fully submerged orifice where the discharge Q flows under the head difference of $(H_1 - H_2) = H$ as shown in the figure.

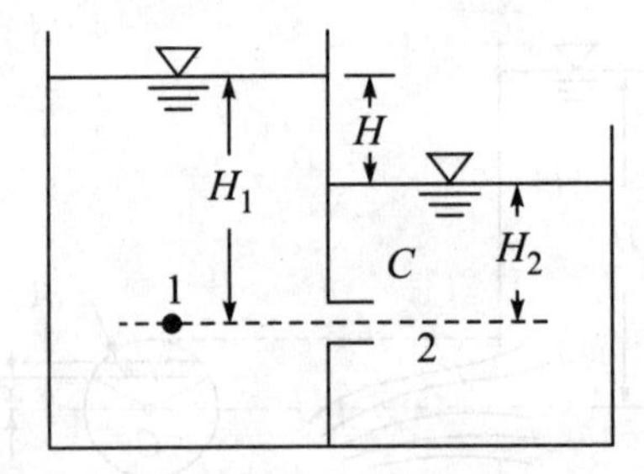

Figure 8.5 Discharge through fully submerged orifice.

Applying Bernoulli's equation at 1 and 2, i.e. at centre of the Vena Contracta, and neglecting losses:

$$Z_1 + \frac{p_1}{w} + \frac{V_1^2}{2g} = \frac{p_2}{w} + Z_2 + \frac{V_2^2}{2g}$$

Here $Z_1 = Z_2$, V_1 is small, $\frac{V_1^2}{2g}$ is neglected.

$$\therefore \quad \frac{V_2^2}{2g} = \frac{p_1}{w} - \frac{p_2}{w} = H_1 - H_2$$

$$\therefore \quad V_2 = \sqrt{2g(H_1 - H_2)} = \sqrt{2gH}$$

The above expression shows that Torricelli's formula is still applicable.

$$\therefore \quad Q_{\text{act}} = C_d\, a\sqrt{2g(H_1 - H_2)} = C_d\, a\sqrt{2gH} \tag{8.10}$$

8.8 DISCHARGE THROUGH A PARTIALLY SUBMERGED ORIFICE

Figure 8.6 shows the flow situation under a partially submerged orifice. This can happen if the size of the orifice is large. Here the upper part of the orifice discharges to the atmosphere, i.e. as a free orifice, and the lower part behaves as a submerged orifice. Let Q_1 and Q_2 be the discharges in the free and submerged portions, respectively. Total discharge:

$$Q = Q_1 + Q_2$$

$$Q_1 = \frac{2}{3}\, C_{d_1}\, b\sqrt{2g}\,(H^{3/2} - H_2^{3/2})$$

and

$$Q_2 = C_{d_2}\, b(H_1 - H_2)\,\sqrt{2gH}$$

where b is the breadth of the orifice.

$$\therefore \quad Q = \frac{2}{3}\, C_{d_1}\, b\sqrt{2g}\,(H^{3/2} - H_2^{3/2}) + C_{d_2}\, b(H_1 - H)\,\sqrt{2gH} \tag{8.11}$$

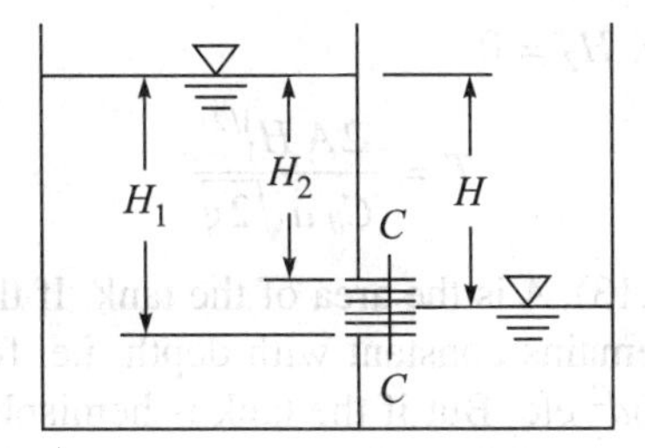

Figure 8.6 Flow through a partially submerged orifice.

8.9 TIME OF EMPTYING A TANK WITH NO INFLOW BY AN ORIFICE AT BOTTOM

Figure 8.7 shows a tank of area A with an orifice of area a at the bottom. The head of water in the tank initially at $t = 0$ was H_1. Find the time required to empty the tank up to level H_2. Let, at any instant, the head of water above the orifice be h. Let the water level fall through a distance dh in time dt seconds. Discharge in dt sec = $A\ dh$, where A is the area of tank, i.e.

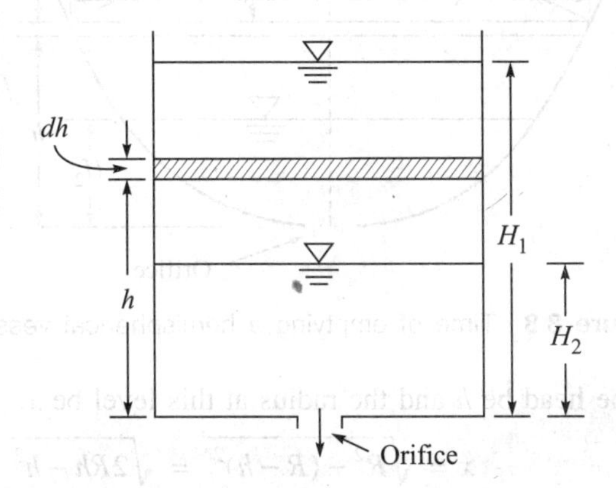

Figure 8.7 Time of emptying a tank with no inflow.

$$Adh = C_d\, a\sqrt{2gh}\, dt$$

$$\therefore \quad dt = \frac{-A}{C_d\, a\sqrt{2g}} h^{-1/2} dh$$

If T is the time required to empty the tank from level H_1 to H_2:

$$\int dt = T = \int_{H_2}^{H_1} \frac{A}{C_d\, a\sqrt{2g}} h^{-1/2} dh$$

$$T = \frac{2A}{C_d\, a\sqrt{2g}} (H_1^{1/2} - H_2^{1/2}) \tag{8.12}$$

If the tank is emptied fully, $H_2 = 0$

$$T = \frac{2A\,H_1^{1/2}}{C_d\,a\sqrt{2g}} \tag{8.13}$$

In Equations. (8.12) and (8.13), A is the area of the tank. If the tank is rectangular or square or circular in shape, the area remains constant with depth, i.e. for rectangular it will be $b \times d$, square b^2, and for circular $(\pi/4)d^2$ etc. But if the tank is hemispherical or conical or cylindrical in shape, with axis horizonal, area of flow A is not constant with the depth.

8.9.1 When the Area is Hemispherical

Figure 8.8 shows a hemispherical vessel of radius R to be emptied from the head H_1 to H_2 in T seconds through the orifice at the bottom.

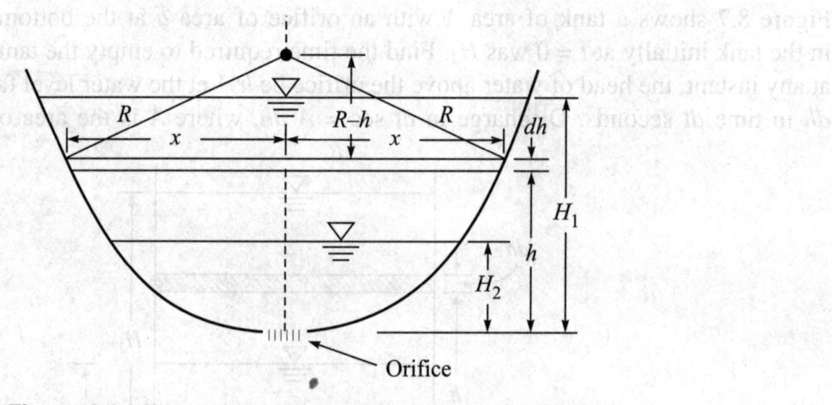

Figure 8.8 Time of emptying a hemispherical vessel.

At any instant, let the head be h and the radius at this level be x.

$$\therefore \quad x = \sqrt{R^2 - (R-h)^2} = \sqrt{2Rh - h^2}$$

$\therefore$ Area A at level $h = \pi x^2$

$$A = \pi\left[\sqrt{(2Rh - h^2)}\right]^2$$

$$A = \pi h(2R - h)$$

Quantity discharged in time $dt = A \cdot dh$

$$= \pi h(2R - h)\ dh \tag{A}$$

$$A = C_d\,a\sqrt{2gh}\ dt$$

$$\therefore \quad \pi h(2R - h)dh = C_d\,a\sqrt{2gh}\ dt$$

$$\therefore \quad dt = \frac{\pi}{C_d\,a\sqrt{2g}}\,h^{1/2}\ (2R - h)\ (dh)$$

If T is the time taken to empty the tank from level H_1 to H_2, integrating the above expression:

$$T = \frac{-\pi}{C_d\, a\sqrt{2g}}\left[\frac{4}{3}Rh^{3/2} - \frac{2}{5}h^{5/3}\right]_{H_2}^{H_1}$$

$$\therefore \quad T = \frac{2\pi}{C_d\, a\sqrt{2g}}\left[\frac{2}{3}R(H_1^{3/2} - H_2^{3/2}) - \frac{1}{5}(H_1^{5/2} - H_2^{5/2})\right] \tag{8.14}$$

If the tank is emptied completely, $H_2 = 0$

$$\therefore \quad T = \frac{2\pi}{C_d\, a\sqrt{2g}}\left[\frac{2}{3}RH_1^{3/2} - \frac{1}{5}H_1^{5/2}\right] \tag{8.15}$$

If tank was initially full and has to be emptied fully:

$$H_1 = R, \quad H_2 = 0$$

$$\therefore \quad T = \frac{2\pi}{C_d\, a\sqrt{2g}}\left[\frac{2}{3}R^{5/2} - \frac{1}{5}R^{5/2}\right]$$

$$\therefore \quad T = \frac{14\pi R^{5/2}}{15\, C_d\, a\sqrt{2g}} \tag{8.16}$$

8.9.2 When Vessel is Horizontal Cylindrical

Figure 8.9 shows a horizontal cylindrical vessel of length L and radius R with an orifice at the bottom. The tank has to be emptied from H_1 to H_2 in time t. Let at any instant surface be a height h from orifice and let in time dT surface falls through a height dh as shown in Figure 8.9(a).

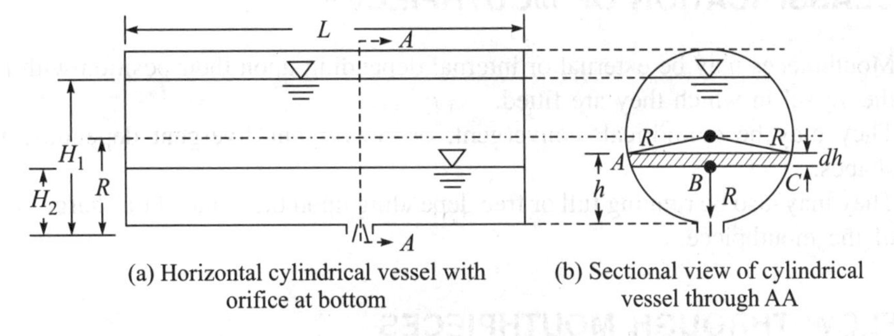

Figure 8.9 Time of emptying a horizontal cylindrical vessel.

Surface area at height $h = L \times AC = L\,(2AB) = L[2\{R^2 - (R - h)^2\}]$

$$\text{Surface area} = 2L\sqrt{2Rh - h^2}$$

$\therefore$ Volume of water drained out in time dt = Surface area

$$\text{at } A_c \times dh = 2L\sqrt{2Rh - h^2}\; dh \tag{A}$$

Also the volume of water drained out by the orifice in time dt when the level is at h

$$= C_d\, a\sqrt{2gh}\, dt \qquad \text{(B)}$$

Now (A) = (B) gives:

$$C_d\, a\sqrt{2g}\, h^{1/2}\, dt = 2L\sqrt{2Rh - h^2}\, dh$$

$$\therefore \quad dt = \frac{2L}{C_d\, a\sqrt{2g}}(\sqrt{2Rh - h^2})\, h^{-1/2}(dh)$$

Integrating between the limit of h from H_1 to H_2 and simplifying:

$$T = \frac{4L}{3C_d\, a\sqrt{2g}}[(2R - H_2)^{3/2} - (2R - H_1)^{3/2}] \qquad (8.17)$$

If emptied completely $H_2 = 0$

$$T = \frac{4L}{3C_d\, a\sqrt{2g}}[(2R)^{3/2} - (2R - H_1)^{3/2}] \qquad (8.18)$$

If initially the vessel was full and has been emptied, $H_1 = 2R$

$$\therefore \quad T = \frac{4L(2)^{3/2}\, R^{3/2}}{3C_d\, a\sqrt{2g}}$$

$$T = \frac{8\sqrt{2}\, LR^{3/2}}{3C_d a\sqrt{2g}} \qquad (8.19)$$

8.10 CLASSIFICATION OF MOUTHPIECES

(i) Mouthpieces may be external or internal depending upon their position with respect to the vessel in which they are fitted.

(ii) They may be cylindrical, convergent, or convergent-divergent depending upon the shapes.

(iii) They may also be running full or free depending upon the value of discharge at the outlet of the mouthpiece.

8.11 FLOW THROUGH MOUTHPIECES

8.11.1 When Mouthpiece is External and Cylindrical

An external cylindrical mouthpiece of length approximately 2 to 3 times of its diameter is fitted externally as shown in Figure 8.10. Its area is a and it is discharging under a constant head H. Water entering the mouthpiece forms a Vena Contracta at CC. After CC, the jet slowly fill the mouthpiece and the jet of water runs full equal to the diameter of the mouthpiece.

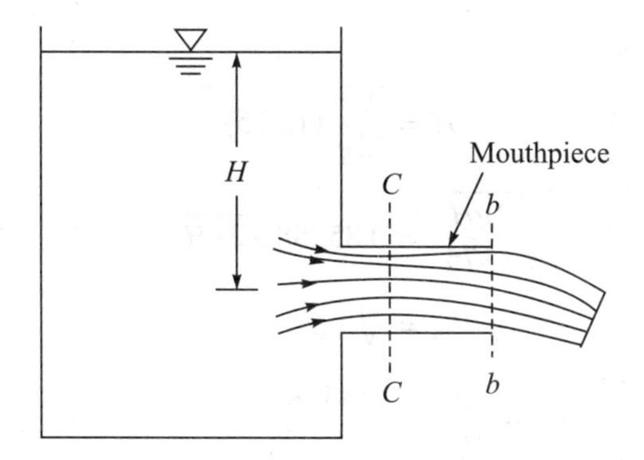

Figure 8.10 Flow through an external cylindrical mouthpiece.

If a, a_c and C_c are the area of the mouthpiece, area at CC and coefficient of contraction, respectively, then:

$$a_c = C_c a$$

Assuming $C_c \simeq 0.62$, $a_c = 0.62a$

If V_c and V are the velocity at Vena Contracta and velocity of the mouthpiece beyond section CC,

Continuity gives $a_c V_c = aV$

$$\therefore \quad V_c = \frac{a}{a_c} V = \frac{V}{0.62}$$

Applying Bernoulli's equation at tank water level at bb of the mouthpiece:

$$\frac{p_a}{w} + \frac{V_a^2}{2g} + Z_1 = \frac{p_b}{w} + \frac{V^2}{2g} + Z_b + h_L$$

$$\frac{p_a}{w} = 0 \because \text{atmospheric} = \frac{p_b}{w}$$

$$V_a = 0$$

$$\therefore \quad Z_1 - Z_b = \frac{V^2}{2g} + h_L$$

or

$$H = \frac{V^2}{2g} + h_L \qquad \text{(A)}$$

$\therefore$ h_L is the loss from CC to bb due to expansion

$$\therefore \quad h_L = \frac{(V_c - V)^2}{2g} = \left(\frac{V}{0.62} - V\right)^2 \Big/ 2g$$

$$h_L = 0.375 \frac{V^2}{2g}$$

Substituting h_L in (A):

$$H = \frac{V^2}{2g}(1.375)$$

$$\therefore \quad V = \sqrt{\frac{2gH}{1.375}} = 0.85288\sqrt{2gH}$$

Again

$$V_{th} = \sqrt{2gH}$$

$$\therefore \quad C_v = \frac{V}{V_{th}} = \frac{0.85288\sqrt{2gH}}{\sqrt{2gH}} = 0.85288$$

Since the cross-sectional area of the jet emerging from the mouthpiece is equal to the area of mouthpiece, hence C_c of the mouthpiece is 1.

$$\therefore \quad C_d = C_c \cdot C_d = 1 \times 0.85288 = 0.85288$$

However, the C_v for the mouthpiece, considering friction in mouth, becomes $\simeq$ 0.82

$$\therefore \quad C_d = 1 \times 0.82 = 0.82$$

$$\therefore \quad Q = C_d\, a\sqrt{2gH} = 0.82a\sqrt{2gH} \tag{8.20}$$

Equation (8.20) shows that C_d for mouthpiece (external) is more than the standard orifice of the same diameter, and discharge through mouthpiece is more than the orifice of the same diameter.

8.11.2 When Mouthpiece is Convergent–Divergent

In Figure 8.11, a convergent–divergent mouthpiece is used to discharge water under a head of *H*. In this type of mouthpiece, as there is no sudden enlargement of the jet, loss of energy due to sudden enlargement is eliminated. Coefficient of discharge for this mouthpiece is taken to be unity.

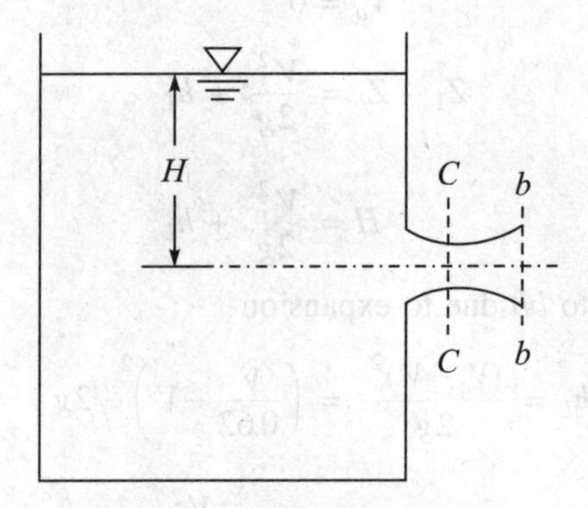

Figure 8.11 Flow through convergent and divergent mouthpiece.

Applying Bernoulli's equation to the free surface of the tank and at section CC, taking datum through the centre line of the mouthpiece:

$$\frac{p}{w} + \frac{V^2}{2g} + Z = \frac{p_c}{w} + \frac{V_c^2}{2g} + Z_c$$

$$Z_c = 0$$

$$Z = H$$

$$\frac{p}{w} = \text{atmospheric head} = H_a, \quad \frac{p_c}{w} = H_c, \quad V = 0$$

$$\therefore \quad H_a + 0 + H = H_c + \frac{V_c^2}{2g} + 0 \quad \text{or} \quad (H_a + H) = \left(H_c + \frac{V_c^2}{2g}\right) \quad \text{(A)}$$

$$\therefore \quad \frac{V_c^2}{2g} = (H_a + H - H_c)$$

$$\therefore \quad V_c = \sqrt{2g(H_a + H - H_c)}$$

Again applying Bernoulli's equation at CC and bb, taking the same datum line:

$$\frac{p_c}{w} + \frac{V_c^2}{2g} + z_c = \frac{p_2}{w} + \frac{V_b^2}{2g} + z_b$$

$$Z_c = Z_b$$

$$\frac{p_2}{w} = \frac{p}{w} = H_a$$

$$\therefore \quad \left(H_c + \frac{V_c^2}{2g}\right) = H_a + \frac{V_b^2}{2g}$$

From (A), Substitute $\left(H_c + \frac{V_c^2}{2g}\right)$:

$$H_a + H = H_a + \frac{V_b^2}{2g}$$

$$\therefore \quad V_b^2 = 2gH$$

$$\therefore \quad V_b = \sqrt{2gH}$$

Applying continuity $a_c V_c = a_b V_b$

or

$$\frac{a_b}{a_c} = \frac{V_c}{V_b} = \frac{\sqrt{2g(H_a + H - H_c)}}{\sqrt{2gH}} = \sqrt{\frac{H_a}{H} + 1 - \frac{H_c}{H}}$$

$$= \sqrt{1 + \frac{H_a - H_c}{H}} \quad (8.21)$$

Discharge Q is given by:

$$Q = a_c\sqrt{2gH} \tag{8.22}$$

Equation (8.21) is $\dfrac{a}{a_c} = \sqrt{1+\dfrac{H_a - H_c}{H}}$ $\quad\because a_b = a$(area of mouthpiece)

If the liquid is water, the limiting value of suction pressure at Vena Contracta $(H_a - H_c) = 7.8$, hence the maximum value of the ratio is:

$$\frac{a}{a_c} = \sqrt{1+\frac{7.8}{H}} \tag{8.23}$$

8.12 TIME OF FLOW FROM ONE VESSEL TO ANOTHER

In Figure 8.12, liquid from vessel A_1 flows to vessel A_2 through an orifice of area a. The initial head difference between A_1 and A_2 is H_1. The orifice is drowned. It is required to find time t taken to reduce the difference of the water level from H_1 to H_2.

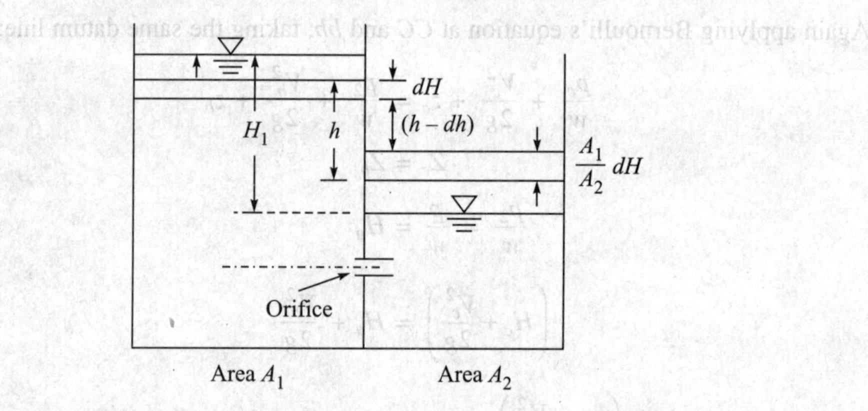

Figure 8.12 Flow of liquid from one vessel to another.

As shown in Figure 8.12, liquid from A_1 flow to A_2. At any instant, let the difference in the liquid level be h and let a small quantity Q flow through the orifice in time dt. This will cause the liquid surface to fall in A_1, by dH which, in turn will cause the liquid level in A_2 to rise by $\left(\dfrac{A_1}{A_2}dH\right)$.

$\therefore$ Difference of liquid level after time $dt = h - \left(dH + \dfrac{A_1}{A_2}dH\right)$

$$= h - dH\left(1+\frac{A_1}{A_2}\right)$$

Hence change in liquid level in time dt is $= dH\left(1+\frac{A_1}{A_2}\right)$

It is the change of liquid levels in time dt that equals the head that causes the flow. Let this head be dh

$$\therefore \qquad dh = dH\left(1+\frac{A_1}{A_2}\right) \quad \text{or} \quad dH = \frac{dh}{1+\frac{A_1}{A_2}}$$

Again if Q is the discharge through the orifice at this instant, then the total volume of liquid passing through the orifice in time dt is $Q\cdot dt$

But
$$Q = C_d a\sqrt{2gh}$$

$$\therefore \qquad Q\cdot dt = C_d a\sqrt{2gh}\, dt.$$

This Qdt or $C_c a\sqrt{2gh}\ dt$ will be equal to the volume emptied from A_1, equal to $A_1\, dH$. Equating these two values, we get:

$$-AdH = C_d a\sqrt{2gh}\, dt \quad (-v_e \text{ indicates that } dH \text{ decreases when } dt \text{ increases})$$

$$\therefore \qquad dt = -\frac{AdH}{C_d a\sqrt{2gh}} = -\frac{A_1 dH}{C_d a\left(1+\frac{A_1}{A}\right)\sqrt{2gh}}$$

Integrating:
$$\int_0^t dt = -\int_{H_1}^{H_2}\frac{A_1 dh}{C_d a\left(1+\frac{A_1}{A_2}\right)\sqrt{2gh}}$$

$$t = \frac{2A_1(H_1^{1/2}-H_2^{1/2})}{C_d\, a\left(1+\frac{A_1}{A_2}\right)\sqrt{2g}} \qquad (8.24)$$

If
$$A_1 = A_2 = A:$$

$$t = \frac{2A(H_1^{1/2}-H_2^{1/2})}{C_d\, a\cdot 2\cdot\sqrt{2g}} = \frac{A(H_1^{1/2}-H_2^{1/2})}{C_d\, a\sqrt{2g}} \qquad (8.25)$$

8.13 FLOW THROUGH INTERNAL OR BORDA'S MOUTHPIECES

In Figure 8.13, Borda's mouthpieces fitted internally are shown. In Figure 8.13(a), the length of the mouthpieces is about the size of the diameter of the pipe, hence the jet, after contraction, does not touch the sides and is running free. In Figure 8.13(b), the length of mouthpieces is about 2 to 3 times its diameter, so the jet, after contraction, expands and the jet flows out by touching the side of the mouthpiece and it is called running full.

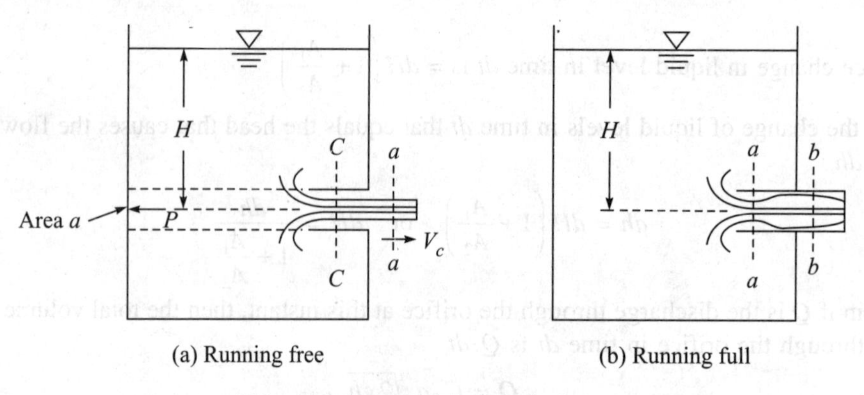

(a) Running free (b) Running full

Figure 8.13 Borda's mouthpieces.

Consider Figure 8.13(a), i.e. the jet running free. It forms a Vena Contracta. Let a be the area of the mouthpiece and a_c and V_c be the area and velocity of the emerging jet at CC respectively. Then mass of liquid flowing out of the mouthpiece = $\frac{w a_c V_c}{g}$ and the velocity of this liquid changes from 0 in the tank to V_c in the jet. Rate of change of momentum = $\frac{w a_c V_c}{g}(V_c - 0)$ = $\left(\frac{w a_c V_c}{g}\right) V_c$. If the mouthpiece were to be prolonged, it would have met the wall with an area of a where pressure force $P = wH \times a = waH$. This pressure causes the flow with a force, i.e. momentum $\left(\frac{w a_c V_c}{g}\right) V_c$

$$\therefore \qquad waH = \frac{w a_c V_c^2}{g}$$

$$\therefore \qquad \frac{a_c}{a} = \frac{gH}{V_c^2} \qquad \text{(A)}$$

Applying Bernoulli's equation at the free surface of the tank, at aa:

$$H_a + H = H_a + \frac{V_c^2}{2g}$$

$$\therefore \qquad V_c^2/2g = H \qquad \text{(B)}$$

Combining Equations (A) and (B),

$$\frac{a_c}{a} = \frac{g}{V_c^2} \cdot \frac{V_c^2}{2g} = \frac{1}{2}$$

$$\therefore \qquad a_c = 0.5\ a$$

i.e. coefficient of contraction of Borda's mouthpiece running free = 0.5.

However, some loss of energy is considered, then $V_c = 0.98\sqrt{2gH}$

or $$\frac{V_c^2}{2g} = 0.96 \text{ H}$$

Then $$C = \frac{0.5}{.96} = 0.52$$

C_v of Borda's mouthpiece running free = 1 and C_c = 0.5 or 0.52

$\therefore$ $$C_d = 0.5 \text{ (or } 0.52)$$

When running full, i.e. Figure 8.13(b), where after forming Vena Contracta, the coefficient of contraction at the outlet is 1, since the jet runs full equal to diameter of mouthpiece i.e. no contraction. At the entrance, the portion is just like running free, hence C_c here = 0.5.

Continuity gives:

$$a_c V_c = aV$$

$$V_c = \frac{a}{a_c} V = \frac{V}{0.5} \tag{i}$$

Consider the Vena Contracta at *aa* and *bb* just outside the mouthpiece and applying Bernoulli's equation at free surface and at *bb*:

$H_a + H = H_a + \dfrac{V^2}{2g} + h_L$, where h_L is the loss between *aa* and *bb*. Between *aa* and *bb*, the flow pattern is similar to sudden expansion. Hence $h_L = \dfrac{(V_c - V)^2}{2g} = \dfrac{\left(\dfrac{V}{0.5} - V\right)^2}{2g} = \dfrac{V^2}{2g}$

Substituting the value h_L:

$$H_a + H = H_a + \frac{V^2}{2g} + \frac{V^2}{2g}$$

or $$\frac{V^2}{g} = H$$

$\therefore$ $$V = \sqrt{gH} \tag{ii}$$

But theoretical $V_{th} = \sqrt{2gH}$

$\therefore$ $$C_v = \frac{V}{V_{th}} = \frac{\sqrt{gH}}{\sqrt{2gH}} = \frac{1}{\sqrt{2}} = 0.707$$

Since C_c at the outlet is equal to 1:

$$C_d = C_v \cdot C_c = 1 \times 0.707$$

$$C_d = 0.707$$

But in actual practice, Borda's mouthpiece running full is:

$$C_d = 0.75$$

This may be due to the fact that the loss of energy is slightly less than sudden expansion.

In order to obtain pressure at the Vena Contracta, apply Bernoulli's equation at free surface and at Vena Contracta:

$$H_a + H = H_c + \frac{V_c^2}{2g}$$

or

$$H_c = H_a + H - \frac{V_c^2}{2g}$$

But $V_c = 2V$ and $V^2 = gH$ by (i) and (ii):

$$H_c = H_a + \frac{V^2}{g} - \frac{(2V)^2}{2g} = H_a + \frac{V^2}{g} - \frac{4V^2}{2g} = H_a - \frac{V^2}{2g} = H_a - H$$

$$\therefore \quad H_e = H_a - H$$

which shows that at Vena Contracta, pressure is less than atmosphere. If the absolute pressure at Vena Contracta is 2.5 m of water, maximum value of H upto which the flow through this mouthpiece has to take place:

$$2.5 = 10.3 - H$$

$$\therefore \quad H = 7.8 \text{ m}$$

EXAMPLE 8.6 Find the discharge through a rectangular orifice 3 m wide and 2 m deep fitted to a water tank. The water level in the tank is 4 m above the top level of the orifice. Take $C_d = 0.62$.

Solution: Here $H_1 = 4$ m

$H_2 = (H_1 + 2) \text{ m} = (4 + 2) \text{ m} = 6 \text{ m}$

$C_d = 0.62$

This is the case of a large orifice.

$\therefore$ Discharge $Q = C_d\, b\sqrt{2g}\,(H_2^{3/2} - H_1^{3/2})$

(without considering velocity of approach) $Q = [0.62 \times 2\sqrt{2 \times 9.81}\,(6^{3/2} - 4^{3/2})] \text{ m}^3/\text{sec}$

$Q = 36.783 \text{ m}^3/\text{sec}$ **Ans.**

EXAMPLE 8.7 Find the discharge through a fully submerged orifice of width 2 m if the difference of water on both sides of the orifice is 0.8 m. The heights from the top and bottom of the orifice are 2.5 m and 3 m, respectively. Take $C_d = 0.6$.

Solution: $b = 2$ m, $H = 0.8$ m, $H_1 = 2.5$ m, $H_2 = 3$ m, $C_d = 0.6$

$\therefore$ Q (through a fully submerged orifice) $= C_d \times b \times (H_2 - H_1)\sqrt{2gH}$

$Q = (0.62 \times 2 \times .5 \times \sqrt{2 \times 9.81 \times 0.8}) \text{ m}^3/\text{sec}$

$= 2.4563 \text{ m}^3/\text{sec}$ **Ans.**

EXAMPLE 8.8 A rectangular orifice 1.5 m wide and 1.2 m deep is fitted in one side of a large tank. The water level on one side of the orifice is 2 m above the top edge of the orifice, while on the other side of the orifice, the water level is 0.4 m below the top edge. Calculate the discharge through the orifice if $C_d = 0.62$.

Solution: This is a partially submerged orifice. Width of orifice $b = 1.5$ m, H_1 = height of water level from bottom of the orifice

$$H = 2 + 1.2 = 3.2 \text{ m}$$

$$H_2 = \text{Height of water level above the top edge} = 2 \text{ m}$$

$$H = \text{Difference of water level on both sides.}$$

$$= (2 + .4) = 2.4 \text{ m}$$

$$C_{d_1} = C_{d_2} = C_d = 0.62$$

$$\therefore \quad Q = Q_1 + Q_2 = \frac{2}{3} C_a\, b\sqrt{2g}\, (H^{3/2} - H_2^{3/2}) + C_d\, b(H_1 - H)\, \sqrt{2gH}$$

$$= \frac{2}{3} \times 0.62 \times 1.5\, \sqrt{2 \times 9.81}\, (2.4^{3/2} - 2^{3/2}) + 0.62 \times 1.5\, (3.2 - 2.4)\, \sqrt{2 \times 9.81 \times 2.4}$$

$$= (2.44317 + 5.10537) \text{ m}^3\text{/sec}$$

$$= 7.5485 \text{ m}^3\text{/sec}$$

Ans.

EXAMPLE 8.9 A circular tank of diameter 3 m contains water upto a height of 4 m. An orifice of 0.4 m diameter is provided at the bottom to empty the tank. Find the time: (i) to empty the tank from 4 m to 2 m. (ii) for completely emptying the tank. Take $C_d = 0.6$.

Solution: $A = \frac{\pi}{4}(3)^2 = 7.06856 \text{ m}^2$

$$a = \frac{\pi}{4}(0.4)^2 = 0.125663 \text{ m}^2$$

Time to empty the tank from 4 m to 2 m

$$T = \frac{2A}{C_d\, a\sqrt{2g}}(\sqrt{H_1} - \sqrt{H_2}) = \frac{2 \times 7.06858}{0.6 \times 0.125663 \times \sqrt{2 \times 9.81}}(4^{1/2} - 2^{1/2})$$

$$T = 24.7966 \text{ secs.}$$

Time for emptying the tank completely, $H_2 = 0$

$$T = \frac{2A\sqrt{H_1}}{C_d\, a\sqrt{2g}} = \frac{2 \times 7.06858 \times 4^{1/2}}{0.62 \times 0.125663 \times \sqrt{2 \times 9.81}}$$

$$T = 81.93 \text{ sec}$$

Ans.

EXAMPLE 8.10 A hemispherical tank of diameter 4 m contains water upto a height of 1.5 m. An orifice of diameter 0.05 m is provided at the bottom. Find the time required by the water: (i) to fall from 1.5 m to 1 m, (ii) for completely emptying the tank.

Solution: $D = 4$ m, $\therefore R = 2$ m, $d = 0.05$ m $\therefore a = \frac{\pi}{4}(0.05)^2 = 0.001963$ m^2

$H_1 = 1.5$ m, $H_2 = 1.0$, $C_d = 0.6$

(i) $\therefore$ From Equation (8.14), $T = \frac{2\pi}{C_d\, a\sqrt{2g}}\left[\frac{2}{3}R\,(H_1^{3/2} - H_2^{3/2}) - \frac{1}{5}(H_1^{5/2} - H_2^{5/2})\right]$

$$T = \frac{2\pi}{0.6 \times 0.001963 \times \sqrt{2 \times 9.81}}\left[\frac{2}{3} \times 2\,(1.5^{3/2} - 1^{3/2}) - \frac{1}{5}(1.5^{5/2} - 1^{5/2})\right]$$

$T = 921.4$ sec $=$ 15 minutes 21.4 sec **Ans.**

(ii) When fully emptied $H_2 = 0$

$$T = \frac{2\pi}{C_d\, a\sqrt{2g}}\left[\frac{2}{3}R\,H_1^{3/2} - \frac{1}{5}H_1^{3/2}\right]$$

$$T = \frac{2\pi}{0.6 \times 0.001963 \times \sqrt{2 \times 9.81}}\left(\frac{2}{3} \times 2 \times 1.5^{3/2} - \frac{1}{5} \times 1.5^{5/2}\right)$$

$T = 2286.33$ sec $=$ 38 minute 6.33 secs **Ans.**

8.14 CONCLUSION

Orifices and mouthpieces have lot of applications like flow of oil from oil tankers, flow of water from big tanks used for drip and sprinkler irrigation, flow from water supply tank etc. Therefore, velocities, discharges through orifices under different conditions, coefficient of contraction, velocity and discharge, time of emptying vessel through an orifice etc. have been presented in details. Numerical examples are solved to show the applications of theoretical equations. Few problems are given to students to solve themselves and a list of reference is included.

PROBLEMS

8.1 Find the discharge through an orifice of 2 cm diameter under a head of 1.5 m. Take $C_d = 0.6$. **(Ans.** 1.024 litre/sec)

8.2 An external mouthpiece is fitted to the side of a large vessel. The head over the mouthpiece is 4 m. Find the discharge of the mouthpiece if its diameter is 10 cm. Assume $C_d \sim 0.855$

(Ans. 0.0595 m^3/sec)

8.3 The co-ordinate of a point of a jet issuing through an orifice is (48.5 cm, 4.9 cm) measured horizontally and vertically from centre of the Vena Contracta, i.e. the horizontal and vertical distance of the point in the jet. The head of water in the tank is 1.5 m. Find the coefficient of velocity. **(Ans.** 0.982).

8.4 A rectangular orifice 1 m wide and 1.5 m deep is discharging water from a vessel. The top edge of the orifice is 0.8 m below the water surface in the vessel. Calculate the discharge through the orifice if $C_d = 0.6$. Also calculate the p.e. error if the orifice is treated as a small orifice. (**Ans.** 1.058%)

8.5 A circular tank of diameter 1.5 m contains water upto a height of 4 m. An orifice of 40 mm diameter is provided at its bottom. If $C_d = 0.62$, find the height of water above the orifice after 10 seconds (cm).

8.6 A hemispherical tank of diameter 4 m contains water upto a height of 2 m. An orifice of diameter 50 mm is provided at the bottom. Find the time required by the water: (i) to fall from 2 m to 1 m, (ii) for completely emptying

(**Ans.** (i) 30 min 14.34 secs (ii) 53 minutes).

8.7 A convergent–divergent mouthpiece having, throat diameter 60 mm is discharging water under constant head of 3 m. Determine the maximum outlet diameter for maximum discharge. Find maximum discharging also. Take atmospheric pressure head = 10.3 m of water and separation pressure 2.5 m of water absolute. Q_{max} = 0.02169 m^3/sec d = 8.2647 cm.

8.8 An internal mouthpiece of 100 mm diameter is discharging water under a constant head of 5 m. Find the discharge through the mouthpiece, when: (i) the mouthpiece is running free, (ii) the mouthpiece is running full. (**Ans.** (i) 38.8 lits/sec (ii) 54.86 lits/sec)

REFERENCES

1. Bansal, R.K., *Fluid Mechanics and Hydraulic Machines*, Laxmi Publications, New Delhi, 1983.
2. Modi, P.N. and S.M. Seth, *Hydraulics and Fluid Mechanics including Hydraulic Machines*, Standard Book House, New Delhi, 1977.
3. Rouse, H., *Fluid Mechanics for Hydraulic Engineers*, McGraw-Hill Inc., 1938.
4. Rouse, H., *Elementary Mechanics of Fluid*, John Wiley & Sons. Inc., New York, 1946.
5. Streeter, V.L. and E.B. Waylie, *Fluid Mechanics*, McGraw-Hill, New York, 1983.
6. Munson, B.R., D.F. Young, and T.H. Okiishi, *Fundamental of Fluid Mechanics*, Wiley Text Books, 2001.

Chapter 9

Flow Over Notches and Weirs

9.1 INTRODUCTION

A notch is an opening provided in the side of a tank or a vessel such that the liquid surface in the tank is below the top edge of the opening. It is made of a metallic plate and is usually provided in a narrow channel, i.e. a laboratory flume in order to measure the discharge. Thus a notch is used for measuring discharges.

A weir is a concrete or masonry structure built across a river, stream or canal. A weir is used to: (i) raise the water level upstream, (ii) release excess water to the flow over the structure, and (iii) measure the discharge of the stream.

The sheet of water flowing through a notch or over the weir is called 'nappe' or 'vein'. The bottom edge of a notch or weir is called 'crest' or 'sill' and its height above this crest is called 'crest height'.

9.2 CLASSIFICATION OF NOTCHES AND WEIRS

Notches are classified according to their shape as rectangular, triangular, trapezoidal and stepped notches, and according to the effect of the sides on the nappe as notches with end contractions and notches without end contractions.

Weirs are also classified on the basis of the shape of opening, the crest, the effect of the sides on the nappe and the nature of discharge, as rectangular, triangular, trapezoidal, sharp-crested, broad-crested, narrow-crested, ogee-shaped weirs, weirs with end contractions, weir without end contractions, submerged and free weirs.

Thus the discharge equation over the notch or weir of same type is the same.

9.3 DISCHARGE OVER RECTANGULAR NOTCH OR WEIR

Let L be the length of crest of weir or notch

H is the height water upstream causing the flow

To compute discharge Q over the weir or notch, consider a small strip dh at a depth h from top of the water level as shown in Figure 9.1(a). If dQ is the discharge through this elemental strip,

$$dQ = C_d(Ldh)\sqrt{2gh} \quad \because dA = Ldh,\ V = \sqrt{2gh} \text{ and } C_d = \text{coefficient of discharge.}$$

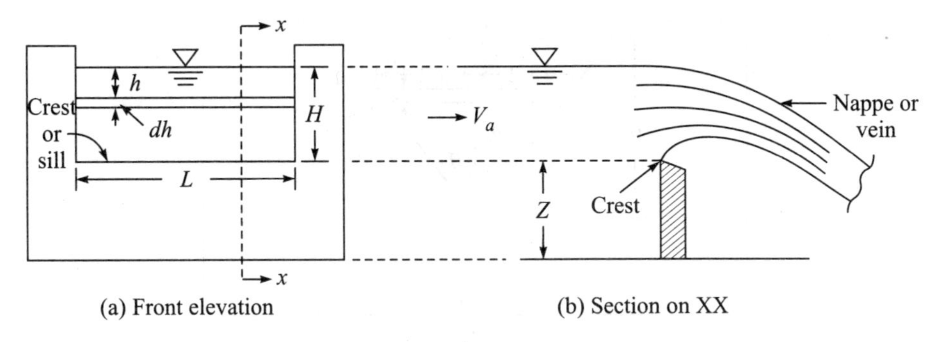

Figure 9.1 Flow over rectangular sharp-crested weir or notch.

The total discharge is

$$\int dQ = CdL\sqrt{2g}\int_0^H h^{1/2} dh$$

$$\therefore \qquad Q = \frac{2}{3} C_d \sqrt{2g}\, LH^{3/2} \tag{9.1}$$

If the velocity of the approach is V_a (the small velocity upstream as shown in Figure 9.1(b) approaching the weir or notch, then the integration limit of the head will be from $\frac{V_a^2}{2g}$ to $\left(H + \frac{V_a^2}{2g}\right)$

then

$$Q = \frac{2}{3} C_d \sqrt{2g}\, L\left[\left(H + \frac{V_a^2}{2g}\right)^{3/2} - \left(\frac{V_a^2}{2g}\right)^{3/2}\right] \tag{9.2}$$

usually V_a is small and $\frac{V_a^2}{2g}$ is much smaller and hence normally, the approach velocity head $\left(\frac{V_a^2}{2g}\right)$ is neglected.

9.4 DISCHARGE OVER TRIANGULAR NOTCH OR WEIR

Let the weir or notch make an angle θ at the vertex as shown in Figure 9.2. Consider a small strip dh at a depth h from the top of the water level. If x is the width of the water at depth h, then

$$\frac{x}{2}\Big/(H - h) = \tan \theta/2$$

$$\therefore \qquad \frac{x}{2(H-h)} = \tan \theta/2$$

or

$$x = 2(H - h) \tan \theta/2$$

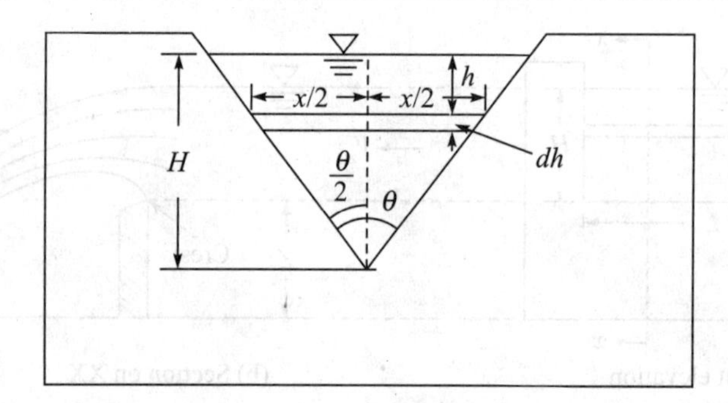

Figure 9.2 Flow over triangular notch or weir.

Area of the strip = $xdh = 2(H - h) \tan \theta/2 \, dh$

Velocity in the strip = $\sqrt{2gh}$

$$\therefore \quad dQ = C_d \, 2(H - h) \tan (\theta/2) \sqrt{2gh}$$

Integrating $Q = 2C_d \tan (\theta/2) \sqrt{2g} \int_0^H (H - h) h^{1/2} \, dh$

Simplifying:

$$Q = \frac{8}{15} C_d \sqrt{2g} \tan (\theta/2) H^{5/2} \tag{9.3}$$

If the vertex angle is 90°, $\theta/2 = 45°$ $\therefore$ $\tan 45° = 1$ $\therefore$ $Q = \frac{8}{15} C_d \sqrt{2g} H^{5/2}$ (9.4)

If the velocity of approach V_a is considered:

$$Q = \frac{8}{15} C_d \sqrt{2g} \tan (\theta/2) \left[\left(H + \frac{V_a^2}{2g} \right)^{5/2} - \left(\frac{V_a^2}{2g} \right)^{5/2} \right] \tag{9.5}$$

9.5 DISCHARGE OVER TRAPEZOIDAL NOTCH OR WEIR

Figure 9.3 shows the trapezoidal notch or weir which is a combination of a rectangular weir or notch with width L and a triangular notch or weir with vertex angle θ. Hence, the discharge over such a notch or weir may be obtained by adding discharges of two different types.

$\therefore$ Q = Flow of rectangular part + Flow of triangular part

$$\therefore \quad Q = \frac{2}{3} Cd_1 L \sqrt{2g} \, H^{3/2} + \frac{8}{15} Cd_2 \sqrt{2g} \tan (\theta/2) H^{5/2} \tag{9.6}$$

If $Cd_1 = Cd_2$ (assumed)

$$Q = C_d \sqrt{2g} \, H^{3/2} \left[\frac{2}{3} L + \frac{8}{15} \tan (\theta/2) H \right] \tag{9.7}$$

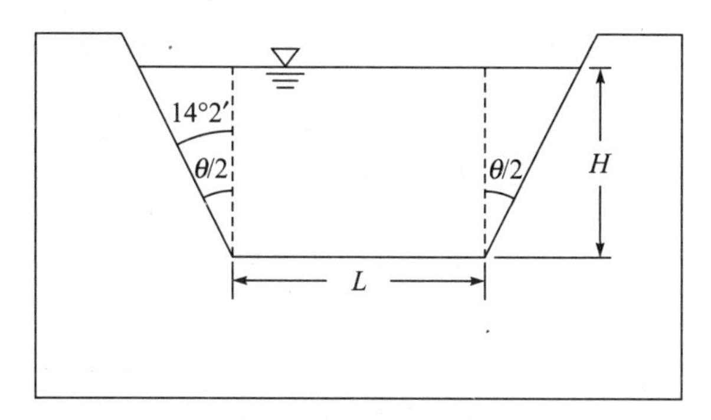

Figure 9.3 Flow over trapezoidal notch or weir.

In 1887, an Italian engineer, Cipolletti adopted the value of $\theta/2 = 14°2'$ which gives $\tan\left(\frac{\theta}{2}\right) = \tan 14°2' = \frac{1}{4}$. Thus, if a trapezoidal weir or notch $\frac{\theta}{2} = 14°2'$, it is called a Cipolletti weir. If $\frac{\theta}{2} = 14°$, the increase in the flow in the triangular portion is equal to the decrease in the rectangular equation due to two end contractions. Thus, $Q = Q_R + Q_T = \frac{2}{3} C_d \sqrt{2g}\,(L - 0.2H)\, H^{3/2} + \frac{8}{15} C_d \sqrt{2g}\, H^{5/2} \tan 14°$

where $(L - 0.2H)$ is the effective length due to two end contractions

$$Q = \frac{2}{3} C_d \sqrt{2g}\,(L - 0.2H)\, H^{3/2} + \frac{8}{15} C_d \sqrt{2g}\; H^{5/2} \times \frac{1}{4}$$

$$Q = \frac{2}{3} C_d \sqrt{2g}\; H^{3/2} \left[L - \frac{H}{5} + \frac{H}{5}\right]$$

$$Q = \frac{2}{3} C_d \sqrt{2g}\, LH^{3/2} \tag{9.8}$$

which is equivalent the Equation (9.1) for discharges in rectangular weir or notch. On the basis of his own experiments, he proposed that the term $\left(\frac{2}{3} C_d \sqrt{2g}\right)$ is equal to 1.866 when $C_d = 0.632$.

Hence he proposed that $$Q = 1.866\, LH^{3/2} \tag{9.9}$$

and his $C_d = 0.632$.

9.6 FLOW OVER BROAD-CRESTED WEIR

Most commonly, when the crest is broad, it is called a broad-crested weir. But according to recent mathematical definition if $0.1 \le \frac{H}{B} \le 0.35$, it is called broad-crested.

Applying Bernoulli's equation at 1.1 and at 2.2 in Figure 9.4 (neglecting velocity of approach) and taking crest as the datum, we get:

$$H = h + \frac{V^2}{2g}$$

$$\therefore \quad V = \sqrt{2g(H-h)}$$

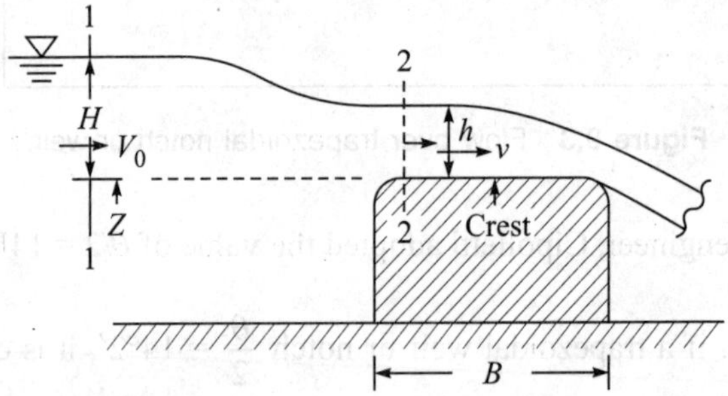

Figure 9.4 Flow over broad-crested weir.

If L is the length of the crest,

$$Q_{th} = (Lh)\sqrt{2g(H-h)}$$ where h is the depth of water over the crest.

$$Q = C_d\, Lh\sqrt{2g(H-h)} \tag{9.10}$$

In order to determine the discharge, only two heads H and h are to be measured. For Q to be maximum, a definite relationship between H and h may be obtained. Differentiating Q with respect to h and equating to zero, we get:

$$\frac{dQ}{dh} = C_d L\sqrt{2g}\left[\sqrt{H-h} - \frac{h}{2\sqrt{H-h}}\right] = 0$$

$$\therefore \quad \sqrt{H-h} - \frac{h}{2\sqrt{H-h}} = 0$$

or

$$\frac{2(H-h)-h}{2\sqrt{H-h}} = 0$$

$$\therefore \quad 2H - 2h - h = 0$$

$$2H = 3h$$

$$\therefore \quad h = \frac{2}{3}H \tag{9.11}$$

which is the condition for maximum Q over the weir.

Later it will be shown that h is the critical depth of the flow (For reference to critical depth, see Chapter 10).

9.7 FLOW OVER SUBMERGED WEIR

Figure 9.5 shows the submerged weir in which the crest of the weir is submerged by a depth H_2 and the level of the water upstream above the weir is H_1. This submerged weir may be divided into two parts. The portion between the head difference $(H_1 - H_2)$ is treated as a free weir while the portion with head H_2 may be taken as being submerged. If Q_1 and Q_2 are the discharges for these two parts,

$$Q = Q_1 + Q_2 = \frac{2}{3} Cd_1 L \sqrt{2g}\ (H_1 - H_2)^{3/2} + Cd_2\ (L \times H_2)\ \sqrt{2g(H_1 - H_2)} \qquad (9.12)$$

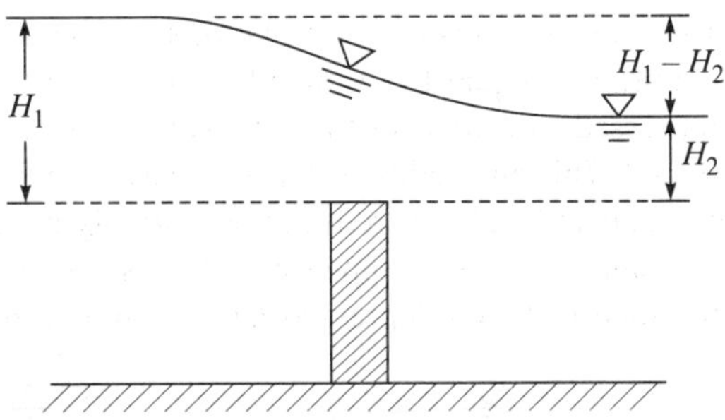

Figure 9.5 Flow over submerged weir.

9.8 FRANCIS FORMULA: END CONTRACTIONS

J.B. Francis, from his experiments, proposed that the end walls of the weir shown in Figure 9.6 influence the nappe to contract which is called 'effect on length' by end contractions. He found from his experiment that this contraction in each end wall is equal to $\frac{1}{10}$ times of H or equal 0.1 H. If there are two walls, effective $L_{eff.}$ = (L – 0.2 H) and if n ends, L_{eff} = (L – 0.1 nH)

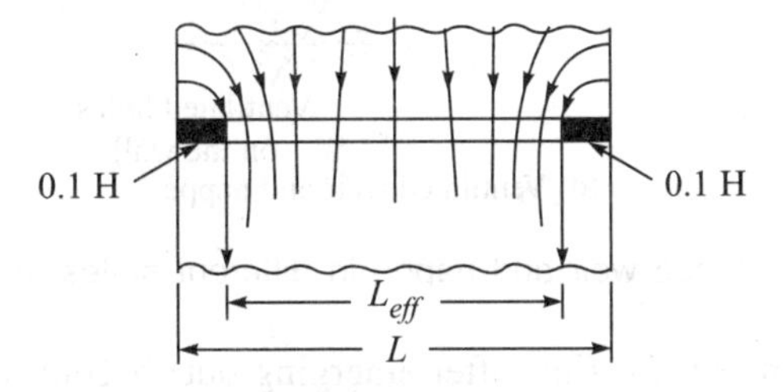

Figure 9.6 End contraction.

Thus the discharge equation becomes:

$$Q = \frac{2}{3} C_d \sqrt{2g}\ (L - 0.1\ \text{nH})H^{3/2} \qquad (9.13)$$

If C_d = 0.623, Q = 1.84 (L – 0.1 nH) $H^{3/2}$ (9.13a)

If the velocity of approach is taken into account:

$$Q = \frac{2}{3} C_d \sqrt{2g}\,(L - 0.1\,\text{nH}_1)\left[H_1^{3/2} - \left(\frac{V_a^2}{2g}\right)^{3/2}\right]$$

where $$H_1 = \left(H + \frac{V_a^2}{2g}\right) \tag{9.14}$$

9.9 VENTILATION OF WEIRS

The suppressed weir and its nappe in states with ventilation have been shown in Figure 9.7. The crest length of the weir is normally equal to the width of the canal or channel and the nappe emerging out from this suppressed weir touches the side walls of the channel. The air below the nappe and the walls are trapped. This air gradually carried away with flowing water and pressure in this shape (i.e. between nappe and walls) is reduced. Eventually negative pressure (i.e. below the atmosphere) is developed and as a result nappe starts depressing i.e. coming near the outside wall of the weir and such nappe is called depressed nappe (shown in Figure 9.7(b))

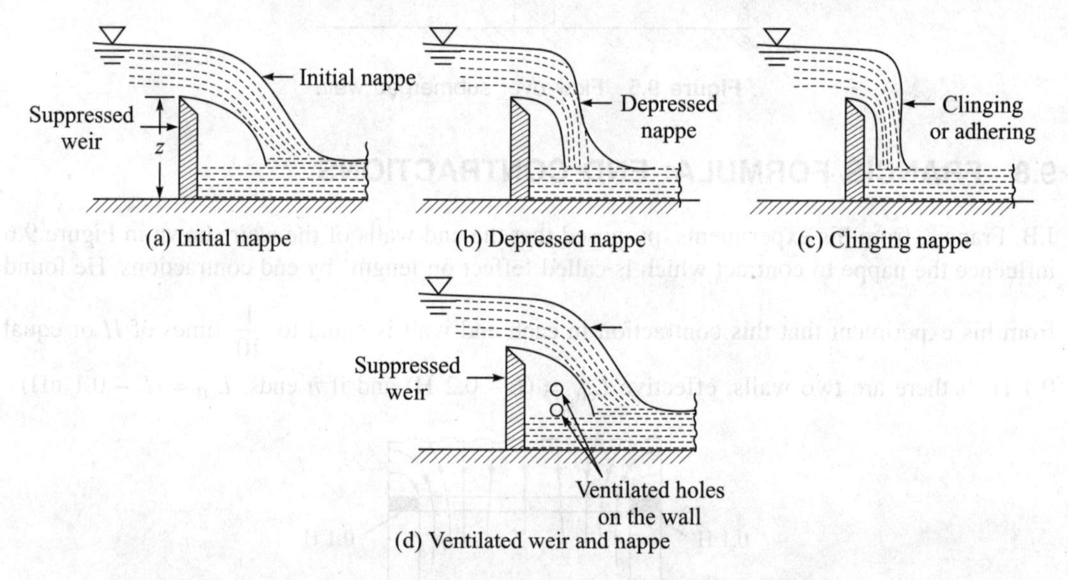

Figure 9.7 Suppressed weir and nappe in different states and with ventilation.

The initial nappe (Figure 9.7a) (just after emerging out) becomes a depressed nappe where pressure falls in the area below the nappe and the walls. Discharge increases due to low pressure in the downstream side. With further withdrawal of air from this space, i.e. when no air is left in this space, nappe addresses the downstream surface of the wall as shown in Figure 9.7(c). This nappe is called clinging or adhering nappe which further increases the discharge more than the actual discharge. The analytical equations derived for suppressed weir are based on the assumption that pressure below the nappe and the wall always remains atmospheric. Maintain the state of atmospheric pressure, ventilation holes are made on the wall as shown in

Figure 9.7(d). These holes are called ventilation holes and the weir is called ventilated weir. Experiments show that ventilated area may be about 0.5% of $(L \times Z)$, where L is the crest length, and Z is the depth of the weir.

9.10 SUTRO WEIR OR PROPORTIONAL WEIR

The discharge equation of a weir in general is expressed as $Q \propto H^n$, where the n value is equal to $\frac{3}{2}$ for rectangular type and equal to $\frac{5}{2}$ for triangular type of weir. Proportional or sutro weir means that Q varies linearly with H, i.e. $Q \propto H$. It is called Sutro weir who first developed this weir where Q varies linearly with H. Figure 9.8 shows this proportional or Sutro weir.

Sutro analytically obtained the relationship for the shape of this weir profile having $Q \propto H$ as

$$x \propto y^{-1/2} \tag{A}$$

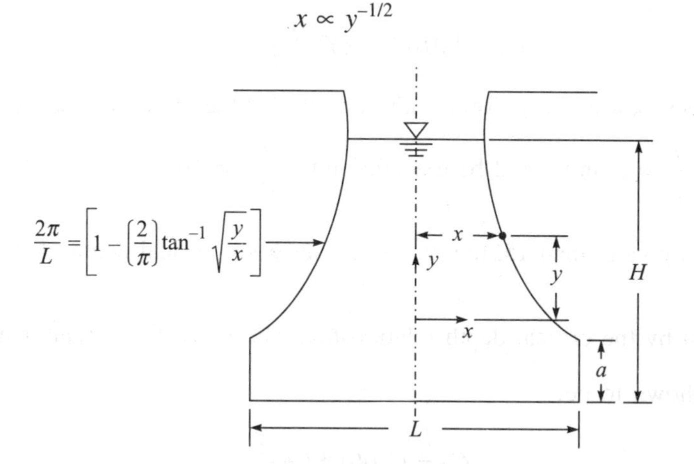

Figure 9.8 Sutro or proportional weir.

The above relationship (A) shows that as $y \to 0$, $x \to \infty$, which means that the width of the weir aperture becomes infinite at the crest. In order to overcome this practical limitation, Sutro modified the weir profile to obtain a finite width at the crest. This weir has been shown in Figure 9.8.

He gave the hyperbolic curve of weir profile having the following equation:

$$\frac{2x}{L} = \left[1 - \frac{2}{\pi}\tan^{-1}\sqrt{\left(\frac{y}{a}\right)}\right] \tag{9.15}$$

where a and L are the height and width of the small rectangular aperture forming the base of the weir.

The discharge equation given by this weir is:

$$Q = C_d\, L\, (2ga)^{1/2}\left(H - \frac{a}{3}\right) \tag{9.16}$$

C_d of this weir varies from 0.6 to 0.65.

This proportional weir is very useful in chemical dosing and sampling.

9.11 BAZIN'S FORMULA AND REHBOCK'S FORMULA

In 1886, H.E. Bazin in France gave the following formula of discharge based on his experiment:

$$Q = \left(0.405 + \frac{0.003}{H_1}\right) \sqrt{2g}\, LH_1^{3/2} \tag{9.17}$$

where $H_1 = \left(H + c\frac{Vd}{2g}\right)$, c is a constant, and the mean value given by him is 1.6.

Similarly T. Rehbocks (1929) on the basis of the suppressed rectangular weir, proposed the following empirical formula:

$$Q = \frac{2}{3}\left(0.611 + 0.08\frac{H}{z}\right) \sqrt{2g}\, LH^{3/2} \tag{9.18}$$

where
$$C_d = \left(0.611 + 0.08\frac{H}{z}\right) \tag{9.19}$$

Hunter Rouse indicated that Rehbock's formula for the weir may be used with reasonable accuracy upto $\frac{H}{z} = 5$ and could be extended upto $\frac{H}{z} = 10$.

When $\frac{H}{z}$ is greater than 15, the weir becomes a sill critical depth of flow approximately equal to $(z + H)$ by the critical depth relationship. The value C_d in Equation $Q = \frac{2}{3} C_d \sqrt{2g}$ $LH^{3/2}$, can be shown to be:

$$C_d = (1.06)\left(1 + \frac{z}{H}\right)^{1.5} \tag{9.20}$$

9.12 FLOW OVER STEPPED NOTCH

A stepped notch is a combination of rectangular notches as shown in Figure 9.9.

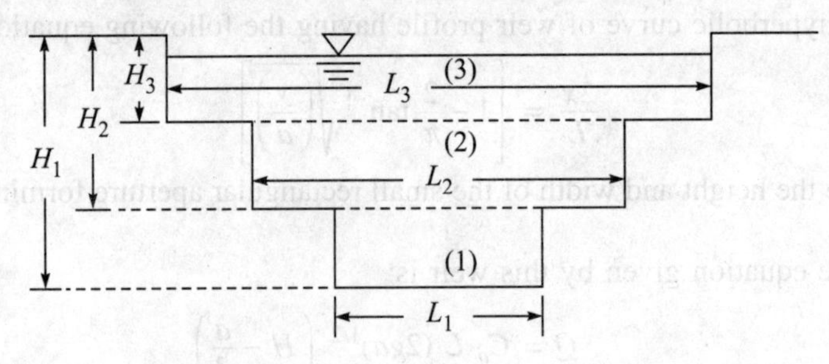

Figure 9.9 Flow over stepped notch.

Let L_1 be the crest length or notch 1. Similarly L_2 and L_3 are the crest lengths of notches 2 and 3. H_1, H_2, H_3 are heads of the water in notch (1), (2) and (3).

If Q_1, Q_2, Q_3 are the discharges of the three notches and C_d is the same coefficient of discharge for these notches:

$$Q = Q_1 + Q_2 + Q_3 = \frac{2}{3} C_d L_1 \sqrt{2g}\, [H_1^{3/2} - H_2^{3/2}] + \frac{2}{3} C_d L_2 \sqrt{2g}\, [H_2^{3/2} - H_3^{3/2}]$$

$$+ \frac{2}{3} C_d \sqrt{2g}\, [H_3^{3/2}] \tag{9.21}$$

EXAMPLE 9.1 Find the discharge of water flowing over a rectangular notch of 3 m length when the constant head over the notch is 40 cm. Take $C_d = 0.6$.

Solution: If Q is the discharge,

$$Q = \frac{2}{3} 2dL \sqrt{2g}\, H^{3/2}, \text{ here } C_d = 0.6,\ L = 3 \text{ m},\ H = 0.4 \text{ m}$$

$$\therefore \quad Q = \left[\frac{2}{3} \times 0.6 \times 3\sqrt{2 \times 9.81}\, (.4)^{3/2}\right] \text{m}^3\text{/sec}$$

$$Q = 1.34468 \text{ m}^3\text{/sec}$$ **Ans.**

EXAMPLE 9.2 Determine the height of rectangular weir of length 5 m to be built across a rectangular channel. The maximum depth of water upstream is 1.5 m and the discharge is 1.5 m^3/sec. Take $C_d = 0.6$ m, neglect end contractions.

Solution:

$$\Rightarrow \quad Q = \frac{2}{3} C_d L H^{3/2} \sqrt{2g}$$

$$\Rightarrow \quad 1.5 = \frac{2}{3} \times 0.6 \times 5\, H^{3/2} \times \sqrt{2 \times 9.81}$$

$$\therefore \quad H = 0.30606 \text{ m.}$$

$$\therefore \quad Z = 1.5 - H = 1.5 - 0.30606 \text{ m.}$$

$$\therefore \quad Z = 1.19394 \text{ m}$$ **Ans.**

EXAMPLE 9.3 A rectangular channel 1.5 m wide has a discharge of 200 lits/sec which is measured by a right-angled V-notch. Find the position of apex of the notch from the bed of the channel if maximum depth of water is not to exceed 1 m. Take $C_d = 0.62$.

Solution: Q = 200 lits/sec = 0.20 m^3/sec

Depth of water in the channel = 1 m, $\theta = 90°$

Let H be the head on the triangular notch,

$$Q = \frac{8}{15} C_d \sqrt{2g} \tan(\theta/2)\, H^{5/2}$$

$$0.20 = \frac{8}{15} \times 0.62 \times \sqrt{2 \times 9.81} \tan 45° \times H^{5/2}$$

$$\therefore \quad H = 0.450937 \text{ m.}$$

$\therefore$ Position of apex from the bed = Depth of water in channel – H

$$= 1 - 0.450937$$

$$= 0.549043 \text{ m}$$ **Ans.**

EXAMPLE 9.4 Find the discharge through a trapezoidal notch which is 1.2 m wide at the top and 0.5 m at the bottom and is 40 cm in height. The head of water on the notch is 30 cm. Assume C_d for the rectangular notch to be 0.62 and for the triangular notch to be 0.6.

Solution: $H = 0.3$ m, $Cd_1 = 0.62$, $Cd_2 = 0.6$

$Q = Q_1$ (rectangular) + Q_2 (triangular), Depth of triangular notch = 0.4 m

$$= \frac{2}{3} Cd_1\, L\sqrt{2g}\, H^{3/2} + \frac{8}{15} Cd_2 \sqrt{2g}\ \tan\theta/2\ H^{5/2}$$

To find $\theta/2$, $\tan\theta/2 = \dfrac{(1.2 - 0.5)/2}{0.4} = \dfrac{0.35}{0.4} = 0.875$

$$Q = \left(\frac{2}{3} \times 0.62 \times \sqrt{2 \times 9.81} \times .3^{3/2} + \frac{8}{15} \times 0.6 \times \sqrt{2 \times 9.81 \times 0.875 \times 0.3^{5/2}}\right) \text{ m}^3/\text{sec}$$

$Q = 0.22$ m³/sec **Ans.**

9.13 EFFECT OF COMPUTED DISCHARGE DUE TO MEASUREMENT OF HEAD

The equation of discharge in rectangular notch is:

$$Q = \frac{2}{3} Cd \sqrt{2g}\ LH^{3/2}$$

$$\frac{dQ}{dH} = \left(\frac{2}{3} C_d \sqrt{2g}\ L\right) \cdot \frac{3}{2} H^{1/2}$$

$$\therefore \quad dQ = \left(\frac{2}{3} C_d \sqrt{2g}\ L\right) \cdot \frac{3}{2} H^{1/2} dH$$

Let dQ and dH be the error measurement of the discharge and the head.

$$\therefore \quad \frac{dQ}{Q} = \frac{\left(\frac{2}{3} C_d \sqrt{2g}\ L\right) \frac{3}{2} H^{1/2}\, dH}{\left(\frac{2}{3} C_d \sqrt{2g}\ L\right) H^{3/2}}$$

$$\therefore \quad \left(\frac{dQ}{Q}\right) = \frac{3}{2}\left(\frac{dH}{H}\right)$$

i.e. $$\frac{dQ}{Q} \times 100 = 1.5 \times 100 \frac{dH}{H}$$

i.e. an error of 1% in the discharge is equal to 1.5 error in the head of a rectangular channel.

Similarly Q for triangular weir a notch is:

$$Q = \frac{8}{15} C_d \sqrt{2g} \ \tan \theta/2 \ H^{5/2}$$

$$dQ = \left(\frac{8}{15} C_d \sqrt{2g} \tan \theta/2\right) \frac{5}{2} \ H^{3/2} \ dH$$

$$\therefore \quad \frac{dQ}{Q} = \frac{\left(\frac{8}{15} C_d \sqrt{2g} \tan \theta/2\right) \frac{5}{2} H^{3/2} \, dH}{\left(\frac{8}{15} \sqrt{2g} \tan \theta/2\right) H^{5/2}}$$

$$\therefore \quad \frac{dQ}{Q} = \frac{5}{2} \cdot \frac{dH}{H}$$

i.e. an error of 1% in discharge is equivalent to an error of 2.5% in head measurement.

EXAMPLE 9.5 A rectangular notch 50 cm long is used to measure a discharge of 40 litre/sec of water. An error of 2 mm is made in measuring the head over the notch. Calculate the percentage error in discharge. Take $C_d = 0.60$.

Solution: $Q = \frac{2}{3} Cd \, L \sqrt{2g} \, H^{3/2}$, $Cd = 0.6$, $Q = 40$ liters/sec $= 40{,}000$ cm^3/sec, $L = 50$ cm

Error in head $dH = 2$ mm $= 0.2$ cm

$$\therefore \quad 40{,}000 = \frac{2}{3} \times 0.6 \times 50 \sqrt{2 \times 9.81} \ H^{3/2}$$

$$H = 12.68 \text{ m.}$$

$$\frac{dQ}{Q} = \frac{3}{2} \frac{dH}{H} = \frac{3}{2} \cdot \frac{0.2}{12.68} \times 100 = 2.366\% \qquad \textbf{Ans.}$$

EXAMPLE 9.6 Water flows in a rectangular channel 1.2 m wide and 0.8 m deep. Find the discharge over the weir of the crest of 70 cm if the head of water over the crest of weir is 25 cm. Take $Cd = 0.6$. Neglect end contractions but consider the velocity of approach.

Solution: Discharge without velocity of approach is:

$$Q = \frac{2}{3} C_d \, L \sqrt{2g} \, H^{3/2}, \ Cd = 0.6, \ L = 70 \text{ cm} = 0.7 \text{ m}, \ H = 25 \text{ cm} = 0.25 \text{ m}$$

$$\therefore \quad Q = \left[\frac{2}{3} \times 0.6 \times .7 \times \sqrt{2 \times 9.81} \, (.25)^{3/2}\right] \text{m}^3\text{/sec}$$

$$Q = 0.155 \text{ m}^3\text{/sec.}$$

$$\text{Velocity of approach } V_a = \frac{Q}{\text{area of channel}} = \frac{0.155}{1.2 \times 0.8} \text{m/sec}$$

$$V_a = 0.1615 \text{ m/sec}$$

$$\therefore \quad \text{Additional head } h_a = \frac{V_a^2}{2g} = \frac{(0.1615)^2}{2 \times 9.81} = 0.001329 \text{ m}$$

$$\therefore \quad \text{Total head} = (0.25 + .001329) \text{ m} = 0.251329 \text{ m.}$$

$$Q = \frac{2}{3} Cd\, L\sqrt{2g}\ [0.251329^{3/2} - 0.001329^{3/2}] \text{ m}^3/\text{sec}$$

$$Q = \frac{2}{3} \times .6 \times .7 \times \sqrt{2 \times 9.81}\ [0.251329^{3/2} - 0.001329^{3/2}] \text{ m}^3/\text{sec}$$

$$Q = 0.1562 \text{ m}^3/\text{sec}$$ **Ans.**

EXAMPLE 9.7 The head of water over a rectangular weir is 50 cm. The length of the crest of the weir with end contractions suppressed is 1.4 m. Find the discharge using: (i) Francis formula, (ii) Basin formula.

Solution: Assuming $C_d = 0.623$, the Francis formula of discharge is Equation [9.12a]

$$Q = 1.84\ (L - 0.1\ \text{nH})\ H^{3/2}$$

$$Q = (1.84 \times 1.4 \times 0.5^{3/2}) \text{ m}^3/\text{sec}$$

$$Q = 0.91075 \text{ m}^3/\text{sec}$$ **Ans.**

The Basin formula gives Equation [9.16]

$$Q = \left(0.405 + \frac{0.003}{H}\right)\sqrt{2g}\ LH^{3/2}$$

$$Q = \left[\left(0.405 + \frac{0.003}{5}\right)\sqrt{2 \times 9.81} \times 1.4 \times 0.5^{3/2}\right] \text{ m}^3/\text{sec}$$

$$Q = 0.9011 \text{ m}^3/\text{sec}$$ **Ans.**

EXAMPLE 9.8 Find the discharge over a Cipolletti weir of length 1.8 m when the head over the weir is 1.2 m. Take $C_d = 0.632$.

Solution: The discharge equation over the Cipolletti weir is (Equation 9.8)

$$Q = 1.866\ LH^{3/2} \text{ when } C_d = 0.632$$

$$Q = (1.866 \times 1.8 \times 1.2^{3/2}) \text{ m}^3$$

$$Q = 4.415 \text{ m}^3/\text{sec}$$ **Ans.**

EXAMPLE 9.9 The height of water on the upstream and downstream sides of a submerged weir of length 3.5 m is 30 cm, and 15 cm, respectively. If C_d values for the free and submerged portions are 0.6 and 0.8, respectively, find the discharges over the weir.

Solution: The discharge Q of a submerged weir (Equation 9.11) is:

$$Q = Q_1 + Q_2 = \frac{2}{3} Cd_1\, L\sqrt{2g}\ (H_1 - H_2)^{3/2} + Cd_2(L \times H_2)\sqrt{2g\,(H_1 - H_2)}$$

$$Cd_1 = 0.6,\ C_2 = 0.8,\ L = 3.5 \text{ m},\ H_1 = 30 \text{ cm} = 0.3 \text{ m},\ H_2 = 15 \text{ cm} = 0.15 \text{ m}$$

$$\therefore \quad Q = \left[\frac{2}{3} \times 0.6 \times 3.5 \times \sqrt{2 \times 9.81}\,(0.3 - 0.15)^{3/2} + 0.8 \times 3.5 \times 1.5 \times \sqrt{2 \times 9.81(.3 - .15)}\right] \text{m}^3/\text{sec}$$

$Q = 1.0807 \text{ m}^3/\text{sec}$ **Ans.**

EXAMPLE 9.10 A rectangular weir 6 m long is divided into three boys by two vertical points each 3 cm wide. Find the discharge when the head in 45 cm.

Solution: Using the Francis formula with end contractions:

$$Q = 1.84\ (L - 0.1\ \text{nH})\ H^{3/2}$$

Here $L = (6 - 2 \times .3) = 5.4$ m, since two vertical parts, end contractions $n = 6$ nos.

$$\therefore \quad Q = [1.84\ [5.4 - 0.1 \times 6 \times 0.45] \times (0.45)^{3/2}]\ \text{m}^3/\text{sec}$$

$$Q = 2.8494 \text{ m}^3/\text{sec} \simeq 2.85 \text{ m}^3/\text{sec}$$ **Ans.**

9.14 CONCLUSION

The use of weir and notch has already mentioned in the section 9.1 (i.e. INTRODUCTION) of this chapter. Different equations of measuring the discharges of the different types of weirs and notches have been derived in the different sections of the chapter. Such equations are quite frequently required by the field engineers to find the design discharge, height and width of the weirs (or notches). All the existing types of weir, their contraction effects, ventilation etc. have been presented. Numerical examples on each and every types of weir are solved and few other problems are listed for the students to solve. A list of references of some other existing textbooks on the subject is also attached at the end for further comparison of the author's presentation.

PROBLEMS

9.1 Find the discharge over a triangular notch of angle 60° when the head over the notch is 0.2 m. Take $C_d = 0.6$. (**Ans.** 0.0164 m^3/sec)

9.2 Find the discharge over a rectangular weir of length 80 m. The head over the weir is 1.2 m. The velocity of approach is given as 1.5 m/sec. Take $C_d = 0.6$. (**Ans.** 208.11 m^3/sec)

9.3 A broad crested weir of length 40 m has a 40 cm height of water above the crest. Find the maximum discharge. Take $C_d = 0.6$, and neglect velocity of approach. (**Ans.** 10.352 m^3/sec)

9.4 Determine the height of a rectangular weir of length 6 m, constructed across a rectangular channel. The depth of water upstream of the weir is 1.328 m and the discharge is 2 m^3/sec. Take $C_d = 0.6$, neglect end contractions. (**Ans.** 1 m)

9.5 Water flows over a rectangular weir 2 m wide at a height of 0.15 m, which passes through a right-angled triangular notch. Taking C_d for the rectangular and triangular weirs to be 0.62 and 0.59, respectively, find the depth over the triangular notch. (**Ans.** 0.47146 m)

9.6 Find the rate of flow over a triangular notch of 45° for a head of 0.4 m taking $C_d = 0.6$. If this flow is to be maintained within ± 0.2%, what are the limiting values of the head?

(**Ans.** 0.8911 m^3/sec, $H \pm dH = 0.4 \pm .0032 = 0.4032$ m and 0.3968 m)

9.7 The maximum flood discharge of a stream is 1600 m^3/sec, which has to pass over a weir. A road bridge is provided over the weir which divides it into a number of openings, each with a span of 6 m. Find the number of openings needed in order to ensure that the head of water over the crest of weir does not exceed 1.75 metres. The velocity of approach is 3 m/sec.

(*Hints:* Find Q of one opening considering V_a and end contractions, then the number of openings = Total discharge/discharge of one opening. **Ans.** 27 Nos.)

REFERENCES

1. Jaeger, C., *Engineering Fluid Mechanics*, Blackie, Glasgow, 1956.
2. Ramamruthum, S., *Hydraulics, Fluid Mechanics and Fluid Machines*, Dhanpat Rai and Sons, Delhi, 1980.
3. Olson, R.M., *Engineering Fluid Mechanics*, International Text Book Company, Seranton, Pennsylvania, 1967.
4. Modi, P.N. and S.M. Seth, *Hydraulics and Fluid Mechanics including Hydraulic Machines*, Standard Book House, New Delhi, 1977.
5. Streeter, V.L. and E.B. Waylie, *Fluid Mechanics*, McGraw-Hill, New York, 1983.
6. Mott, R.L., *Applied Fluid Mechanics*, Prentice Hall, Upper Saddle River, New York, 1999.

Chapter 10

Open Channel Flow

10.1 INTRODUCTION

The passage in which the liquid is not completely enclosed by a solid boundary but has a free surface exposed to the atmosphere is called 'open channel'. The flow of liquid in this open channel is called 'open channel flow'. It flows under atmospheric pressure due to the component of gravity with a free surface. Since this flow is always associated with a free surface, it is also called free surface flow. Figure 10.1 is an example of an open channel flow.

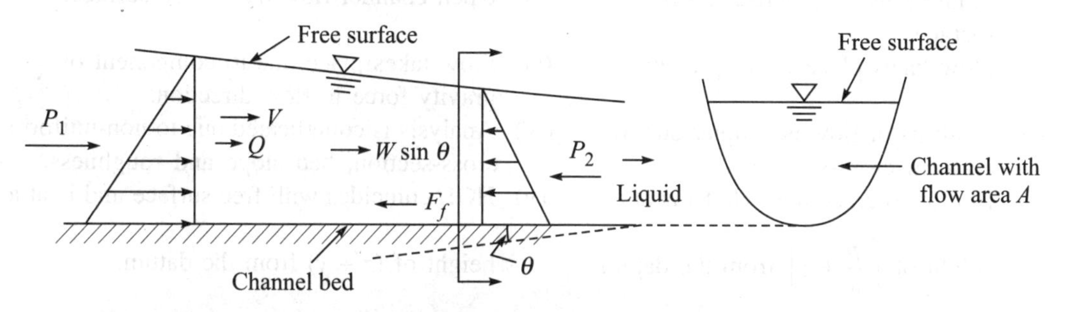

Figure 10.1 Open channel flow.

The examples of open channel are flows in natural rivers, streams, rivulets, drains, irrigation canals, sewers, culverts with a free surface, pipes not running full, and streets after heavy rainfall.

The different forces that govern the open channel flow are component of gravity ($W \sin \theta$) due to bed slope (θ), inertia force, surface tension (almost negligible), viscous (small for water when the flow is turbulent), resistance due to friction on the bed (F_f), etc.

10.2 DIFFERENCES IN OPEN CHANNEL FLOW AND PIPE FLOW

Figure 10.2 represents the flow in pipe and open channel to show the basic differences in both the flows. These differences are given below in tabular form.

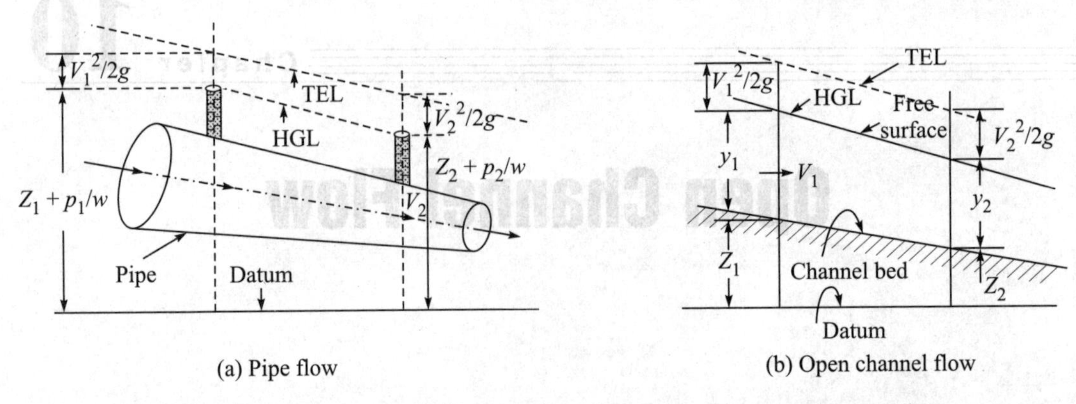

(a) Pipe flow

(b) Open channel flow

Figure 10.2 Basic differences between pipe flow and open channel flow.

Pipe Flow	*Open Channel Flow*
(i) The height of Total Energy Line (TEL) is $\left(\frac{p}{w} + z + \frac{V^2}{2g}\right)$	(i) The height of TEL flow from the datum is $\left(z + y + \frac{V^2}{2g}\right)$
(ii) Liquid runs full, there is no free surface.	(ii) Open channel flow has a free surface.
(iii) Flow takes place under pressure.	(iii) Flow takes place due to component of gravity force in flow direction.
(iv) Analysis of flow is simpler due to uniform cross-section.	(iv) Analysis is complicated due to non-uniform cross-section, bed slope and roughness.
(v) Hydraulic grade line (HGL) is at a height of $\left(\frac{p}{w} + z\right)$ from the datum.	(v) HGL coincides with free surface and is at a height of $(z + y)$ from the datum.
(iv) Surface tension force is dominant if the diameter is small.	(iv) Surface tension force is negligible.

10.3 TYPES OF CHANNELS

Channels are classified on the following bases:

(a) On the basis of whether channel is man made or not as natural or artificial.

(b) On the basis of shape as rectangular, triangular, trapezoidal, parabolic, exponential, circular, wide rectangular, compound and natural (irregular).

(c) On boundary characteristics as rigid or mobile boundary channel.

(d) On cross-section as prismatic (cross-section and slope remain constant) and non-prismatic.

10.4 TYPES OF OPEN CHANNEL FLOW

Open channel flow may be classified in many types and described in the various ways described below.

(a) *Steady flow:* when flow parameters do not change with time, i.e.

$$\frac{dQ}{dt} = 0, \quad \frac{\partial V}{\partial t} = 0 \quad \frac{\partial y}{\partial t} = 0, \text{ etc.}$$

(b) *Unsteady flow:* when flow parameters change with time, i.e.

$$\frac{dQ}{dt} \neq 0, \quad \frac{\partial V}{\partial t} \neq 0, \quad \frac{\partial y}{\partial t} \neq 0, \text{ etc.}$$

(c) *Uniform flow:* when flow parameters do not change with space, i.e.

$$\frac{\partial Q}{\partial x} = 0, \quad \frac{\partial V}{\partial x}, \frac{\partial y}{\partial x} = 0, \text{ etc.}$$

The uniform flow may be steady or unsteady.

(d) *Non-uniform flow:* when parameters do change with space, i.e.

$$\frac{\partial Q}{\partial x} \neq 0, \quad \frac{\partial V}{\partial x} \neq 0, \quad \frac{\partial y}{\partial x} \neq 0, \text{ etc.}$$

Again non-uniform or varied flow may be steady or unsteady.

(e) *Varied flow:* may also be rapidly varied flow, gradually varied flow and spatially varied flow. If the depth of flow changes rapidly, in a comparatively short distance, it is known as rapidly varied flow. If the depth changes gradually, it is gradually varied flow. If the discharge in the channel varies along the length due to the addition or withdrawal of water, the flow is spatially varied flow.

(f) *Super-critical, critical and sub-critical flow:* depending on Froude's number (F_r) of flow which is the ratio of inertia force to gravity force. If $F_r > 1$, the flow is super-critical, $Fr = 1$, the flow is critical and if $F_r < 1$, the flow is sub-critical.

(g) *Laminar, turbulent and transitional flow:* based on Reynolds number (R_e) of the flow which is the ratio of inertia force to viscous force. If Reynolds number $R_e < 2000$, the flow is laminar in the pipe flow. In case of open channel, there is no definite upper limit, yet R_e values from 500 to 12,500 flow changes from laminar to turbulent. In between these two states of flow, the flow is transitional. In pipe $R_e = \frac{VD}{v}$ and in open channel, diameter D is replaced by $4R$ where R = Area of flow/wetted perimeter, thus R_e (in open channel) $= \frac{4VR}{v}$.

(h) *One-dimensional (1-D), two-dimensional (2-D) and three-dimensional (3-D) flow* based on the coordinates system.

(i) *Homogenous and stratified flow* based on the Richardson number, i.e. change of density and velocity with depth.

(j) *Hydraulically smooth and hydraulically rough flow* based on the roughness on the bed and thickness of the laminar sub-layer.

10.5 GEOMETRIC AND FLOW PARAMETERS IN OPEN CHANNEL

The depth of flow y at a section is the vertical distance between the water surface and the bed. In most cases, this terminology is used. The actual depth d at a section is the distance of water perpendicular to the bed of the channel. If θ is the slope of the bed, the relationship between y and d is:

$$y = \frac{d}{\cos\theta}$$

$$y\cos\theta = d \tag{10.1}$$

θ is normally small, $\cos\theta \simeq 1$

$$\therefore \quad y = d \tag{10.1a}$$

The top width (T) is the width of the channel section at the free surface. The bottom width of the rectangular and trapezoidal channel is represented by B. The area of flow or water area A is the cross-sectional area of the flow normal to the flow direction. The wetted perimeter P is the length of the line of intersection of the channel wetted surface with cross-section of the flow normal to the direction of the plane. All these parameters are shown in Figure 10.3.

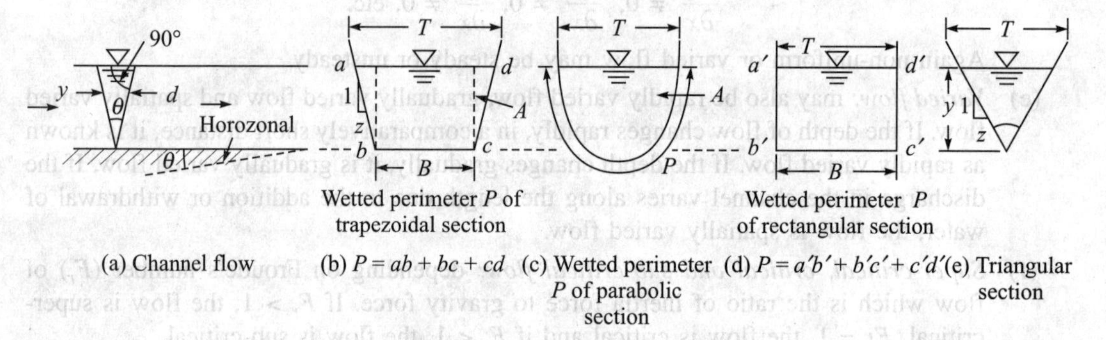

Figure 10.3 Open channel sections showing A, B, T, P, side slope zH : 1v., etc.

Hydraulic radius $\left(\text{Hydraulic mean depth for pipe, i.e. } m = \dfrac{A}{P} = \dfrac{(\pi/4)D^2}{AD} = \dfrac{D}{4}\right)$ R of open channel is:

$$R = \frac{A}{P} \tag{10.2}$$

Hydraulic depth is:

$$D = \frac{A}{T} \tag{10.3}$$

Section factor Z_n for normal depth computation is:

$$Z_n = AR^{2/3} \tag{10.4}$$

Section factor Z_e for critical depth computation is:

$$Z_c = A\sqrt{\frac{A}{T}} \tag{10.5}$$

Discharge or flow rate is:

$$Q = AV \text{ (length}^3\text{/time).} \tag{10.6}$$

Conveyance $K = \frac{Q}{\sqrt{s_o}}$, where $K = AC\sqrt{R}$ with Chezy's equation and $K = (AR^{2/3})/n$ with Manning's equation.

10.6 UNIFORM FLOW

A flow in open channel is said to be uniform if the depth of the flow is the same in every section of the channel, (Figure 10.4), i.e. $y_1 = y_2 = y_n$ and this constant depth y_n is called the normal depth. Besides the constant depth, uniform flow has a few more characteristics. Here $V_1 = V_2 = V$, $Q_1 = Q_2 = Q$, $A_1 = A_2 = A$ i.e., $\frac{dy}{dx} = 0$, $\frac{dV}{dx} = 0$, $\frac{dQ}{dx} = 0$, $\frac{dA}{dx} = 0$, etc. Since the depth remains constant, the free surface is parallel to the bed, i.e. bed slope s_b = Water Surface slope s_w. Again applying Bernoulli's equation, $Z_1 + y_1 + \frac{V_1^2}{2g} = Z_2 + \frac{V_2^2}{2g} + y_2$ + Losses

$$y_1 = y_2,\ V_1 = V_2,\ Z_1 - Z_2 = \text{losses } (h_f)$$

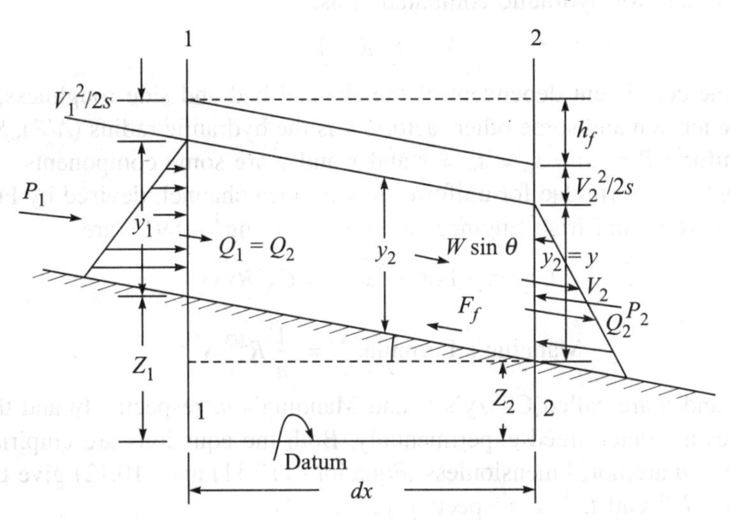

Figure 10.4 Channel reach *dx* to define uniform flow.

Dividing by *dx:*

$$\frac{Z_1 - Z_2}{dx} = \frac{h_f}{dx}$$

$$\frac{dz}{dx} = \frac{h_f}{dx}$$

$$\therefore \quad s_b = s_f$$

$$\therefore \quad s_b = s_w = s_f \tag{10.7}$$

Writing the momentum equation:

$$P_1 - P_2 + W \sin\theta - F_f = \frac{wQ}{g}(V_2 - V_1) \tag{10.8}$$

where $P_1 = w\, A_1 \bar{x}_1$ (hydrostatic force at 1-1)

$P_2 = w\, A_1 \bar{x}_2$ (hydrostatic force at 2-2)

and $P_1 = P_2$ since $A_1 = A_2$, $\bar{x}_1 = \dfrac{y}{2}$, $\bar{x}_2 = \dfrac{y}{2}$

$$V_2 = V_1$$

$\therefore$ Equation (10.8) becomes $W \sin\theta = F_f$ (10.9)

Equation (10.9) shows that in uniform flow component of gravity $W \sin\theta$ is exactly balanced by force of friction F_f and this is the real condition of establishment of uniform flow in open channel.

10.7 AVERAGE VELOCITY EQUATIONS IN UNIFORM FLOW

In uniform flow, the average velocity of flow V is expressed approximately by the so-called uniform flow formula for hydraulic computation as:

$$V = C_1 R^x s_f^y \tag{10.10}$$

where C_1 is some coefficient dependent of the channel bed and side roughness, magnitude of velocity, surface tension and some other factors, R is the hydraulic radius (A/P), S_f is the energy slope and in uniform flow $s_f = s_b = s_w = s$ and x and y are some components.

The most practical formulae for uniform flow in open channel, devised by French engineer Antonic Chezy (1969), and Irish Engineer Roberts Manning[3] (1891), are:

$$\text{Chezy's Formula: } V = C\sqrt{Rs} \tag{10.11}$$

$$\text{Manning's Formula: } V = \frac{1}{n} R^{2/3}\, s^{1/2} \tag{10.12}$$

Wherein C and n are called Chezy's C and Manning's n, respectively and their values for different surfaces are determined experimentally. Both the equations are empirical as the two coefficients C and n are not dimensionless. Equations (10.11) and (10.12) give the dimensions of C and n as $L^{1/2}T^{-1}$ and $L^{-1/3}T$, respectively.

The relationship between C and n is:

$$C = \frac{1}{n} R^{1/6} \tag{10.13}$$

Swiss engineers Ganguillet and Kutter in 1969, and French hydraulic engineer Bazin in 1897 also developed such empirical formulae which are not in use now. The most popular and widely used formula even today is Mannings formula. Ven Te Chow has given a table of n values for different surface roughnesses in his book *Open Channel Hydraulics*. His values are still used by investigators in the study of open channels even till today.

Another resistance coefficient is Darcy's friction factor f, a dimensionless coefficient and it is used in pipe flow. A group of hydraulicians are of the opinion that this friction factor f can be also used in open channel satisfactorily when pipe diameter is replaced by four times of hydraulic radius as

$$R = \frac{A}{P} = \frac{(\pi/4)D^2}{\pi D} = \frac{D}{4} \quad \therefore D = 4R \tag{10.14}$$

If this concept is used in Darcy's equation for open channel,

then
$$h_f = \frac{fLV^2}{2gD} = \frac{fLV^2}{2g(4R)} = \frac{fLV^2}{8gR}$$

$$\therefore \quad V^2 = \frac{8gR}{f}\left(\frac{h_f}{L}\right) = \frac{8gR}{f}s_f = \frac{8gR}{f}s_b \quad \because s_f = s_b = s_w = s \text{ in uniform flow.}$$

$$\therefore \quad V = \sqrt{\frac{8g}{f}}\sqrt{R\,s_g} \tag{10.15}$$

Comparing Equation (10.15) with Equations (10.4) and (10.12),

$$C = \frac{1}{n}R^{1/6} = \sqrt{\frac{8g}{f}} \tag{10.16}$$

10.8 ECONOMIC OR EFFICIENT SECTION IN OPEN CHANNEL

A section of an open channel is said to be most economical or efficient when it carries the maximum discharge Q for a given cross-sectional area, resistance coefficients like C or n or f and bed slope s_b. From continuity (i.e. $Q = AV$), it is evident if the area is constant, Q is maximum if V is maximum. Again from Chezy's or Manning's equation, V is maximum if R is maximum for a given value of resistance coefficient and bed slope. R is given by $R = \frac{A}{P}$, if A is constant, R becomes maximum if wetted perimeter P is minimum. This condition of minimum wetted perimeter P is utilized to find the economic sections of the channel.

(a) *Economic Section of Rectangular Channel*

In Figure 10.5, different economic sections like rectangular, triangular, trapezoidal and circular channel are shown. These are described below.

(i) **Rectangular Channel**

$$A = By \quad \therefore B = \frac{A}{y}$$

$$P = B + 2y = \frac{A}{y} + 2y$$

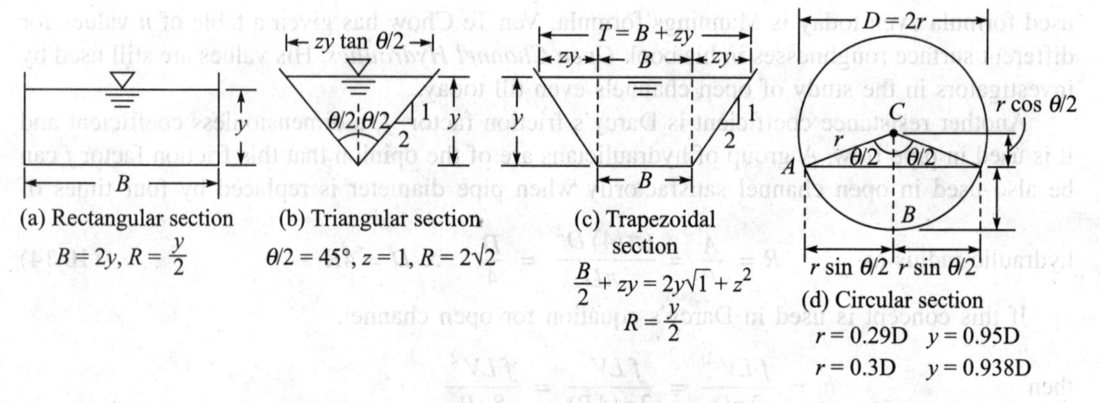

Figure 10.5 Different economic channel sections.

differentiating P with respect to y and equating with zero for a minimum,

$$\frac{dP}{dy} = A(-1)\,\frac{1}{y^2} + 2 = 0$$

$$-\frac{A}{y^2} + 2 = 0$$

$$-\frac{By}{y^2} + 2 = 0$$

$$-\frac{By}{y^2} = 2$$

$$\therefore \qquad By = 2y^2$$

$$\therefore \qquad B = 2y \quad \text{or} \quad y = \frac{B}{2} \tag{10.17}$$

(ii) **Triangular Channel**

$$A = \frac{1}{2}(2y \tan\theta/2)\, y = y^2 \tan\theta/2$$

$$\therefore \qquad y = \sqrt{\frac{A}{\tan\theta/2}}$$

$$P = 2y \sec\theta/2 = 2\sqrt{\frac{A}{\tan\theta/2}} \cdot \sec\theta/2$$

$$\therefore \qquad \frac{dP}{dy} = 2\sqrt{A}\left[\frac{\sec\theta/2\tan\theta/2}{\sqrt{\tan\theta/2}} - \frac{\sec^2\theta/2}{2(\tan(\theta/2))^{3/2}}\right] = 0 \quad \text{for } P \text{ to be minimum}$$

$$\frac{\sec\theta/2\cdot\tan\theta/2}{\sqrt{\tan\theta/2}} - \frac{\sec^3\theta/2}{2(\tan(\theta/2))^{3/2}} = 0$$

$$\sec\ \theta/2[2\ \tan^2\ \theta/2 - \sec^2\theta] = 0$$

$$\sec \theta/2 \neq 0$$

$$\therefore \quad 2 \tan^2\theta/2 - \sec^2\theta/2 = 0$$

$$\therefore \quad \sqrt{2} \tan \theta/2 = \sec \theta/2$$

$$\therefore \quad \sqrt{2} \frac{\sin \theta/2}{\cos \theta/2} = \frac{1}{\cos \theta/2}$$

$$\therefore \quad \sqrt{2} \sin \theta/2 = 1$$

$$\therefore \quad \sin \theta/2 = \frac{1}{\sqrt{2}}$$

$$\therefore \quad \theta/2 = 45° \qquad (10.18)$$

i.e. the angle subtended by the triangular channel is 90° for economic condition or in side slope

zH : 1V, $z = 1$ and $R = \dfrac{A}{P} = \dfrac{y^2 \tan 45°}{2y \sec 45°}$

$$R = \frac{y^2 \times 1}{2y\sqrt{2}} = \frac{y}{2\sqrt{2}} \qquad (10.19)$$

(iii) **Trapezoidal Channel**

$$A = By + zy^2$$

$$\therefore \quad B = \frac{A}{y} - zy$$

$$T = B + 2zy$$

$$P = B + 2y\sqrt{1+z^2} = \frac{A}{y} - zy + 2y\sqrt{1+z^2}$$

$$\frac{dP}{dy} = -\left(\frac{A}{y^2}\right) - z + 2\sqrt{Hz^2} = 0$$

$$\frac{A}{y^2} + z = 2\sqrt{1+z^2}$$

$$\frac{(B+zy)\,y}{y^2} + z = 2\sqrt{1+z^2}$$

$$\frac{B+zy}{y} + z = 2\sqrt{1+z^2}$$

$$B + zy + yz = zy\sqrt{1+z^2}$$

$$\frac{B+2zy}{2} = y\sqrt{1+z^2}$$

$$\left(\frac{B}{2} + zy\right) = y\sqrt{1+z^2} \tag{10.20}$$

Equation (10.20) is the condition for economic section, i.e. half of top width (i.e.) $\left(\frac{B}{2} + zy\right)$ is equal to the length of one slanting side $\left(y\sqrt{1+z^2}\right)$.

and

$$R = \frac{A}{P} = \frac{(B+zy)\,y}{B+2y\sqrt{1+z^2}}$$

Substituting the value of B from Equation (10.20)

$$R = \frac{y}{2} \tag{10.21}$$

(iv) **Circular Channel:** If A is the area of flow, from Figure 10.5 (d),

$$A = \text{Area } ABDC - \text{triangle } ACD$$

$$A = (\theta/2)r^2 - \frac{1}{2}\, 2r \sin \theta/2 \; r \cos \theta/2$$

$$A = \frac{r^2}{2}(\theta - 2 \sin \theta/2 \cdot \cos \theta/2)$$

$$A = \frac{r^2}{2}(\theta - \sin \theta) \tag{A}$$

$$\frac{dA}{d\theta} = \frac{r^2}{2}(1 - \cos \theta) \tag{B}$$

$$P = r\theta \quad \therefore \quad \frac{dP}{d\theta} = r \tag{C}$$

$$Q = AV = \text{A.C}\left(\frac{A}{P}\right)^{1/2} s_b^{1/2} = C\sqrt{s_b}\,\frac{A^{3/2}}{P^{1/2}} = C\sqrt{s_b}\left(\frac{A^3}{P}\right)^{1/2}$$

For a given value of Chezy's C and s_b, Q is max^m if $\left(\frac{A^3}{P}\right)$ is maximum. Differentiating $\left(\frac{A^3}{P}\right)$, with respect to θ and equating with zero for a maximum:

$$\frac{d\left(\frac{A\theta}{P}\right)}{d\theta} = \frac{P3A^2\frac{dA}{d\theta} - A^3\frac{dP}{d\theta}}{P^2} = 0$$

gives $$3P\frac{dA}{d\theta} - A\frac{dP}{d\theta} = 0 \quad \text{(D)}$$

Substituting A, $\frac{dA}{d\theta}$ and $\frac{dP}{d\theta}$ from (A) and (B) and from (C) in (D)

$$3r\theta\cdot\frac{r^2}{2}(1 - \cos\theta) - \frac{r^2}{2}(\theta - \sin\theta) = 0$$

Simplifying: $$2\theta - 3\theta\cos\theta + \sin\theta = 0$$

Solving this equation, when $\theta \simeq 308°$

$$\therefore \quad y = r + r\cos\theta(180° - \theta/2)$$

$$y \simeq 1.8988\, r$$

$$\therefore \quad y \simeq 0.95\, D \quad (10.22)$$

Similarly when Manning's equation is used, the condition for max^m discharge or economic condition, it can be shown that:

$$y \simeq 0.938\, D \quad (10.23)$$

EXAMPLE 10.1 Find the velocity of flow and discharge through a rectangular channel 5 m wide and 2 m water depth. The bed slope of the channel is 1/3000 and Chezy's coefficient $C = 50$

Solution: Chezy's equation $V = C\sqrt{Rs_b}$

$$C = 50,\ s_b = \frac{1}{3000},\ R = \frac{A}{P} = \frac{5\times 2}{5+2+2} = \frac{10}{9}\text{ m}$$

$$\therefore \quad V = 50\sqrt{\frac{10}{9}\times\frac{1}{3000}} = 0.96225 \text{ m/sec}$$ **Ans.**

$$\therefore \quad Q = AV = 5 \times 2 \times 0.96225 = 9.6225 \text{ m}^3\text{/sec}$$ **Ans.**

EXAMPLE 10.2 A flow of water flows at 150 litres per sec in a rectangular channel of width 70 cm and depth of flow 40 cm. Uniform flow occurs at a certain bed slope with Chezy's C equal to 60. Find the bed slope and conveyance of the channel.

Solution: $Q = 150$ litres/sec – 0.15 m^3/sec

$B = 70$ cm $= 0.7$ m

$y = 40$ cm $= 0.4$ m

$C = 60 \quad \therefore A = By = (0.7 \times 0.4) = 0.28 \text{ m}^2$

$Q = A(C\sqrt{(A/P)s_b}) \quad P = B + 2y = 0.7 + 2 \times .4 = 1.5$ m

$$0.15 = \left(60\sqrt{\frac{.28}{1.5} - s_b}\right) \times 0.28$$

$$\therefore \quad s_b = 1/2341.546$$ **Ans.**

$$\text{Conveyance } K = \frac{Q}{\sqrt{s_b}} = \frac{AC\sqrt{Rs_b}}{\sqrt{s_b}} = AC\sqrt{R} = 0.28 \times 60 \times \sqrt{\left(\frac{.28}{1.5}\right)}$$

$\therefore$ $K = 7.2584 \text{ m}^3/\text{sec}$ **Ans.**

EXAMPLE 10.3 Find the discharge through a trapezoidal channel of bottom width 6 m side slope 1 horizontal 3 vertical $\left(\frac{1}{3}\text{H} : 1\text{ V i.e. } z = \frac{1}{3}\right)$. The depth of flow is 3 m and Chezy's $C = 60$ and $s_b = \frac{1}{5000}$.

Solution: $B = 6$ m, side slope $\frac{1}{3}$H : 1 V i.e. $z = \frac{1}{3}$ (ZH : 1)

Depth of flow $y = 3$ m, $C = 60$

$\therefore$ Area of flow $A = (B + zy)y = \left(6 + \frac{1}{3} \times 3\right) \times 3 = 21 \text{ m}^2$

$$P = B + 2y\sqrt{1 + z^2} = 6 + 2 \times 3\sqrt{1 + \left(\frac{1}{3}\right)^2} = 11.1622 \text{ m}$$

$$\therefore \quad R = \frac{A}{P} = \frac{21}{11.1622} = 1.88135 \text{ m}$$

$$\therefore \quad Q = AV = A \cdot (C\sqrt{Rs_b}) = \left(21 \times 60 \times \sqrt{1.88135 \times \frac{1}{5000}}\right) \text{m}^2/\text{sec}$$

$Q = 24.441 \text{ m}^3/\text{sec}$ **Ans.**

EXAMPLE 10.4 Water flows through a rectangular channel of 60 cm width and 30 cm depth is 100 litres/sec. If Manning's $n = 0.013$, find the bottom slope s_b and conveyance K of the channel.

Solution: $Q = 100$ litres/sec $= 0.1 \text{ m}^3/\text{sec}$

$B = 60$ m $= 0.6$ m

$y = 30$ cm $= 0.3$ m

$\therefore$ $A = B \cdot y = 0.6 \times 0.3 = 0.18 \text{ m}^2$

$$n = 0.013,\ P = 0.6 \times 2 \times .3 = 1.2 \text{ m},\ R = \frac{A}{\rho} = \frac{0.18}{1.2} = 0.15 \text{ m}$$

We have $Q = AV = (By)\,\frac{1}{n}\,(R)^{2/3}\, s_b^{1/2}$

$$0.1 = 0.18 \times \frac{1}{0.013}\,(0.15)^{2/3}\, s_b^{1/2}$$

$\therefore$ $s_b = 1/1527.964 \simeq 1/1528$ **Ans.**

$$\text{Conveyance } K = \frac{Q}{\sqrt{s_b}} = \frac{AR^{2/3}}{n}$$

or $$K = \left(\frac{0.1}{\sqrt{1/1528}}\right) \text{ m}^3/\text{sec} = 3.90896 \text{ m}^3/\text{sec}$$ **Ans.**

EXAMPLE 10.5 Find the bed slope of a trapezoidal channel of bed width 6 m, depth of water 3 m. side slope 3H : 4V, when discharge flowing through the channel is 30 m^3/sec. Take Manning's n = 0.0158. (See Figure 10.6).

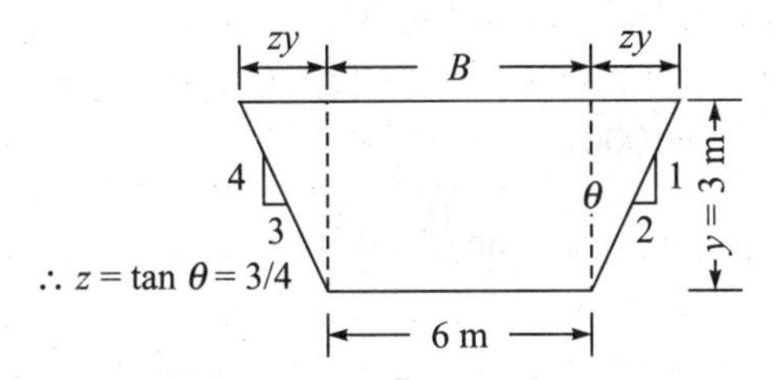

Figure 10.6 Visual of Example 10.5.

Solution: B = 6 m, y = 3 m, side slope 3H : 4V i.e. $Z = \frac{3}{4}$

Q = 30 m^3/sec, n = 0.0158

$$A = (B + zy)y = \left[\left(6 + \frac{3}{4} \times 3\right) \times 3\right] \text{m}^2 = 24.75 \text{ m}^2$$

$$P = B + 2y\sqrt{1+z^2} = \left(6 + 2 \times 3\sqrt{1+\left(\frac{3}{4}\right)^2}\right) \text{m} = 13.5 \text{ m}$$

$$\therefore \quad R = \frac{A}{P} = \left(\frac{24.75}{13.5}\right) \text{m} = 1.8333 \text{ m}$$

We know $Q = AV = (24.75) \times \frac{1}{n}(R)^{2/3}\, s_b^{1/2}$

$$30 = 24.75 \times \frac{1}{0.0158} \times (1.8333)^{2/3}\, s_b^{1/2}$$

$$s_b = \frac{1}{6114.26}$$ **Ans.**

EXAMPLE 10.6 Find the discharge through a triangular channel that makes an angle at the vertex equal to 60° with the depth of water 4 m, Take s_b = 1/1000 and Manning's n = 0.0182. (See Figure 10.7).

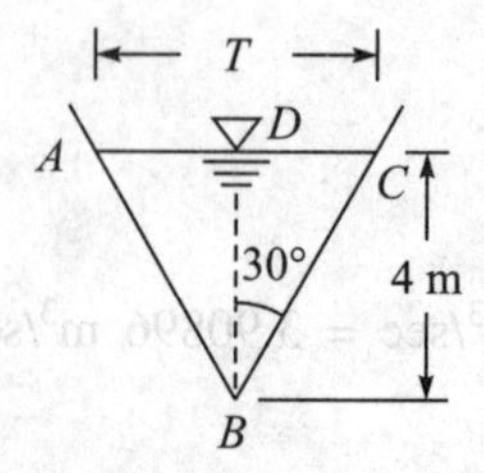

Figure 10.7 Visual of Example 10.6.

Solution: $y = 4$ m

θ = Vertex angle = 60° $\therefore\ \theta/2 = 30°$

$$n = 0.0182, \quad s_b = \frac{1}{1000}$$

$$T = AD + DC = 2DC = \left(2\tan\frac{\theta}{2} \times 4\right)\text{m}$$

$$\therefore \quad \text{Area } A = \frac{1}{2}Ty = \frac{1}{2}(2\tan 30° \times 4) \times 4 = 9.2376 \text{ m}^2$$

$$P = AB + BC = 2BC = 2\sqrt{BD^2 + DC^2}$$

$$P = 2\sqrt{4^2 + (4\tan 30°)^2} = 9.2376 \text{ m}$$

$$\therefore \quad R = \frac{A}{P} = \frac{9.2376}{9.2376} = 1 \text{ m}$$

$$\therefore \quad Q = A \cdot V = 9.2376 \times \frac{1}{0.0182}(1)^{2/3}\left(\frac{1}{1000}\right)^{1/2}$$

$$\therefore \quad Q = 16.067 \text{ m}^3\text{/sec}$$ **Ans.**

EXAMPLE 10.7 Find the diameter of a circular channel laid at a slope of 1 in 10,000, carrying a discharge 1000 litres/sec when running half full. Take Manning's n = 0.02.

Solution: Q = 1000 litres/sec = 1.0 m^3/sec

$n = 0.02$

$s_b = 1/10{,}000$

$y = D/2$, if diameter = D.

$$A = \frac{\pi}{4}D^2 \times \frac{1}{2} = \frac{\pi D^2}{8} \text{ (half full)}$$

$$P = \frac{\pi D}{2} \text{ (half the circumference)}$$

$$R = A/P = \frac{\pi D^2 \times 2}{8 \times \pi D} = \frac{D}{4}$$

$$\therefore \quad Q = AV = A \times \frac{1}{n} (R)^{2/3} (s_b)^{1/2}$$

$$1 = Q = \frac{\pi D^2}{8} \cdot \frac{1}{n} \left(\frac{D}{4}\right)^{2/3} \left(\frac{1}{10,000}\right)^{1/2}$$

$$1 = \frac{\pi}{8} \times \frac{1}{0.02} \times \left(\frac{1}{10000}\right)^{1/2} \times \frac{D^{8/3}}{4^{2/3}}$$

$$1 = 0.077921 \; D^{8/3}$$

$$\therefore \quad D = 2.604 \text{ m} \qquad \textbf{Ans.}$$

EXAMPLE 10.8 A rectangular channel carries water at the rate of 500 litres/sec when the bed slope is 1/3000. Find the most economical dimensions of the channel when (i) $C = 60$ (ii) $n = 0.015$.

Solution: Discharge $Q = 500$ lits/sec $= 0.5$ m^3/sec

$$s_b = 1/3000$$

Economical section is $B = 2y$ and $R = \dfrac{y}{2}$

$$\therefore \quad A = B \cdot y = 2y \cdot y = zy^2$$

$$\text{(i)} \quad Q = AV = (2y)^2 \, C\sqrt{Rs_b} = 2y^2 \times 60 \times \left(\frac{y}{2}\right)^{1/2} \times \left(\frac{1}{3000}\right)^{1/2}$$

or

$$0.5 = 2 \times 60 \times \left(\frac{1}{2}\right)^{1/2} \times \left(\frac{1}{3000}\right)^{1/2} y^{5/2}$$

$$0.5 = 1.54919 \; y^{5/2}$$

$$y = 0.63613 \text{ m} \qquad \textbf{Ans.}$$

$$\therefore \quad B = 2y = 1.27226 \text{ m} \qquad \textbf{Ans.}$$

$$\text{(ii)} \quad Q = AV = A\frac{1}{n}(R)^{2/3}(s_b)^{1/2} = 2y^2 \times \frac{1}{0.015} (y/2)^{2/3} \left(\frac{1}{3000}\right)^{1/2}$$

$$\therefore \quad 0.5 = 2 \times \frac{1}{0.015} \times \left(\frac{1}{2}\right)^{2/3} \times \left(\frac{1}{3000}\right)^{1/2} y^{8/3}$$

$$0.5 = 1.5336 \; y^{8/3}$$

$$\therefore \quad y = 0.65685 \text{ m} \qquad \textbf{Ans.}$$

$$\therefore \quad B = 2y = 1.3137 \text{ m} \qquad \textbf{Ans.}$$

EXAMPLE 10.9 A trapezoidal channel has a side slope 1 vertical and 2 horizontal and bed slope 1 in 2000. The area of the section 42 m^2. Find the dimension of the channel if a section is an economic section. Determine the discharge of the channel if (i) $C = 60$ (ii) $n = 0.02$.

Solution: $A = 42 \text{ m}^2$, $s_b = \dfrac{1}{2000}$, $C = 50$, $n = 0.012$, side slope is 1 horizontal 2 vertical, i.e. $\dfrac{1}{2}H$: 1V. i.e. $z = \dfrac{1}{2}$.

Condition for economical section in trapezoidal channel is:

half of the top width = length of slanting side

i.e.
$$\left(\frac{B}{2} + zy\right) = y\sqrt{1+z^2}$$

or
$$\frac{B}{2} + \frac{1}{2}y = y\sqrt{1+\left(\frac{1}{z}\right)^2}$$

$$\frac{B+y}{2} = 1.118\ y$$

$$\therefore \quad B + y = 2.236\ y \quad \therefore\ B = 1.236\ y$$

$$\text{Area} = (B + zy)y = By + zy^2 = 1.236\ y^2 + \frac{1}{2}\ y^2 = 1.736\ y^2$$

$$\therefore \quad 42 = 1.736\ y^2$$

$$\therefore \quad y = 4.9187 \text{ m} \qquad \textbf{Ans.}$$

$$\therefore \quad B = (1.236 \times 4.9187)\text{m}$$

$$B = 6.0795 \text{ m} \qquad \textbf{Ans.}$$

For economic section, $R = \dfrac{y}{2}$, i.e. $R = \dfrac{4.9187}{2}$ m = 2.45935 m

(i) when $C = 60$

$$Q = AC\sqrt{Rs_b} = \left(42 \times 60\sqrt{2.45935 \times \frac{1}{2000}}\right) \text{m}^3/\text{sec}$$

$$Q = 88.368 \text{ m}^3/\text{sec} \qquad \textbf{Ans.}$$

(ii) when $n = 0.02$

$$Q = A\frac{1}{n}\ (R)^{2/3}\ (s_b)^{1/2} = 42 \times \frac{1}{0.02}\ (2.45935)^{2/3} \left(\frac{1}{2000}\right)^{1/2}$$

$$Q = 85.556 \text{ m}^3/\text{sec} \qquad \textbf{Ans.}$$

EXAMPLE 10.10 A trapezoidal channel with side slope 1H : 1V has been designed to convey 12.2 m³/sec at a velocity 1.9 m/sec so that the amount of concrete lining is minimum. Calculate the area of lining per metre length of the channel.

Solution: $Q = 12.2 \text{ m}^3/\text{sec}$, $V = 1.9$ m/sec

Side slope 1H : 1V i.e. $z = 1$

Area of flow $A = \dfrac{12.2}{1.9} = 6.421 \text{ m}^2$

If lining area is to be minimum, wetted perimeter P is minimum, i.e. it is a case of economic section.

If B is the bottom width, y is the depth, then

$$\frac{B}{2} + zy = y\sqrt{1+z^2}$$

or
$$\frac{B}{2} + 1 \times y = y\sqrt{1+1^2} = y\sqrt{2}$$

$$B = 2y(\sqrt{2} - 1) = 2(1.4142 - 1)y = 0.8284\ y$$

$$A = (B + zy)y = (0.8284 + 1 \times y)y = 1.8284\ y^2$$

$$6.421 = 1.8284\ y^2$$

$\therefore$
$$y = 1.874\ \text{m}$$

$\therefore$
$$B = (1.874 \times 0.8284)\text{m} = 1.5524\ \text{m}$$

$$\text{Wetted perimeter } P = B + 2y\sqrt{1+z^2} = (1.5524 + 2 \times 1.874\sqrt{1+1^2})\text{m}$$

$$P = 8.22845\ \text{m}$$

$$\text{Area of wetted perimeter/unit length} = (P \times 1)\ \text{m}^2$$

$$= (8.22845 \times 1) = 8.22845\ \text{m}^2$$ **Ans.**

EXAMPLE 10.11 A triangular channel is economic. The discharge is 0.04 m³/sec when the depth of flow is 0.225 m. If $C = 50$, calculate the slope.

Solution: The channel is economic, i.e. the vertex angle is 90° and side slope is 45°. i.e. $z = 1$, $y = 0.225$ m

$\therefore$
$$\text{Area } A = zy^2 = 1 \times 0.225^2 = 0.050625\ \text{m}^2$$

Wetted $P = 2\sqrt{2}\ y$

and
$$R = \frac{A}{P} = \frac{zy^2}{C\sqrt{2}y} = \frac{y^2}{2\sqrt{2}y} = \frac{y}{C\sqrt{2}}\ \text{m}$$

$\therefore$
$$Q = AC\sqrt{Rs_b}$$

$$0.04 = 0.050625 \times 50\sqrt{\frac{0.225}{2\sqrt{2}}}\ s_b^{1/2}$$

$\therefore$
$$s_b^{1/2} = \frac{0.04}{.050625 \times 50 \times 0.282}$$

$$s_b = \frac{1}{393}$$ **Ans.**

EXAMPLE 10.12 A rectangular channel 2 m wide and has water depth of 0.5 m. It is laid in a bed slope of 0.0004. If Manning's $n = 0.012$, find C and f.

Solution: $R = \dfrac{A}{P} = \dfrac{2 \times .5}{2 + 2 \times .5} = \dfrac{1}{3}$ m

$\therefore \quad C = \dfrac{1}{n} R^{1/6} = \dfrac{1}{0.012}\left(\dfrac{1}{3}\right)^{1/6} = 69.39 \text{ m}^{1/3}/\text{sec}$

We have $C = \sqrt{\dfrac{8g}{f}} \quad \therefore \sqrt{f} = \sqrt{\dfrac{8g}{C^2}} = \left(\dfrac{8 \times 9.81}{69.39^2}\right)^{1/2}$

$\therefore \quad f = 0.0163$ **Ans.**

10.9 COMPUTATION OF UNIFORM FLOW DEPTH

The depth in uniform flow is called normal depth. The computation of this normal depth except in a few situations like wide rectangular and triangular channel section cannot be done directly.

As such, computation of normal depth y_n is implicit, i.e. a direct solution does not exist. Therefore, trial and error, graphical method or method of iteration with the help of a small computer programme will be necessary to compute this normal depth.

In order to examine the above situation of computation, we may start with a trapezoidal channel of bottom with B, side slope zH : 1V with known Q, s_b. Uniform flow formulae of velocity is taken through Manning's equation which is widely used even in the present-day analysis of open channel flow.

Let the normal depth of flow be y_n which has to be computed.

Now Area $A = (B + zy_n)y_n$, $P = B + 2y_n\sqrt{1+z^2}$

$$\therefore \quad R = \frac{A}{P} = \left(\frac{(B + zy_n)\, y_n}{B + 2y_n\sqrt{1+z^2}}\right)$$

Writing discharge equation with Manning's equation:

$$Q = A \cdot \frac{1}{n} (R)^{2/3} s_b^{1/2}$$

$$Q = \frac{(B + zy_n)\, y_n}{n}\left[\frac{(B + zy_n)\, y_n}{B + 2y_n\sqrt{1+z^2}}\right]^{2/3} s_b^{1/2} \tag{10.24}$$

In the above equation, Q, n, B, z and s_b are known, but solution of y_n is implicit, i.e. no direct solution exists.

Equation (10.24) may be made dimensionless, using non-dimensional normal depth $Y_n = \dfrac{y_n}{B}$

i.e. $y_n = BY_n$

Now replacing y_n by BY_n, Equation (10.24) may be written as:

$$\frac{Qn}{s_b^{1/2}} = \frac{B^{10/3}\left[(1 + zY_n)\, Y_n\right]^{5/3}}{B^{2/3}\left[1 + 2Y_n\sqrt{1+z^2}\,\right]^{2/3}}$$

or

$$\left(\frac{Qn}{B^{8/3}\, s_b^{1/2}}\right) = \frac{\left[(1+zY_n)Y_n\right]^{5/3}}{[1+2Y_n\sqrt{1+z^2}\,]^{2/3}} \tag{10.25}$$

where $\left(\frac{Qn}{B^{8/3}\, s_b^{1/2}}\right)$ is the non-dimensional discharge and Y_n is the non-dimensional normal depth.

(a) *When the Channel is Triangular*

In Equation (10.24), $B = 0$

∴ Equation (10.24) reduces to:

$$Q = \frac{(zy_n)^{5/3}}{n(2y_n\sqrt{1+z^2})^{2/3}}\, s_b^{1/2}$$

$$\frac{Qn}{s_b^{1/2}} = \frac{z^{5/3}\, y_n^{10/3}}{2^{2/3}\, y_n^{2/3}\,(1+z^2)^{1/3}}$$

$$\left[\frac{2^{2/3}\cdot Qn\,(1+z^2)^{1/3}}{s_b^{1/2}\, z^{5/3}}\right] = y_n^{8/3}$$

$$\therefore \qquad y_n = \left[\frac{4^{1/3}\, Qn\,(1+z^2)^{1/3}}{z^{5/3}\, s_b^{1/2}}\right]^{3/8} \tag{10.26}$$

Equation (10.26) shows that a direct solution for normal depth in case of a triangular channel exists.

(b) *When the Channel is Wide Rectangular*

A channel is wide rectangular when $B >> y$ (i.e. Breadth >> depth of the channel).

and $\qquad P \simeq B, \quad R = \frac{A}{P} = \frac{A}{B} = \frac{By_n}{B} = y_n$

i.e. $\qquad R \simeq y.$

$$Q = AV = A\,\frac{1}{n}\,(R)^{2/3} s_b^{1/2} = By_n \cdot \frac{1}{n}\,(y_n)^{2/3} s_b^{1/2}$$

$$\therefore \qquad \frac{Qn}{B\, s_b^{1/2}} = y_n^{5/3}$$

$$\therefore \qquad y_n = \left(\frac{Qn}{B\, s_b}\right)^{3/5} \tag{10.27}$$

Hence a direct solution of normal depth exists.

(c) *When the Channel is Rectangular*

In Equation (10.25), $z = 0$

$\therefore$ Equation (10.23) becomes:

$$\left(\frac{Qn}{B^{8/3}\,s_b^{1/2}}\right) = \frac{Y_n^{5/3}}{(1+Y_n)^{2/3}} \qquad (10.28)$$

The solution of Y_n is implicit, i.e. there is no direct solution.

Barr and Das[21] (ICE, London, 1986) had developed an approximate direct solution of normal depth for rectangular channel which gives fairly good results. Their equation is:

$$Y_n = \frac{y_n}{B} = \left(\frac{Qn}{B^{8/3}\,s_b^{1/2}}\right)^{3/5}\left[1+0.855\left(\frac{Qn}{B^{5/3}\,s_b^{1/2}}\right)^{3/5}\right] \qquad (10.29)$$

Similarly for a trapezoidal channel also, Barr and Das[21] gave another approximate direct solution for normal depth in the trapezoidal channel, which may be written as:

$$Y_n = \frac{y_n}{B} = \left(\frac{Qn}{B^{8/3}\,s_b^{1/2}}\right)^{3/5}\left[1-\frac{0.69}{1+\dfrac{3}{\left(\dfrac{Qn}{B^{8/3}\,s_b^{1/2}}\right)^{3/5}\cdot z}}\right] \qquad (10.30)$$

In the trial and error solution, it is required to assume a value of y_n for iteration. If the direct solutions given by Barr and Das in Equations (10.29) and (10.30) are taken as the first assumed values of y_n for rectangular and trapezoidal channel, the solution converges to an exact y_n in one or two trials.

EXAMPLE 10.13 A rectangular channel 6 m wide carries a discharge of 5 m^3/sec at a slope of 0.006. Compute normal depth when Manning's $n = 0.014$.

Solution: $Q = AV = By_n \dfrac{1}{n}\left(\dfrac{By_n}{B+2y_n}\right)^{2/3}(s_b)^{1/2}$

$$\therefore \quad 5 = 6y_n \times \frac{1}{0.014}\left(\frac{6y_n}{6+2y_n}\right)^{2/3}(0.006)^{1/2} \qquad (A)$$

Solution of y_n is implicit.

To start the trial and error method,

Assume a trial value of y_n by Barr and Das Equation (10.29)

i.e.

$$Y_n = \frac{y_n}{B} = \left(\frac{Qn}{B^{8/3}\,s_b^{1/2}}\right)^{3/5}\left[1+0.855\left(\frac{Qn}{B^{8/3}\,s_b^{1/2}}\right)^{3/5}\right]$$

$$\frac{y_n}{6} = \left(\frac{5\times 0.014}{6^{8/3}\times(0.006)^{1/2}}\right)^{3/5}\left[1+0.855\left(\frac{5\times 0.014}{6^{8/3}\times(0.006)^{1/2}}\right)^{3/5}\right]$$

$$y_n = 0.33585 \text{ m}$$

Substituting $y_n = 0.33585$ m in (A)

$$Q \simeq 5.01 \text{ m}^3/\text{sec} \simeq \text{L.H.S.}$$

$$\therefore \quad y_n \simeq 0.33585 \text{ m.}$$ **Ans.**

EXAMPLE 10.14 A trapezoidal channel of 10 m bottom width, side slope 1.5H : 1V, has a bed slope of 0.0003. The value of Manning's $n = 0.012$. If the depth of flow is 3 m, compute the velocity and discharge.

Solution: $z = 1.5,\ s_b = 0.0003,\ B = 10 \text{ m},\ n = 0.012,\ y = 3 \text{ m}$

$$A = (B + zy)y = [(10 + 1.5 \times 3) \times 3] \text{ m}^2 = 43.5 \text{ m}^2$$

$$P = B + 2y\sqrt{1+z^2} = [10 + 2 \times 3 \times \sqrt{1+1.5^2}\,] \text{ m} = 20.81665 \text{ m}$$

$$R = \frac{A}{P} = \frac{43.5}{20.81665} \text{ m} = 2.089673 \text{ m}$$

$$V = \frac{1}{n} R^{2/3} s_b^{1/2} = \left[\frac{1}{0.012}(2.089673)^{2/3}(0.0003)^{1/2}\right] \text{m} = 2.3592 \text{ m/sec}$$

$$\therefore \quad Q = AV = (43.5 \times 2.3592) \text{ m}^3/\text{sec} = 102.625 \text{ m}^3\text{sec.}$$ **Ans.**

EXAMPLE 10.15 In a wide rectangular channel if the normal depth is increased by 20%, find the increase in discharge.

Solution: Wide rectangular channel, $B >>> y_n$

$$\therefore \quad P \simeq B \quad \therefore R = \frac{A}{B} = y_n$$

$$\therefore \quad Q_1 = A_1V_1 = (B_1Y_{n1})\left(\frac{1}{n}\right)(Y_{n1})^{2/3}(s_b)^{1/2} \qquad \text{(A)}$$

$\because$ If depth increases by 20%, $Y_2 = 1.2Y_{n1}$

$$\therefore \quad Q_2 = A_2V_2 = (B_1 \times 1.2Y_{n1})\left(\frac{1}{n}\right)(1.2Y_{n1})^{2/3} s_b^{1/2} \qquad \text{(B)}$$

$$\therefore \quad \frac{Q_1}{Q_2} = \frac{(B_1y_{n1})\frac{1}{n}(Y_{n1})^{2/3}(s_b)^{1/2}}{(B \times 1.2Y_{n1})\left(\frac{1}{n}\right)(1.2Y_{n1})^{2/3} s_b^{1/2}}$$

$$\therefore \quad \frac{Q_1}{Q_2} = \frac{Y_{n1}}{1.2Y_{n1}}\left(\frac{Y_{n1}}{1.2Y_{n1}}\right)^{2/3}$$

$$\frac{Q_1}{Q_2} = \frac{1}{1.2}\left(\frac{1}{1.2}\right)^{2/3} = 1/1.335$$

$$\therefore \quad Q_2 = 1.335\ Q_1$$

i.e. the discharge increases by 35.5%. **Ans.**

10.10 SPECIFIC ENERGY, SPECIFIC FORCE AND CRITICAL DEPTH COMPUTATION

The concept of specific energy was first proposed by Boris A. Bakhmeteff[2,5] in 1912. His concept has made the study of open channel flow analysis easier and simpler.

We have already shown that total energy or head of flowing in a channel is:

$H = Z + Y + \dfrac{V^2}{2g}$ where energy is measured with respect to a datum line. If energy is measured with respect to the channel bottom, i.e. $z = 0$, the energy is called specific energy and is denoted by E.

$$\therefore \quad E = y + \frac{V^2}{2g} \tag{10.31}$$

Equation (10.31) may be written as:

$$E = y + \frac{Q^2}{2g\,A^2} \tag{10.32}$$

i.e. $V = \dfrac{Q}{A}$.

Equation (10.32) shows that for a given discharge, specific energy E is the function of depth y only ($\because$ Area A is also a function of depth). If depth of y is plotted against E for a given discharge, a specific energy curve or diagram is obtained (Figure 10.8).

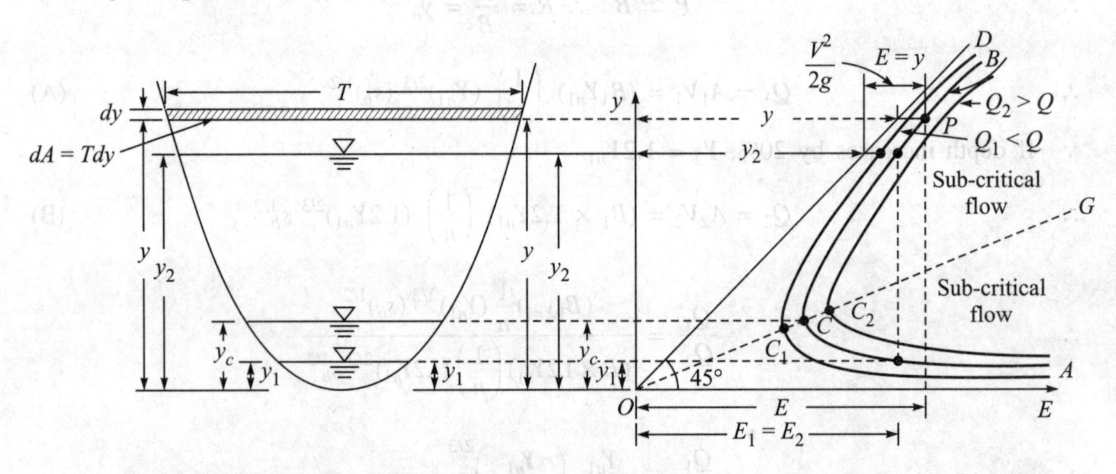

Figure 10.8 Specific energy curves for different given discharges Q, Q_1 and Q_2.

The curves with different discharges like Q, $Q_2 > Q$, $Q_1 < Q$ have two limbs. Consider the curve with Q (in the middle) which has two limbs AC and CB. The line OD is drawn at 45° to the horizontal which has the equation $E = y$. At point P, the ordinate represents y and abscissa represents $\left(y + \dfrac{V^2}{2g}\right)$.

The curves have some characteristics. They are asymptotic to the abscissa and to the line $E = y$. As the depth increases from low depth, i.e. from super-critical depth, the E value decreases upto a minimum value at C and after that the E value increases. This minimum depth at C is called depth y_c where specific energy E is minimum, which will be shown later. For two different depths y_1 and y_2, the same specific energy is obtained, and these two depths y_1 and y_2 with the same specific energy are called alternate depths. If the discharge increases ($Q_2 > Q$), or decreases ($Q_1 > Q$), the points of minimum specific energy, i.e. p_2 and p_1 on specific energy curves also follow their alignment towards right or left which is shown in the figure. The point joining C_1, C, C_2 ..... by a line OG represents the critical state of flow. The zone below this line is the super-critical flow zone and that above is the sub-critical flow zone.

The condition for minimum specific energy at a given value of Q may be obtained to be the critical state of flow as follows:

$$E = y + \frac{V^2}{2g} = y + \frac{Q^2}{2g\,A^2}$$

As Q is constant, by differentiating E with respect to y for minimum and equating it with zero, we get:

$$\frac{dE}{dy} = 1 + (-2)\frac{Q^2}{2g\,A^3}\frac{dA}{dy} = 0$$

from Figure 10.8, $dA = Tdy$

$$\therefore \quad 1 + (-2)\frac{Q^2\,T}{2g\,A^3} = 0$$

$$\therefore \quad \frac{Q^2\,T}{gA^3} = 1 \tag{10.33}$$

which is the equation representing the condition of minimum specific energy. But the expression $\frac{Q^2\,T}{gA^3}$ represents the (Froude's number)^2 which may be shown as:

$$\frac{Q^2\,T}{gA^3} = \frac{(Q^2)}{g(A^2)(A/T)} = \frac{V^2}{gD} \quad \because \frac{Q^2}{A^2} = V^2, \ \frac{A}{T} = D$$

but

$$\frac{V^2}{gD} = \left(\frac{V}{\sqrt{gD}}\right)^2 = F_r^2 \quad \because \frac{V}{\sqrt{gD}} = F_r$$

$$\frac{Q^2\,T}{gA^3} = F_r^2 = 1 \ \therefore \ F_r = 1, \text{ which means that the flow is critical.}$$

Again the same condition of critical flow may be obtained for a given value of E when Q is the maximum.

Specific force may be explained from the momentum [Equation (10.8)].
Equation (10.8) is:

$$P_1 - P_2 + W \sin\theta - F_f = \frac{w}{g} Q(V_2 - V_1) \tag{10.8}$$

If the channel is short, horizontal and prismatic, F_f and $W \sin\theta$ are neglected.

$\therefore$ Equation (10.8) becomes $P_1 - P_2 = \dfrac{w}{g} Q(V_2 - V_1)$

$$\frac{wQV_1}{g} + wA_1\bar{x}_1 = \frac{w}{g} QV_2 + wA_2\bar{x}_2 \quad \because P_1 = wA_1\bar{x}_1$$

and $$\because P_2 = wA_2\bar{x}_2$$

$\therefore$ $$\frac{QV_1}{g} + A_1\bar{x}_1 = \frac{Q}{g} V_2 + A_2\bar{x}_2$$

or $$\frac{Q}{g}\cdot\left(\frac{Q}{A_1}\right) + A_1\bar{x}_1 = \frac{Q}{g}\cdot\frac{Q}{A_2} + A_2\bar{x}_2$$

or $$\frac{Q^2}{gA_1} + A_1\bar{x}_1 = \frac{Q^2}{gA_2} + A_2\bar{x}_2 \tag{10.34}$$

The two sides of Equation (10.34) are analogous and hence may be reduced to a general form for a short channel section as:

$$F = \frac{Q^2}{gA} + A\bar{x} \tag{10.35}$$

The first term represents the momentum of flow for a unit time per unit weight, while the second term is also a force per unit weight of water. Thus F represents a force per unit weight per unit time and F is called specific force and Equation (10.35) is the specific force equation. Equation (10.35) was first formulated by Bresse[8], (1860) in Paris.

Specific force curve can similarly be obtained by plotting F against depth y as shown in Figure 10.9. Here also when specific force is minimum, depth is critical. Two depths y_1 and y_2 give the same specific force and these two depths are the depth before and after the hydraulic jump, respectively, that will be discussed later.

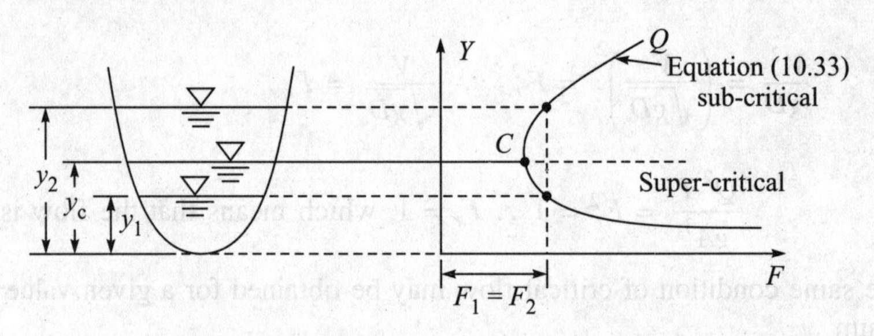

Figure 10.9 Specific force curve diagram.

The critical depth y_e which occurs at minimum E or F may be computed. Direct computation is possible for the channel of rectangular, triangular and parabolic channels. But for the channel of trapezoidal channel, computation is implicit. Trial and error method of solution is required.

When the channel is rectangular,

Critical flow condition is $$\frac{Q^2T}{gA^3} = 1 \qquad (10.33)$$

for rectangular channel $T = B$.

$$\therefore \quad \frac{Q^2 B}{g(By_c)^3} = 1$$

or $$\left(\frac{Q^2}{B^2}\right)\frac{B}{By_c^3} = 1$$

or $$\frac{g^2}{g\, y_c^3} = 1 \quad \left(\text{Unit discharge of} = \frac{Q}{B}\right)$$

$$\therefore \quad y_c^3 = \frac{g^2}{g}$$

$$\therefore \quad y_c = \left(\frac{g^2}{g}\right)^{1/3} \qquad (10.36)$$

Thus critical depth y_c in a rectangular channel is obtained for a given discharge by Equation (10.33) as follows:

$$E_c = y_c + \frac{V_c^2}{2g} = y_c + \frac{Q^2}{2g\,A_c^2} = y_c + \frac{(Q^2)}{2g\,(B^2)y_c^2} = y_c + \frac{(g^2)}{2\,(g)\,y_c^3}$$

$$E_c = y_c + \frac{(g)^2}{2\,(g)\,y_c^2} = Y_c + \frac{y_c^3}{2y_c^2} = y_c + \frac{y_c}{2} = \frac{3}{2}y_c$$

$$\therefore \quad E_c = \frac{3}{2}y_c \qquad (10.37)$$

When the channel is triangular:

Substituting value of T and A of triangular channel in Equation (10.31):

$$\frac{Q^2\,2zy_c}{g\,(zy_c^2)^3} = 1$$

Simplifying: $$y_c = \left(\frac{2Q^2}{gz^2}\right)^{1/5} \qquad (10.38)$$

Similarly, $$E_c = \frac{5}{4} y_c \tag{10.39}$$

When the channel is parabolic:

$$T = K y_c^{1/2},\ A = \frac{2}{3} y_c T = \frac{2}{3} y_c K y_c^{1/2} = K \frac{2}{3} y_c^{3/2}$$

Substituting T and A in $\frac{Q^2 T}{g A^3} = 1$ and simplifying:

$$y_c = \left(\frac{27 Q^2}{8 g K^2} \right)^{1/4} \tag{10.40}$$

Similarly, $$E_c = \frac{4}{3} y_c. \tag{10.41}$$

When the channel is trapezoidal:

$$T = B + 2 z y_c,\ A = (B + z y_c) y_c$$

Substituting T and (A) in $\frac{Q^2 T}{g A^2} = 1$,

$$\frac{Q^2 (B + 2 z y_c)}{g[(B + z y_c)\, y_c]^3} = 1 \tag{10.42}$$

For known Q, B, g and z, values of solution y_c is implicit. Solution of y_c can be obtained by trial and error or by method of iteration by a small computer programme.

10.11 HYDRAULIC JUMP

In the field of open channel flow, hydraulic jump is one of the most important topics that has been extensively investigated and studied. The first description of hydraulic jump was given by Leonardo da Vinci (1452–1519). After a gap of three hundred years, it was Bidone, an Italian engineer who first investigated it experimentally in 1818. This had led Belanger (1828) to distinguish between sub-critical to super-critical slopes as he observed that in a steep slope channel, the jump is produced by a barrier in uniform flow. Thereafter, comprehensive studies were undertaken by many investigators. Outstanding works about hydraulic jump can be credited to Bresse (1860), Darcy and Bazin (1863), Matzke (1936), Escande (1938), Rouse[12] (1958), and many others. In honour of Bidone[4], hydraulic jump in Italy is named "il salto di Bidone" (jump of Bidone).

The theory of jump developed in early days was confirmed only for horizontal frictionless channel where the weight of water upon the behaviour of a jump was ignored in the analysis.

Hydraulic jump is defined as a local phenomenon of sudden or abrupt rise of water when the flow condition changes from sub-critical to super-critical. This sudden rise of water is accompanied by turbulent rollers. Since the depth of the water changes rapidly, in a relatively short length of reach of the channel, it is classified as Rapidly Varied Flow (RVF).

It occurs frequently below a regulating sluice gate, at the point of over fall spillway section of dam, when steep channel slopes suddenly turns flat, when there is an under water obstruction, downstream of narrow section, in irrigation canal fall etc. Figure 10.10 shows the occurrence of a hydraulic jump in practical field.

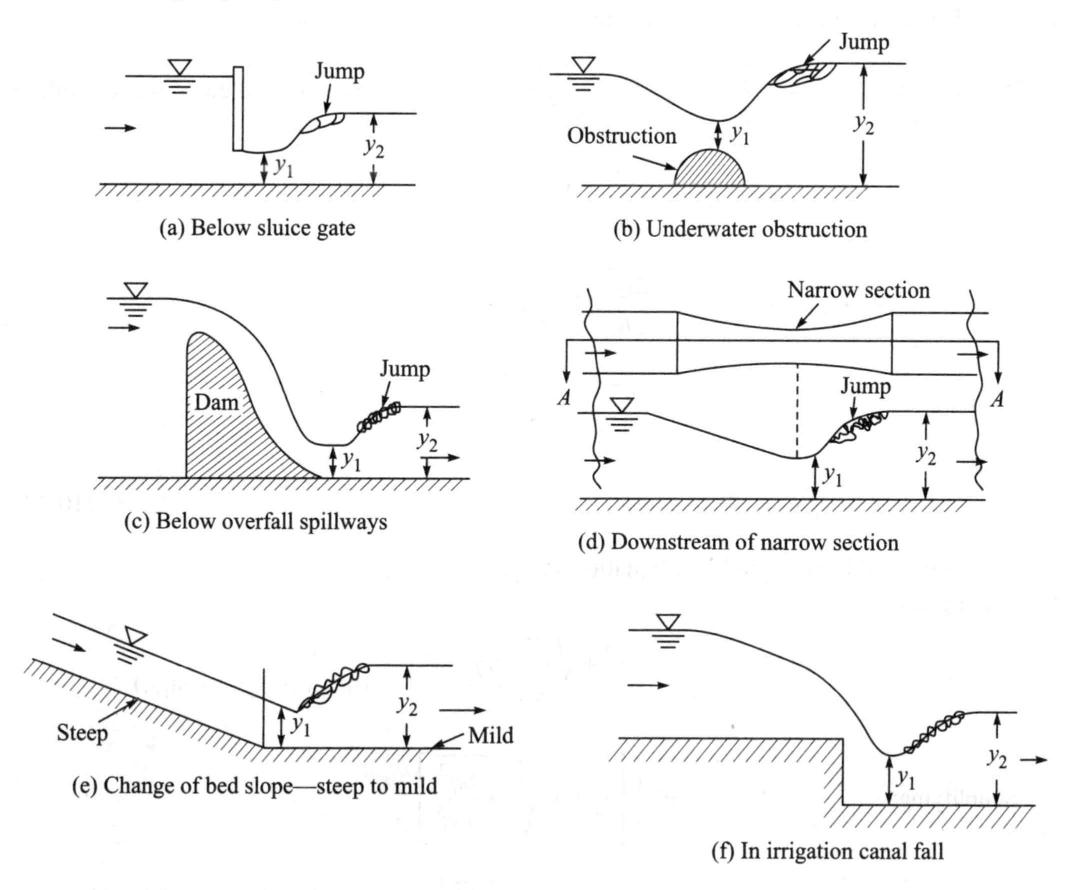

Figure 10.10 Occurrences of hydraulic jump in different field situations.

This hydraulic jump is used for dissipating energy below over fall spillway, in a canal fall as an energy dissipator to protect the bed against erosion, for increasing the weight of an apron to reduce uplift force under masonry structure, to mix chemicals for water purification in a water plant, to raise the water level in an irrigation canal, to indicate the presence of super-critical flow; to remove air pockets in a water supply line, to aerate water for city water supply, to transfer gases in chemical processes, and as flow measuring devices, among other applications. Out of all the above uses, the use of energy dissipator below hydraulic structures like dams etc. is of the most important.

Analytical analysis of hydraulic jump is presented here only for rectangular channel. We start with specific force [Equation (10.34)].

i.e.
$$\frac{Q^2}{gA_1} + A_1\bar{x}_1 = \frac{Q^2}{gA_2} + A_2\bar{x}_2 \qquad (10.43)$$

For a rectangular channel with bottom width B, Equation (10.34) may be written as:

$$\frac{Q^2}{g\,By_1} + By_1 \cdot \frac{y_1}{2} = \frac{Q^2}{gBy_2} + By_2 \cdot \frac{y_2}{2}$$ (where y_1 and y_2 are the depths before and after the jump, respectively)

Dividing by $\frac{2}{B}$, $$y_1^2 + \frac{2Q^2}{gB^2 y_1} = y_2^2 + \frac{2Q^2}{gB^2 y_2}$$ (They are called sequent depths.)

or

$$y_2^2 - y_1^2 = \frac{2Q^2}{gB^2}\left(\frac{1}{y_1} - \frac{1}{y_2}\right)$$

$$(y_2 + y_1)\,(y_2 - y_1) = \frac{2Q^2}{gB^2} \cdot \left(\frac{y_2 - y_1}{y_1 y_2}\right)$$

$$y_1 y_2 (y_2 + y_1) = \frac{2q^2}{g} \quad \therefore \frac{Q^2}{B^2} = q^2$$

$$y_1 y_2^2 + y_1^2 y_2 - \frac{2q^2}{g} = 0 \tag{10.44}$$

Equation (10.44) is a quadratic equation in y_2,.
Solving for y_2.

$$y_2 = \frac{-y_1^2 + \sqrt{y_1^4 + 8q^2 y_1/g}}{2y_1}$$ (neglecting $-ve$ sign)

Simplifying: $$y_2 = \frac{1}{2}\left[y_1 + y_1\sqrt{1 + \frac{8q^2}{gy_1^3}}\right]$$

or

$$\frac{y_2}{y_1} = \frac{1}{2}\left[-1 + \sqrt{1 + Fr_1^2}\right] \tag{10.45}$$

$$\because \quad \frac{8q^2}{gy_1^3} = 8\left(\frac{q^2}{g}\right)\frac{1}{y_1^3} = \frac{8V_1^2 y_1^2}{[gy_1^3]} = 8\left(\frac{V_1}{\sqrt{gy_1}}\right)^2 = 8Fr_1^2.$$

$\because$ $\left(\frac{V_1}{\sqrt{gy_1}}\right)$ = Froude's number of flow at depth y_1 of the rectangular channel.

Similarly solving quadratic equation y_1 Equation (10.44):

$$\frac{y_1}{y_2} = \frac{1}{2}\left[-1 + \sqrt{1 + Fr_2^2}\right] \tag{10.46}$$

The term $\left(\dfrac{y_2}{y_1}\right)$ in Equation (10.45) is called the sequent depths ratio for initial Froude's number Fr_1 in a horizontal frictionless rectangular channel and is known as the Belanger (1828) momentum equation. For a high value of $Fr_1 > 8$, Equation (10.45) approximated to be:

$$\frac{y_2}{y_1} = 1.41\, Fr_1^2 \tag{10.47}$$

In a hydraulic jump, a considerable energy is lost.

If E_1 and E_2 are the specific energy values before and after the jump, respectively, $E_1 - E_2 = \Delta E$ (loss of energy)

$$\therefore \quad \Delta E = \left(y_1 + \frac{V_1^2}{2g}\right) - \left(y_2 + \frac{V_2^2}{2g}\right)$$

$$= (y_1 - y_2) + \left[\frac{Q^2}{2g(By_1)^2} - \frac{Q^2}{2g(By_2)^2}\right]$$

$$= (y_1 - y_2) + \frac{q^2}{2g}\left(\frac{1}{y_1^2} - \frac{1}{y_2^2}\right)$$

$$= (y_1 - y_2) + \frac{q^2}{g} \cdot \frac{1}{2} \cdot \frac{(y_2 + y_1)(y_2 - y_1)}{y_1^2\, y_2^2} \tag{A}$$

From Equation (10.44), $$\frac{q^2}{g} = y_1 y_2 \left(\frac{y_2 + y_1}{2}\right) \tag{B}$$

Substituting $\dfrac{q^2}{g}$ value from (B) in (A):

$$\Delta E = (y_1 - y_2) + y_1 y_2 \left(\frac{y_2 + y_1}{2}\right) \cdot \frac{1}{2} \cdot \frac{(y_2 + y_1)(y_2 - y_1)}{y_1^2\, y_2^2}$$

$$= -(y_2 - y_1) + \frac{(y_2 + y_1)^2 (y_2 - y_1)}{4 y_1 y_2}$$

$$= (y_2 - y_1)\left[-1 + \frac{(y_2 + y_1)^2}{4 y_1 y_2}\right]$$

$$= (y_2 - y_1)\left[\frac{-4 y_1 y_2 + y_2^2 + y_1^2 + 2 y_2 y_1}{4 y_1 y_2}\right]$$

$$= (y_2 - y_1)\, \frac{(y_2 - y_1)^2}{4 y_1 y_2}$$

$$\therefore \quad \Delta E = \frac{(y_2 - y_1)^3}{4 y_1 y_2} \tag{10.48}$$

which gives the loss of energy in jump. The term (y_2/y_1) is equal to the height of jump h_j. Length of jump L_j cannot be exactly evaluated. However L_j varies from 5 to 7 terms of h_j.

Further analytical analysis in different channel sections, such as in a sloping channel is beyond the scope of this chapter. Here only preliminary knowledge of jump is presented.

10.12 STEADY GRADUALLY VARIED FLOW

The development of theory of steady gradually varied flow dates to the eighteenth century. The contributors in the early stages are Daniel Bernoulli, Bussinesq and Ponceplet. The next contributors are Belanger, Bresse, Bakhmeteff and Jaeger, out of which the contribution of Belanger is believed to be outstanding.

In order to derive the dynamic equation, consider the profile of gradually varied flow in an elementary length dx of the channel (Figure 10.11) where θ is small.

$$\text{Total head of energy } H = z + y + \frac{V^2}{2g}$$

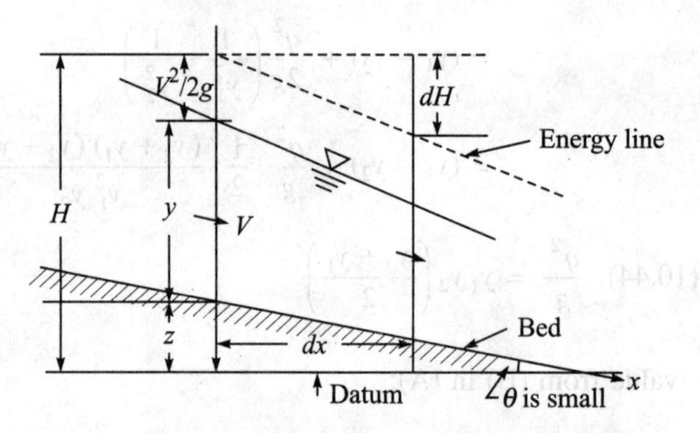

Figure 10.11 Sketch to derive dynamic equation.

Taking the bottom as x-axis and differentiating with respect to x:

$$\frac{dH}{dx} = \frac{dz}{dx} + \frac{dy}{dx} + \frac{d\left(\frac{V^2}{2g}\right)}{dx}$$

In the above equation, $\frac{dH}{dx}$ is the slope of energy line. $\therefore$ $\frac{dH}{dx} = -s_f$ (friction slope is $-ve$ as H decreases as x increases, $\frac{dz}{dx} = -s_b$, as x increases, z decreases.

Hence the above equation is written as:

$$s_b - s_f = \frac{dy}{dx} + \frac{d}{dy}\left(\frac{V^2}{2g}\right)\frac{dy}{dx}$$

$$s_b - s_f = \frac{dy}{dx}\left[1 + \frac{d}{dy}\left(\frac{V^2}{2g}\right)\right] \quad \text{(A)}$$

To simplify the term:

$$\frac{d}{dy}\left(\frac{V^2}{2g}\right),$$

$$\frac{d}{dy}\left(\frac{Q^2/A^2}{2g}\right) = \frac{Q^2}{2g}\left(-2\frac{dA}{A^3 dy}\right) \quad \text{But } \frac{dA}{dy} = T$$

$$= \frac{-Q^2 T}{g\,A^3}$$

∴ (A) becomes:

$$s_b - s_f = \frac{dy}{dx}\left[1 - \frac{Q^2 T}{gA^3}\right]$$

$$\therefore \quad \frac{dy}{dx} = s_b - s_f \bigg/ \left(1 - \frac{Q^2 T}{gA^3}\right) \quad (10.49)$$

which is the dynamic equation of gradually varied flow.

The term $\frac{Q^2 T}{gA^3} = F_r^2$ (already shown)

∴ Equation (10.49) may also be written as:

$$\frac{dy}{dx} = \frac{s_b - s_f}{1 - F_r^2} \quad (10.50)$$

when the channel is wide rectangular $B = T$, $R \simeq y$

$$\therefore \quad \frac{Q^2 T}{gA^3} = \frac{Q^2 B}{g(B^2)\,y^3 \cdot B} = \frac{(q^2)}{(g)\,y^3} = \frac{y_c^3}{y^3} = \left(\frac{y_c}{y}\right)^3 \text{(A)} \quad \because \left(\frac{q^2}{g}\right) = y_c^3$$

When Manning's equation is used, $Q = (By)\frac{1}{n}\,(y^{2/3})\,s_f^{1/2}$ (B)

when the flow is uniform, $y = y_n$ (normal depth)

$$\therefore \quad Q = (By_n)\frac{1}{n}\,(y_n)^{2/3}\,s_b^{1/2} \quad \text{(C)}$$

Equating (C) and (B): $B_Y^{5/3}\,\frac{1}{n}\,s_f^{1/2} = By_n^{5/3}\,\frac{1}{n}\,s_b^{1/2}$

$$\therefore \quad \left(\frac{s_f}{s_b}\right) = \left(\frac{y_n}{y}\right)^{10/3} \quad \text{(C)}$$

Equation (10.49) is written as:

$$\frac{dy}{dx} = \frac{s_b\left(1 - \frac{s_f}{s_b}\right)}{1 - \frac{Q^2 T}{gA^3}} \quad \text{(D)}$$

Substituting $\left(\frac{s_f}{s_b}\right)$ from (C) and $\left(\frac{Q^2 T}{gA^3}\right)$, from (A) in (D):

$$\frac{dy}{dx} = s_b \frac{1 - (y_n/y)^{10/3}}{1 - (y_c/y)^3} \quad (10.51)$$

Similarly when Chezy's equation is used, the dynamic equation in wide rectangular channel can be shown to be:

$$\frac{dy}{dx} = \frac{1 - \left(\frac{y_n}{y}\right)^3}{1 - \left(\frac{y_c}{y}\right)^3} \quad (10.52)$$

The surface profiles of gradually varied flow may be backwater profiles or drawdown profiles. Based on slopes (horizontal, mild, steep, critical and adverse slopes) and the zones of profiles, they are further classified into 12 different profiles. Details of these analyses are not included in this chapter. Interested readers are advised to study books on open channel flow.

The profiles are computed by solving the dynamic Equation (10.49). Unfortunately, the equation cannot be solved analytically. Therefore, the equation is solved for computation of surface profiles by approximate methods like step method, graphical integration method, integration by varied flow functions, numerical integration method, etc. The present trend of solution of the equation or computation of profiles are by different numerical methods like trapezoidal integration, Runge-Kutto's, Milines, Pickwards, Tailor series expansion, Newton-Raphson or finite difference methods with the help of a high speed computer. Thus the computation of profiles or solution of the equation is not discussed in the chapter.

10.13 CONTINUITY AND DYNAMIC EQUATIONS OF UNSTEADY GRADUALLY VARIED FLOW

The continuity equation for unsteady flow may be derived by considering the conservation of mass in infinitesimal space dx between two channel sections. In unsteady flow, discharge Q changes with distance with a rate $\frac{\partial Q}{\partial x}$ and the depth changes with time with a rate $\frac{\partial y}{\partial t}$. Hence the change of discharge through space in time dt is $\left(\frac{\partial Q}{\partial x}\right) dx \cdot dt$. The corresponding change in channel storage in space in time dt is $(T\, dx)\frac{\partial y}{\partial t} dt$ which is equal to $\frac{\partial A}{\partial t} \cdot dx \cdot dt \because \partial A = T \partial y$.

Since water is incompressible, the net change in discharge plus the change in storage should be equal to zero, i.e.

$$\left(\frac{\partial Q}{\partial x}\right) dx \cdot dt + \left(\frac{\partial A}{\partial t}\right) dx \cdot dt = 0$$

Dividing by $dx \cdot dt$,

$$\frac{\partial Q}{\partial x} + \frac{\partial A}{\partial t} = 0 \tag{10.53}$$

which is the equation of continuity in unsteady open channel flow.

The dynamic equation of unsteady open channel flow was deduced by A.J.C. Barre de Saint-Venant (1873) in Italy (Paris) and this equation is known as the Saint-Venant equation. It is used popularly till today to solve the unsteady flow in different situations.

The derivation of this equation is presented here by energy approach.

Writing energy equation in two sections of the reach of length ∂x:

$$z_1 + y_1 + \frac{V_1^2}{2g} = z_2 + y_2 + \frac{V_2^2}{2g} + \text{Losses} \tag{A}$$

Losses in unsteady flow are due to friction plus loss of energy head for acceleration produced. Thus Losses = $h_f + h_a$

But $$s_f = \frac{h_f}{\partial x} \quad \therefore\ h_f = s_f\, \partial x$$

To find the loss of energy head h_a,

In unsteady flow $\frac{\partial V}{\partial t}$ exists, i.e. acceleration is produced by causing a force to be produced. Production of force will cause loss of energy and this loss, in head form per unit weight, is h_a.

$\therefore$ Force produced due to this acceleration/unit time = $\left(\frac{\text{mass}}{\text{unit time}}\right) \times$ acceleration

$$= \left(\frac{1}{g}\right)\left(\frac{\partial V}{\partial t}\right)$$

Work done by this force = force × distance

$= \frac{1}{g}\frac{\partial V}{\partial t} dx$ and this work done is loss of energy head due to acceleration = h_a

$$\therefore \quad h_a = \frac{1}{g}\frac{\partial V}{\partial t}\partial x$$

Substituting these two losses in (A):

$$z_1 + y_1 + \frac{V_1^2}{2g} = z_2 + y_2 + \frac{V_2^2}{2g} + s_f\,\partial x + \frac{1}{g}\frac{\partial V}{\partial t}\partial x$$

or

$$(y_2 - y_1) + (z_2 - z_1) + \left(\frac{V_2^2}{2g} - \frac{V_1^2}{2g}\right) + s_f\,\partial x + \frac{1}{g}\frac{\partial V}{\partial t}\partial x = 0$$

or $$\partial y + \partial z + \partial\left(\frac{V^2}{2g}\right) + s_f\,\partial x + \frac{1}{g}\frac{\partial V}{\partial t}\,\partial x = 0$$

or $$\frac{\partial y}{\partial x} + \frac{\partial z}{\partial x} + \frac{\partial\left(\frac{V^2}{2g}\right)}{\partial x} + s_f + \frac{1}{g}\frac{\partial V}{\partial t} = 0$$

or $$g\frac{\partial y}{\partial x} + g\frac{\partial z}{\partial x} + \frac{2V\,\partial V}{2\partial x} + g s_f + \frac{\partial V}{\partial t} = 0$$

$\therefore$ $$\frac{\partial z}{\partial x} = -s_b,$$

$\therefore$ $$\frac{\partial V}{\partial t} + V\frac{\partial V}{\partial x} + g\left(\frac{\partial y}{\partial x} - s_b + s_f\right) = 0 \qquad (10.54)$$

which is the dynamic equation of unsteady open channel flow. If the flow is steady, $\frac{\partial V}{\partial t} = 0.$

Substituting $\frac{\partial V}{\partial t} = 0$ in Equation (10.54) and simplifying the dynamic equation steady gradually varied flow, i.e. Equation (10.49) may be obtained, i.e.

$$\frac{dy}{dx} = \frac{s_b - s_f}{1 - \frac{Q^2 T}{gA^3}} \qquad (10.49)$$

The solutions of these two unsteady flows [Equations (10.53) and (10.54)] is analytically impossible as they constitute a set of hyperbolic partial differential equations. Therefore, these equations are solved by different numerical solutions. Continued research on these two equations in different situations by different numerical solutions has been investigated till the twenty-first century. The different numerical methods that have been used are method of characteristics, finite difference, both explicit and implicit, and finite element methods. Some of the numerical schemes like Lax diffusive, Lamda, Gabeetti, Maecormack, and Preissimann are popular both in conservative and non-conservative forms, but the discussions of these are beyond the scope of this chapter.

EXAMPLE 10.16 Find the specific energy of water flowing through a rectangular channel of width 4 m with discharge 8 m³/sec. The depth of flow of the channel is 2.5 m.

Solution: $B = 4$ m, $y = 2.5$ m, $Q = 8$ m³/sec

$\therefore$ Velocity $V = \frac{Q}{A} = \left(\frac{8}{4\times 2.5}\right)$ m/sec $= 0.8$ m/sec.

Specific energy $= E = y + \frac{V^2}{2g} = \left(2.5 + \frac{0.8^2}{2\times 9.81}\right)$ m of water

$E = 2.53262$ m of water **Ans.**

EXAMPLE 10.17 A rectangular channel carries 18 m^3/sec. The width of the channel is 6 m. Find the critical depth, critical velocity and value of minimum specific energy.

Solution: $Q = 18$ m^3/sec, $B = 6$ m,

$$\therefore \quad \frac{Q}{B} = q \text{ unit discharge (m}^3\text{/sec/m).}$$

$$\therefore \quad q = \frac{18}{6} = 3 \text{ m}^3\text{/sec/m.}$$

For rectangular channel, the equation of critical depth is $y_c = \left(\dfrac{q^2}{g}\right)^{1/3}$

$$\therefore \quad y_c = \left(\frac{3^2}{9.81}\right)^{1/3} \text{ m}$$

$$y_c = 0.97168 \text{ m} \qquad \textbf{Ans.}$$

Now $q = y_c V_c$ where V_c is the critical velocity.

$$\therefore \quad V_c = \frac{3}{0.97168} \text{ m/sec} = 3.0874 \text{ m/sec} \qquad \textbf{Ans.}$$

Minimum specific energy occurs when the depth is critical and we have the following equation for rectangular channel:

$$E_{\min} = \frac{3}{2} y_c = (\frac{3}{2} \times 0.97168) \text{ m of water}$$

$$E_{\min} = 1.4575 \text{ m of water.} \qquad \textbf{Ans.}$$

EXAMPLE 10.18 Show that the relation between alternate depth y, and y_2 in a rectangular channel can be expressed by

$$y_c^3 = \frac{2y_1^2 y_2^2}{(y_1 + y_2)}$$

Solution: If y_1 and y_2 are alternate depths, the specific energy values in these two depths are the same.

$$\therefore \quad E = y_1 + \frac{V_1^2}{2g} = y_2 + \frac{V_2^2}{2g}$$

$$\text{or} \quad (y_1 - y_2) = \frac{V_2^2}{2g} - \frac{V_1^2}{2g} = \frac{Q^2}{2g(By_2)^2} - \frac{Q^2}{2g(By_1)^2}$$

$$\text{or} \quad (y_1 - y_2) = \frac{1}{2g} \cdot \left(\frac{Q^2}{B^2}\right)\frac{1}{Y_2^2} - \frac{1}{2g}\left(\frac{Q^2}{B^2}\right)\frac{1}{Y_1^2}$$

or $$(y_1 - y_2) = \frac{1}{2}\left(\frac{q^2}{g}\right)\frac{1}{Y_2^2} - \frac{1}{2}\left(\frac{q^2}{g}\right)\frac{1}{Y_1^2} \quad \because \left(\frac{Q}{B}\right)^2 = q^2$$

or $$(y_1 - y_2) = \frac{y_c^3}{2y_2^2} - \frac{y_c^3}{2y_1^2} \quad \because \left(\frac{q^2}{g}\right)^{1/3} = y_c$$

or $$(y_1 - y_2) = y_c^3\left(\frac{y_1^2 - y_2^2}{2y_2^2\,y_1^2}\right)$$

or $$(y_1 - y_2) = y_c^3\,\frac{(y_1 + y_2)\,(y_1 - y_2)}{2y_1^2\,y_2^2}$$

or $$1 = y_c^3\,\frac{(y_1 + y_2)}{2y_1^2\,y_2^2}$$

$$\therefore \quad y_c^3 = \frac{2y_1^2\,y_2^2}{(y_1 + y_2)} \quad \text{Shown.}$$

EXAMPLE 10.19 The specific energy in a 5 m wide rectangular channel is 4 m of water. The discharge through the channel is 20 m^3/sec. Determine the alternate depths.

Solution: An analytical solution of alternate depths involves a solution of the cubic equation and hence a rigorous trial and error solution is required. It may, however, be solved by the following graphical procedure:

$$E = y + \frac{V^2}{2g} = y + \frac{Q^2}{2gA^2} = y + \frac{Q^2}{2gB^2y^2} = y + \frac{q^2}{2gy^2}$$

$$\therefore \quad 4 = y + \frac{4^2}{2gy^2} \quad \because g = \frac{Q}{B} = \frac{20}{5} = 4 \text{ m}^3\text{/sec/m}$$

$$\therefore \quad 4 = y + \frac{16}{2 \times 9.81 y^2} \quad \text{(A)} \quad E = 4 \text{ m}, B = 5.$$

Find first critical depth $y_c = \left(\frac{q^2}{g}\right)^{1/3} = \left(\frac{4^2}{9.81}\right)^{1/3} = 1.177$ m.

Now plot E Vs y (See Figure 10.12).

Find values of E at different values of y from Equation (A).

y	E
$y = y_c = 1.177$	1.765 $\because E = \frac{3}{2}y_c$
$y = 0.5$	3.7619
$y = 0.45$	4.477

$y = 0.6$	2.865
$y = 2$	2.20
$y = 3.5$	3.57
$y = 4.0$	4.05
$y = 3.95$	4.002

$\therefore$ At $y = 3.95$ m, $E \cong 4.0$ m

The corresponding super-critical depth when $y = 3.95$ m from the graph is = 0.48 m.

$\therefore$ The alternate depths are 0.48 m and 3.95 m. **Ans.**

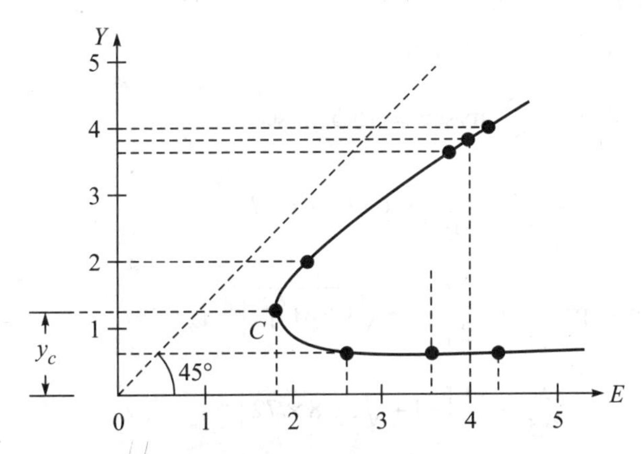

Figure 10.12 Visual of Example 10.17 to compute alternate depths.

EXAMPLE 10.20 A rectangular channel flowing with super-critical depth ($F_{r_1} = 8.5$) is used as an energy dissipator by creating a hydraulic jump. If the energy loss in the jump is 5 m of water, find the sequent depths, i.e. depths before and after the jump.

Solution: The channel is rectangular. If y_2 is the depth after the jump and y_1 the depth before the jump, loss of energy $\Delta E = \dfrac{(y_2 - y_1)3}{4y_1y_2}$ (10.48)

To have a jump, we have the equation:

$$\frac{y_2}{y_1} = \frac{1}{2}\left[-1+\sqrt{1+8Fr_1^2}\right] \tag{10.42}$$

$$Fr_1 = 8.5 \quad \text{(given)}$$

$$\therefore \quad \frac{y_2}{y_1} = \frac{1}{2}\left[-1+\sqrt{1+8\times 8.5^2}\right] = 11.5308 \text{ m.}$$

$$\therefore \quad \left.\begin{aligned} y_2 &= 11.5308\, y_1 \\ \text{or} \quad y_1 &= 0.0867\, y_2 \end{aligned}\right\} \tag{B}$$

Substituting y_1 in (A):

$$\Delta E = 5 = \frac{(y_2 - 0.0867\,y_2)^3}{4\times 0.0867\times y_2}$$

$$5 = \frac{y_2^3(1-0.0867)^3}{y_2 \times 0.3472} = \frac{0.0761548\, y_2^2}{0.3472} = 2.1934\, y_2^2$$

$\therefore$ $y_2 = 1.5098$ m. **Ans.**

$\therefore$ $y_1 = 0.0868 \times 1.5098$ m $= 0.131$ m **Ans.**

EXAMPLE 10.21 A spillway having a downstream horizontal apron of depth 0.5 m flows with unit discharge 7.75 m^3/sec/m. What tail water depth is needed to have a hydraulic jump? Find the length of the jump and the energy loss.

Solution: $q = 7.75$ m^3/sec/m, $y_1 = 0.5$ m

$$V_1 y_1 = q \therefore V_1 = \frac{7.75}{0.5} \text{ m/sec} = 15.5 \text{ m/sec}$$

$$\therefore \quad Fr_1 = \frac{V}{\sqrt{g y_1}} = \frac{15.5}{\sqrt{9.81 \times 0.5}} = 6.9986 \simeq 7$$

Using the relationship, $\dfrac{y_2}{y_1} = \dfrac{1}{2}\left[-1+\sqrt{1+8Fr_1^2}\right]$

$$\frac{y_2}{0.5} = \frac{1}{2}\left[-1+\sqrt{1+8\times 72}\right]$$

$y_2 = 4.706$ m. **Ans.**

Length of jump L_j is taken 5 to 7 times h_j (i.e. $y_2 - y_1$)

Taking average $L_j = \left(\dfrac{5+7}{2}\right) h_j = 6(y_2 - y_1) = 6(4.706 - .5)$

$\therefore$ $L_j = 25.236$ m **Ans.**

Loss of energy $\Delta E = \dfrac{(y_2 - y_1)^3}{4 y_1 y_2} = \left(\dfrac{(4.706-0.5)^3}{4\times 4.706\times 0.5}\right)$ m

$\Delta E = 7.905$ m of water **Ans.**

EXAMPLE 10.22 A steady gradual flow occurs in a rectangular channel of breadth 10 m. Bed slope is 1 in 4000. Water flows with a velocity of 1.0 m/sec. Initial depth of water is 1.5 m. Energy slope is 0.00004. Find the water surface slope.

Solution: Water surface slope is $\dfrac{dy}{dx}$.

The dynamic equation of steady gradually varied flow is:

$$\frac{dy}{dx} = \frac{S_b - S_f}{1 - \dfrac{Q^2 T}{g A^3}}$$

$$s_b = 1/4000,\ s_f = 0.00004,\ T = B\ 10\ \text{m},\ y = 1.5\ \text{m}$$

$$A = By = (10 \times 1.5)\text{m}^2 = 15\ \text{m}^2,\ V = 1\ \text{m/sec}$$

$$Q = AV = 15 \times 1 = 15\ \text{m}^3/\text{sec}.$$

$$\therefore \quad \frac{dy}{dx} = \frac{\dfrac{1}{4000} - 0.00004}{1 - \dfrac{15^2 \times 10}{9.81 \times 15^3}} = 0.0002253 \qquad \textbf{Ans.}$$

EXAMPLE 10.23 While measuring the discharge in a river with unsteady flow, the depth y was found to increase at a rate of 0.06 m/hour. The surface width of the river is 30 m and discharge at this section is 35 m^3/sec. Estimate the discharge at section 1 km upstream.

Solution: The continuity equation in unsteady flow gives:

$$\frac{\partial Q}{\partial x} + \frac{\partial A}{\partial t} = 0 \quad \text{But } \partial A = T\ dy$$

$$\therefore \quad \frac{\partial Q}{\partial x} + T\frac{\partial Y}{\partial t} = 0$$

$$\frac{Q_2 - Q_1}{\partial x} = -T\frac{\partial Y}{\partial t} \quad \text{But } T\frac{\partial Y}{\partial t} = 30 \times \frac{0.6}{60 \times 60} = \frac{0.06}{120}\ \text{m}^2/\text{sec}$$

If Q_1 is the discharge at upstream 1 km distance,

$$Q_1 = Q_2 + \frac{T\partial y}{\partial t}\cdot\partial x = 35 + \frac{0.06}{120}\ (1 \times 1000)$$

$$\because \quad \partial x = 1\ \text{km} = 1000\ \text{m}\ \ Q_2 = 35\ \text{m}^3/\text{sec}$$

$$Q_1 = 35.5\ \text{m}^3/\text{sec} \qquad \textbf{Ans.}$$

10.13 CONCLUSION

Open channel flow itself is a vast subject in which people have been writing books. This chapter has been presented as an introduction to open channel flow for the beginners. Almost all the topics of open channel flow have been mentioned, equations have been derived, but have not been continued further. But the Dynamic equations of unsteady continuity and momentum, are new addition in the book than in general books on Hydraulics and Fluid Mechanics. Approximate Direct solutions of normal depth in uniform flow computation by Barr and Das[21] (1986) for both rectangular and trapezoidal channel is another new part added in the chapter.

The details of theory of Open Channel Flow developed since ancient time by Bernoulli, Euler, Chezy, Manning[3] Bazin and Darcy[19], Whitham[23], Dressler[24], Chow[1], Henderson[13], Stoker[14], Rouse[12] and many more till date which cannot be shown just in one chapter. But the theory presented here, numerical solved examples about 21 numbers, help the student to learn many things about open channel flow. Student can practice to apply the theory in the given 10 problems. References given will help the ambitious undergraduate and P.G. students a lot for further study.

The works of Saint-Venant[6], Belanger[7], Bussinesq[10], Bakmeteff[5], Bidone[4], Manning[3], Ritter[22] in early stages and further works by Chow[1], Jaeger[11], Rouse[12], Stoker[14], Henderson[13], Choudhury[17], French[18] Ranga-Raju[16], Subramanya[15], Barr[20,21] Ligget[25], Brutsaert[26] and few more in the subsequent years are important to be referred.

PROBLEMS

10.1 Find the discharge in a trapezoidal channel of bottom width 6 m, side slope 1H : 3V. The depth of flow is 3 m and the Chezy's $C = 60$. The slope of the bed is 1 in 5000. (**Ans.** 23.26 m^3/sec)

10.2 Find the discharge in a triangular channel having a vertex angle of 60°. Take $C = 50$ and bed slope 1/1500. The depth of flow is 6 m. (**Ans.** 32.864 m^3/sec)

10.3 A trapezoidal channel with slopes 1H : 1V is designed for discharge of 9 m^3/sec at velocity 1.5 m/sec so that the amount of concrete lining on the bed and the sides is the minimum. Calculate the area of the lining required per metre length of the channel. (**Ans.** 6.62 m^2)

10.4 A power canal of trapezoidal has to be excavated through hard clay at the least cost. Determine the dimensions of the channel. Data for the channel are: $Q = 15$ m^3/sec, $s_b = 1/2000$, $n = 0.02$. (**Ans.** $B = 2.96$ m, $y = 2.6$ m)

10.5 Find the discharge through a circular pipe of diameter 4 m if the depth of water in the pipe is 1.33 m. The pipe is laid at a slope of 1 in 1500. Take $C = 60$. (**Ans.** 4.89 m^3/sec)

10.6 A rectangular channel of width 60 cm carries discharge equal to 100 litres/sec. If Chezy's $C = 56$, find the bed slope in uniform flow depth 30 cm. Find also the conveyance K of the channel. (**Ans.** 1/1525, 3.9 m^3/sec)

10.7 The specific energy for a 6 m wide rectangular channel is 5 m of water. If the discharge of water through the channel is 24 m^3/sec, determine the alternate depths of flow. (**Ans.** 4.831 m, 1.457 m)

10.8 The depth of flow of water at a certain section of a rectangular channel of 5 m wide is 0.6 m. The discharge is 15 m^3/sec. If a hydraulic jump takes place on the downstream side, find the depth after the jump. (**Ans.** 1.474 m)

10.9 Find the slope of the free water surface in a rectangular channel of width 15 m having a depth of flow 4 m. The discharge through the channel is 40 m^3/sec. Take $s_b = 1/1000$ and $n = 0.023465$. (**Ans.** 1/5430)

10.10 In the measurement of discharge in a river, it was found that the depth increases at a rate of 0.5 m/hour. If the discharge at that section is 15 m^3/sec, and surface width is 15 m, estimate the discharges at a section 1.2 km upstream. (**Ans.** 17.5 m^3/sec)

REFERENCES

1. Chow, Ven Te, *Open Channel Hydraulics*, McGraw-Hill Inc., 1959.
2. Bakhmeteff, B.A., *Hydraulics of Open Channels*, McGraw-Hill Inc., 1932.

3. Manning, R., "On the flow of water on open channel and pipes" Trans. *ICE*, Ireland, Col. 20, pp. 161–207, 1831.
4. Bidone, G., "Observation on height of hydraulic jump", A report presented in December 12, 1819, *Meeting of the Royal Academy of Science*, Turin, 1819.
5. Bakhmeteff, R.A., *Varied Flow in Open Channels*, St. Petersburg, Russia, 1912.
6. Saint-Venant, A.J.C. Barre'de, "Theory of the nonpermanent movement of water with application to the floods of rivers and to the introduction of tides with their beds", *Competes rendus seances de l'Academic des science*", Vol. 73, pp 147–154 and 234–240, 1871.
7. Blasius, H., "The law of similitude for friction in fluids", *Forschungsheft des vereins deutcher Ingenieure*, No. 131, Bertin, 1913.
8. Bresse, J.A. Ch., "Course in Applied Mechanics", Part 2, *Hydraulics*, Hallet–Bachelier Paris, 1960.
9. Belanger, J.B., *Summary of Lectures*, Paris, 1838.
10. Boussinesq, "Essay of the theory of water flow", Memmoires prisentis par divers savants a' l'Academic des sciences, Paris, Vol. 23, Ser. 2, No. 1, pp. 1–680, 1877.
11. Jaeger, C., "Engineering Fluid Mechanics" translated from German by P.O. Wolf, Blackie and Sons Ltd., London and Glasgow, 1956.
12. Rouse, H. and Ince, S., "History of Hydraulics", Iawa institute of Hydraulic Research, Iawa city Iawa, 1958.
13. Henderson, F.M., *Open Channel Flow*, Macmillan, New York, 1966.
14. Stoker, J.J., *Water Waves*, Wiley Interscience, New York, 1957.
15. Subramanya, K., *Flow in Open Channels*, Tata McGraw-Hill, New Delhi, 1988.
16. Rang-Raju, K.G., *Flow Through Open Channels*, Tata McGraw-Hill, New Delhi, 1981.
17. Choudhury, M.H., *Open Channel Flow*, Practice-Hall of India, New Delhi, 1994.
18. French, R.H., *Open Channel Hydraulics*, McGraw-Hill, Singapore, 1980.
19. Darcy, H. and H. Bazin, "Experimental research on back water and wave propagation", Vol. II of *Hydraulic Researches*, Academic des Sciences, Paris, 1865.
20. Barr, D.I.H., "Two additional methods of direct solution of Colebrook–White Function" *Proc. ICE*, Part 2, London, June, 1976.
21. Barr, D.I.H. and M.M. DAS, "Direct Solution of Normal Depth using Mannings Equation" *Proc. ICE*, London, Part 2, 81, pp. 315–333, Sept., 1986.
22. Ritter, A., "Die Fortpflanzung der Wasserwellen", *Z. ver. deut. Ing*. 36. 1892.
23. Whitham, G.B., "Effect of hydraulic resistance in the dam-break problem", *Proc. Roy. Society of London*, Series A, Vol. 227, 1955.
24. Dressler R.F., "Hydraulic resistance effect upon dam-break functions", *Jour. of Res., Nat'l Bur. of Standard S.* Vol. 49, No. 3, 1952.
25. Ligget, *Basic Equations of Unsteady Flow*, Edited by K. Mahmood and V. Yevjerech, Vol. 1, Chapter 2, water Resources Publication, Fort Collins, Co., 1969.
26. Brutsaert, W.H., "Saint-Venant equations experimentally varied" Jour. Hy. Div., *ASCE*, Vol 97, No Hy 9, 1971.

Chapter 11

Laminar Flow

11.1 INTRODUCTION

The definition of laminar flow has already been given in Chapter 5. It occurs at a low velocity of flow. Viscous force predominates over the inertia force. Therefore, laminar flow is also called viscous flow. The ratio of inertia force to viscous force is called the Reynolds number (R_e).

i.e. $$R_e = \frac{\text{Inertia force}}{\text{Viscous force}} = \frac{\text{Mass} \times \text{Acceleration}}{\text{Shear stress} \times \text{Area}} = \frac{\rho \times \text{Volume} \times \dfrac{\text{Velocity}}{\text{Time}}}{\left(\mu \dfrac{dV}{dy}\right) A}$$

or $$R_e = \frac{\rho \times \dfrac{\text{Volume}}{\text{Time}} \times \text{Velocity}}{\dfrac{dV}{dy} A} = \frac{\rho (Q) V}{\dfrac{dV}{dy} A} = \frac{\rho (AV) V}{\left(\dfrac{V}{L}\right) A} = \frac{\rho L^2 V^2}{\dfrac{V}{L} L^2}$$

$$R_e = \frac{\rho L^2 V^2}{\mu V L}$$

$$R_e = \frac{\rho L V}{\mu} = \frac{LV}{\dfrac{\mu}{\rho}} = \frac{VL}{\nu} \tag{11.1}$$

If the value of R_e is less than 2000 in the pipe flow, the flow is laminar or viscous. Thus viscosity μ (or ν) plays a main role and this viscosity of fluid induces relative motion within the fluid as the fluid forms layers, one over the other, which, in turn, gives rise to shear stresses (λ). The magnitude of this viscous shear is maximum at the boundary and gradually goes on decreasing with an increase in the distance from the boundary. The shear stresses so produced result in the development of a resistance to flow. In order to overcome this resistance to flow, the pressure drops from section to section in the direction of the flow. Hence, pressure gradient exists in laminar flow. An expression relating shear stress λ and pressure gradient $\dfrac{\partial p}{\partial x}$ in laminar flow is being developed to help analyse the various situations of laminar flow like laminar flow between two parallel plates, Hagen–Poiseville flow, Couette flow, Darcy's law, Stoke's law, application of laminar flow (i.e. lubrication mechanics), dash-pot mechanism and measurement of viscosity, among others. All these will be discussed in this chapter.

11.2 RELATION BETWEEN SHEAR STRESS AND PRESSURE GRADIENT

Consider an infinitesimal fluid element of sides dx, dy and dz as shown in Figure 11.1.

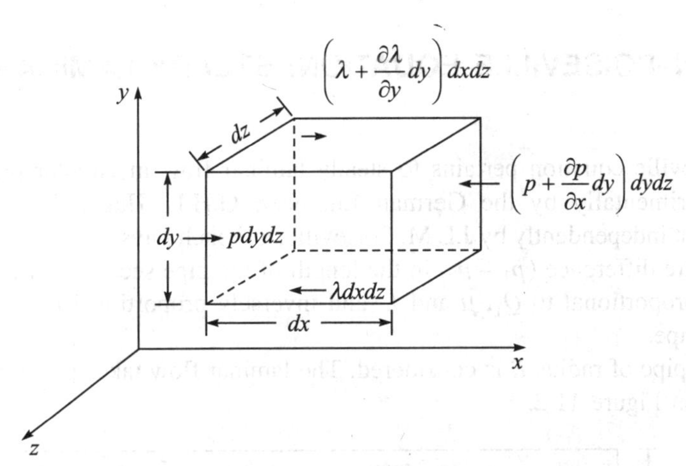

Figure 11.1 Sketch to establish relation between pressure and shear stress.

$$\text{The net shearing force from Figure 11.1} = \left(\lambda + \frac{\partial \lambda}{\partial y} dx\right) dxdz - \lambda dxdz$$

$$= \left(\frac{\partial \lambda}{\partial y}\right) dx \cdot dy \cdot dz$$

$$\text{Net pressure force on the element} = pdydx - \left(p + \frac{\partial p}{\partial x} dx\right) dydz$$

$$= \left(\frac{-\partial p}{\partial x}\right) dx \cdot dydz$$

Summation of these net two forces for steady motion (i.e. acceleration is zero), i.e. resultant force = 0.

$$\therefore \quad \left(\frac{\partial \lambda}{\partial y}\right) dxdydz + \left(-\frac{\partial p}{\partial x}\right) dx \cdot dy \cdot dz = 0$$

$$\therefore \quad \left(\frac{\partial \lambda}{\partial y}\right) - \left(\frac{\partial p}{\partial x}\right) = 0$$

$$\therefore \quad \frac{\partial \lambda}{\partial y} = \frac{\partial p}{\partial x} \tag{11.2}$$

According to Newton's equation of viscosity, [i.e. Equation (11.3)]

$\lambda = \mu \dfrac{\partial v}{\partial y}$, where v is the velocity at a distance y from the boundary.

Substituting λ in Equation (11.2), the following differential equation is obtained:

$$\mu \frac{\partial^2 v}{\partial y^2} = \frac{\partial p}{\partial x} \tag{11.3}$$

11.3 HAGEN–POISEVILLE EQUATION: STEADY LAMINAR FLOW IN PIPE

The Hagen–Poiseville equation pertains to steady laminar flow in circular pipes. It was first determined experimentally by the German Engineer, G.H.L. Hagen in 1839 and almost simultaneously but independently by J.L.M. Poiseville, a French physician in 1840. The equation shows that pressure difference $(p_1 - p_2)$ in the length of the pipe section L with steady uniform flow is directly proportional to Q_1, μ and L, and inversely proportional to D^4 where D is the diameter of the pipe.

A horizontal pipe of radius R is considered. The laminar flow takes place in the pipe in the direction shown in Figure 11.2.

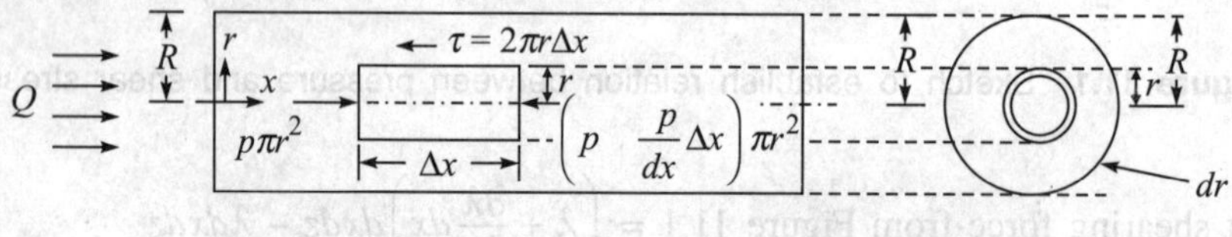

Figure 11.2 Laminar flow through circular pipe.

A fluid element of Δx, with radius r, sliding in the cylindrical fluid element of radius $(r + dr)$. As the flow is steady laminar, there is no acceleration, and the summation force in the cylindrical fluid element is zero, i.e. $p\pi r^2 - \left(p + \dfrac{\partial p}{\partial x}\Delta x\right)\pi r^2 - \lambda\, 2\pi r\, \Delta x = 0$

$$-\left(\frac{\partial p}{\partial x}\Delta x\right)\pi r^2 - \lambda 2\pi r \Delta x = 0$$

$$-\frac{\partial p}{\partial x} r - 2\lambda = 0$$

$$\lambda = \left(-\frac{\partial p}{\partial x}\right)\frac{r}{2} \tag{11.4}$$

Equation (11.4) shows that shear stress varies linearly along the radius of the pipe. At centre $r = 0$, $\therefore$ $\lambda = 0$ and at the pipe wall $r = R$.

$$\therefore \qquad \lambda_0 = \left(-\frac{\partial p}{\partial x}\right)\frac{R}{2} \tag{11.5}$$

A negative sign indicates the decrease of p as x increase i.e. in the direction of the flow. Equation (11.4) holds good for both laminar and turbulent flow as in the derivation, no assumption is made regarding the nature of the flow.

Applying Bernoulli's equation in two sections of the pipe.

$$\frac{p_1}{w} + \frac{V_1^2}{2g} + z_1 = \frac{p_2}{w} + \frac{V_2^2}{2g} + z_2 + h_f$$

Here $V_1 = V_2$, $z_1 = z_2$, (horizontal pipe)

$$\therefore \quad \frac{p_1}{w} - \frac{p_2}{w} = h_f \quad \therefore p_1 - p_2 = wh_f$$

$$\therefore \quad \frac{p_1 - p_2}{L} = \frac{wh_f}{L}$$

But

$$\left(\frac{p_1 - p_2}{L}\right) = \left(-\frac{dp}{dx}\right)$$

$$\therefore \quad \left(-\frac{dp}{dx}\right) = \left(\frac{p_1 - p_2}{L}\right) = \frac{wh_f}{L} \tag{11.6}$$

$\therefore$ Equation (11.4) is now written as:

$$\lambda = \frac{wh_f}{2L} r \tag{11.6a}$$

In Newton's equation of viscosity, $\lambda = \mu \frac{\partial v}{\partial y}$ where v is the velocity at a distance y from boundary. In this case $y = R - r$ and $dy = -dr$

$$\therefore \quad \lambda = -\mu \frac{\partial v}{\partial r} \tag{11.7}$$

Substituting this shear stress from Equation (11.7) in Equation (11.4):

$$-\mu \frac{\partial v}{\partial r} = -\frac{\partial p}{\partial x} \cdot \frac{r}{2}$$

$$\therefore \quad \frac{\partial v}{\partial r} = \frac{1}{\mu} \cdot \frac{\partial p}{\partial x} \cdot \frac{r}{2} \tag{11.8}$$

The drop in pressure depends on the distance x and is independent of r, hence the pressure gradient $\left(\frac{\partial p}{\partial x}\right)$ is constant. Now integrating Equation (11.8), we get:

$$v = \frac{1}{4\mu}\left(\frac{\partial p}{\partial x}\right) r^2 + c \tag{11.9}$$

The constant of integration C in Equation (11.9) is obtained by putting the boundary condition $r = R$, and $v = 0$, i.e. the velocity at the boundary is zero.

$$\therefore \quad C = -\frac{1}{4\mu}\left(\frac{\partial p}{\partial x}\right) R^2$$

Equation (11.9) becomes:

$$v = \frac{1}{4\mu}\left(-\frac{\partial p}{\partial x}\right)(R^2 - r^2) \tag{11.10}$$

The maximum velocity occurs when $r = 0$

$$\therefore \quad v_{\max} = \frac{1}{4\mu}\left(-\frac{\partial p}{\partial x}\right)R^2 \tag{11.11}$$

$$v_{\max} = \frac{1}{16\mu}\left(-\frac{\partial p}{\partial x}\right)D^2 \tag{11.11a}$$

Further Equation (11.10) is written as:

$$v = \frac{1}{4\mu}\left(-\frac{\partial p}{\partial x}\right)R^2 - \frac{1}{4\mu}\left(-\frac{\partial p}{\partial x}\right)r^2$$

$$v = v_{\max}\left[1 - \left(\frac{r}{R}\right)^2\right] \tag{11.12}$$

Figure 11.3 shows the distribution of velocity and shear stress across a pipe section. The velocity distribution is parabolic (Equation 11.12) and shear stress distribution is linear (Equation 11.4).

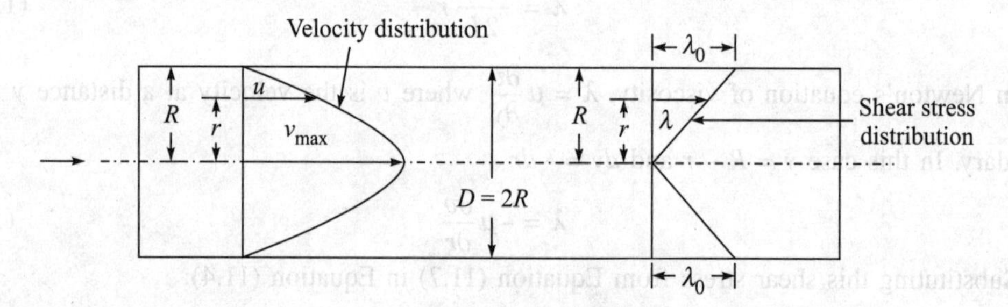

Figure 11.3 Velocity and shear stress distribution in pipe for laminar flow.

To find an expression of discharge Q flowing through the pipe is obtained as follows:

If dQ is the discharge that flows through the elementary rising of thickness dr in Figure 11.2, then $dQ = vdA = v(2\pi r)dr$.

$$\therefore \quad dQ = \frac{1}{4\mu}\left(-\frac{\partial p}{\partial x}\right)(R^2 - r^2)(2\pi r)\,dr$$

Integrating: $Q = \dfrac{\pi}{2\mu}\left(-\dfrac{\partial p}{\partial x}\right)\displaystyle\int_0^R (R^2 - r^2)r\,dr$

$$\therefore \quad Q = \frac{\pi}{8\mu}\left(-\frac{\partial p}{\partial x}\right)R^4 \tag{11.13}$$

If D is the diameter, $R = \frac{D}{2}$,

$$\therefore \qquad Q = \frac{\pi}{128\mu}\left(-\frac{\partial p}{\partial x}\right)D^4 \qquad (11.14)$$

The mean velocity of flow is $V = \frac{Q}{A} = \frac{Q}{\pi R^2} = \frac{Q}{\pi/4D^2}$

$$\therefore \qquad V = \frac{1}{32\mu}\left(-\frac{\partial p}{\partial x}\right)D^2 \qquad (11.15)$$

Comparing Equations (11.11a) and (11.15), the relationship of average V and v_{max} is:

$$V = \frac{1}{2}v_{max} \qquad (11.16)$$

The distance from the centre of the pipe (i.e. the value of r) where the local velocity v is equal to the average velocity, may be obtained by equating V from Equation (11.16) to the Equation (11.12)

$$V = \frac{1}{2}v_{max} = v_{max}\left[1-\left(\frac{r}{R}\right)^2\right]$$

or

$$1 - \left(\frac{r}{R}\right)^2 = \frac{1}{2}$$

$$\therefore \qquad \left(\frac{r^2}{R^2}\right) = \frac{1}{2}$$

$$\therefore \qquad r = \frac{R}{\sqrt{2}} \qquad (11.17)$$

i.e. at $r = \frac{R}{\sqrt{2}}$, or the radial distance $\frac{R}{\sqrt{2}}$ the mean velocity occurs.

Again Equation (11.15) gives

$$\left(-\frac{\partial p}{\partial x}\right) = \frac{32\mu V}{D^2}$$

or

$$\left(\frac{p_1 - p_2}{L}\right) = \frac{32\mu V}{D^2}$$

$$\therefore \qquad p_1 - p_2 = \frac{32\mu VL}{D^2} \qquad (11.18)$$

or

$$p_1 - p_2 = \frac{32\mu}{D^2}\cdot\left(\frac{Q}{(\pi/4)\,D^2}\right)L \qquad \because V = \frac{Q}{(\pi/4)\,D^2}$$

$$\therefore \qquad p_1 - p_2 = \frac{128\mu QL}{\pi D^4} \qquad (11.18a)$$

Equation (11.18a) is the Hagen–Poiseuille equation for laminar flow in a circular pipe. The most important feature of the Hagen–Poiseuille equation is that it involves viscosity, discharge, length and diameter for difference of pressure between two points in a length L of the pipe. There is no empirical coefficient or experimental factor. If h_f represents the drop of pressure head, i.e. Equation (11.6), then from Equation (11.18)

$$h_f = \frac{p_1 - p_2}{w} = \frac{32\mu VL}{wD^2}$$

$$\therefore \quad h_f = \frac{32\mu VL}{\rho g D^2} = \frac{2 \times 32\, LV^2}{\left(\frac{\rho VD}{\mu}\right)(2gD)} = \frac{64}{R_e}\left(\frac{LV^2}{2gD}\right)$$

$$\therefore \quad h_f = \frac{64}{R_e}\left(\frac{LV^2}{2gD}\right) \tag{11.19}$$

The Darcy–Weisbach equation is (Equation (7.3)):

$$h_f = f\left(\frac{LV^2}{2gD}\right) \tag{7.3}$$

Comparing Equations (7.3) and (11.19), we get

$$f = \frac{64}{R_e} \tag{11.20}$$

which gives the equation of friction factor for laminar flow in the pipe.

Further from Equation (11.6), we get:

$$\frac{p_1 - p_2}{w} = h_f \quad \text{and} \quad h_f = \frac{fLV^2}{2gD}$$

and from Equation (11.5), $\lambda_0 = \left(-\frac{\partial p}{\partial x}\right)\frac{R}{2} = \left(\frac{p_1 - p_2}{L}\right)\frac{D}{4}$.

Substituting $(p_1 - p_2)$:

$$\lambda_0 = \left(\frac{wh_f}{L}\right)\frac{D}{4} = \frac{w\,(fLV^2)\cdot D}{4\,(2gD)\,L}$$

or

$$\lambda_0 = \frac{wfV^2}{8g} = \frac{\rho g \cdot fV^2}{8g} = \frac{\rho f V^2}{8}$$

or

$$\frac{\lambda_0}{\rho} = \frac{f}{8}V^2$$

$$\therefore \quad \sqrt{\frac{\lambda_0}{\rho}} = \sqrt{\frac{f}{8}}\,V \tag{11.21}$$

The expression $\sqrt{\dfrac{\lambda_0}{\rho}}$ has the dimension of velocity, which is called shear or friction velocity, denoted normally by V_* and hence:

$$\sqrt{\frac{\lambda_0}{\rho}} = \sqrt{\frac{f}{8}}\, V = V_* \tag{11.22}$$

In steady uniform flow, certain power is required to overcome the resistance to flow. This resistance to flow is offered by the pressure gradient $\left(-\dfrac{\partial p}{\partial x}\right)$ in the opposite direction of the force of resistance. Pressure gradient is thus the average force per unit volume of fluid. Again power is the rate of doing work. If A and L are the area and length of the pipe, respectively, the total force of resistance is equal to $\left(-\dfrac{\partial p}{\partial x}\right) AL$. If P is the power, then:

$$P = \left(-\frac{\partial p}{\partial x}\right)(AL)\cdot V = \left(\frac{p_1 - p_2}{L}\right) AL\cdot V = (p_1 - p_2)\,(AV)$$

$$\therefore \qquad P = (p_1 - p_2)Q \tag{11.23}$$

$$\because \qquad Q = AV, \left(\frac{-\partial p}{\partial x}\right) = \frac{p_1 - p_2}{L}$$

11.4 LAMINAR FLOW BETWEEN TWO PARALLEL PLATES AT REST

Two plates at rest are shown in Figure 11.4, in which they are placed at a distance B apart through which the laminar flow of fluid takes place. Laminar flow is flowing as shown in the figure from left to right i.e. in the x direction. Consider a fluid element of length Δx, thickness Δy and of width equal to unity. The lower face of the element is at a distance y from the lower plate, where the velocity is v, and the upper face of the elemental fluid is at a distance $(y + \Delta y)$ where the velocity is $(v + \partial v)$. The upper face moves with higher velocity and the lower face moves with lower velocity, and hence it tends to retard its motion. λ is the shear stress at the lower face and $\left(\lambda + \dfrac{\partial \lambda}{\partial y}\Delta y\right)$ at the upper face.

Summing up the pressure and shear forces acting on the element and equating with zero for laminar flow (since there is no acceleration), we get:

$$p\cdot\Delta y \times 1 - \left(p + \frac{\partial p}{\partial x}\Delta x\right)\cdot\Delta y \times 1 - \lambda\,\Delta x \times 1 + \left(\lambda \frac{\partial \lambda}{\partial y}\Delta y\right)\cdot \Delta x \times 1 = 0$$

$$-\left(\frac{\partial p}{\partial x}\Delta x\right)\Delta y + \left(\frac{\partial \lambda}{\partial y}\Delta y\right)\Delta x = 0$$

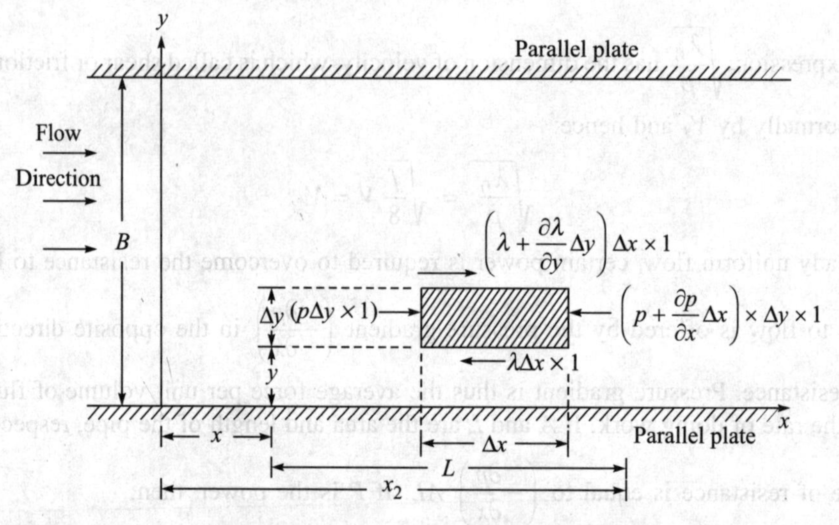

Figure 11.4 Laminar flow between two plates: both plates are at rest.

Dividing by $\Delta x \cdot \Delta y$:

$$\frac{-\partial p}{\partial x} + \frac{\partial \lambda}{\partial y} = 0$$

$$\therefore \qquad \frac{\partial \lambda}{\partial y} = \frac{\partial p}{\partial x} \tag{11.2}$$

i.e. the same Equation (11.2) is obtained.

To find the velocity distribution between the two plates, replace λ by Newton's equation of viscosity $\left(\text{i.e. } \lambda = \mu \frac{\partial^2 v}{\partial y}\right)$ in Equation (11.2) to get:

$$\mu \frac{\partial^2 v}{\partial y^2} = \frac{\partial p}{\partial x} \tag{11.3}$$

Since $\left(\frac{\partial p}{\partial x}\right)$ is independent of y, integrating the above expression twice with respect to y to obtain the equation of velocity, we get:

$$v = \frac{1}{\mu}\left(\frac{\partial p}{\partial x}\right)\frac{y^2}{2} + C_1 y + C_2 \tag{11.3a}$$

where C_1 and C_2 are constant of integration.

To evaluate C_1 and C, put the boundary condition

at $\qquad y = 0,\ v = 0 \text{ and } y = B,\ v = 0$

which gives $C_2 = 0$

and $\qquad C_1 = \frac{-B}{2\mu}\left(\frac{\partial p}{\partial x}\right) = \frac{B}{2\mu}\left(-\frac{\partial p}{\partial x}\right)$

By substituting C_2 and C_1, the velocity expression becomes,

$$v = \frac{1}{2\mu}\left(-\frac{\partial p}{\partial x}\right)(By - y^2) \tag{11.24}$$

Equation (11.24) gives the velocity distribution curve for laminar flow between two parallel plates, which is parabolic.

Differentiating v in Equation (11.24) with respect to y and equating with zero for maximum, we get:

$$\frac{\partial v}{\partial y} = \frac{1}{2\mu}\left(-\frac{\partial p}{\partial x}\right)(B - 2y) = 0$$

$$\therefore \quad B - 2y = 0 \qquad \because \frac{1}{2\mu}\left(-\frac{\partial p}{\partial x}\right) \neq 0$$

$$\therefore \quad y = \frac{B}{2}$$

Thus at $y = \frac{B}{2}$, i.e. at half the distance between them, the velocity is maximum as shown in Figure 11.5.

$$\therefore \quad v_{\max} = \frac{1}{2\mu}\left(-\frac{\partial p}{\partial x}\right)\left[B\cdot\frac{B}{2} - \left(\frac{B}{2}\right)^2\right]$$

$$\therefore \quad v_{\max} = \frac{B}{8\mu}\left(-\frac{\partial p}{\partial x}\right) \tag{11.25}$$

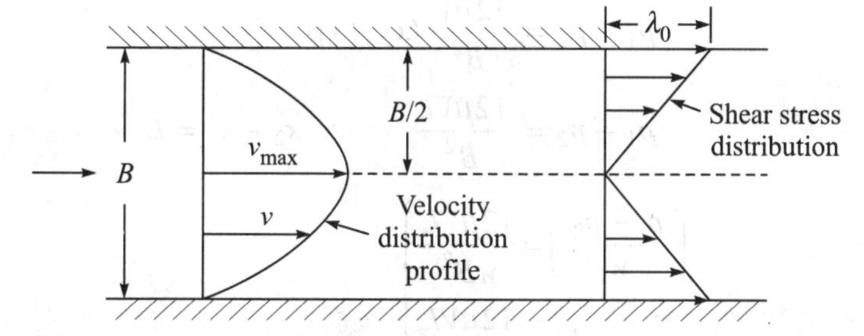

Figure 11.5 Velocity and shear stress distribution between two plates.

If dq is the discharge passing through the strip of dy per unit width, then $dq = v(dy \times 1) = vdy$

Integrating:

$$q = \int_0^B v\,dy = \int_0^B \frac{1}{2\mu}\left(-\frac{\partial p}{\partial x}\right)(By - y^2)\,dy$$

$$q = \frac{1}{2\mu}\left(-\frac{\partial p}{\partial x}\right)\left[\frac{By^2}{2} - \frac{y^3}{3}\right]_0^B$$

$$q = \frac{1}{2\mu}\left(-\frac{\partial p}{\partial x}\right)\left[\frac{B^3}{2} - \frac{B^3}{3}\right]$$

$$q = \frac{1}{2\mu}\left(-\frac{\partial p}{\partial x}\right)\left[\frac{B^3}{6}\right]$$

$$q = \frac{B^3}{12\mu}\left(-\frac{\partial p}{\partial x}\right) \tag{11.26}$$

where q is the discharge between the two plates per unit width of the plates. If V is the average velocity of flow, then:

$$(B \times 1)V = q$$

$$\therefore \quad V = \frac{q}{B} = \frac{B^2}{12\mu}\left(-\frac{\partial p}{\partial x}\right) \tag{11.27}$$

which is two-third of v_{max} (i.e. from Equation 11.25), again the pressure gradient from Equation (11.27) is:

$$\left(-\frac{\partial p}{\partial x}\right) = \frac{12\mu V}{B^2}$$

or

$$-\partial p = \frac{12\mu V}{B^2}\partial x$$

$$\int_{p_1}^{p_2}(-\partial p) = \frac{12\mu V}{B^2}\int_{x_1}^{x_2} dx$$

or

$$p_1 - p_2 = \frac{12\mu V}{B^2}(x_2 - x_1)$$

or

$$p_1 - p_2 = \frac{12\mu VL}{B^2} \qquad \because x_2 - x_1 = L$$

or

$$\left.\begin{aligned} \left(\frac{p_1 - p_2}{w}\right) &= \frac{12\mu VL}{wB^2} \\ \text{or} \quad h_f &= \frac{12\mu VL}{wB^2} \end{aligned}\right\} \tag{11.28}$$

Equation (11.28) shows that the drop of pressure head in the laminar flow varies linearly with velocity.

To find the distribution of shear stress (λ), substitute the equation of v (i.e. Equation 11.24) in Newton's equation of viscosity $\left(\text{i.e., } \lambda = \mu\frac{\partial v}{\partial y}\right)$,

$$\lambda = \mu\frac{\partial}{\partial y}\left[\frac{1}{2\mu}\left(-\frac{\partial p}{\partial x}\right)(By - y^2)\right]$$

or
$$\lambda = \frac{\mu}{2\mu}\left(-\frac{dp}{\partial x}\right)(B - 2y)$$

$\therefore$
$$\lambda = \left(-\frac{\partial p}{\partial x}\right)\left(\frac{B}{2} - y\right) \tag{11.29}$$

Equation (11.29) shows that shear stress varies linearly with y, i.e. the distance from the boundary. At $y = B/2$, $\lambda = 0$,

and at $y = 0$, $\lambda_0 = \left(-\frac{\partial p}{\partial x}\right)\frac{B}{2}$ (11.30)

Figure (11.5) also shows shear stress distribution along with velocity distribution.

11.5 COUETTE FLOW

It is a laminar flow between two parallel flat plates wherein the lower plate is fixed and the upper plate moves with a uniform velocity at a distance B as shown in Figure 11.6. M.F.A. Couette was the first to analyse this laminar flow which is why it is called Couette flow.

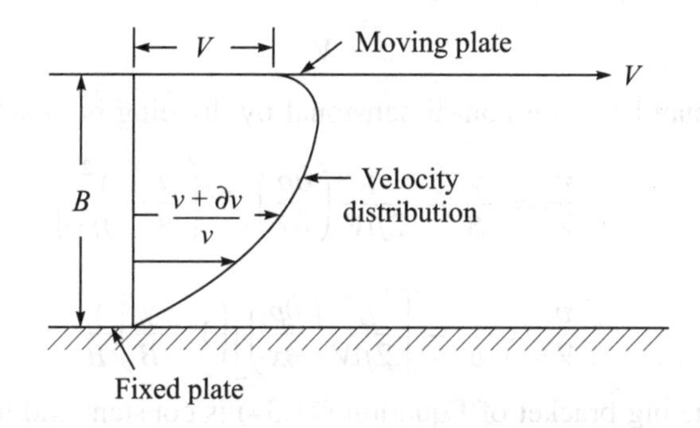

Figure 11.6 Velocity distribution in Couette flow.

The general Equation (11.3a) of laminar flow between two fixed plates is applicable here. Hence:

$$v = \frac{1}{\mu}\left(\frac{\partial p}{\partial x}\right)\frac{y^2}{2} + C_1 y + C_2 \tag{11.3a}$$

The boundary conditions in Couette flow are:

$$v = 0 \text{ at } y = 0$$

$\therefore$
$$C_2 = 0$$

$$v = V \text{ at } y = B$$

$\therefore$
$$V = \frac{1}{\mu}\left(\frac{\partial p}{\partial x}\right)\frac{B^2}{2} + C_1 B$$

$$\therefore \quad C_1 = \frac{V}{B} - \frac{B}{2\mu}\left(\frac{\partial p}{\partial x}\right)$$

$$\therefore \quad v = \frac{1}{\mu}\left(\frac{\partial p}{\partial x}\right)\frac{y^2}{2} + \frac{VY}{B} - \frac{By}{2\mu}\left(\frac{\partial p}{\partial x}\right)$$

$$\therefore \quad v = \frac{V}{B}y - \frac{1}{2\mu}\left(\frac{\partial p}{\partial x}\right)(By - y^2) \tag{11.31}$$

If $V = 0$, the upper plate is fixed and Equation (11.31) reduces to Equation (11.24), i.e.

$$v = \frac{1}{2\mu}\left(-\frac{\partial p}{\partial x}\right)(By - y^2) \tag{11.24}$$

If there is no pressure gradient, Equation (11.31) reduces to

$$v = \frac{V}{B}y \tag{11.32}$$

which shows that the velocity distribution is linear. This is a case of plain Couette flow or simple shear flow.

When $y = B$, Equation (11.31) becomes:

$$v = V \tag{11.33}$$

Equation (11.31) may be made non-dimensional by dividing both sides by V as follows:

$$\frac{v}{V} = \frac{y}{B} - \frac{1}{2\mu V}\left(\frac{\partial p}{\partial x}\right)B^2\left(\frac{y}{B} - \frac{y^2}{B^2}\right)$$

or

$$\frac{v}{V} = \frac{y}{B} - \left[\frac{B^2}{2\mu V}\left(\frac{\partial p}{\partial x}\right)\right]\left(1 - \frac{y}{B}\right)\frac{y}{B} \tag{11.34}$$

The term within the big bracket of Equation (11.34) is constant and may be called the non-dimensionless pressure gradient P_n. Then Equation (11.34) is written as:

$$\frac{v}{V} = \frac{y}{B} + P_n\left(1 - \frac{y}{B}\right)\frac{y}{B} \tag{11.34a}$$

where $P_n = \dfrac{B^2}{2\mu V}\left(-\dfrac{\partial p}{\partial x}\right)$. For different values of P_n, non-dimensional Equation (11.34a) is plotted $\dfrac{y}{B}$ against $\dfrac{v}{V}$ to obtain a family of velocity distribution curves as shown in Figure 11.7.

The Couette flow with a pressure gradient has its application in the hydrodynamic theory of lubrication which will be discussed later.

To find the distribution of shear stress at any section, Equation (11.31) for velocity v is substituted in Newton's equation of viscosity, i.e.

$$\lambda = \mu\frac{d}{dy}\left[\frac{V}{B}y - \frac{1}{2\mu}\left(\frac{\partial p}{\partial x}\right)(By - y^2)\right]$$

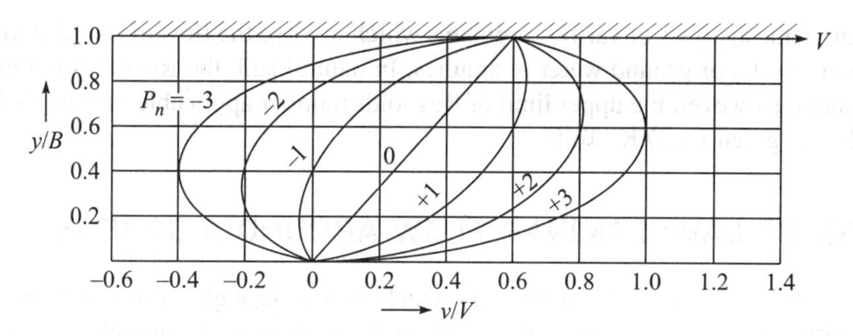

Figure 11.7 Non-dimensional velocity curves in Couette flow.

or
$$\lambda = \mu\left[\frac{V}{B} - \frac{1}{2\mu}\left(\frac{\partial p}{\partial x}\right)(B - 2y)\right]$$

or
$$\lambda = \mu\frac{V}{B} + \left(\frac{-\partial p}{\partial x}\right)\left(\frac{B}{2} - y\right) \tag{11.35}$$

where shear stress λ varies linearly with the distance from the boundary of the fixed plate.

11.6 DARCY'S LAW: LAMINAR FLOW THROUGH POROUS MEDIA

In 1856, Henri Darcy, a French engineer, derived the equation relating average velocity V, coefficient of permeability and hydraulic gradient for laminar flow through porous media. This equation is called Darcy's law. The flow of water or any liquid through soil or any media takes place at low velocity. This media is called porous media. As the water flows or seeps with a low velocity, the flow is laminar and the Reynolds number is very small. On the basis of his experimental evidence, Darcy stated that the discharge through this porous media (i.e., soil) was directly proportional to the head loss h_f and the area of flow section, and inversely proportional to the length L of the soil sample.

In other words,

$$Q \propto \frac{h_f}{L} \cdot A$$

or
$$\frac{Q}{A} \propto \frac{h_f}{L}$$

or
$$V \propto \frac{h_f}{L}$$

$\therefore$
$$V = K\left(\frac{h_f}{L}\right) \tag{11.36}$$

which is Darcy's equation, where $\left(\frac{h_f}{L}\right)$ is the hydraulic gradient and constant K is the coefficient of permeability (i.e., the ability of the porous media to pass through it), having the dimension of velocity. Darcy's law or equation given in Equation (11.36) through porous media is quite useful in the analysis of various problems which involve the flow of liquid at low velocity through porous media or ground water hydraulics. In natural soil, the Reynolds number of flow is less than unity. However, the upper limit of Reynolds number applicable to Darcy's law ranges from 1 to 10, as given by D.K. Todd.

11.7 STOKES LAW: LAMINAR FLOW AROUND A SPHERE

Laminar flow also exists around a solid body which moves through a fluid of infinite extent in which viscosity is the only fluid property affecting the motion of the particle. The accelerating effect is very small and is therefore, neglected. Practical examples of such a falling body through a fluid with some velocity are sand grains falling towards the bottom in water at rest, dust particles falling on the ground surface through air, etc. When the body or the particle moves through the fluid medium, it experiences a resistance which acts in the opposite direction to the flow direction of the body. This resistance is called Drag Force (F_D).

G.G. Stokes analytically developed an expression of the F_D experienced by a sphere of diameter D, moving with velocity V in a fluid medium with dynamic viscosity μ. His analytical equation is:

$$F_D = 3\,\pi\mu\, VD \tag{11.37}$$

This is called Stokes law and it is experimentally verified to be valid for Reynolds number less than 0.2.

When the sphere falls through the fluid under its own weight at a constant velocity V, then the buoyant force plus the resistance (F_D) to its motion must be equal to its weight. Thus it may be written as:

$$\left(\frac{\pi}{6}D^3\right)w_f + 3\pi\mu\, VD = \left(\frac{\pi}{6}D^3\right)w_s$$

where w_f and w_s are the specific weights of the fluid and of the sphere, respectively.

or

$$3\pi\mu VD = \frac{\pi}{6}D^3\,(w_s - w_f)$$

$\therefore$

$$V = \frac{D^2}{18\mu}(w_s - w_f) \tag{11.38}$$

The velocity given by Equation (11.38) is called 'terminal fall velocity of the sphere'.

If V is the velocity of the body in the fluid medium of density ρ, A is the projected area of the body perpendicular to flow direction, then F_D is expressed as:

$$F_D = C_d\,A\left(\frac{\rho V^2}{2}\right) \tag{11.39}$$

where C_d is called the coefficient of drag.

For the sphere, $F_D = 3\pi\mu VD$, projected area $A = \frac{\pi}{4}D^2$

$\therefore$

$$F_d = C_d\left(\frac{\pi}{4}D^2\right)\left(\frac{\rho V^2}{2}\right)$$

or
$$3\pi\mu\, VD = C_d\left(\frac{\pi}{4}D^2\right)\left(\frac{\rho V^2}{2}\right)$$

$\therefore$
$$C_d = 24\left(\frac{\mu}{\rho VD}\right)$$

$\therefore$
$$C_d = \frac{24}{R_e} \tag{11.40}$$

which gives the drag coefficient of the sphere on the basis of Stokes' law.

11.8 APPLICATION OF LAMINAR FLOW: LUBRICATION MECHANICS

One of the most significant applications of the laminar flow of viscous fluid is the lubrication of various types of bearings. Reynolds, who initiated the theory of lubrication, showed that if a thin layer of viscous fluid is maintained between two parallel or nearly parallel surfaces, the sliding of surfaces, one over the other, becomes very smooth without any direct contact between the surfaces. This is the working principle of various bearings with lubricating oil as viscous fluid. Although high velocities in the lubricating system are involved, the thickness of the film of lubricating fluid is so small, that the Reynolds number produced remains far below the critical value for laminar flow. Thus, laminar flow is always assumed to be applied in the analysis of lubrication mechanics. The common types of bearing in which lubrication mechanics is applied are slipper and journal bearing.

11.8.1 Slipper Bearing

A slipper bearing in its simplest form consists of a slipper or a side block moving over the horizontal bearing plate as shown in Figure 11.8.

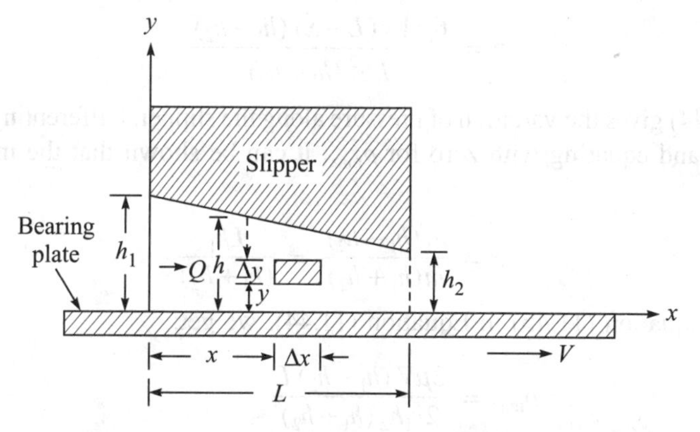

Figure 11.8 Slipper bearing simplified sketch.

The plane surface of the slipper is kept slightly inclined over the horizontal bearing plate. The principle of Couette flow between the bearing plate and slipper is applied here. Considering a small fluid element of Δx and Δy per unit width, equating the summation of pressure and shear force, the same Equation (11.2) is obtained, i.e.

$$\frac{\partial \lambda}{\partial y} = \frac{\partial p}{\partial x} \tag{11.2}$$

Using Newton's Equation (11.2), the same Equation (11.3) is obtained, i.e.

$$\frac{\partial p}{\partial x} = \mu \frac{\partial^2 v}{\partial y^2} \tag{11.3}$$

Integrating Equation (11.3) and putting the boundary condition from Figure (11.8): i.e. $y = 0$, $v = V$ and $y = h$, $v = 0$ and evaluating the constant, the equation for v is obtained as:

$$v = V\left(1 - \frac{y}{h}\right) + \frac{h^2}{2\mu}\left(\frac{\partial p}{\partial x}\right)\left(\frac{y}{h}\right)\left(\frac{y}{h} - 1\right) \tag{11.41}$$

Considering discharge dQ through elementary strip dy per unit breadth,

$$dQ = v dy$$

Integrating after substituting v from Equation (11.41), and simplifying:

$$Q = \frac{Vh}{2} - \frac{1}{12\mu}\left(\frac{\partial p}{\partial x}\right)h^3 \tag{11.42}$$

Solving for $\frac{\partial p}{\partial x}$, $\frac{\partial p}{\partial x} = \frac{12\mu}{h^3}\left(\frac{Vh}{2} - Q\right)$ (11.43)

From Figure (11.8), $h = h_1 - \frac{x}{L}(h_1 - h_2) = h_1 - mx$

where $m = \frac{h_1 - h_2}{L}$

Integrating Equation (11.43), the pressure distribution along the slipper after simplication is:

$$p = \frac{6\mu Vx(L - x)(h_1 - h_2)}{Lh^2 (h_1 + h_2)} \tag{11.44}$$

Equation (11.44) gives the variation of pressure along the slipper. Differenting Equation (11.44) with respect to x and equating with zero for p_{max}, it can be shown that the maximum pressure occurs at x, i.e.

$$x = \frac{h_1 (h_1 - h_2)}{m(h_1 + h_2)} = \frac{Lh_1}{(h_1 + h_2)} \tag{11.45}$$

Substituting Equation (11.45) in Equation (11.44), we get:

$$p_{max} = \frac{3\mu V (h_1 - h_2) L}{2h_1 h_2 (h_1 + h_2)} \tag{11.46}$$

It is due to this p_{max}, in the film, that the bearing supports the load.

11.8.2 Journal Bearing

It consists of a sleeve which is partially or completely wrapped around a rotating journal. Journal bearing is designed to support the radial load. Its rubbing surfaces are separated by a lubricant film.

Figure 11.9(a) shows the journal at rest and in that position, there is a metal-to-metal contact at the lowest point. When the journal begins to rotate in the clockwise direction, it moves up the bearing in an anti-clockwise direction as shown in Figure 11.9(b). The lubricant then adheres to the surface of the journal which now rolls on a lubricated surface. At the time of sufficient speed of journal, the lubricant will automatically be drawn between the surfaces, thus floating the journal in the bearing and the centre of the bearing moves to the left of the centre of the bearing as shown in Figure 11.9(c). The convergent film of the lubricant is established and positive pressure is thus developed, which can support the load.

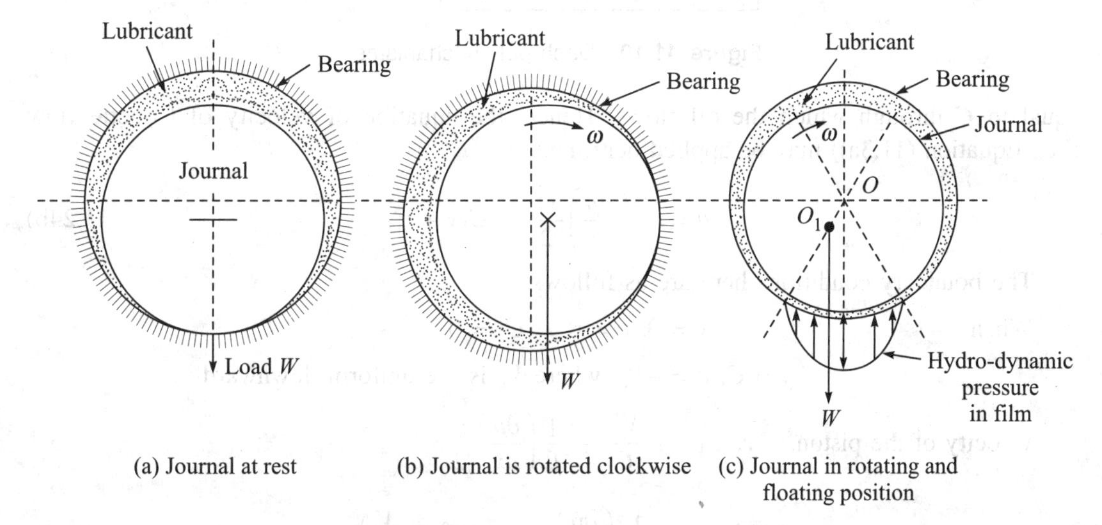

Figure 11.9 Journal bearing: Mechanisms of load bearing.

11.9 DASH-POT MECHANISMS

It is mechanism used to dampen the vibrations of machines. This is achieved by using fluid (i.e. oil) of high viscosity. A simple dash-pot mechanism shown in Figure 11.10, consists of a piston that moves in a concentric cylinder containing viscous oil. The diameter of the cylinder is slightly greater than that of the piston. The piston whose movement is to be restrained is connected to the machine. When the piston moves downwards, the viscous oil moves upward through the annular space between the piston and the cylinder with a laminar flow. This flow occurs due to an increase in the pressure of oil below the piston. Similarly, the upward movement of the piston decreases the pressure of oil below the piston and the oil moves downwards with a laminar flow.

If force F is applied downward, the oil flows upwards. If the clearance space between the piston and cylinder is small and equal to C, the piston and cylinder are considered to be two flat plates of breadth equal to the circumferential length of the pisiton πD, length L and distance apart

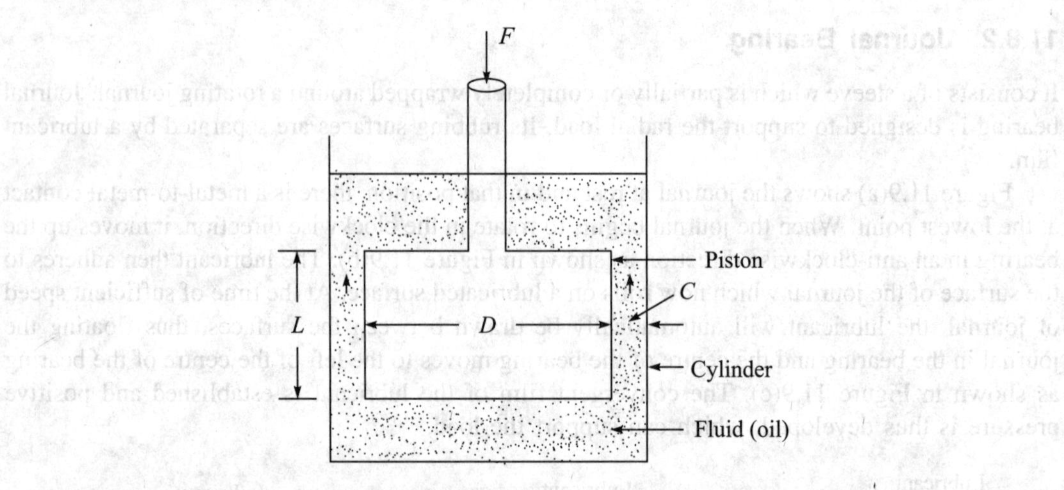

Figure 11.10 Dash-pot mechanisms.

equal to C through which the oil flows. Hence, the equation of velocity of Couette flow (i.e., Equation (11.3a)) may be applied here, i.e.

$$v = \frac{1}{\mu}\left(\frac{\partial p}{\partial x}\right)\frac{y^2}{2} + C_1 y + C_2 \tag{11.24b}$$

The boundary conditions here are as follows:

When $y = 0, v = 0 \therefore C_2 = 0$

$y = C, v = -V_u$, where V_u is the uniform downward velocity of the piston. $\therefore C_1 = -\frac{V_u}{C} - \frac{1}{\mu}\left(\frac{\partial p}{\partial x}\right)\frac{C}{2}$

$$\therefore \quad v = \frac{1}{2\mu}\left(\frac{\partial p}{\partial x}\right)(y - Cy^2) - \frac{V_u y}{C} \tag{11.47}$$

If the velocity of piston V_u is much smaller than the average velocity of oil through the clearance space, Equation (11.47) becomes (neglecting V_u),

$$v = \frac{1}{2\mu}\left(\frac{\partial p}{\partial x}\right)(y^2 - Cy)$$

Considering the flow dQ through an elemental strip of breath πD and thickness dy,

$$dQ = v\,(\pi D)\,dy$$

$$dQ = \pi D\left[\frac{1}{2\mu}\left(\frac{\partial p}{\partial x}\right)(y^2 - Cy)\right]dy$$

or

$$Q = \frac{\pi D}{2\mu}\left(\frac{\partial p}{\partial x}\right)\int_0^C (y^2 - Cy)\,dy$$

$$Q = \frac{\pi D}{2\mu}\left(\frac{\partial p}{\partial x}\right)\left(\frac{C^3}{3} - \frac{C^3}{2}\right)$$

$$Q = \left[\frac{\pi D}{12\mu}\left(-\frac{\partial p}{\partial x}\right)C^3\right] \tag{11.48}$$

Q given in Equation (11.48) is equal to the amount of oil displaced by the piston,

i.e.
$$\frac{\pi D}{12\mu}\left(-\frac{\partial p}{\partial x}\right)C^3 = \frac{\pi D^2}{4}\cdot V_u \tag{11.48a}$$

But
$$\frac{\partial p}{\partial x} = \frac{-p}{L}$$

$$\therefore \quad \frac{\pi D}{12\mu}\left(\frac{p}{L}\right)C^3 = \frac{\pi D^2}{4}\cdot V_u$$

$$\therefore \quad p = \frac{12 D V_u \mu L}{4C^3} = \frac{3 D V_u \mu L}{C^3} \tag{11.49}$$

Due to this difference of pressure, an upward force will be exerted on the piston which opposes the downward motion and is given by:

$$F_v = p\cdot\frac{\pi D^2}{4} = \frac{3 D V_u \mu L}{C^3} \times \frac{\pi D^2}{4} = \frac{3\pi D^3 V_u \mu L}{4C^3}$$

$$F_v = \frac{3\pi D^3 V_u \mu L}{4C^3}$$

Since the shear force (or skin friction drag) on the piston wall is small, it is neglected.

$$\therefore \quad F = F_v = \frac{3}{4}\pi\mu\, V_u\, L\left(\frac{D}{C}\right)^3 \tag{11.50}$$

Equation (11.50) shows that force F is dependent on the $\left(\frac{D}{C}\right)$ ratio. If the value of $\left(\frac{D}{C}\right)$ is large, the velocity of the piston V_u is smaller.

11.10 MEASUREMENT OF VISCOSITY

In laminar flow, viscosity of fluid or the Reynolds number plays an important role and measurement viscosity of the fluid is also important. Various methods of measuring the viscosity of the fluid are based on the principles of Newton's law of viscosity, the Hagen–Poiseville equation, and the Stokes law. Following are some of the methods or viscometers used to measure the viscosity:

(i) Capillary tube method or viscometer
(ii) Falling sphere viscometer

(iii) Sabolt viscometer
(iv) Redwood viscometer
(v) Rotating cylinder viscometer
(vi) Coaxial cylinder viscometer or rotating cylinder method
(vii) Orifice type viscometer.

Two very simple methods of measuring viscosity based on the Hagen–Poiseville and Stokes equations are described below.

Capillary Tube Viscometer

It is based on the Hagen–Poiseville equation. The viscosity of liquid is calculated by measuring the head difference in the capillary tube of length L.

Figure 11.11 shows a constant head tank, a piezometer, capillary tube of diameter D with laminar flow, and a discharge measuring tank.

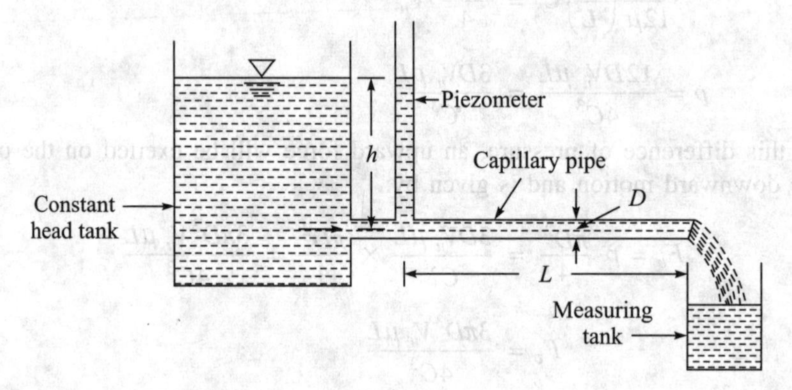

Figure 11.11 Capillary tube viscometer.

The Hagen–Poiseville equation is [i.e. Equation (11.18)]:

$$p_1 - p_2 = \frac{128\mu QL}{\pi D^4}$$

Dividing by $w = \rho g$, we get:

$$\frac{p_1 - p_2}{w} = \frac{128\mu QL}{w\pi D^4}$$

$$\therefore \quad h = \frac{128\mu QL}{\rho g \cdot \pi D^4}$$

$$\therefore \quad \mu = \frac{\pi \cdot \rho g D^4 h}{128 QL} \tag{11.51}$$

In Equation (11.50) Q is measured in the measuring tank. Measurements of h, D and L, viscosity of the fluid is obtained from Equation (11.50).

Falling Sphere Viscometer

Figure 11.12 shows the falling sphere viscometer with a constant temperature bath, a cylindrical vertical container with the fluid whose viscosity is to be measured, a sphere of steel and a sphere releasing mechanism. It is based on Stokes' Law. The ball is released as shown. It is allowed to fall freely. The terminal velocity V of the fall is obtained by measuring time t in length of L by $V = \dfrac{L}{t}$.

Then Stokes' equation is $\mu = \dfrac{D^2}{18V(W_s - W_l)}$ (11.38)

where D is the diameter of the sphere, and w_s and w_l are the specific weights of the steel ball and the liquid in the cylindrical container, respectively.

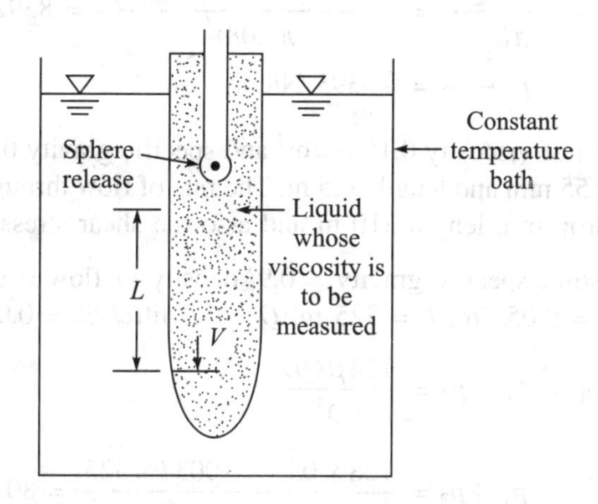

Figure 11.12 Falling sphere viscometer.

EXAMPLE 11.1 A crude oil of viscosity 0.9 Poise and specific gravity 0.8 is flowing through a horizontal pipe of length 15 m, and diameter 80 mm. Calculate the difference of pressure at the two ends of the pipe if 50 kg of oil is collected in a tank in 10 seconds.

Solution: $\mu = 0.9$ Poise $= \dfrac{0.9}{10} = 0.09$ Ns/m^2

Specific gravity = 0.8

$\therefore$ Density of liquid = 800 kg/m^3

Diameter of pipe D = 80 mm = 0.08 m

Length of the pipe L = 15 m

If Q is the discharge in m^3/sec, then:

$$\rho Q = \frac{50}{10}$$

or

$$800 \times Q = \frac{50}{10}$$

$$\therefore \quad Q = \frac{50}{10 \times 800} = 0.00625 \text{ m}^3/\text{sec.}$$

$$\therefore \quad V = \frac{Q}{A} = \frac{0.00625}{\frac{\pi}{4}(.08)^2} = 1.2434 \text{ m/sec}$$

$$\therefore \quad \text{Reynolds number } R_e = \frac{\rho VD}{\mu} = \frac{800 \times 1.2434 \times 0.08}{0.09} = 884.195$$

$R_e < 2000$, hence the flow is laminar.

Using the Hagen–Poiseville Equation [Equation (11.18)], we get:

$$p_1 - p_2 = \frac{128\mu QL}{\pi D^4} = \frac{128 \times .09 \times 0.00625 \times 15}{\pi (.08)^4} = 8392.996 \text{ N/m}^2$$

$$p_1 - p_2 = 0.8393 \text{ N/cm}^2 \qquad \textbf{Ans.}$$

EXAMPLE 11.2 An oil of viscosity 0.15 Ns/m^2 and specific gravity 0.9 is flowing through a circular pipe of diameter 55 mm and length 325 m. The rate of flow through the pipe is 3.7 litres/sec. Find the pressure drop in a length 310 m and also the shear stress at the wall.

Solution: $\mu = 0.15$ Ns/m^2, specific gravity = 0.9, density of flow = 900 kg/m^3,
$D = 55$ mm $= 0.055$ m, $L = 325$ m, $Q = 3.7$ litre/sec $= 0.0037$ m^3/sec

$$\text{Pressure drop} = p_1 - p_2 = \frac{128\mu QL}{\pi D^4} \qquad (11.18)$$

$$p_1 - p_2 = \frac{128 \times 0.15 \times 0.0037 \times 325}{\pi \times (0.055)^4} = 803129.69 \text{ N/m}^2$$

$$\therefore \quad p_1 - p_2 = \frac{803129.69}{10^4} = 80.313 \text{ N/cm}^2 \qquad \textbf{Ans.}$$

$$R_e = \frac{\rho VD}{\mu} = \frac{900 \times (0.0037) \times 0.055}{(\pi \times .0532) \times 0.15} = 128.48 < 2000$$

So the flow is laminar, hence the Hagen–Poiseville Equation (11.18) is applicable here. The shear stress at the wall is:

$$\lambda_0 = \left(-\frac{\partial p}{\partial x}\right)\frac{R}{2}, \text{ i.e. Equation (11.5)}$$

$$\lambda_0 = \left(\frac{p_1 - p_2}{L}\right)\frac{R}{2}$$

$$\lambda_0 = \left(\frac{80.313}{325 \times 100}\right)\left(\frac{55}{10 \times 2}\right) \text{ N/cm}^2 \quad \because D = 55 \text{ mm} = \frac{55}{10} \text{ cm}$$

$$\therefore \quad R = \frac{D}{2} = \left(\frac{55}{2 \times 10}\right)$$

$$\lambda_0 = 0.006795715 \text{ N/cm}^2$$

$$\lambda_0 = 67.957 \text{ N/m}^2$$

Ans.

EXAMPLE 11.3 A fluid of viscosity 5 Poise and specific gravity 1.2 is flowing through a pipe of diameter 10 cm. The maximum shear stress at the pipe wall is given to be 147.15 N/m^2. Find the pressure gradient, average velocity and Reynolds number of flow.

Solution: $\mu = 0.5 \text{ Poise} = \frac{5}{10} \text{ Ns/m}^2 \qquad \because 1 \text{ Poise} = \frac{1}{10} \text{ Ns/m}^2$

$$\mu = 0.5 \text{ N/m}^2$$

Specific gravity = 1.2 $\therefore$ Density = (1.2 × 1000) kg/m^3

$$D = 10 \text{ cm} = 0.1 \text{ m}$$

$$\lambda_0 = 147.15 \text{ N/m}^2$$

We know that

$$\lambda_0 = \left(-\frac{\partial p}{\partial x}\right)\frac{R}{2} = \left(-\frac{\partial p}{\partial x}\right)\frac{D}{4}$$

or

$$147.15 = \left(-\frac{\partial p}{\partial x}\right)\frac{0.1}{4}$$

$$\therefore \quad \frac{\partial p}{\partial x} = -\left(\frac{147.15 \times 4}{0.1 \text{ m}}\right)^{\text{N/m}^2} = -5886.0 \text{ N/m}^3$$

Ans.

$$\text{Average velocity } V = \frac{1}{32\,\mu}\left(-\frac{\partial p}{\partial x}\right)D^2 \text{ [Equation (11.15)]}$$

$$\therefore \quad V = \left[\frac{1}{32 \times 0.5}(5886)\,(.1)^2\right] \text{m/sec}$$

$$V = 3.6787 \text{ m/sec}$$

Ans.

$\therefore$ Reynolds number

$$R_e = \frac{\rho V D}{\mu} = \frac{1.2 \times 1000 \times 3.6787 \times 0.1}{0.5}$$

$$R_e = 882.88$$

Ans.

EXAMPLE 11.4 The flow of glycerine through a horizontal pipe of diameter 0.1 m is 0.01 m^3/sec. If μ of glycerine is 8 Poise, what power is required per kilometre of line to overcome the viscous resistance to the flow of glycerine?

Solution: L = 1 km = 1000 m, D = 0.1 m, Q = 0.01 m^3/sec

$$\mu = 8 \text{ Poise} = \frac{8}{10} \text{ Ns/m}^2 = 0.8 \text{ Ns/m}^2$$

$$\text{Velocity} = V = \frac{Q}{A} = \frac{0.01}{\frac{\pi}{4}(0.1)^2} = 1.2732 \text{ m/sec}$$

$$\text{Loss of head} = h_f = \frac{32\mu VL}{\rho g D^2}$$

Power required = Wh_f where W is the weight of water flowing per sec

and $$W = (\rho g Q)$$

$$\therefore \quad \text{Power required} = (\rho g Q \times h_f) = \left(\rho g Q \times \frac{32\mu VL}{\rho g D^2}\right) \text{watts.}$$

$$= \frac{32 Q\mu VL}{D^2} = \frac{32 \times 0.01 \times .8(1.2732) \times 1000}{(.1)^2} \text{watts.}$$

$$= 32594.93 \text{ watts.}$$

$$= 32.594 \text{ kW}$$

Ans.

EXAMPLE 11.5 The oil of viscosity 0.02 Ns/m^2 is flowing between two stationary parallel plates 1 m wide, placed at a destance of 0.01 m apart. The maximum velocity at the middle is 2 m/sec. Find the pressure gradient along flow direction, average velocity, discharge and shear at the wall.

Solution: $\mu = 0.02$ Ns/m^2, $B = 0.01$ m, $b = 1$ m

$$V_{max} = 2 \text{ m/sec}$$

We have the Equation (11.25) as follows:

$$V_{max} = \frac{B^2}{8\mu}\left(-\frac{\partial p}{\partial x}\right)$$

$$2 = \frac{(.01)^2}{8 \times 0.02}\left(-\frac{\partial p}{\partial x}\right)$$

$$\therefore \quad \frac{\partial p}{\partial x} = -3200 \text{ N/m}^2 \text{ per m.}$$

Ans.

$$\text{Average velocity } V = \frac{B^2}{12\mu}\left(-\frac{\partial p}{\partial x}\right) \quad \text{[Equation (11.27)]}$$

$$\therefore \quad V = \left[\frac{(0.01)^2}{12 \times 0.02}(3200)\right] \text{m/sec}$$

$$V = 1.333 \text{ m/sec}$$

Ans.

$$\text{Discharge } Q = \text{area of flow} \times V = (B \times b)V$$

$$Q = (0.01 \times 1 \times 1.333) \text{ m}^3\text{/sec}$$

$$Q = 0.01333 \text{ m/sec}$$ **Ans.**

Shear stress at the wall is $\lambda_0 = \left(-\dfrac{\partial p}{\partial x}\right)\dfrac{B}{2}$ [Equation (11.30)]

$$\lambda_0 = (3200) \times \frac{.01}{2} \text{ N/m}^2$$ **Ans.**

$$\lambda_0 = 16 \text{ N/m}^2$$ **Ans.**

EXAMPLE 11.6 Water is flowing through a pipe of 15 cm diameter with Darcy's friction factor $f = 0.05$. The shear stress at a point 4 cm from the pipe wall is 0.01962 N/cm². Calculate the shear stress at the pipe wall and the pressure gradient.

Solution: D = 15 cm = 0.5 m, $f = 0.05$

Shear stress λ at $r = 0.04$ m is

equal to 0.01962 N/cm² = 0.01962×10^4 N/m²

Find the Reynolds number R_e in the equation $f = \dfrac{64}{R_e}$ for

laminar flow, i.e.
$$0.05 = \frac{64}{R_e}$$

$\therefore$
$$R_e = 1280$$

$\therefore$ the flow is laminar.

Shear stress in laminar flow is $\lambda = \left(-\dfrac{\partial p}{\partial x}\right)\dfrac{r}{2}$

$\therefore$
$$0.01962 = -\left(\frac{\partial p}{\partial x}\right)\cdot\frac{4}{2}$$

$\therefore$
$$\left(\frac{\partial p}{\partial x}\right) = -\left(\frac{0.01962 \times 2}{4}\right) = -0.00981 \text{ N/m}^3$$ **Ans.**

Shear stress $\lambda \propto r$

as $\left(\dfrac{\partial p}{\partial x}\right)$ remains constant.

$\therefore$
$$\frac{x}{r} = \frac{\lambda_0}{R} \quad \therefore \quad R = \frac{D}{2} = \frac{15}{2} = 7.5 \text{ cm}$$

$\therefore$
$$\frac{0.01962}{4} = \frac{\lambda_0}{7.5}$$

$\therefore$
$$\lambda_0 = \left(\frac{0.01962 \times 7.5}{4}\right) \text{ N/cm}^2$$

$$\lambda_0 = 0.0367875 \text{ N/cm}^2$$ **Ans.**

11.11 CONCLUSION

Presentation given in this chapter has shown that laminar or viscous flow has a lot of practical importance. It has been found that pressure gradient and shear stress gradient are equal. Hagen–Poiseville equation, Stokes law, Darcy's Law[1], application of laminar flow in fabrication mechanisms, dash-pot mechanism have been found to be very important in laminar flow. Fluids with high viscosity play an important role in different flow situations of fluid mechanics. Analytical analyses of different situations are explained with facts and figures. Few solved examples are presented for better understanding of the students and some problems with answer are listed for students to solve. Presentations by Weisbach[2], and Blasius[3] on both laminar and turbulent flow is worth to be referred. Todd[4] book on "Ground Water Hydrology" is another example of detailed analysis of Laminar flow in porous media.

PROBLEMS

11.1 A laminar flow occurs in a pipe of 10 cm. The maximum velocity is 2 m/sec. Find the average velocity and the radius at which it occurs. Also calculate the velocity at 3 cm from the wall of this pipe. (**Ans.** 1 m/sec, r = 3.555 cm, 1.68 m/sec)

11.2 Laminar flow occurs between two fixed parallel plates. The plates are placed at a distance of 8 cm apart. The maximum velocity between the plate is 1.5 m/sec. If the viscosity of the fluid is 1.962 Ns/m^2, determine the pressure gradient, shear stress on the plates, and discharge per metre width of the plates.

(**Ans.** –3678.7 (N/m^2)/m 147.15 N/m, 0.08 m^3/sec/m.)

11.3 Water flows through two parallel fixed plates placed at a distance of 2 mm apart. If the viscosity of the fluid flowing between the plates is 0.01 Poise, and the average velocity is 0.4 m/sec, determine the pressure drop per unit length, maximum velocity, and shear stress at walls of the plates.

(**Ans.** 1199.7 N/m^2 per m, 0.6 m/sec, 1.199 N/m^2)

11.4 The viscosity of oil of specific gravity 0.8 is measured by a capillary tube of diameter 4 cm. The difference of pressure between two points placed 1.5 m apart is 0.3 m of water. The mass of oil collected in a measuring tank is 40 kg in 120 seconds. Find the viscosity of the oil.

(**Ans.** 2.36 Poise)

11.5 A sphere of diameter 3 mm balls falls 100 mm in 1.5 seconds in a viscous liquid. The density of the sphere is 7000 kg/m^3 and of the liquid is 800 kg/m^3. Find the viscosity of the liquid. (**Ans.** 45.61 Poise)

11.6 Oil of viscosity 0.02 Ns/m^2 is flowing between two stationary parallel plates placed at a distance of 10 mm apart. The width of the plates is 2 m. The velocity midway is 2 m/sec. Calculate the pressure gradient, average velocity and discharge through the plates.

(**Ans.** $\frac{dp}{dx}$ = – (3200 N/m^2)/m V = 1.333 m/sec, Q = 0.01333 m^3/sec)

REFERENCES

1. Darcy, H., "Experimental researches on the flow of water in pipes", *Competes rendus des se'ances de l'Academic des sciences*, Vol. 38, pp. 1109–1121, June 26, 1954.
2. Weisbach, J., *Text Book of Engineering Mechanics*, Brunswick, Germany, 1845.
3. Blasius, H., "The Law of similitude for frictions in fluid", *Forschungsheft des vereins deutscher Ingenieure*, No. 131, Berlin, 1913.
4. Todd, D.K., *Ground Water Hydrology*, John Wiley & Sons, New York, 2001.
5. Modi, P.N. and S.M. Seth, *Hydraulics and Fluid Mechanics including Hydraulic Machines*, Standard Book House, New Delhi, 1977.
6. Streeter, V.L. and E.B. Waylie, *Fluid Mechanics*, McGraw-Hill, New York, 1983.
7. Chow, V.T., *Open Channel Hydraulics*, McGraw-Hill, New York, 1959.
8. Bansal, R.K., *Fluid Mechanics and Hydraulic Machines*, Laxmi Publication, New Delhi.

Chapter 12

Turbulent Flow

12.1 INTRODUCTION

In 1883 Osborne Reynolds established the concept of laminar, transition and turbulent flow with the help of an experiment. In his experiment, he used a constant head tank of water, a small tank containing dye, a glass tube having a bell-mouthed entrance and a regulating valve. The regulating valve can regulate velocity through the glass tube. A liquid dye having the same specific gravity of water was introduced from the dye tank at the middle of the pipe. When velocity is low, the dye filament in the glass tube was parallel to the glass tube and it is the case of laminar flow. When velocity is slowly increased, the filament becomes wavy, and it becomes a situation of transitional flow with any further increase of velocity, the wavy dye filament is broken up and diffused in water, i.e. fluid elements of the dye move in a random path. It is a case of turbulent flow. It is finally obtained when the Reynolds number $R_e < 2000$, the flow remains laminar, $2000 < R_e < 4000$, the flow is transitional, $R_e > 4000$, the flow is turbulent. Head loss in turbulent flow varies with V^n and in turbulent flow, n varies from 1.75 to 2. Generally n is taken as 2. It was already shown in pipe flow chapter that $h_f = \left(\frac{fL}{2gD}\right) V^2$.

12.2 REYNOLDS SHEAR STRESS IN TURBULENT FLOW

J. Boussinesq, a French mathematician in 1877, developed an expression of shear stress in turbulent flow in analogy to laminar shear stress as:

$$\lambda = \eta \frac{dv}{dy} = \rho \in \frac{dv}{dy} \qquad (12.1)$$

where η (eta) has been introduced by Boussinesq as eddy (or apparent or virtual) viscosity, analogous to μ and

$$\in = \frac{\eta}{\rho} \text{ (or } \eta = \rho\in) \qquad (12.2)$$

is called eddy kinematic viscosity.

v is the velocity at a distance y from the boundary. When viscous shear stress is included, the total shear stress is:

$$\lambda = \mu \frac{dv}{dy} + \eta \frac{dv}{dy} \tag{12.3}$$

The magnitude of η is zero when the flow is laminar but increases to several thousand than that of μ when the flow is turbulent. The Boussinesq hypothesis on η or E is of limited use.

In 1886, Reynolds developed an equation for shear stress in turbulent flow. He considered two fluid layers, A and B at small distance apart shown in Figure 12.1(a), having average velocities $\overline{V}_A$ and $\overline{V}_B$, respectively. He assumed that $\overline{V}_A > \overline{V}_B$ and that the relative velocity of layer A with respective to B is $(\overline{V}_A - \overline{V}_B) = V_x$, where V_x is the fluctuating velocity component in x direction due to turbulence [Figure 12.1(b)]. It may be stated that if layer B is stationary, layer A moves with a velocity V_x. Similarly in y-direction, let the fluctuating velocity be V_y. Now over a surface area A perpendicular to y, V_y is uniformly distributed, then the mass of fluid transferred from the bottom layer to the top layer is (ρAVy). This mass moves with a velocity V_x in x-direction and hence the transfer of momentum is $(\rho AV_y)V_x$, thus resulting a tangential force on each layer. The corresponding turbulent shear stress exerted on fluid layer is:

$$\lambda = \frac{\text{force}}{A} = \frac{\rho A v_y \cdot v_x}{A}$$

$$\therefore \quad \lambda = \rho \, v_y \, v_x \tag{12.4}$$

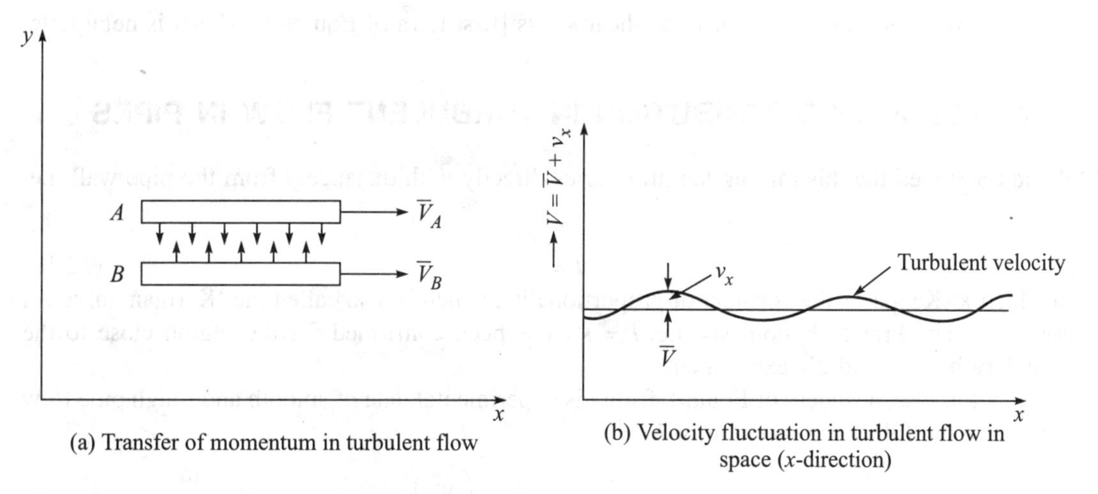

(a) Transfer of momentum in turbulent flow

(b) Velocity fluctuation in turbulent flow in space (x-direction)

Figure 12.1 Transfer of momentum and velocity fluctuation in turbulent flow.

Since v_x and v_y are varying, λ will also vary. Taking the time average on both sides of Equation (12.4), it is written as:

$$\lambda = \rho \, \overline{v_x v_y} \tag{12.5}$$

Thus Equation (12.5) is called Reynolds stress equation in turbulent flow.

12.3 PRANDTL MIXING LENGTH HYPOTHESIS

L. Prandtl, a German engineer, made an important advance in this direction in 1925 and his contribution is known as 'Prandtl mixing length theory or hypothesis'. According to him, the mixing length l is the distance traversed by the fluid particles of a layer in exchanging the momentum transfer. He further related this mixing length l with v_x, v_y and $\bar{V}$ through the following equation:

$$v_x = l\frac{d\bar{v}}{dy} \quad \text{and} \quad v_y = l\frac{d\bar{v}}{dy} \tag{12.6}$$

and hence $\overline{v_x \cdot v_y}$ can be written as:

$$\overline{v_x \cdot v_y} = l^2\left(\frac{dv}{dy}\right)^2 \tag{12.7}$$

From Equations (12.5) and (12.7), we obtain:

$$\bar{\lambda} = \rho l^2\left(\frac{d\bar{v}}{dy}\right)^2 \tag{12.8}$$

Thus the shear stress of Equation (12.3) becomes:

$$\bar{\lambda} = \mu\frac{d\bar{v}}{dy} + \rho l^2\left(\frac{d\bar{v}}{dy}\right)^2 \tag{12.9}$$

However, viscous or laminar flow shear stress [first term of Equation (12.9)] is negligible.

12.4 VELOCITY DISTRIBUTION IN TURBULENT FLOW IN PIPES

Prandtl assumed that his mixing length l varies directly with distance y from the pipe wall, i.e.

$$l \propto y$$

$$\therefore \quad l = \kappa y \tag{12.10}$$

in which κ (Kappa) is the constant of proportionality, which is also called the 'Karman universal constant'. The Prandtl hypothesis, i.e. $l = \kappa y$ has been confirmed for the region close to the boundary by Nikuradse's experiment.

J. Nikuradse, a student of Prandtl, from his experimental data of smooth and rough pipe flow data, gave the following equation of l:

$$l = 0.4\, y - 0.44\left(\frac{y^2}{R}\right) \tag{12.11}$$

where R is the radius of the pipe. If y is very close to the boundary, $\frac{y^2}{R}$ is negligible and Equation (12.11) becomes:

$$l = 0.4\, y \tag{12.12}$$

Thus, the value κ (Kappa), i.e. the Karman universal constant is 0.4. Substituting Equation (12.10) in Equation (12.8),

$$\lambda = \rho \kappa y^2 \left(\frac{dv}{dy}\right)^2 \tag{12.13}$$

For convenience, bars denoting time average are omitted in Equation (12.13). Now shear stress close to the boundary is λ_0 and $\kappa = 0.4$.

$$\therefore \quad \lambda_0 = \rho(0.16)\, y^2 \left(\frac{dv}{dy}\right)^2$$

$$\text{or} \quad \frac{\lambda_0}{\rho} = 0.16\, y^2 \left(\frac{dv}{dy}\right)^2$$

$$\text{or} \quad \sqrt{\frac{\lambda_0}{\rho}} = 0.4\, y \frac{dv}{dy}$$

$$\text{or} \quad V_* = 0.4\, y \frac{dv}{dy} \quad \because \sqrt{\frac{\lambda_0}{\rho}} = V_* = \text{shear velocity}$$

$$\therefore \quad \frac{1}{V_*}\, dv = \frac{1}{0.4\, y}\, dy$$

Integrating $\dfrac{v}{V_*} = 2.5 \log_e y + C'$, where C' is the constant of integration

$$\text{or} \quad \frac{v}{V_*} = 2.5 \log_e \left(\frac{y}{C}\right) \tag{12.14}$$

where C is another constant.

Equation (12.14) is known as the Prandtl–Karman universal logarithmic velocity distribution law. Changing the base of the log,

$$V = (2.5 \times 2.3)\, V_* \log_{10} \left(\frac{y}{C}\right)$$

$$V = 5.75\, V_* \log_{10} \left(\frac{y}{C}\right) \tag{12.15}$$

Nikuradse obtained from his experiment that the form smooth pipe turbulent flow, $C = \dfrac{\nu}{9V_*}$, where ν is the Kinematic viscosity.

$\therefore$ Equation (12.15) becomes

$$v = 5.75\, V_* \log_{10} \left(\frac{y}{\frac{\nu}{9} V_*}\right)$$

or
$$\frac{v}{V_*} = 5.75 \log_{10}\left(\frac{V_* y}{\nu}\right) + 5.50 \tag{12.16}$$

which is applicable to a smooth pipe surface.

When the pipe is rough, Nikuradse gave $C = \frac{k_s}{30}$, where k_s is the Niguradse's equivalent sand roughness size.

∴ Equation (12.15) becomes

$$v = 5.75\, V_* \log_{10}\left(\frac{y}{k_s/30}\right)$$

or
$$\frac{v}{V_*} = 5.75 \log_{10}\left(\frac{y}{k_s}\right) + 8.50 \tag{12.17}$$

which gives the velocity distribution in rough pipes.

12.5 RESISTANCE LAWS IN SMOOTH AND ROUGH PIPES

When fluid flows through pipes, resistance to the flow is offered by frictional resistance. In an earlier chapter it was shown that due to this frictional resistance, the energy of the flowing fluid is lost and that lost of energy has been given by Darcy's equation, i.e. $h_f = \frac{fLV^2}{2gD}$. The correct assessment of this friction factor is essential for the design of the pipe. This correct assessment depends on the state of the flow and roughness of the pipe. It was found that this friction factor f depends on the Reynolds number and roughness number, i.e.

$$f = \phi\left[\left(\frac{\rho VD}{\mu}\right), \left(\frac{D}{k_s}\right)\right] \tag{12.18}$$

where $\frac{\rho VD}{\mu} = \frac{VD}{\nu} = R_e$ is the Reynolds number, D is the diameter of the pipe and k_s is the average height of the pipe roughness and (D/k_s) is a non-dimensional roughness number.

When the flow is laminar or viscous, roughness has no effect on the flow and hence $f = \frac{64}{R_e}$ (as already shown in Chapter. 11, equation number 11.20)

When the flow is turbulent, f depends both on $\left(\frac{\rho VD}{\mu}\right)$ and $\left(\frac{D}{k_s}\right)$ values.

In 1911, Blasius, after studying the data of Saph and Schoder, gave the following empirical relationship for the smooth turbulent flow for friction factor:

$$f = \frac{0.316}{(R_e)^{1/4}} \tag{12.19}$$

Belter equations for assessment of resistance coefficient, i.e. friction factor in pipe for both smooth and rough turbulent flow, have been given by Prandtl and Karman, based on Nikuradse's experimental data for smooth turbulent flow, it is given to be:

$$\left.\begin{aligned} \frac{1}{\sqrt{f}} &= 2\log_{10}(R_e\sqrt{f}/2.51) \\ \text{or} \qquad \frac{1}{\sqrt{f}} &= 2\log_{10}(R_e\sqrt{f}) - 0.8 \end{aligned}\right\} \tag{12.20}$$

For rough turbulent flow, the friction is independent of R_e and the equation is given by:

$$\frac{1}{\sqrt{f}} = 2\log_{10}(3.71\ D/k_s) \quad \text{or} \quad \frac{1}{\sqrt{f}} = 2\log_{10}\left(\frac{D}{k_s}\right) + 1.139 \tag{12.21}$$

Colebrook and White combined both smooth turbulent resistance law or Equation (12.20) and rough turbulent resistance law or Equation (12.21) into one equation to represent the resistance coefficient f for a commercial pipe. This equation so combined is known as the Colebrook–White equation which is in the form of:

$$\frac{1}{\sqrt{f}} = -2\log\left[\frac{k_s}{3.71D} + \frac{2.51}{R_e\sqrt{f}}\right] \tag{12.22}$$

The significance of Equation (12.22) is that if the pipe is rough, the second term may be neglected, and Equation (12.22) is reduced to rough turbulent Equation (12.21). Similarly, if the pipe is smooth, first written the bracket is ignored, and the equation becomes the smooth turbulent Equation (12.20). It is seen that equation f for smooth turbulent flow from Equation (12.20) is implicit. A trial and error solution is essential. Barr has given an approximate direct solution of f from this smooth turbulent equation as:

$$\frac{1}{\sqrt{f}} = 1.785\ \log_{10} R_e - 1.424 \tag{12.23}$$

Equation (12.23) gives a very close value of f as obtained through the trial and error solution from Equation (12.20). To start trial and error for solution f from Equation (12.20), evaluate f directly from Equation (12.23), put this value in Equation (12.20) as the first assumed to start the trial and error. Within one or two iteration f valve can be obtained.

Similarly the Colebrook–White equation is again implicit for a solution of f. Barr has developed an approximate direct solution f for Equation (12.22) as:

$$\frac{1}{\sqrt{f}} = -2\log_{10}\left[\left(\frac{k_s}{3.71D}\right) + \left(\frac{5.186}{R_e^{0.89}}\right)\right] \tag{12.24}$$

The value of f is obtained directly from Equation (12.24) may be used to evaluate the first assumed value for a trial and error solution of Equation (12.22).

12.6 MOODY'S OR STANTON–PANNELL DIAGRAM TO EVALUATE f

L.F. Moody and T.E. Stanton–J.R. Pannell developed a three parameters plot $\left(f - R_e - \frac{D}{k}\right)$ independently almost at the same time. Therefore, this three parameters plot is known as both Moody's diagram or the Stanton–Pannell diagram in different countries. Its diagram is drawn on a log-log scale in Figure 12.2.

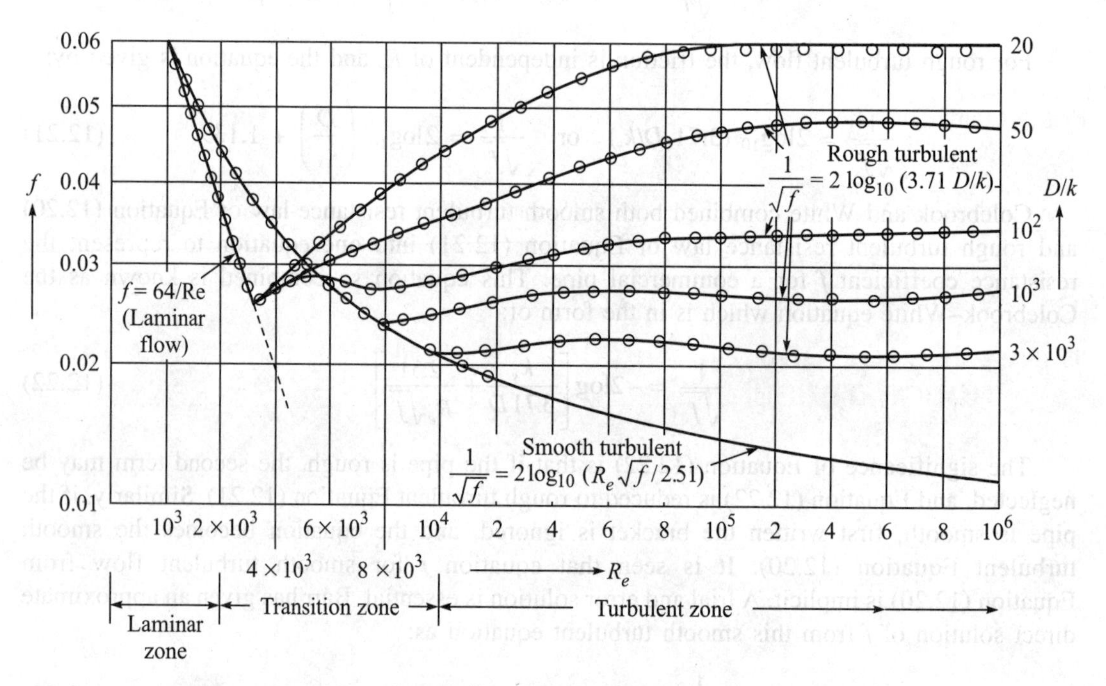

Figure 12.2 Moody's or Stanton–Pannell diagram.

From Figure 12.2, if R_e and D/k of the pipe are known, the f value may obtained, depending on whether the flow is laminar or turbulent and the pipe is smooth or rough.

12.7 HYDRODYNAMICALLY SMOOTH AND ROUGH SURFACES

In general, all boundaries are rough. If k is the average height of roughness and this roughness projection is completely submerged in laminar sub-layer (∂_b) of the boundary layer (which will be discussed later in a separate chapter) as shown in Figure 12.3(a), then the surface is called a hydro dynamically smooth surface. Again, if the roughness projection k is more than the laminar sub-layer (∂_b), then it is called a hydrodynamically rough surface [see Figure 12.3(b)].

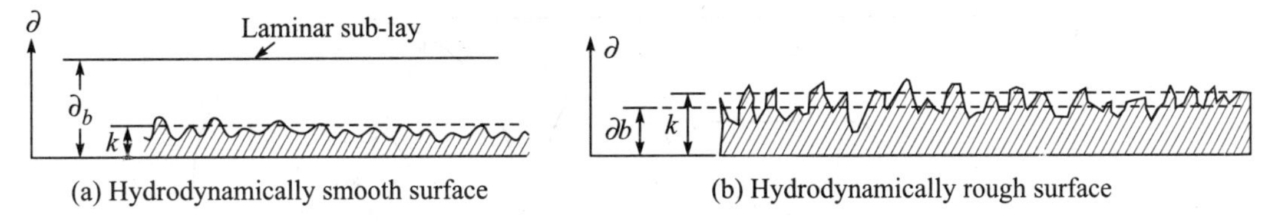

Figure 12.3 Definition of hydrodynamically smooth and rough surfaces.

Mathematically, it has been found from Nikuradse's experiment that if $\dfrac{k}{\partial_b} < 0.25$, it is called hydrodynamically smooth and if $\dfrac{k}{\partial_b} > 6.0$, it is hydrodynamically rough. For $0.25 < \dfrac{k}{\partial_b} < 6.0$, the boundary is in transition. And in terms of roughness, if the Reynolds number $R_* = \dfrac{V_* k}{\nu} < 4$, the boundary is smooth and if $R_* > 6.0$, the boundary is rough.

12.8 AVERAGE OR MEAN VELOCITY IN PIPE FOR SMOOTH AND ROUGH TURBULENT FLOW

For a smooth pipe, the general equation of velocity distribution is Equation (12.16), i.e. $\dfrac{v}{V_*} = 5.75 \log_{10}\left(\dfrac{V_* y}{\nu}\right) + 5.50$

For the pipe, $y = R - r$.

$$\therefore \qquad v = V_*\left[5.75 \log_{10}\left(\frac{V_*}{\nu}(R-r)\right)\right] + 5.50 \qquad \text{(A)}$$

To determine the average velocity: V in a pipe, consider Figure 12.4,

$$V = \frac{Q}{\pi R^2} = \frac{1}{\pi R^2}\int_A v dA$$

$$V = \frac{1}{\pi R^2}\int_0^R v(2\pi r) dr$$

$$V = \frac{1}{V_* \pi R^2}\int_0^R \left[2\pi r \times 5.75 \log_{10}\frac{V_*(R-r)}{\nu} + 5.50\right] dr$$

On integration and simplification, the following equation of average velocity V is obtained:

$$V = V_*\ 5.75 \log_{10}\left(\frac{V_* R}{\nu}\right) + 1.75 \qquad (12.25)$$

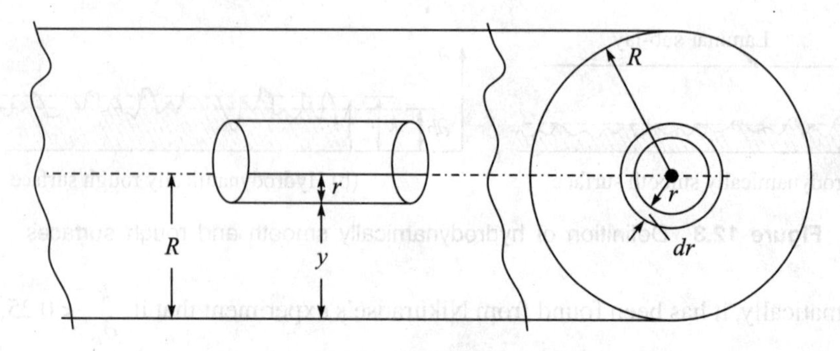

Figure 12.4 Elementary ring of thickness dr at a radial distance r.

Similarly for a rough pipe, starting with Equation (12.17) and replacing $y = (R - r)$, after integration, the average velocity V is:

$$V = V_* \, 5.75 \log_{10}\left(\frac{R}{k_s}\right) + 4.75 \tag{12.26}$$

Subtracting Equation (12.25) from Equation (12.16), we get:

$$5.75 \log_{10}\left(\frac{V_* y}{\nu}\right) + 5.5 - V_* \times 5.75 \log_{10}\left(\frac{V_* R}{y}\right) + 1.75$$

$$= \frac{\upsilon}{V_*} - V$$

or $$\frac{\upsilon - V}{V_*} = 5.75 \log_{10}\left[\left(\frac{V_* y}{\nu}\right)\Big/\left(\frac{V_* R}{y}\right)\right] + 5.5 - 1.75$$

or $$\frac{\upsilon - V}{V_*} = 5.75 \log_{10}\left[\frac{V_* y}{\nu} \times \frac{\nu}{V_* R}\right] + 3.75$$

$$\therefore \quad \frac{\upsilon - V}{V_*} = 5.75 \log_{10}\left(\frac{y}{R}\right) + 3.75 \tag{12.27}$$

Similarly subtracting Equation (12.26) from Equation (12.17), we get:

$$5.75 \log_{10}\left(\frac{y}{k_S}\right) + 8.50 - 5.75 \, V_* \log_{10}\left(\frac{R}{k_S}\right) + 4.75$$

$$= \frac{\upsilon}{V_*} - V$$

$$\therefore \quad \frac{\upsilon - V}{V_*} = 5.75 \log_{10}\left[\frac{y}{k_s} \times \frac{k_s}{R}\right] + 8.50 - 4.75$$

$$\therefore \quad \frac{\upsilon - V}{V_*} = 5.75 \log_{10}\left(\frac{y}{R}\right) + 3.75 \tag{12.27}$$

Thus the same Equation (12.27) is obtained.

This shows that the Karman–Prandtl equation of velocity, when referred to average velocity, becomes identical, i.e. Equation (12.27), which is independent of roughness k_s for both smooth and rough pipes. Since at $y = R$, v is equation of v_{max}, Equation (12.27) becomes:

$$\frac{v_{max} - V}{V_*} = 5.75 \log_{10}\left(\frac{R}{R}\right) + 3.75$$

or

$$\frac{v_{max} - V}{V_*} + 0 + 3.75$$

$$\therefore \quad \frac{v_{max} - V}{V_*} = 3.75 \tag{12.28}$$

EXAMPLE 12.1 For turbulent flow in a pipe, find the value of distance y from a wall where the point velocity becomes the average velocity V.

Solution: The relation between point velocity v and average velocity V for both smooth and rough turbulent flow is given by Equation (12.27), i.e:

$$\frac{v - V}{V_*} = 5.75 \log_{10}\left(\frac{y}{R}\right) + 3.75$$

when $v = V$,

$$5.75 \log_{10}\left(\frac{y}{R}\right) + 3.75 = 0$$

$$\log_{10}\left(\frac{y}{R}\right) = \frac{-3.75}{5.75} = \bar{1}.347826$$

$$\therefore \quad \left(\frac{y}{R}\right) = 0.22275$$

$$\therefore \quad y = 0.22275\, R$$

Ans.

EXAMPLE 12.2 For turbulent flow in pipe for both smooth and rough pipes, find the value of $\frac{v_{max}}{V}$ in terms of friction factor f.

Solution: Taking Equation (12.27):

$$\frac{v - V}{V_*} = 5.75 \log_{10}\left(\frac{y}{R}\right) + 3.75$$

At $y = R$, $v = V_{max}$

$$\therefore \quad \frac{v_{max} - V}{V_*} = 3.75 \tag{A}$$

In Chapter 11, Equation (11.22) is:

$$V_* = \sqrt{\frac{f}{8}}\; V = \sqrt{\frac{\lambda_0}{\rho}}$$

$\therefore$ (A) becomes $$\frac{v_{max} - V}{V\sqrt{\frac{f}{8}}} = 3.75$$

$$v_{max} - V = 3.75\, V\sqrt{\frac{f}{8}}$$

or $$\frac{v_{max}}{V} - 1 = 3.75\sqrt{\frac{f}{8}}$$

$\therefore$ $$\frac{v_{max}}{V} = \frac{3.75}{\sqrt{8}}\sqrt{f} + 1$$

$\therefore$ $$\frac{v_{max}}{V} = (1.3258\sqrt{f} + 1)$$ **Ans.**

EXAMPLE 12.3 A pipeline carrying water has an average roughness height $k_s = 0.20$ mm. What type of boundary it is? The shear stress at the wall is 7.84 N/m^2. Take $\nu = 0.01$ stokes.

Solution: $k_s = 0.20$ mm $= 0.20 \times 10^{-3}$m, $\lambda_0 = 7.848$ N/m^2
$\nu = 0.01$ stokes $= 0.01$ cm^2/sec $= 0.01 \times 10^{-4}$ m^2/sec

ρ of water = 1000 kg/m^3 $\therefore$ $$V_* = \sqrt{\frac{\lambda_0}{\rho}} = \sqrt{\frac{7.848}{1000}} = 0.08858 \text{ m/sec}$$

$\therefore$ Roughness or the shear Reynolds number $$R_* = \frac{V_* k_s}{\nu}$$

$$R_* = \frac{0.08858 \times 0.20 \times 10^{-3}}{0.01 \times 10^{-4}}$$

$$R_* = 17.717$$

$4 < R_* < 60$, the flow is transitional

Transitional flow. **Ans.**

EXAMPLE 12.4 Water is flowing with a rough turbulent flow through a pipe of diameter 60 cm with a discharge of 0.6 m^3/sec. The roughness k_s is 0.003 m. Find the power lost in a 1 km length of the pipe.

Solution: $D = 60$ cm $= 0.6$ m, $Q = 0.6$ m^3/sec, $k_s = 0.003$ m
$L = 1$ km $= 1000$ m

For rough turbulent flow,

$$\frac{1}{\sqrt{f}} = 2\log_{10}\left(3.71\frac{D}{k_s}\right) = 2\log_{10}\left(3.71 \times \frac{0.6}{.003}\right)$$

$$\frac{1}{\sqrt{f}} = 5.7408$$

$\therefore$ $$f = 0.030342$$

$\therefore$ Head lost due to friction $h_f = \dfrac{fLV^2}{2gD}$

$$h_f = \frac{0.030342 \times 1000}{2 \times 9.81 \ \times .6}\left(\frac{0.6}{\frac{\pi}{4}(.6)^2}\right)^2 \text{ m}$$

$$h_f = 11.6067 \text{ m}$$

Power lost $P = \dfrac{\rho g Q\, h_f}{1000} = \dfrac{1000 \times 9.81 \times 0.6 \times 11.6067}{1000}$ kw

P = 68.317 kw **Ans.**

EXAMPLE 12.5 A pipe with diameter 10 cm and roughness size 0.0025 m carries water at a velocity of 2 m/sec. If ν of water is 10^{-6} m^2/sec, find friction factor f assuming that the flow is governed by

(1) Only smooth turbulent law
(2) Only rough turbulent law
(3) Both smooth and rough turbulent law.

Solution: Reynolds number $R_e = \dfrac{VD}{\nu} = \dfrac{2 \times 0.1}{10^{-6}} - 0.2 \times 10^6 = 2 \times 10^5$

(1) When the flow is smooth turbulent:

The Prandtl–Karman equation of smooth turbulent is Equation (12.20), i.e. $\dfrac{1}{\sqrt{f}} = 2 \log_{10}$ $[R_e\sqrt{f}/2.51]$, where the solution f is implicit.

To start trial and error for solution of f, first calculate f by Barr approximate direct solution equation, i.e. Equation (12.33)

$$\frac{1}{\sqrt{f}} = 1.785 \log_{10} R_e - 1.424$$

or
$$\frac{1}{\sqrt{f}} = 1.785 \log_{10} (2 \times 10^5) - 1.424$$

$\therefore$
$$f = 0.015476$$

Assume that f = 0.015476 as the first assumed value to start trial and error in Equation (12.20), i.e.

$$\frac{1}{\sqrt{0.015476}} = 2\log_{10} (2 \times 10^5\sqrt{0.015476}/2.51)$$

$$8.03833 = 7.99237$$

$$\text{L.H.S.} \simeq \text{R.H.S.}$$

$\therefore$
$$f = 0.0155$$ **Ans.**

(2) When the flow is rough turbulent, i.e. by Equation (12.21)

$$\frac{1}{\sqrt{f}} = 2\log_{10}\left(3.71\frac{D}{k_s}\right)$$

$$\frac{1}{\sqrt{f}} = 2\log_{10}\left(3.71\times\frac{0.4}{0.0025}\right)$$

$$f = 0.053$$ **Ans.**

(3) When both smooth and rough turbulent flows are considered, use the Colebrook–White equation, i.e. Equation (12.22)

$$\frac{1}{\sqrt{f}} = -2\log_{10}\left[\frac{k_s}{3.71D}+\frac{2.51}{R_e\sqrt{f}}\right]$$

Again the solution of f is implicit. Take the Barr approximate direct solution to start trial and error in Equation (12.22), i.e. Equation (12.24),

which is $$\frac{1}{\sqrt{f}} = -2\log_{10}\left[\frac{k_s}{3.71D}+\frac{5.186}{R_e^{0.89}}\right]$$

$$\therefore \quad f = 0.0533$$

Putting this $f = 0.0533$ in the Colebrook–White equation, i.e. Equation (12.22), we get:

$$\frac{1}{\sqrt{.0533}} = -2\log_{10}\left[\frac{.0025}{3.71\times 0.1}+\frac{2.51}{200000\times\sqrt{0.0533}}\right]$$

$$4.33148 = 4.33018$$

$$\text{LHS} \simeq \text{RHS}$$

$$\therefore \quad f = 0.0533$$ **Ans.**

12.9 CONCLUSION

Most of the flows of fluid in real situation are turbulent and unsteady. Flow of flood waves in river, flow in the wake region past a submerged body, efflux of smoke from a Chimney, storms over the earth surface, flow past an overfall spillway, hydraulic jump are some of the examples of turbulent flow. Prandtl, von Karman, Reynolds, Stanton–Pannel, Moody, Darcy, Weisbach, Nikuradse, Blasius, Schlichting, Colebrook and White, Barr have shown that resistance to flow in turbulent flow is quite different than that of laminar flow. Therefore, theories presented by Reynolds, Prandtl[4], von Karman[3], Darcy–Weisbach[1,9], Stanton–Pannel, Moody, Nikuradse[8], Colebrook and White[10], Barr[11] and others have been presented in this chapter for both smooth and rough surface under turbulent flow condition.

Numerical examples have been solved to show the application of those theories, A list of references is included at the end of the chapter.

PROBLEMS

12.1 Determine the wall shearing stress in a pipe with turbulent flow when the diameter of the pipe is 10 cm. The velocities at the pipe centre and at a distance of 3 cm from pipe centre are 2 m/sec and 1.5 m/sec, respectively.

(**Ans.** 47.68 N/m^3)

12.2 A smooth pipe with turbulent flow is 800 m long and 8 cm diameter. It carries a discharge of 0.008 m^3/sec. Calculate the loss of head, λ_0, centre line velocity, velocity and shear stress at 3 cm from pipe wall. ν of liquid = 0.015 stokes. Calculate f from the Blasius equation $f = \dfrac{0.316}{R_e^{1/4}}$.

(**Ans.** h_f = 23.9246 m, λ_6 = 5.8675 N/m^2, V_{max} = 1.88 m/sec, λ at 3 cm = 1.466875 N/m^2, v at 3 cm = 1.8216 m/sec)

12.3 A pipe with smooth turbulent flow and diameter 30 cm carries liquid with centre line velocity of 2 m/sec and the velocity at 10 cm from centre is 1.6 m/sec. Find V_*, V and Q.

(**Ans.** V_* = 0.1458 m/sec, V = 1.45325 m/sec, Q = 0.102724 m^3/sec)

12.4 A pipe of diameter 15 cm and roughness size 0.003 m carries water at a velocity 4 m/sec. If ν of the water is 10^{-6} m^2/sec, find the friction factor assuming that the flow is governed by:

(I) Only the smooth turbulent law

(II) Only the rough turbulent law

(III) Both the smooth and rough turbulent law.

REFERENCES

1. Darcy, H., "Experimental researches on the flow of water in pipes", *Competes rendus des seances de l'Academic des Sciences*, Vol. 38, pp. 1109–1121, June 26, 1854.
2. Blasius, H., "The law of similitude for friction in fluids", *Forchungsheft des Vereins deutscher Ingenieure*, No 131, Berlin, 1913.
3. Von Karman, T., "Mechanical similitude and turbulence", *Proc. of 3rd International Congress for Applied Mechanics*, Stockholm, Vol. 1, pp. 85–93, 1930.
4. Prandtl, L. "Mechanics of Viscous Fluids" in W.F. Durand (Ed.) *Aerodynamic Theory*, Springer-Verlag, Berlin, Vol. III, div. G., 1935.
5. Dryden, H.L., "Transition for laminar to turbulent flow", Princeton University Press, section of *Turbulent Flow and Heat Transfer*, Edited by C.C. Lin, 1959.
6. Schlichting, H., *Boundary Layer Theory*, translated by J. Kestin, McGraw-Hill Inc., New York, 1960.
7. Stanton, T.E. and J.R. Pannell, "Similarity of motion in relation to surface friction of fluids", Philosophical Trans., *Royal Society of London*, Vol. 214A, pp. 199–224, 1914.

8. Nikuradse, J., "Laws of Turbulent Flow in Smooth Pipes", *Forschungsheft des Vereins deutschor, Ingenieure*, No 356, Berlin, 1932.
9. Weisbach, J., *Textbook of Engineering Mechanics*, Brunswick, Germany, 1845.
10. Colebrook, C.F. and C.M. White, "Experiments with fluid friction in roughened pipes", *Proc. Royal Society of London*, Series A, Vol. 161, pp. 367–381, 1938.
11. Barr, D.I.H. "Two additional methods of direct solution of Colebrook–White function", *Proc. ICE*, London, Part 2, June, 1976.

Chapter 13

Boundary Layer in Incompressible Flow

13.1 INTRODUCTION

Ludwig Prandtl, the father of modern fluid mechanics in Germany, introduced for the first time in 1904 the concept of a thin boundary layer next to the body in which viscous effects were concentrated. Beyond that thin boundary layer, viscous effects were considered to be negligible and the flow could be treated as that of an ideal fluid. The formation of boundary layer over a stationary cylinder placed in a flowing fluid with velocity V_s is shown in Figure 13.1.

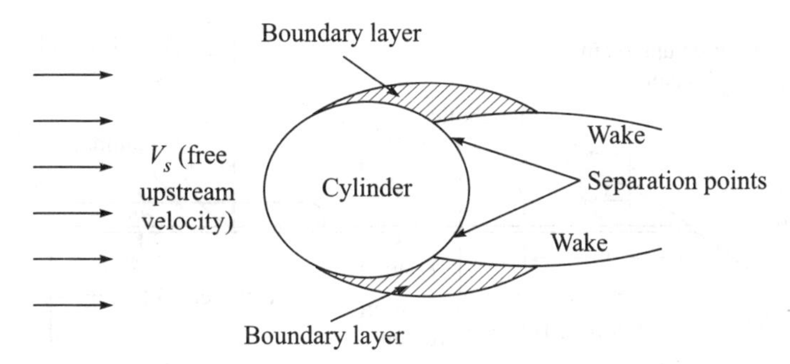

Figure 13.1 Flow of real fluid past a cylinder showing thin boundary layer on surface.

Inside the boundary, the velocity gradient exists and hence there is a shear stress exist, whereas outside the boundary layer, no velocity gradient exists and hence there is no shear stream.

The boundary layer theory is of extreme importance in modern fluid mechanics and aerodynamics, and is also essential for an understanding of convective head transfer. An exhaustive treatment or analysis of the boundary layer theory has been given by Prandtl and H. Schlichting.

13.2 DESCRIPTION OF THE BOUNDARY LAYER

The development of Boundary Layer (B.L.) can perhaps be best described by a study of the flow of fluid along a flat plate (Figure 13.2) or plate or body moving in stationary fluid. Consider the uniform flow of an incompressible fluid at a free upstream V_s approaching the plate. When the fluid reaches the leading edge of the plate, large shear stresses are set up near the plate surface and the fluid particles at the plate surface are brought to rest, while those at a short distance normal to the plate are retarded because of viscous shear. The region of retarded flow where the velocity gradient exists is called Boundary Layer (B.L.) and its thickness is designated as ∂. For some longitudinal distance x_c, the flow within the B.L. is laminar. Downstream of this point, the B.L. becomes unstable to some distance and eventually becomes turbulent. If the free upstream velocity V_s is increased, X_e is decreased such that $V_s x_c$ remains essentially constant. The ratio of $V_s x_c$ and Kinematic viscosity is the Reynolds number. Beyond this point, there is instability and the flow within the boundary layer is turbulent and this is called turbulent B.L. Outside the boundary layer thickness ∂, the velocity gradient does not exist, hence the velocity is equal to free upstream velocity V_s. The thickness of boundary ∂ increases as the distance x from leading edge increases. Between the plate and turbulent boundary layer, there is a thin laminar layer. This is called the *laminar* sub-layer whose thickness is designated ∂_b. This is reasonable to expect, since the inertia force is less because of low velocity near the plate surface. Also the viscous force is predominant and complete mixing of fluid particles is inhibited due to the presence of the plate change of the boundary layer from laminar to turbulent, which depends on the Reynolds number $R_e = \dfrac{V_s x}{\nu}$, where x is the distance from the leading edge. For practical purposes, this R_e is approximately equal to 4×10^5. The velocity profiles within the B.L. are shown. In laminar B.L., it is parabolic and in turbulent B.L. it is logarithmic.

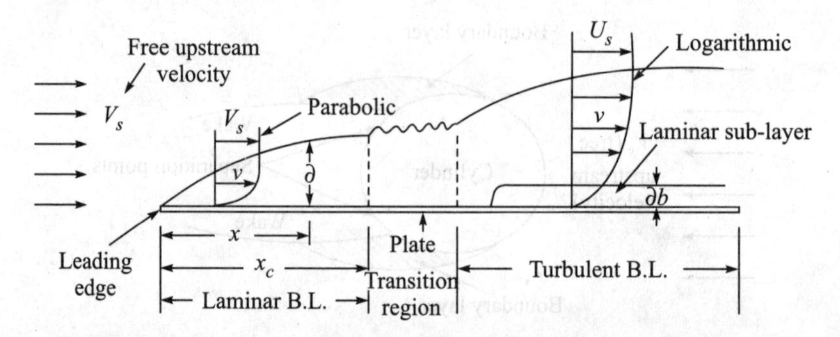

Figure 13.2 Boundary layer and velocity distribution along a flat plate.

13.3 THICKNESS OF THE BOUNDARY LAYER

The velocity within the B.L. increases from zero at the boundary surface to the velocity of free upstream V_s asymptotically. Therefore, the thickness of B.L. is defined as that distance from the boundary in which the velocity reaches 99% of V_s i.e. when $v = 0.99\ V_s$. This distance y is considered as the thickness of the B.L. (δ). This δ is called the nominal thickness of B.L. For greater accuracy, B.L. thickness is defined in terms of some mathematical expressions which are the measures of the effect of boundary layer on the flow. There are three such definitions of the boundary layer. They are:

(i) Displacement thickness (∂_d)
(ii) Momentum thickness (∂_m)
(iii) Energy thickness (∂_e)

Displacement Thickness (∂_d)

This is defined as the distance by which boundary surface would have to be displaced outwards (Figure 13.3) so that the total actual discharge would be the same as that of an ideal (frictionless) fluid passing the displaced boundary.

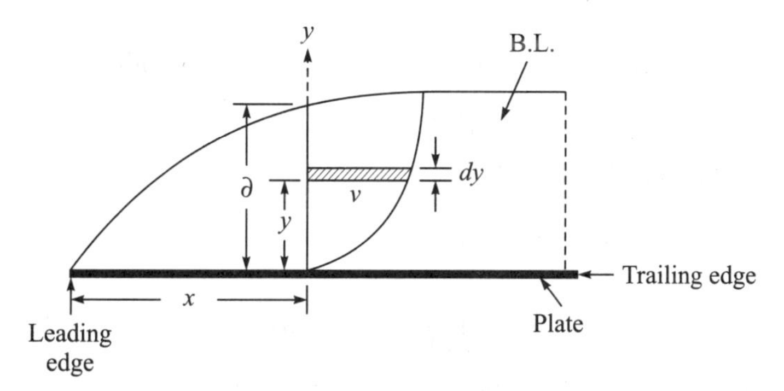

Figure 13.3 Sketch to define displacement thickness.

To obtain the mathematical expression, consider Figure 13.3.
Let at a distance x from the leading edge, ∂ be the thickness of B.L.
Let at a distance y from the boundary at this distance x, velocity within the B.L. be v.
Consider the unit width of the plate.
Let dy is the elemental strip.
Discharge through this strip = Area × Velocity = $(1 \times dy)v = vdy$
If the plate would not have been present, discharge through the same strip = $(1 \times dy) \times V_s$ = $V_s dy$.
Reduction of discharge through the strip due to the presence of the plate

$$= V_s dy - v\,dy = (V_s - v)\,d_y$$

$\therefore$ Total reduction of discharge = $\int_0^{\partial} (V_s - v)dy$ (A)

If ∂_d be the displacement thickness, then by definition:

$$V_s\,\partial_d = \int_0^{\partial} (V_s - v)\,dy$$

$$\therefore \qquad \partial_d = \int_0^{\partial} \left(1 - \frac{v}{V_s}\right) dy \qquad (13.1)$$

which is the equation of displacement thickness.

Momentum Thickness (∂_m)

Similarly momentum of the flow through the strip

$= \text{Mass through the strip} \times \text{Velocity through the strip}$

$= (\rho v dy)v = \rho v^2 dy$

Momentum of the strip in absence of the plate or boundary

$= (\rho v dy) V_s$

$\therefore$ Loss of momentum/sec $= \rho v V_s dy - \rho v^2 dy = \rho v (V_s - v) dy$

Total loss of momentum within the B.L.

$$= \int_0^{\partial} \rho v (V_s - v) dy \quad \text{(A)}$$

If ∂_m is the momentum thickness, by definition, momentum with velocity V_s

$$(\rho\ V_s\ \partial_m) V_s = \rho\ \partial_m\ V_s^2 \quad \text{(B)}$$

Now equating (A) and (B), we get:

$$\rho \partial_m\ V_s^2 = \int_0^{\partial} \rho v (V_s - v) dy$$

$$\therefore \quad \partial_m = \int_0^{\partial} \frac{v}{V_s}\left(1 - \frac{v}{V_s}\right) dy \quad (13.2)$$

which is the equation of momentum thickness.

Energy Thickness (∂_e)

Similarly, by equating the total loss of kinetic energy to the K.E. with velocity V_s through energy thickness ∂_e, we get:

$$\frac{1}{2}(\rho\ \partial_e\ V_s) V_s^2 = \int_0^{\partial} \frac{1}{2}[\rho v (V_s^2 - v^2)] dy$$

$$\partial_e = \int_0^{\partial} \frac{v}{V_s}\left(1 - \frac{v^2}{V_s^2}\right) dy \quad (13.3)$$

which is the mathematical expression for energy thickness.

EXAMPLE 13.1 If the velocity profile within the B.L. is linear, i.e.

$$\frac{v}{V_s} = \frac{y}{\partial}, \text{ find } \partial_d, \partial_m \text{ and } \partial_e.$$

Solution: $\dfrac{v}{V_s} = \dfrac{y}{\partial}$ (given)

$$\partial_d = \int_0^{\partial}\left(1 - \frac{v}{V_s}\right) dy = \int_0^{\partial}\left(1 - \frac{y}{\partial}\right) dy = \partial - \frac{1}{\partial}\cdot\frac{\partial^2}{2} = \frac{\partial}{2}$$

$$\partial_m = \int_0^{\partial} \frac{v}{V_s}\left(1 - \frac{v}{V_s}\right) dy = \int_0^{\partial} \frac{y}{\partial}\left(1 - \frac{y}{\partial}\right) dy = \frac{1}{\partial}\cdot\frac{\partial^2}{2} - \frac{1}{\partial^2}\cdot\frac{\partial^3}{3} = \frac{\partial}{6}$$

$$\partial_e = \int_0^{\partial} \frac{v}{V_s}\left(1 - \frac{v^2}{V_s^2}\right) dy = \int_0^{\partial} \frac{y}{\partial}\left(1 - \frac{y^2}{\partial^2}\right) dy = \frac{1}{\partial}\cdot\frac{\partial^2}{2} - \frac{1}{\partial^3}\cdot\frac{\partial^4}{4} = \frac{\partial}{4}$$

13.4 DRAG FORCE DUE TO B.L.

Consider Figure 13.3 and let b be the width of the plate.

The total retarding force at a distance x from the leading edge, i.e. the drag force = rate of change of momentum.

Rate of change of momentum of quantity of discharge through the strip = $\rho(bdy)v\ (V_s - v)$

Total drag force on the plate for length x from the leading edge

$$= F_{d_x} = \int_0^\delta \rho v b(V_s - v)\,dy$$

or

$$F_{D_x} = \int_0^\partial \rho V_s^2\, b\left[\frac{v}{V_s} - \left(\frac{v}{V_s}\right)^2\right]dy \tag{13.4}$$

If λ_0 is the boundary shear intensity at a distance x from the leading edge, the drag resistance for elemental length dx

$$= d\,F_{d_x} = \lambda_0(b \cdot dx)$$

$\therefore$

$$\lambda_0 b = \frac{d\,F_{dx}}{dx} \tag{13.5}$$

13.5 LAMINAR B.L. ANALYSIS

In the laminar boundary layer, the following velocity distribution laws are followed:

(i) $\dfrac{v}{V_s} = 2\left(\dfrac{y}{\partial}\right) - \left(\dfrac{y}{\partial}\right)^2$ (ii) $\dfrac{v}{V_s} = \dfrac{3}{2}\left(\dfrac{y}{\partial}\right) - \dfrac{1}{2}\left(\dfrac{y}{\partial}\right)^3$ (iii) $\dfrac{v}{V_s} = 2\left(\dfrac{y}{\partial}\right) - 2\left(\dfrac{y}{\partial}\right)^3 + \left(\dfrac{y}{\partial}\right)^4$

Now the expression for boundary layer thickness ∂, boundary shear stress λ_0 and drag coefficient may be obtained for a prescribed velocity profile.

Let us take the velocity profile (i) i.e. $\dfrac{v}{V_s} = 2\left(\dfrac{y}{\partial}\right) - \left(\dfrac{y}{\partial}\right)^2$ (A)

The drag force for a distance x from the leading edge, i.e. Equation (13.4) is:

$$F_{d_x} = \int_0^\partial \rho V_s^2\, b\left[\frac{v}{V_s} - \left(\frac{v}{V_s}\right)^2\right]dy$$

$$= \int_0^\partial \rho V_s^2\, b\frac{v}{V_s}\left[1 - \frac{v}{V_s}\right]dy$$

Substituting $\dfrac{v}{V_s}$ from (A):

$$F_{d_x} = \int_0^\partial \rho V s^2\, b\left[2\left(\frac{y}{\partial}\right) - \left(\frac{y}{\partial}\right)^2\right]\left[1 - 2\frac{y}{\partial} + \left(\frac{y}{\partial}\right)^2\right]dy$$

Integrating and simplifying:

$$F_{d_x} = \frac{2}{15}\rho b V_s^2 \partial \tag{B}$$

Again using Equation (13.5), we get:

$$\lambda_0 b = \frac{d}{dx}(Fd_x) = \frac{d}{dx}\left(\frac{2}{15}\rho b V_s^2 \delta\right)$$

$$\lambda_0 = \frac{2}{15}\rho V_s^2 \frac{d\partial}{\partial x} \tag{C}$$

We know that $\lambda_0 = \mu\left(\dfrac{\partial v}{dy}\right)_{y=0}$ (D)

But v from (A), $v = \left[2\left(\dfrac{y}{\partial}\right) - \left(\dfrac{y}{\partial}\right)^2\right] V_s$

$\therefore$

$$\frac{\partial v}{dy} = \left[\frac{2}{\partial} - \frac{2y}{\partial^2}\right] V_s$$

$$\left.\frac{dv}{dy}\right|_{y=0} = \frac{2V_s}{\partial}$$

(D) becomes $\lambda_0 = \dfrac{2\mu V_s}{\partial}$ and equating with (C);

$$\frac{2\mu V_s}{\partial} = \frac{2}{15}\rho V_s^2 \frac{d\partial}{dx}$$

or

$$\partial d\partial = 15\frac{\mu}{\rho V_s}dx$$

Integrating:

$$\frac{\partial^2}{2} = \frac{15\mu x}{\rho Vs} + C$$

$$\text{at } x = 0, \partial = 0 \therefore C = 0$$

$\therefore$

$$\partial^2 = 30\frac{\mu x}{\rho Vs} = 30\left(\frac{\mu}{\rho Vsx}\right)x^2 = \frac{30x^2}{R_{ex}}$$

$\therefore$

$$\partial = \frac{\sqrt{30}x}{\sqrt{R_{ex}}}$$

$\therefore$

$$\partial = \frac{5.477x}{\sqrt{R_{ex}}} \tag{13.6}$$

We get: $$\lambda_0 = \frac{2\mu V_s}{\partial} = \frac{2\mu V_s}{\frac{5.477\, x}{\sqrt{R_{ex}}}}$$

$$\therefore \quad \lambda_0 = 0.365 \frac{\mu V_s}{x}\sqrt{R_{ex}} \tag{13.7}$$

$$F_D = \frac{2}{15}\rho b\, V_s^2 \left(\frac{5.477\, x}{\sqrt{R_{ex}}}\right) = C_d\, \rho \frac{V_s^2}{2}\, bx$$

$$\therefore \quad C_d = \frac{1.46}{\sqrt{R_{ex}}} \tag{13.8}$$

Similarly it can be shown for the velocity profile (ii) $\frac{3}{2}\left(\frac{y}{\partial}\right) - \frac{1}{2}\left(\frac{y}{\partial}\right)^3 = \frac{v}{V_s}$,

$$\partial = 4.64 \frac{x}{\sqrt{R_{ex}}} \tag{13.9}$$

$$\lambda_0 = 0.323 \frac{\mu Vs}{x}\sqrt{R_{ex}} \tag{13.10}$$

$$C_d = \frac{1.292}{\sqrt{R_{ex}}} \tag{13.11}$$

and for the velocity profile (iii) $\frac{v}{V_s} = 2\left(\frac{y}{\partial}\right) - 2\left(\frac{y}{\partial}\right)^3 + \left(\frac{y}{\partial}\right)^4$,

$$\partial = 5.835 \frac{x}{\sqrt{R_{ex}}} \tag{13.12}$$

$$\lambda_0 = 0.343 \frac{\mu Vs}{x}\sqrt{R_{ex}} \tag{13.13}$$

$$C_d = \frac{1.371}{\sqrt{R_{ex}}} \tag{13.14}$$

The Blasius solution gave $$\partial = \frac{4.91 x}{\sqrt{R_{ex}}} \tag{13.15}$$

and $$C_d = \frac{1.328}{\sqrt{R_{ex}}} \tag{13.16}$$

EXAMPLE 13.2 The velocity profile within a laminar B.L. is given by $\frac{v}{V_s} = \frac{2y}{\partial} - \left(\frac{y}{\partial}\right)^2$.

Determine the boundary layer thickness ∂, λ_0 and drag force on one side of the plate at a distance of 1 m from the leading edge. The plate is 1.5 m long and 1.2 m wide. Take $\mu = 0.01$ poise of water and the free upstream velocity is 0.25 m/sec.

Solution: Here $x = 1$ m, $L = 1.5$ m, $b = 1.2$ m, $V_s = 0.25$ m/sec

$$\mu = 0.01 \text{ poise} = \frac{0.01 \text{ Ns}}{10 \text{ m}^2} = 0.001 \text{ Ns/m}^2$$

$$\rho = 1000 \text{ kg/m}^3$$

At a distance of 1 m from the leading edge,

$$R_{e_x} = \frac{\rho V_s x}{\mu} = \frac{1000 \times 0.25 \times 1}{0.001} = 2.5 \times 10^5$$

$\therefore$ The thickness of B.L. at x, $\delta = 5.477 \frac{x}{\sqrt{R_{ex}}}$

$$\partial = 5.477 \times \frac{1}{\sqrt{2.5 \times 10^5}} = 0.010954 \text{ m} \qquad \because x = 1$$

$$\partial = 1.0954 \text{ cm}$$ **Ans.**

$$\lambda_0 = 0.365 \frac{\mu V_s}{x} \sqrt{R_{ex}}$$

$$\lambda_0 = 0.365 \times \frac{0.001 \times 0.25}{1} \sqrt{2.5 \times 10^5} = 0.045625 \text{ N/m}^2$$ **Ans.**

At the trailing edge $x = L = 1.2$ m

$$R_{e_L} = \frac{\rho VsL}{\mu} = \frac{1000 \times .25 \times 1.2}{0.001} = 3 \times 10^5$$

$$\therefore \qquad C_d = \frac{1.46}{\sqrt{R_{ex}}} = \frac{1.46}{\sqrt{3 \times 10^5}} = 2.665 \times 10^{-3} = 0.002665$$

$\therefore$ Drag force $F_D = \frac{1}{2} C_d \, \rho V_s^2 \, A$

$$F_D = \frac{1}{2} \times .002665 \times 1000 \times (0.25)^2 \times (1.5 \times 1.2)$$

$$F_D = 0.15 \text{ N}$$ **Ans.**

EXAMPLE 13.3 The velocity profile in a laminar flow is given by:

$$\frac{v}{V_s} = \frac{3}{2}\left(\frac{y}{\partial}\right) - \frac{1}{2}\left(\frac{y}{\partial}\right)^3$$

A plate 2 m long and 1.4 m wide is placed in flowing water whose upstream velocity is 0.2 m/sec. If viscosity μ = 0.01 poise, find the boundary layer thickness at a distance of 1.5 m from the leading edge. Also find λ_0 and F_D on both sides of the plate.

Solution: x = 1.5 m, L = 2 m, b = 1.4 m, V_s = 0.2 m/sec, μ = 0.01 poise = 0.001 Ns/m^2

Now $$R_{e_x} = \frac{\rho V_s x}{\mu} = \frac{1000 \times 0.2 \times 1.5}{0.001} = 3 \times 10^5$$

$\therefore$ $$\partial = \frac{4.64 \times 1.5}{\sqrt{3 \times 10^5}} = 0.012707 \text{ m} = 12.707 \text{ mm}$$ **Ans.**

$$\lambda_0 = 0.323 \frac{\mu V s}{x}\sqrt{R_{ex}} = 0.323 \times \frac{0.001 \times 0.2}{1.5}\sqrt{3 \times 10^5}$$

$$\lambda_0 = 0.023588 \text{ N/m}^2$$ **Ans.**

$$C_d = \frac{1.292}{\sqrt{R_{ex}}} = \frac{1.292}{\sqrt{3 \times 10^5}} = 0.00235885$$

$$F_D = \frac{1}{2} C_d \rho V_s^2 (2A) \quad \because 2A \text{ since both sides.}$$

$$F_D = \left[\frac{1}{2} \times 0.00235885 \times 1000 \times (.2)^2 \times (2 \times 2 \times 1.4)\right] \text{N}$$

$$F_D = 0.2641912 \text{ N.}$$ **Ans.**

13.6 TURBULENT BOUNDARY LAYER ANALYSIS

Prandtl suggested that the velocity within the turbulent B.L. varies with the seventh root of distance from the wall or plate, i.e.

$$\frac{v}{V_s} = \left(\frac{y}{\partial}\right)^{1/7} \tag{13.17}$$

In Figure 13.4, velocity profile within the turbulent B.L. is considered.

This Equation (13.17) cannot be applied at wall surface as the velocity gradient from Equation (3.17) is

$$\frac{dv}{dy} = \frac{V_s}{7\partial^{1/7} y^{6/7}}$$

when y = 0 at the surface

$$\frac{dv}{dy} \rightarrow \propto \quad \text{and} \quad \lambda_0 = \mu \frac{dv}{dy} \rightarrow \propto$$

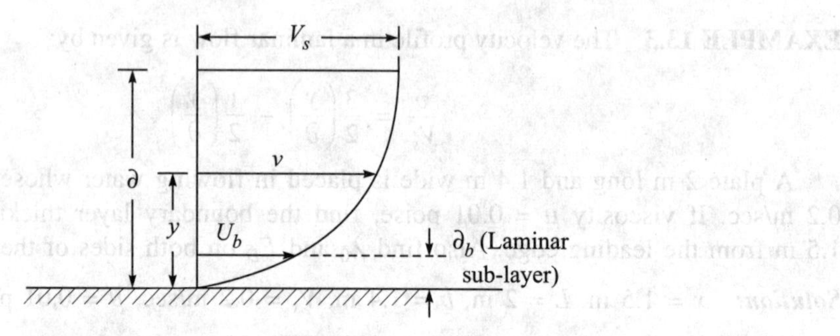

Figure 13.4 Velocity profile in turbulent B.L. with laminar sub-layer.

However, near the wall, the laminar sub-layer is formed and the velocity distribution in the laminar sub-layer is assumed to be laminar.

Like Prandtl, Blasius also suggested the same velocity profile of the seventh root when $5 \times 10^5 < R_e < 10^7$, on the basis of his experimental data.

Blasius gave the following empirical equation of boundary shear:

$$\lambda_0 = 0.0225\ \rho V_s^2 \left(\frac{\nu}{V_s \partial}\right)^{1/4} \tag{13.18}$$

where ∂ is the thickness of turbulent B.L.

Applying the momentum principle within a boundary layer, a flat surface may be applied to a region of unit width of infinitesimal length and equation for shear stress λ_0 to flow the past flat plate with no pressure gradient ($\because$ V_s is constant) is obtained to be

$$\lambda_0 = \rho \frac{d}{dx} \int_0^{\partial} (V_s - v)\, v dy$$

$$= \rho V_s^2 \frac{d}{dx} \int_0^{\partial} \left(1 - \frac{v}{V_s}\right) \frac{v}{V_s} dy \tag{13.19}$$

Equating Equations (13.18) and (13.19) and substituting $\frac{v}{V_s} = \left(\frac{y}{\partial}\right)^{1/7}$, then integrating and separating the variables, we get:

$$\partial^{1/4}\, d\partial = 0.232 \left(\frac{\nu}{V_s}\right)^{1/4} dx$$

Simplifying
$$\frac{\partial}{x} = \frac{0.371}{\left(\frac{u_s x}{\nu}\right)^{1/5}}$$

$\therefore$
$$\partial = \frac{0.371}{(R_{ex})^{1/5}} x \tag{13.20}$$

Similarly
$$C_d = \frac{0.072}{(R_{ex})^{1/5}} \tag{13.21}$$

For $10^7 < R_{e_x} < 10^9$, Schlichting gave the following equation of C_d:

$$C_d = \frac{0.455}{(\log_{10} R_{ex})^{2.58}} \tag{13.22}$$

13.7 LAMINAR SUB-LAYER

This is a laminar thin layer formed between the plate and the turbulent B.L. and its thickness is designated as ∂_b.

We know that:

$$\lambda_0 = \mu\left(\frac{dv}{dy}\right)_{y=0} = \mu\left(\frac{v}{y}\right)_{0\le y\le\partial_b}$$

$$\lambda_0 = \mu\frac{v_b}{\partial_b} \tag{A}$$

Equating Equation (A) with is Blasius Equation (13.18), we get:

$$\mu\frac{v_b}{\partial_b} = 0.0225\ \rho V_s^2\left(\frac{\nu}{Vs\partial}\right)^{1/4}$$

or

$$\frac{\partial_b}{\partial} = \frac{v_b}{V_s}\cdot\frac{1}{0.0225}\left(\frac{\nu}{Vs\partial}\right)^{3/4} \tag{A}$$

At $y = \partial_b$, $v = v_b$, $y = \partial$, $v = V_s$ (at the junction of turbulent B.L. and laminar sub-layer)

$\therefore$ Equation (3.17) becomes:

$$\frac{v_b}{V_s} = \left(\frac{\partial_b}{\partial}\right)^{1/7}$$

$$\therefore \quad \frac{\partial_b}{\partial} = \left(\frac{v_b}{V_s}\right)^7 \tag{B}$$

Equating Equations (A) and (B) and simplifying,

$$\frac{\partial_b}{\partial} = \frac{198}{(R_{ex})^{7/10}} \tag{13.22a}$$

Nikuradse's experimental studies had shown that:

$$\partial_b = \frac{11.6\nu}{\sqrt{\lambda_0/\rho}}$$

i.e.

$$\partial_b = \frac{11.6\nu}{V_*} \qquad \because \sqrt{\lambda_0/\rho} = V_* \tag{13.23}$$

13.8 SEPARATION OF BOUNDARY LAYER

Boundary layer is considerably affected by the pressure gradient in the direction of flow.

Consider a flow of fluid above a curved surface in the direction of s as shown in Figure 13.5. As the flow is on the curved surface, it is accelerated over the left hand section until a point C where the pressure is minimum. The velocity gradient in the left of C is positive. As the velocity increases, as s increases, pressure falls, hence the pressure with respect to the distance is negative. Up till C, the flow converges, hence the velocity increases and pressure decreases.

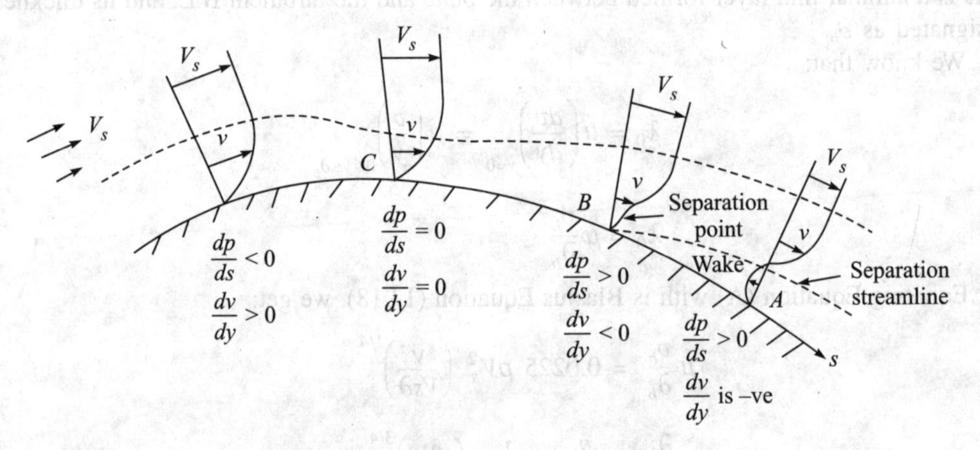

Figure 13.5 Separation of boundary layer.

Beyond point C, the flow enters a diverging area, the velocity decreases, pressure increases somewhere at B, and the velocity close to the boundary is zero. At point A, near the boundary, back flow takes place. Thus, the pressure goes on increasing beyond the point C resulting in decrease of velocity or kinetic energy. When this happens, the boundary layer separates itself from the boundary. This phenomenon is called 'boundary layer separation'. The separation point is followed by a region called 'wake' in which intense eddies exist. In the wake region, energy of flowing fluid decreases and thereby drag force becomes more. Therefore, an attempt has been made to avoid making the wake region smaller to minimise the loss of energy. Also an attempt has been made to adopt suitable boundary layer control measures. These measures consist of removing the retarded flow in the boundary layer by suction or imparting energy to the retarded flow just to avoid the possibility of separation.

EXAMPLE 13.4 The velocity profile within the boundary layer is given by $\frac{v}{V_s} = \left(\frac{y}{\partial}\right)^{1/7}$.

Calculate displacement and energy thickness.

Displacement thickness ∂_d:

Solution:

$$\partial_d = \int_0^{\partial}\left(1 - \frac{v}{V_s}\right)dy$$

$$\partial_d = \int_0^{\partial}\left[1 - \left(\frac{y}{\partial}\right)^{1/7}\right]dy$$

$$\partial_d = \left[y - \frac{y^{8/7}}{(8/7)\,\partial^{1/7}} \right]_0^{\partial} = \frac{\partial}{8}$$ **Ans.**

Energy thickness ∂_e:

$$\partial_e = \int_0^{\partial} \frac{v}{V_s}\left(1 - \frac{v}{V_s}\right) dy$$

$$\partial_e = \int_0^{\partial} \frac{y^{1/7}}{\partial^{1/7}}\left(1 - \frac{y^{1/7}}{\partial^{1/7}}\right) dy$$

$$\partial_e = \frac{7}{72}\,\partial$$ **Ans.**

EXAMPLE 13.5 For the velocity profile given below, state whether the boundary layer has separated or is on the verge of separation or will remain attached with the boundary surface.

(i) $\frac{v}{V_s} = 2\left(\frac{y}{\partial}\right) - \left(\frac{y}{\partial}\right)^2$ (ii) $\frac{v}{V_s} = -2\left(\frac{y}{\partial}\right) + \frac{1}{2}\left(\frac{y}{\partial}\right)^3$ (iii) $\frac{v}{V_s} = \frac{3}{2}\left(\frac{y}{\partial}\right)^2 + \frac{1}{2}\left(\frac{y}{\partial}\right)^3$

Solution: (i) $v = \frac{2V_s}{\partial} y - \frac{V_s}{\partial^2} y^2$

$$\therefore \quad \frac{dv}{dy} = \frac{2V_s}{\partial} - \frac{2V_s}{\partial^2}\,y$$

$$\left.\frac{dv}{dy}\right|_{Y=0} = \frac{2V_s}{\partial}$$ which is positive, so it remains attached

(ii) $v = \frac{-2V_s}{\partial} y + \frac{1}{2\partial^3} y^3$

$$\frac{\partial v}{\partial y} = \frac{-2V_s}{\partial} + \frac{3y^2}{2\partial^3}$$

$$\left.\frac{dv}{dy}\right|_{Y=0} = \frac{-2V_s}{\partial}$$, which is negative, and has separated.

(iii) $v = \frac{3}{2}\frac{V_s}{\partial^2} y^2 + \frac{1}{2}\frac{V_s}{\partial^3} y^3$

$$\frac{dv}{dy} = \frac{3}{2}\frac{V_s}{\partial^2}\cdot 2y + \frac{1}{2}\frac{V_s}{\partial^3}\cdot 3y^2$$

$$\left.\frac{dv}{dy}\right|_{Y=0} = 0$$, Hence this is on the verge of separation.

13.9 CONCLUSION

Extensive works have already been done on boundary layer theory. Schlichting wrote a book on the topic that has been translated into English. Partial differential equations on this theory in 2-dimensional and 3-dimensional flow with unsteady and viscous effect have also been analysed by different investigators. This book is an introduction fluid mechanics to beginners. Therefore, detailed analysis of boundary layer theory is out of the scope of this scope.

Attempts are only made to offer an idea about the formation of boundary layer—both laminar and turbulent, separation of boundary layer, drag force and laminar sub-layer, Preliminary equations of boundary layer theory are derived and few numerical examples are solved to show the application of those theoretical equations. For ambitious readers, Schlichting[2], Prandtl[1,3,11,12,13] are referred for further study.

PROBLEMS

13.1 Find the displacement thickness, energy thickness and momentum thickness for the velocity distribution given by:

$$\frac{v}{V_s} = 2\left(\frac{y}{\partial}\right) - \left(\frac{y}{\partial}\right)^2 \qquad \left(\textbf{Ans. } \frac{\partial}{3}, \frac{2}{15}\partial, \frac{22}{105}\partial\right)$$

13.2 For the following velocity profiles, determine whether the flow has separated or is on the verge of separation or will attach with the surface.

$$\text{(i) } \frac{v}{V_s} = \frac{3}{2}\left(\frac{y}{\partial}\right) - \frac{1}{2}\left(\frac{y}{\partial}\right)^2 \quad \text{(ii) } \frac{v}{V_s} = 2\left(\frac{y}{\partial}\right)^2 - \left(\frac{y}{\partial}\right)^3 \quad \text{(iii) } \frac{v}{V_s} = -2\left(\frac{y}{\partial}\right) + \left(\frac{y}{\partial}\right)^2$$

[**Ans.** (i) will attach, (ii) is on the verge of separation, and (iii) has separated]

13.3 For the velocity profile in a laminar B.L. given as $\frac{v}{V_s} = \frac{3}{2}\left(\frac{y}{\partial}\right) - \frac{1}{2}\left(\frac{y}{\partial}\right)^3$, find the thickness of B.L. and the shear stress at a distance of 1.8 m from the leading edge of the plate. The plate is 2.5 m long and 1.5 m wide, and is placed in water moving with a velocity of 0.15 m/sec. Find the drag on one side of the plate if the viscosity is 0.01 poise.

(**Ans.** 1.6 cm, 0.0014 N/m^2, 0.0889 N)

13.4 Oil with free stream velocity of 2 m/sec flows over a thin plate 2 m wide and 2 m long. Calculate the boundary layer thickness and shearing stress at the trailing edge and determine the total surface resistance of the plate. Take specific gravity of oil to be 0.86 and kinematic viscosity as 10^{-5} m^2/sec.

(*Hints:* Velocity profile not given, then always use Blasicus solution.

$$\partial = \frac{4.91x}{\sqrt{R_{ex}}}, C_d = \frac{1.326}{\sqrt{R_{ex}}}, \lambda_0 = 0.332\frac{\rho V s^2}{\sqrt{R_{ex}}}\Bigg)$$

(**Ans.** ∂ = 15.5 mm λ_0 = 1.805 N/m^2 F_D = 28.88 N)

13.5 Determine the thickness of B.L. at the trailing edge of a smooth plate of length 4 m, and width 1.5 m, when the plate is moving with a velocity of 4 m/sec in stationary air. Take ν of air = 1.5×10^{-5} m^2/sec

(*Hints:* Here $R_{e_L} > 5 \times 10^5$, so the flow is turbulent. Use the turbulent B.L. equation).

Ans. 9.156 cm

REFERENCES

1. Prandtl, L., "Uber Fliissigkeitsbewegung bei sehr kleiner Reibung", *Proceedings, 3rd. International Math. Congress*, Heidelberg, 1904.
2. Schlichting, H., *Boundary Layer Theory*, Translated by J. Kestin, McGraw-Hill Inc., New York, 1960.
3. Prandtl, L., "On Fully Developed Turbulence", *Proc. of the 2nd. International Congress of Applied Mechanics*, Zurich, pp. 64–72, 1926.
4. Prandtl, L., *Outline of Theory of Flow*, Vieweg-Verlag, Brunswick, Germany, 1931.
5. Prandtl, L. and O.G. Tietjens, *Fundamentals of Hydro and Aerodynamics*, McGraw-Hill, New York, 1934.
6. Falkner, V.M. and S.W. Skan, "Some Approximate Solutions of the Boundary Layer Equations", *ARC Reports and Memoranda*, No. 1314, 1930.
7. Wieghardt, K.E.G. "On a Simple Method of Calculating Laminar Boundary Layer", *The Aeronautical Quarterly*, Vol. 5, 1954.
8. Hama, F., "Boundary Layer Characteristics for smooth and Rough surfaces", Trans., *S.N.A.M.E.*, Vol. 62, 1954.
9. Landweber, L., "Generalisation of the Logarithmic Law of Boundary Layer on a Flat Plate", *Schiffstechnik*, Vol. 4, No. 21, 1957.
10. Landweber, L., "Frictional Resistance of Flat Plates in Zero Pressure Gradient", Trans. *S.N.A.M.E.*, Vol. 61, 1953.
11. Prandtl, L., "Mechanics of viscuous fluids" in W.F. Durand (Ed.). *Aerodynamic Theory*, Springer-Verlag, Berlins, 1935.
12. Prandtl, L., *Essential of Fluid Mechanics*, Hafner Publishing Company, New York, 1952.
13. Prandtl, L., "Mechanics of Viscous Fluids" in W.F. Durand (Ed.) *Aerodynamic Theory*, Springer-Verlag, Berlin, 1935.

Chapter 14

Dimensional Analysis and Model Investigation

14.1 INTRODUCTION

Dimensional analysis is a mathematical technique in which the use of dimensions of different parameters of a physical problem is used as an aid to solve many engineering problems.

Model investigation is also a technique used to predict the behaviour or performance of the actual structure or prototype in advance. This investigation is also based on dimensional analysis. Thus, for model investigation, knowledge of dimensional analysis is a pre-requisite.

The uses of dimensional analysis are essential for testing the dimensional homogeneity of any equation of fluid motion, deriving equations expressed in terms of non-dimensional parameters to study the relative significance of each parameter, planning model tests and presenting experimental data to study the actual behaviour of the prototype. For studying the application of dimensional analysis, a preliminary knowledge of dimensions is essential.

14.2 DIMENSIONS AND DIMENSIONAL HOMOGENEITY

The various physical quantities required to describe a phenomenon can be described by a set of quantities which are called fundamental or primary quantities. These primary quantities are mass, length, time and temperature which are designated as M, L, T and θ respectively. Other quantities like area, volume, discharge, velocity, acceleration, energy, force power, frequency, viscosity, density, specific weight, surface tension, etc. are called derived quantities or secondary quantities. Dimension is the expression of derived quantities in terms of primary quantity. As an example, force = mass × acceleration, i.e. Force = $M \cdot \dfrac{L}{T^2} = MLT^{-2}$. Thus, the derived quantity force is expressed in the primary quantities M, L and T. If force is expressed as F instead of M, it is an the F-L-T system. Usually, however, derived quantities are expressed in the H-L-T system.

An equation is said to maintain a dimensional homogeneity if the dimensions of terms on the left hand side are the same as the dimensions of the right hand side. For example, head loss in head of fluid due to friction is $h_f = \dfrac{fLV^2}{2gD}$. Here h_f is head (L), f is non-dimensional ($M^0L^0T^0$), L is the length of the pipe (L), V^2 is (L^2/T^2), g is acceleration $\left(\dfrac{L}{T^2}\right)$, diameter D is length (L) and hence

$$\text{LHS} = L$$

$$\text{RHS} = M^0L^0Y^0 \cdot L \cdot \frac{L^2}{T^2} \times \frac{1}{\dfrac{L}{T^2} \cdot L} = \frac{L^3}{T^2} \times \frac{T^2}{L^2} = L$$

Hence LHS = RHS. Thus, the equation $h_f = \dfrac{fLV^2}{2gD}$ is dimensionally homogeneous.

Consider Chezy's equation:

$$V = C\sqrt{Rs_b}$$

LHS is velocity $= \dfrac{L}{T}$

RHS $R = L$, $s_b = M^0L^0T^0$, C is a resistance coefficient, if C is assumed to be dimensionless, $C = M^0L^0T^0$

$\therefore$ LHS $= \dfrac{L}{T}$, RHS $= M^0L^0T^0 \cdot L^{1/2} \cdot M^0L^0T^0 = L^{1/2}$

$\therefore$ LHS $\neq$ RHS

Thus Chezy's equation is not dimensionally homogenous. In order to maintain dimensional homogeneity, C must have a dimension of $\left(\dfrac{L^{1/2}}{T}\right)$.

The dimensions of some of the physical quantities used in fluid mechanics are given in Table 14.1.

TABLE 14.1 Dimensions of Physical Quantities in M-L-T System

	Physical quantity	*Symbol*	*Dimensions*
(a)	Fundamental		
	Mass	M	M
	Length	L	L
	Time	T	T
(b)	Geometric linear measurement	L	L
	Area	A	L^2
	Volume	$\forall$	L^3
	Curvature	C	$1/L$
	Slope	s	$M^0L^0T^0$

(*Contd.*)

TABLE 14.1 Dimensions of Physical Quantities in M-L-T System (*Contd.*)

	Physical quantity	*Symbol*	*Dimensions*
	Angle	α, θ	$M^0L^0T^0$
	Shape factor	η	$M^0L^0T^0$
(c)	Kinematic		
	Time	T, t	T
	Velocity	V, v	L/T
	Angular velocity	ω	$\frac{1}{T}$
	Acceleration	a, g	$\frac{L}{T^2}$
	Angular acceleration	α	$\frac{1}{T}$
	Discharge	Q	L^3/T
	Unit discharge	q	L^2/T
	Kinematic viscosity	ν	L^2/T
	Circulation	Γ	L^2/T
(d)	Dynamic		
	Force	F	$\frac{ML}{T^2}$
	Weight	W	$\frac{ML}{T^2}$
	Mass density	ρ	M/L^3
	Specific weight	w, γ	$\frac{M}{L^2T^2}$
	Specific gravity	s	$M^0L^0T^0$
	Dynamic viscosity	μ	$\frac{M}{LT}$
	Pressure intensity	p	$\frac{M}{LT^2}$
	Modulus of elasticity	K, E	$\frac{M}{LT^2}$
	Surface tension	σ	$\frac{M}{T^2}$
	Shear stress	λ	$\frac{M}{LT^2}$
	Work or energy	W or E	$\frac{ML^2}{T^2}$

(*Contd.*)

TABLE 14.1 Dimensions of Physical Quantities in M-L-T System (*Contd.*)

	Physical quantity	*Symbol*	*Dimensions*
	Power	P	$\frac{ML}{T^3}$
	Torque	T	$\frac{ML^2}{T^2}$
	Momentum	M	ML/T^{-1}
	Pressure gradient	$\Delta P/L$	M/L^2T^2
(e)	General		
	moment	$F \times L$	$\frac{ML^2}{T^2}$
	Moment of inertia	I	L^4
	Moment of inertia of mass	ML^2	ML^2
	Temperature	t	θ
	Heat conduction	ϕ	$\frac{ML^2}{T^2}$ (Kilocalories)
	Specific heat	ϕ_s	$\frac{L^2}{\theta T}$
	Heat conduction	ϕ_c	$\frac{ML}{T^3\theta}$
(f)	Dimensionless		
	Reynolds number	R_e	
	Fronde number	F_r	
	Euler number	E_u	
	Weber number	W_b	
	Mach number	M_a	
	Friction factor	f	$M^0L^0T^0$
	Slope	s, i	
	K.E. factor	α	
	Momentum correction factor	β	
	Efficiency	η_e	
	Angle	θ	
	Shape factor	η	

14.3 METHODS OF DIMENSIONAL ANALYSIS

When the number of variables in a physical phenomenon are known, the relationship among the variables is determined by the following two methods:

(i) Rayleigh Method
(ii) Buckingham p-theorem

Rayleigh Method

Lord Rayleigh in 1899 proposed this method for determining the effect of temperature on the viscosity of gas. In the Rayleigh method, the functional relationship of some variables is expressed in the form of an exponential equation, which must be dimensionally homogenous. If X is some function of variables X_1, X_2, X_3, ... X_n, the functional equation is written as:

$$X = f(X_1, X_2, X_3, \ldots, X_n) \tag{14.1}$$

Here X is the dependent variable and $X_1, X_2, X_3, \ldots, X_n$ are the independent variables.

$$X = C(X_1^a\, X_2^b\, X_3^c \ldots X_n^n) \tag{14.2}$$

where C is a dimensionless constant and exponents a, b, c, ..., n are evaluated on the basis of the fact that Equation (14.2) is dimensionally homogenous.

For instance, the resistance for F_D of falling sphere in a viscous liquid depends on diameter D of the sphere, its fall velocity V and viscosity of fluid μ.

We can now express the relationship in the form of:

$$F_D = C(D^a,\ V^b,\ \mu^c) \tag{A}$$

Writing in dimensional form:

$$\frac{ML}{T^2} = C\left[L^a \left(\frac{L}{T}\right)^b \left(\frac{M}{LT}\right)^c\right] \quad \text{where } C \text{ is non-dimensional}$$

Equating the powers of M:

$$1 = C \quad \therefore C = 1$$

Equating the powers of L:

$$1 = a + b - c$$

or

$$1 = a + b - 1 \quad \therefore \quad a + b = 2$$

Equating the powers of T:

$$-2 = -b - c \quad \therefore \quad b + c = 2$$

or $\quad b + 1 = 2$

$\therefore \quad b = +1$

$\therefore \quad a + 1 = 2$

$\therefore \quad a = 1$

Substituting a, b, c in (A):

$$F_D = C(D^1\ V^1\ \mu^1) \quad \therefore \quad F_D = C\,\mu VD$$

If constant C is replaced by a constant (3π),

$$F_D = 3\pi\mu VD \text{ which is Stokes' law.}$$

EXAMPLE 14.1 The period of a simple pendulum time T depends on its length L and acceleration due to gravity g. Obtain an expression T by the Rayleigh method.

Solution: $T = f(L, g)$

or $T = C(L^a g^b)$ (A)

or $T = C\left[L^a \cdot \left(\frac{L}{T^2}\right)^b\right]$

Equating powers T, $1 = -2b \therefore b = -\frac{1}{2}$

Equating powers L, $0 = a + b \therefore b = -a$

$$\therefore \quad \frac{-1}{2} = -a$$

$$\therefore \quad a = \frac{1}{2}$$

Substituting a and b in (A):

$$T = C[L^{1/2} \cdot g^{-1/2}] = C\sqrt{\frac{L}{g}}$$

$$\therefore \quad T = C\sqrt{\frac{L}{g}}$$

If the constant C is replaced by 2π,

$\therefore$ $T = 2\pi\sqrt{\frac{L}{g}}$ which is the actual equation of the time period of a simple pendulum.

EXAMPLE 14.2 The critical depth y_c in a triangular channel depends on the discharge Q and acceleration due to gravity.

Solution: $y_c = f(Q, g)$

or $y_c = C(Q^a \cdot g^b)$ (A)

$$L = C\left[\left(\frac{L^3}{T}\right)^a \cdot \left(\frac{L}{T^2}\right)^b\right]$$

Equating powers of L, $1 = 3a + b$

Equating powers of T, $0 = -a - 2b$

$$\therefore \quad a = -2b$$

$$\therefore \quad 1 = 3 \times (-2b) + b$$

$$1 = -5b \quad \therefore \quad b = -1/5$$

$$\therefore \quad a = -2 \times -\frac{1}{5} = \frac{2}{5}$$

Substituting a and (b) in (A):

$$y_c = C(Q^{2/5}\, g^{-1/5}) = C\left[\frac{Q^2}{g}\right]^{1/5}$$

If the constant C is replaced by a constant $\left(\frac{2}{z^2}\right)^{1/5}$ where the side slope is zH : 1V.

$\therefore\ y_c = \left(\frac{2Q^2}{g z^2}\right)^{1/5}$ which is the actual equation of the critical depth in a triangular channel.

EXAMPLE 14.3 The discharge Q through an orifice depends on diameter d of the orifice, head H above the orifice, density ρ and viscosity μ of liquid and acceleration g due to gravity. Show by dimensional analysis that $Q = C_d\, a\sqrt{2gH}$, where C_d is the non-dimensional coefficient of discharge.

Solution:

$$Q = f(\mu, \rho, d, H, g)$$

or

$$Q = C(\mu^a \rho^b \alpha^c H^d g^e) \tag{A}$$

or

$$\left(\frac{L^3}{T}\right) = H^0L^0T^0\left[\left(\frac{M}{LT}\right)^a\left(\frac{M}{L^3}\right)^b (L)^c (L)^d \left(\frac{L}{T^2}\right)^e\right]$$

Equating exponent of M, $0 = a + b \quad \therefore\ b = -a$

Equating exponent of L, $3 = -a + 3b + c + d + e$ (B)

Equating exponent T, $-1 = -a - 2e \quad \therefore\ e = \frac{1}{2} - \frac{a}{2}$

Substituting e and a in B

$$3 = -a + 3(-a) + c + d + \frac{1}{2} - \frac{a}{2}$$

$$\therefore\qquad C = \frac{5}{2} - \frac{3a}{2} - d.$$

Substituting the exponents in (A) in terms a and d,

$$Q = C\left[\mu^a \rho^{-a} d^{\left(5/2-\frac{3a}{2}-d\right)} H^d g^{\left(\frac{1}{2}-a/2\right)}\right]$$

$$Q = C[(d^{5/2}\ g^{1/2})\ (\mu^a\ \rho^{-a}\ d^{-3a/2}\ g^{-a/2})\ (H^d d^{-d})]$$

$$= C\left[(d^2 \cdot d^{1/2} \cdot g^{1/2})\left(\frac{\mu}{\rho\, d^{3/2}\, g^{1/2}}\right)^a\left(\frac{H}{d}\right)^d\right]$$

$$= C\left[-\left(\frac{\mu}{\rho\, d^{3/2}\, g^{1/2}}\right)^a\left(\frac{H}{d}\right)^{d-1/2}(d^2\, H^{1/2}\, g^{1/2})\right]$$

$$= \frac{C}{(\pi/4)\sqrt{2}}\left(\frac{\mu}{\rho\, d^{3/2}\, g^{1/2}}\right)^a\left(\frac{H}{d}\right)^{d-1/2}\cdot\left(\frac{\pi}{4}d^2\right)\sqrt{2gH}$$

$$= a\sqrt{2gH}\ f_1\left[\left(\frac{\mu}{\rho\, d^{3/2}\, g^{1/2}}\right), \left(\frac{H}{d}\right)\right]$$

or $$Q = C_d\, a\sqrt{2gH}$$

where $$C_d = f_1\left[\left(\frac{\mu}{\rho\, d^{3/2}\, g^{1/2}}\right), \left(\frac{H}{d}\right)\right]$$

C_d and both the terms within bracket are nondimensional.

$\therefore$ $$Q = C_d\, a\sqrt{2gH} \quad \text{shown.}$$

Buckingham π-theorem

The Buckingham π-theorem was first stated by Vaschy and proved in increasing generality by Buckingham, and therefore it is known as Buckingham π-theorem. The theorem states that if there are n dimensional variables involved in a phenomenon, which can be completely described by m fundamental quantities or dimensions (such as mass, length and time), then the relationship among the n quantities can always be expressed in terms of exactly $(n - m)$ dimensionless and independent π terms. For example, discharge Q through the orifice depends on H, diameter d, ρ, μ and g. Thus, there are six variables Q, H, d, ρ, μ and g that are required to describe the flow phenomenon through orifice and these six variables can be described by three fundamental quantities (i.e. M, L and T). Hence number π-terms $(n - m)$, i.e. $(6 - 3) = 3$ numbers.

To understand the theorem, we may take Example 14.3, solved by the Rayleigh method.

$$Q = f(\mu, \rho, d, H, g) \tag{A}$$

Here $n = 6$, $m = 3$ (M, L, T)

$\therefore$ No. of π terms $= 6 - 3$.

(A) may also be written as $f_1(Q, \mu, \rho, d, H, g) = C$, where C is a non-dimensional constant.

Let the three π terms be π_1, π_2 and π_3.

$\therefore$ $$f_2(\pi_1, \pi_2, \pi_3) = C_1 \text{ where } C_1 \text{ is a non-dimensional constant.}$$

The π terms are selected in such way that each π term has to choose three repeating variables out of the six and these three repeating variables are such that they, among themselves, contain all three fundamental dimensions (*M.L.T.*).

Let us choose $\rho(M/L^3)$, $d(L)$, $g(L/T^2)$ as repeating variables.

Thus the three π terms will be:

$$\left.\begin{aligned} \pi_1 &= \rho^{a_1}\, d^{b_1}\, g^{c_1}\, Q \\ \pi_2 &= \rho^{a_2}\, d^{b_2}\, g^{c_2}\, \mu \\ \pi_3 &= \rho^{a_3}\, d^{b_3}\, g^{c_3}\, H \end{aligned}\right\}$$

Express π terms dimensionally in terms of the M–L–T system

$\therefore$ $$\pi_1 = H^0 L^0 T^0 = (ML^{-3})^{a_1}\,(L)^{b_1}\,(LT^{-2})^{c_1}\,(L^3T^{-1})$$

Equating exponents of M, $0 = a_1$:

$$L,\ 0 = -3a_1 + b_1 + c_1 + 3$$
$$T,\ 0 = -2c_1 - 1$$

From these three equations, $a_1 = 0, c_1 = -1/2$

$$0 = -3 \times 0 + b_1 + \left(-\frac{1}{2}\right) + 3$$

$$\therefore \quad b_1 = -5/2$$

$$\therefore \quad \pi_1 = r^0 \cdot d^{-5/2} \cdot g^{-1/2}\ Q$$

$$\therefore \quad \pi_1 = \left(\frac{Q}{d^{5/2}\ g^{1/2}}\right) \qquad \text{(A)}$$

Similarly $\pi_2 = M^0L^0T^0 = (ML^{-3})^{a_2}\ (L)^{b_2}\ (LT^{-2})^{c_2}\ (ML^{-1}T^{-1})$

Equating exponents of M, $0 = a_2 + 1 \quad \therefore\ a_2 = -1$

$$L,\ 0 = -3a_2 + b_2 + c_2 - 1$$

$$T,\ 0 = -2c_2 - 1 \quad \therefore\ c_2 = -1/2$$

$$\therefore \quad 0 = -3\ (-1) + b_2 + (-1/2) - 1$$

$$b_2 = -3/2$$

$$\therefore \quad \pi_2 = \rho^{-1}\ d^{-3/2}\ g^{-1/2}\ \mu$$

$$\therefore \quad \pi_2 = \left(\frac{\mu}{\rho\ g^{1/2}\ d^{3/2}}\right) \qquad \text{(B)}$$

Again $\pi_3 = M^0L^0T^0 = (ML^{-3})^{a_3}\ (L)^{b_3}\ (LT^{-2})^{c_3}\ (L)$

Equating exponents M:

$$0 = a_3$$

$$L,\ 0 = -3a_3 + b_3 + c_3 + 1$$

$$T,\ 0 = -2c_3 \quad \therefore\ c_3 = 0$$

$$\therefore \quad 0 = -3 \times 0 + b_3 + 0 + 1$$

$$\therefore \quad b_3 = -1$$

$$\therefore \quad \pi_3 = \rho^0\ d^{-1}\ g^0\ H \quad \therefore\ b_3 = -1$$

$$\therefore \quad \pi_3 = (H/d) \qquad \text{(C)}$$

We may now write $\pi_1 = \phi(\pi_2, \pi_3)$ (D)

or $\phi_1(\pi_1, \pi_2, \pi_3) = K$ where K is a non-dimensional constant.

Writing in (D):

$$\frac{Q}{g^{1/2}\ d^{5/2}} = C_2\ \phi_2\left[\left(\frac{\mu}{\rho\ g^{1/2}\ d^{3/2}}\right), \left(\frac{H}{d}\right)\right] \qquad \text{(E)}$$

If we would have chosen the second repeating variables H instead of d, (E) would have finally been:

$$\frac{Q}{g^{1/2}\ H^{5/2}} = C_3\ \phi_3\left[\left(\frac{\mu}{\rho\ g^{1/2}\ H^{3/2}}\right), \left(\frac{d}{H}\right)\right]$$

$$\therefore \quad Q = g^{1/2} H^{5/2} C_3 \phi_3 \left[\left(\frac{\mu}{\rho g^{1/2} H^{3/2}} \right), \left(\frac{d}{H} \right) \right]$$

$$Q = \left(\sqrt{2gH} \times \frac{1}{2} H^2 \right) \left(\left(\frac{\pi}{4} d^2 \right) \frac{1}{(\pi/4) d^2} \right) C_3 \phi_3 \left[\left(\frac{\mu}{\rho g^{1/2} H^{3/2}} \right), \left(\frac{d}{H} \right) \right]$$

$$Q = a\sqrt{2gH} \cdot \left(\frac{4}{2\pi} \frac{H^2}{d^2} \right) C_3 \phi_3 \left[\left(\frac{\mu}{\rho g^{1/2} H^{3/2}} \right), \left(\frac{d}{H} \right) \right]$$

$$Q = \left\{ C_4 \phi_4 \left[\left(2\frac{H^2}{d^2} \right), \left(\frac{\mu}{\rho g^{1/2} H^{3/2}} \right), \left(\frac{d}{H} \right) \right] \right\} \cdot a\sqrt{2gH}$$

$Q = C_d \, a\sqrt{2gH}$ (F) Equation (F) is the required discharge equation

where C_d (non-dimensional) $= C_4\phi \left[\left(2\frac{H^2}{d^2} \right), \left(\frac{\mu}{\rho g^{1/2} H^{3/2}} \right), \left(\frac{d}{H} \right) \right]$

A study of these two methods shows that if the number of variables is smaller, both the methods are essentially the same. But if the number of variables is more, the Rayleigh method becomes tedious, and the π-theorem can be adopted in this case to avoid complexity.

EXAMPLE 14.4 Capillary rise h depends on density ρ, acceleration g, surface tension σ and radius of the tube r. Show by using the Buckingham π-theorem that $\frac{h}{r} = \phi\left(\frac{\sigma}{\rho g r^2} \right)$

Solution:

$$h = f(\rho g \sigma r)$$

$$n = 5, \ m = 3 \quad \therefore \ \pi \text{ terms} = 5 - 3 = 2$$

$$\pi_1 = \rho^{a_1} g^{b_1} r^{c_1} \sigma$$

$$M^0 L^0 T^0 = \left(\frac{M}{L^3} \right)^{a_1} \left(\frac{L}{T^2} \right)^{b_1} (L)^{c_1} (HT^{-2})$$

Equating power:

$$M, \ 0 = a_1 + 1 \quad \therefore \ a_1 = -1$$

$$L, \ 0 = -3a_1 + b_1 + c_1$$

$$T, \ 0 = -2b_1 - 2 \quad \therefore \ b_1 = -1$$

$$\therefore \quad 0 = -3 \times (-1) + (-1) + c_1$$

$$\therefore \quad c_1 = -2$$

$$\therefore \quad \pi_1 = \rho^{-1} g^{-1} r^{-2} \sigma = \left(\frac{\sigma}{\rho g r^2} \right)$$

$$\pi_2 = \rho^{a_2} g^{b_2} r^{c_2} h$$

$$M^0 L^0 T^0 = (ML^{-3})^{a_2} (LT^{-2})^{b_2} L^{c_2} L$$

Equating power of M, $0 = a_2$:

$$L,\ 0 = -3a_2 + b_2 + c_2 + 1$$
$$T,\ 0 = -2b_2 \quad \therefore\ b_2 = 0$$
$$\therefore \quad 0 = -3 \times 0 + 0 + c + 1$$
$$\therefore \quad c_2 = -1$$
$$\therefore \quad \pi_2 = \rho^0\, g^0\, r^{-1}\, h$$
$$\therefore \quad \pi_2 = \left(\frac{h}{r}\right)$$
$$\therefore \quad \pi^2 = \phi(\pi_1)$$
$$\frac{h}{r} = \phi\left(\frac{\sigma}{\rho g r^2}\right) \quad \text{shown.}$$

EXAMPLE 14.5 The resistance R to the motion of a supersonic aircraft of length L moving with velocity V in the air of density ρ, depends on the viscosity μ and the bulk modulus of elasticity K of air. Using Buckingham's π-theorem, obtain the following expression of R.

$$R = (\rho L^2\, V^2)\ \phi\left[\left(\frac{\mu}{\rho L V}\right), \left(\frac{K}{\rho V^2}\right)\right]$$

Solution: $R = f(\rho, L, V, \mu, K)$

Here $n = 6$, $m = 3$, Hence π terms $= 3$

Choose repeating variables ρ, L, V.

$$\pi_1 = \rho^{a_1}\, L^{b_1}\, V^{c_1}\, R$$
$$\pi_2 = \rho^{a_2}\, L^{b_2}\, V^{c_2}\, \mu$$
$$\pi_3 = \rho^{a_3}\, L^{b_3}\, V^{c_3}\, K$$

Now

$$M^0L^0T^0 = \pi_1 = (ML^{-3})^{a_1}\, (L)^{b_1} \left(\frac{L}{T}\right)^{c_1} \quad (MLT^{-2})$$

Equating the power:

$$M,\ 0 = a_1 + 1 \ \therefore\ a_1 = -1$$
$$L,\ 0 = -3a_1 + b_1 + c_1 + 1$$
$$T,\ 0 = -c_1 - 2 \quad \therefore\ c_1 = -2$$
$$\therefore \quad 0 = -3 \times -1 + b_1 - 2 + 1$$
$$\therefore \quad b_1 = -2$$
$$\therefore \quad \pi_1 = \rho^{-1}\, L^{-2}\, V^{-2}\, R$$
$$\therefore \quad \pi_1 = \frac{R}{\rho\, L^2 V^2}$$

$$\pi_2 = M^0L^0T^0 = (ML^{-3})^{a_2}\, (L)^{b_2}\, (LT^{-1})^{c_2}\, (ML^{-1}T^{-1})$$

Equating the power: $M,\ 0 = a_2 + 1 \quad \therefore a_2 = -1$

$L,\ 0 = -3a_2 + b_2 + c_2 - 1$

$T,\ 0 = -c_1 - 1 \quad \therefore c_2 = -1$

$\therefore \quad 0 = -3 \times -1 + b_2 - 1 - 1$

$\therefore \quad b_2 = -1$

$\therefore \quad \pi_2 = \rho^{-1} L^{-1} V^{-2} \mu$

$$\therefore \quad \pi_2 = \frac{\mu}{\rho LV}$$

$$\pi_3 = M^0L^0T^0 = (\rho)^{a3}\ (L)^{b3}\ (V)^{c3}\ K$$

$$\pi_3 = M^0L^0T^0 = (HL^{-3})^{a3}\ (L)^{b3}\ (LT^{-1})^{c3}\ (HL^{-1}T^{-2})$$

Equating the power: $M,\ 0 = a_3 + 1 \quad \therefore a_3 = -1$

$L,\ 0 = -3a_3 + b_3 + c_3 - 1$

$T,\ 0 = -c_3 - 2 \quad \therefore c_3 = -2$

$\therefore \quad 0 = -3 \times -1 + b_3 - 2 - 1$

$\therefore \quad b_3 = 0$

$\therefore \quad \pi_3 = \rho^{-1}\ L^0\ V^{-2}\ K$

$$\therefore \quad \pi_3 = \frac{k}{\rho V^2}$$

According to the π-theorem:

$$\pi_1 = \phi(\pi_2,\ \pi_3)$$

or

$$\frac{R}{\rho L^2V^2} = \phi\left[\left(\frac{\mu}{\rho LV}\right), \left(\frac{K}{\rho V^2}\right)\right]$$

$$\therefore \quad R = \rho L^2V^2\ \phi\left[\left(\frac{\mu}{\rho LV}\right), \left(\frac{K}{\rho V^2}\right)\right]$$ which is the required expression.

14.4 MODEL INVESTIGATION

The small scale replica of actual structures, machines, rivers, dams, spillways, stilling basins, harbours, ships, submarine, seepage problem areas catchments, submerged objects in fluid flow and some objects in the field of engineering and science is called a model while the actual one is called a prototype. Models are made because scientists, and engineers associated with the design, construction and efficient functioning of the prototypes would want to know in advance how the prototypes would behave when they are constructed. Sometimes, investigators and researchers numerically develop some solutions of problems or equations whose exact solutions are impossible to arrive at through exact mathematics. For comparing the results of the numerical solutions and predicting their applicability, model investigations or model studies are very useful as they help predict the outcome of the prototypes. The tests conducted on a model for generating model data are convenient and economical, because any alterations in design, construction, or operation can be effected easily and without any extra expenditure till the defect is rectified to produce fruitful and meaningful results. However, model test results are useful only if there is a complete similarity between the model and the prototype.

14.5 TYPES OF SIMILARITY: SIMILITUDE

Similitude is defined as the similarity between the model and prototype, i.e. a model and the prototype have similar properties or similarity. The following three types of similarities must be maintained between the model and prototype to facilitate useful study of the model investigation.

(i) Geometric similarity
(ii) Kinematic similarity
(iii) Dynamic similarity

Geometric Similarity

This similarity between the model and the prototype is said to exist if the linear dimensions between the two are equal. If L_m, L_p, B_m, B_p, D_m, D_p, $\forall_m$, $\forall_p$, A_m, A_p, d_m, d_p are the length, breadth, diameter, volume, area, depth of model (with subscript m) and prototype (with subscript p), respectively.

Then:
$$\frac{L_m}{L_p} = \frac{D_m}{D_p} = \frac{B_m}{B_p} = Lr \text{ (i.e. the length ratio)}$$

$$\frac{A_m}{A_p} = \frac{L_m \times B_m}{L_p \times B_m} = Lr \times Lr = Lr^2 \text{ (the area ratio)}$$

$$\frac{\forall_m}{\forall_p} = \frac{L_m \times B_m \times d_m}{L_p \times B_p \times d_p} = Lr \cdot Lr \cdot Lr = Lr^3 \text{ (volume ratio)}$$

If the above ratios of the model and the prototype are maintained, it is said that geometric similarity exists.

Kinematic Similarity

If the ratio of kinematic parameters like velocity, acceleration, time, and discharge at the corresponding points in the model and the prototype are the same, it is said that there is kinematic similarity between the model and the prototype. Since velocity and acceleration are vector quanties, the directions of velocity and acceleration at the corresponding points in the model and prototypes should also be parallel.

Thus:
$$T_r = \frac{T_m}{T_p}, \text{ time scale ratio}$$

$$\frac{V_m}{V_p} = V_r, \text{ velocity scale ratio}$$

$$= (L_m/T_m)\ (L_p/T_p)$$

$$\frac{a_m}{a_p} = a_r, \text{ acceleration scale ratio}$$

$$= \frac{\left(\dfrac{L_m}{T_{m^2}}\right)}{\left(\dfrac{L_p}{T_{p^2}}\right)} = \left(\frac{L_m}{L_p}\right)\left(\frac{T_p^2}{T_{n^2}}\right) = L_r/T_r$$

$$\frac{Q_m}{Q_p} = Q_r = \frac{\left(\dfrac{L_m^3}{T_m}\right)}{\left(\dfrac{L_p^3}{T_p}\right)} = \left(\frac{L_m}{L_p}\right)^3 \cdot \frac{T_p}{T_n} = L_r^3/T_r.$$

Kinematic similarity is attained if the flow net formed by streamlines and the equipotential lines for the model and the prototype are geometrically similar. Changing in the scale, and flow net of the model and the prototype can be superimposed.

Dynamic Similarity

If both geometrical and kinematic similarities exist between the model and the prototype, then dynamic similarity for the model and the prototype is attained. This means that the ratio of all forces acting on homologous points in the model and the prototype are equal. The forces acting on the fluid flow system are:

(a) Inertia force F_i
(b) Viscous force F_v
(c) Gravity force F_g
(d) Surface tension force F_s
(e) Elastic force F_e
(f) Pressure force F_p

14.6 NON-DIMENSIONAL NUMBERS: FORCE RATIO AND MODEL LAWS

Inertia Force Ratio

$$\text{Inertia force} = \text{Mass} \times \text{acceleration}$$

$$= \rho(\text{volume})\,\frac{V}{T}$$

$$= \rho\,\frac{\text{Volume}}{T}\,V = \rho QV = \rho(AV)V = \rho L^2V^2$$

$$\text{Inertia force} = \frac{(Fi)_m}{(Fi)_p} = \frac{\rho_m\, L_n^2\, V_m^2}{\rho_p\, L_p^2\, V_p^2} = \rho_r\, L_r^2\, V_r^2$$

Reynolds Number: *Ratio of Inertia Force to Viscous Force, Reynolds Model Law.*

$$\frac{F_i}{F_v} = \frac{\rho L^2 V^2}{\mu VL} = \frac{\rho LV}{\mu}$$

$$\therefore \quad F_v = \lambda A = \mu\left(\frac{dv}{dy}\right)A = \mu \cdot \frac{L}{TL} \cdot L^2 = \mu\left(\frac{L}{T}\right)L$$

$$= \mu VL \quad \because \quad \frac{L}{T} = V$$

$$\therefore \quad R_e = \frac{\rho LV}{\mu}$$

In order to maintain dynamic similarity:

$(R_{em}) = (R_e)_p$. It is called the Reynolds model law.

Froude's Number: *Ratio of Inertia Force to Gravity Force, Froude's Model Law.*

$$F_i = \rho L^2 V^2$$

F_g = Mass × acceleration due to gravity

$$= \rho L^3 g$$

$$\therefore \quad \frac{F_i}{F_g} = \frac{\rho L^2 V^2}{\rho L^3 g} = \frac{V^2}{gL}$$

But $\frac{V}{\sqrt{gL}}$ is Froude's number $= F_r$

i.e. $$\sqrt{\frac{F_i}{F_g}} = F_r$$

Now $F_{r_m} = F_{r_p}$ for dynamic similarity, which is called Froude's model law.

Weber Number: *Ratio of Inertia Force to Surface Tension Force, Weber Model Law.*

$$F_i = \rho L^2 V^2$$

$$F_s = \sigma L$$

$$\therefore \quad \left(\frac{F_i}{F_s}\right) = \frac{\rho L^2 V^2}{\sigma L} = \frac{\rho LV^2}{\sigma} = \frac{V^2}{\sigma/\rho L} = W_e$$ which is called the Weber number.

For dynamic similarity between the model and the prototype:

$$(We_m) = (W_e)_p$$

Mach Number (M_a): *Ratio of Inertia and Elasticity Force: Mach Model Law.*

$$F_i = \rho L^2 V^2$$

F_e = Bulk modulus of Elasticity × Area

$$= KL^2$$

$$\therefore \qquad \frac{F_i}{F_e} = \frac{\rho L^2 V^2}{K L^2} = \frac{V^2}{K/\rho}$$

or $\sqrt{\dfrac{V^2}{K/\rho}} = \dfrac{V}{\sqrt{K/\rho}}$ But $\sqrt{\dfrac{K}{\rho}}$ represents the velocity of sound in fluid media designated by C.

$\therefore$ $\sqrt{\dfrac{F_i}{F_e}} = \dfrac{V}{C}$ = Mach number (M_a), so-called in honour of the Austrian Philosopher, E. Mach.

For dynamic similarity between the model and the prototype:

$$\left(\frac{V}{C}\right)_m = \left(\frac{V}{C}\right)_p, \text{ which is the Mach model law.}$$

Euler Number (E_u): *Ratio of Inertia and Pressure Force: Euler Model Law.*

$$F_i = \rho L^2 V^2$$

$$F_p = \text{pressure intensity} \times \text{area}$$

$$= pL^2$$

$$\therefore \qquad \frac{F_i}{F_p} = \frac{\rho V^2 L^2}{pL^2} = \frac{V^2}{p/\rho}$$

The square root of $\dfrac{F_i}{F_p}$, i.e. $\left(\dfrac{V}{\sqrt{P/\rho}}\right)$ is called the Euler number E_u, so-called in honour of the Swiss mathematician Euler.

For dynamic similarity between the model and the prototype:

$$\left(\frac{V}{\sqrt{P/\rho}}\right)_m = \left(\frac{V}{\sqrt{P/\rho}}\right)_p.$$

EXAMPLE 14.6 In the model test of a spillway, the discharge and velocity of the flow over the model were 2 m^3/sec and 1.5 m/s, respectively. Calculate V and Q over the prototype spillway if the scale is 1/36.

Solution: For dynamic similarity:

$$(F_r)_P = (F_r)_m \quad \because \text{ gravity force is dominant in the spillway.}$$

$$\frac{V_p}{\sqrt{gL_p}} = \frac{V_m}{\sqrt{gL_m}} \Rightarrow \frac{V_p}{V_m} = \frac{\sqrt{gL_p}}{\sqrt{gL_m}} = \sqrt{\frac{L_p}{L_m}} = \sqrt{\frac{36}{1}}$$

$$\therefore \qquad V_p = \sqrt{\frac{36}{1}} \times V_m = 6 \times 1.5 = 9 \text{ m/sec} \qquad \textbf{Ans.}$$

again $$\frac{Q_p}{Q_m} = \frac{A_p V_p}{A_m V_m} = \left(\frac{L_p}{L_m}\right)^2 \left(\frac{V_p}{V_m}\right) = \left(\frac{36}{1}\right)^2 \left(\frac{9}{1.5}\right)$$

$$\therefore \qquad Q_P = \left(\frac{L_p}{L_m}\right)^2 \left(\frac{9}{6}\right) V_m$$

$$\therefore \qquad Q_P = \left(\frac{36}{1}\right)^2 \times \frac{9}{1.5} \times 2 = 15552 \text{ m}^3/\text{sec} \qquad \textbf{Ans.}$$

EXAMPLE 14.7 A ship 300 m long moves in sea water of density 1030 kg/m^3. A model in the scale of $\frac{1}{100}$ of this ship is to be tested in a wind tunnel. The velocity of air in the wind tunnel is 30 m/sec and the resistance of the model is 60 N. Determine the velocity of the ship in sea water and resistance in sea water. Take ν_{air} = 0.018 Stokes and ν of sea water 0.012 Stokes. ρ_{air} = 1.24 Kg/m^3

Solution:

$$\nu_{air} = 0.018 \text{ Stokes} = 0.018 \times 10^{-4} \text{ m}^2/\text{sec}$$

$$\nu_{sea\ water} = 0.012 \text{ Stokes} = 0.012 \times 10^{-4} \text{ m}^2/\text{sec}$$

for dynamic similarity $(R_e)_p = (R_e)_m$ (shear resistance occurs)

$$\therefore \qquad \left(\frac{V_p L_p}{\nu_{sea\ water}}\right) = \left(\frac{V_m L_m}{\nu_{air}}\right)$$

$$V_p = \frac{\nu_{sea\ water}}{\nu_{of\ air}} \times \left(\frac{L_m}{L_p}\right) V_m = \left(\frac{0.012 \times 10^{-4}}{0.018 \times 10^{-4}} \times \frac{1}{100} \times 30\right) \text{ m/sec}$$

$$V_p = 0.2 \text{ m/sec.}$$

To the F_p,

$$F = \text{mass} \times \text{acceleration} = \rho \times \text{volume} \times V = \rho(AV)V$$

$$= \rho L^2 V^2$$

$$\therefore \qquad \frac{F_p}{F_m} = \frac{\rho_p L_p^2 V_p^2}{\rho_m L_m^2 V_m^2} = \frac{1030}{1.24} \times \left(\frac{100}{1}\right)^2 \left(\frac{0.2}{30}\right)^2$$

$$\therefore \qquad F_p = 60 \times \left(\frac{100}{1}\right)^2 \left(\frac{1030}{1.24}\right) \times \left(\frac{0.2}{30}\right)^2 \text{ N}$$

$$F_p = 22150.537 \text{ N} \qquad \textbf{Ans.}$$

EXAMPLE 14.8 A model in the scale $\frac{1}{15}$ of a boat is towed in water. The prototype is moving in sea water of density 1024 kg/m^3 at a velocity 20 m/sec. Find the corresponding velocity of the model in water and resistance due to wave in the model if the resistance in prototype due to the wave is 600 N.

Solution: For dynamic similarity, $(F_r)_p = (F_r)_m$

$$\frac{V_p}{\sqrt{gL_p}} = \frac{V_m}{\sqrt{gL_m}} \Rightarrow \frac{V_m}{V_p} = \sqrt{\frac{gL_p}{gL_m}} = \sqrt{\frac{L_m}{L_p}}$$

$$\Rightarrow \qquad \frac{V_m}{20} = \sqrt{\frac{1}{15}}$$

$$\therefore \qquad V_m = 20\sqrt{\frac{1}{15}} \text{ m/sec}$$

$$\therefore \qquad V_m = 5.164 \text{ m/sec}$$

$$\frac{(R_w)_p}{(R_w)_m} = \frac{\rho_p L_p^2 V_p^2}{\rho_m L_m^2 V_m^2} = \left(\frac{\rho_p}{\rho_m}\right)\left(\frac{L_p}{L_m}\right)^2\left(\frac{L_p}{L_m}\right) = \left(\frac{\rho_p}{\rho_m}\right)\left(\frac{L_p}{L_m}\right)^3$$

$$\therefore \qquad \frac{600}{(R_w)_m} = \left(\frac{1024}{1000}\right)\left(\frac{15}{1}\right)^3$$

$$\therefore \qquad R_{w_m} = \frac{600}{\left(\frac{1024}{1000}\right)\left(\frac{15}{1}\right)^3} = 0.17361 \text{ N}$$ **Ans.**

EXAMPLE 14.9 The discharge through a weir is 1.5 m^3/sec. Find the discharge through the model if $\left(\frac{L_p}{L_m}\right)_H = 50$ and $\left(\frac{L_p}{L_m}\right)_V = 10$

Solution: We know $\frac{V_p}{V_m} = \sqrt{\frac{29H_p}{29H_m}} = \sqrt{\frac{H_p}{H_m}} = \sqrt{10}$

$$\frac{A_p}{A_m} = \frac{B_p H_p}{B_m H_m} = \left(\frac{L_p}{L_m}\right)_H\left(\frac{H_p}{H_m}\right)_V = \left(\frac{L_p}{L_m}\right)_V = 50 \times 10 = 500$$

$$\frac{Q_p}{Q_m} = \frac{A_p}{A_m}\cdot\frac{V_p}{V_m} = 500\left(\frac{1}{10}\right)^{1/2} = 158.11388$$

$$\therefore \qquad Q_m = \frac{Q_p}{158.11388} = \frac{1.5}{158.11388} = 0.0009468 \text{ m}^3/\text{sec}$$

$$Q_m = 0.94868 \text{ lits/sec}$$ **Ans.**

EXAMPLE 14.10 A model in a concrete channel is constructed in a scale of 1:64. The Manning's n of the channel 0.02, find the n of the model.

Solution: $\frac{L_m}{L_p} = \frac{1}{64}$ or $\frac{L_p}{L_m} = 64 = Lr\ n_p = 0.02,\ n_m = ?$

Using Manning's equation
$$\frac{V_p}{V_m} = \frac{\frac{1}{n_p}(R_p)^{2/3}(s_p)^{1/2}}{\frac{1}{n_m}(R_m)^{2/3}(s_m)^{1/2}} = \left(\frac{n_m}{n_p}\right)\left(\frac{R_p}{R_m}\right)^{2/3}\left(\frac{s_p}{s_m}\right)^{1/2} \qquad \text{(A)}$$

Equation of Froude's model law, $(Fr)_p = (Fr)_m$

$$\frac{V_p}{\sqrt{gL_p}} = \frac{V_m}{\sqrt{gL_m}}$$

$$\therefore \quad \frac{V_p}{V_m} = \sqrt{\frac{L_p}{L_m}} = (Lr)^{1/2} = 64^{1/2} = 8$$

Substituting $\left(\frac{V_p}{V_m}\right)$ in (A):

$$8 = \frac{n_m}{0.02}\left(\frac{L_p}{L_m}\right)^{2/3} \qquad \because \frac{R_p}{R_m} = \frac{L_p}{L_m}, \quad \frac{s_p}{s_m} = 1$$

$$8 = \frac{n_m}{0.02} \times (64)^{2/3}$$

$$\therefore \quad n_m = \frac{8 \times 0.02}{(64)^{2/3}} = 0.01 \qquad \textbf{Ans.}$$

14.7 SCALE EFFECT IN MODEL STUDY

In some model studies, several forces may be predominant. In such cases, it is not practically possible to achieve complete similarity. It becomes difficult to take care of all the dimensionless variables involved in the phenomenon. In such situations, variables which have secondary influence or importance are neglected. However, neglecting these variables causes some discrepancy in the results achieved in case of the model and the prototype after the prototype is constructed. This discrepancy or disturbing effect or influence is called scale effect in model study. For example, assume that in a particular problem both the Reynolds number and the Froude number are predominant. Then, for complete similarity in order to, achieve between the model and the prototype, both the Reynolds model law and Froude model law need to be satisfied simultaneously. Finding the expression to satisfy both the laws:

Reynolds model law gives

$$(R_e)_m = (R_e)_p$$

$$\frac{\rho_m V_m L_m}{\mu_m} = \frac{\rho_p V_p L_p}{\mu_p}$$

or

$$\left(\frac{\rho_m}{\rho_p}\right)\left(\frac{V_m}{V_p}\right)\left(\frac{L_m}{L_p}\right)\left(\frac{\mu_p}{\mu_n}\right) = 1$$

or

$$\left(\frac{V_m}{V_p}\right)\left(\frac{L_m}{L_p}\right) \cdot \left(\frac{\frac{\mu_p}{\rho_p}}{\frac{\mu_m}{\rho_m}}\right) = 1$$

$$\frac{1}{V_r}\left(\frac{1}{L_r}\right)(v_r) = 1 \tag{A}$$

Froude's model law gives

$$(F_r)_m = (F_r)_p$$

$$\frac{V_m}{\sqrt{gL_m}} = \frac{V_p}{\sqrt{gL_p}}$$

$$\therefore \quad \frac{V_m}{V_p} \cdot \sqrt{\frac{gL_p}{gL_m}} = 1$$

$$\left(\frac{1}{V_r}\right)\left(\sqrt{L_r}\right) = 1 \tag{B}$$

Equating (A) and (B):

$$\left(\frac{1}{V_r}\right)\left(\frac{1}{\sqrt{L_r}}\right)(v_r) = \left(\frac{1}{V_r}\right)(L_r)^{1/2}$$

$$\frac{1}{(L_r)^{3/2}} v_r = 1$$

$$\therefore \quad L_r^{3/2} = v_r$$

i.e.
$$\left(\frac{L_p}{L_m}\right)^{3/2} = \left(\frac{v_p}{v_m}\right) \tag{C}$$

Equation (C) shows that once the scale for the model is selected, a liquid of appropriate value of viscosity is required for the model to satisfy Equation (C).

For example, if in both the model and the prototype, the same fluid is used, the right hand side of Equation (C) is 1.

$$\therefore \quad \left(\frac{L_p}{L_m}\right)^{3/2} = 1$$

$$\therefore \quad \frac{L_p}{L_m} = 1$$

i.e. the model size should be equal to the size of the prototype. But it is not practical to have a full size model or values of appropriate viscosities to satisfy Equation (C). Let the model scale be $\frac{1}{100}$ and model be tested in a viscosity v_m. If the prototype is used in the flow of water, Equation (C) gives:

$$\left(\frac{100}{1}\right)^{3/2} = \frac{1\times 10^{-6}}{v_m}$$

$\therefore$ $v_m = 10^{-3}$ m^2/sec, i.e. the viscosity of fluid for the model should be 1000 times higher than others to maintain a scale of $\frac{1}{100}$.

The scale effect in model studies may also be developed in a few situations where in the forces which do not have any role in the prototype can significantly affect in the model. For example, if the depth of water in model flow is very small, i.e. below 1.5 cm, the surface tension effect is dominant. Thus, the results obtained from such models produce a scale effect as the depth of flow in the prototype is free from surface tension effect.

The size of roughness may also produce some scale effect depending on the thickness of the laminar sub-layer and the roughness height of the model. If the model behaves as a hydrodynamically rough surface, the prototype may behave as a hydrodynamically smooth surface.

Due to this scale effect, the experimental data of the model sometimes do not simulate appropriately even with results obtained with a correct numerical solution.

14.8 TYPES OF MODELS

Models may be classified into two types:

(1) Undistorted models
(2) Distorted models

Undistorted models are geometrically similar to their prototypes. The scale ratios of the length, breadth, and height of the models and their prototypes are the same and the basic conditions of perfect similitude are satisfied. Therefore, results in case of undistorted models results are relatively reliable and may be applied to prototypes confidently.

Distorted models, on the other hand, are not identical to their prototypes. Since the basic conditions of perfect similitude are not satisfied, the results obtained from such models are liable to distortion. Such models may have geometrical distortion, hydraulic distortion or material distortion. In distorted models the linear scales of length and depth are different, i.e. the horizontal scales and vertical scales are different. Thus, such models do not bear any resemblance to the configurations of prototypes. Material distortion may arise if the materials uesd in construction are not exactly the same as the materials used in the prototypes. It may not be possible to maintain similitude in respect to some parameters like time, discharge, etc. which eventually may produce hydraulic distortion. These include models of rivers, dams, harbours, and estuaries in which it is difficult to adopt the same scale of length and depth owing to greater length as compared to the depth of the flow, and therefore distorted models have some limitations.

Distortion in velocity distribution and kinetic energy arise because of vertical exaggeration. Different horizontal and vertical scales produce distortion of longitudinal slope, as a result of which the flow details may not be correctly reproduced. River band, erosion, aggradation, and degradation of rivers difficult predict accurately in such models. The wave produced in a model may be different in the prototypes as the sizes of the sand beds are of different sizes in models. Exaggeration of scales may produce a psychological effect on the observer.

EXAMPLE 14.11 A river discharges 3000 m^3/sec. The horizontal scale is 1/1000 and the vertical scale is 1/100. Manning's $n = 0.025$. Calculate the Q_m and n_m.

Solution: $(L_r)_H = \dfrac{1000}{1}$, $(L_r)_V = \dfrac{100}{1}$

$Q_p = 3000$ m^3/sec, $n_p = 0.025$

To find n_m:

$$\frac{Q_p}{Q_m} = \frac{A_pV_p}{A_mV_m} = \frac{(By)_p \dfrac{1}{n_p}(R)_p^{2/3}(s_b)_p^{1/2}}{(By)_m \dfrac{1}{n_m}(R)_m^{2/3}(s_b)_m^{1/2}}$$

$$\frac{Q_p}{Q_m} = \left(\frac{B_p}{B_m}\right)\left(\frac{y_p}{y_m}\right)\frac{n_m}{n_p}\left(\frac{R_p}{R_m}\right)^{2/3}\left(\frac{y_p/L_p}{y_m/L_m}\right)^{1/2}$$

$$\frac{Q_p}{Q_m} = \left(\frac{1000}{1}\right)\left(\frac{100}{1}\right)\frac{n_m}{0.025}\left(\frac{100}{1}\right)^{2/3}\left[\left(\frac{y_p}{y_m}\right)\left(\frac{L_m}{L_p}\right)\right]^{1/2} \quad \text{assume } \frac{R_p}{R_m} = (L_r)_V$$

$$\frac{Q_p}{Q_m} = \frac{1000 \times 100 \times n_m}{0.025}(100)^{2/3}(100)^{1/2}\left(\frac{1}{1000}\right)^{1/2} \qquad \text{(A)}$$

To find Q_m $\dfrac{V_p}{V_m} = \dfrac{\sqrt{2gH_p}}{\sqrt{2gH_m}} = \left(\dfrac{H_p}{H_m}\right)^{1/2} = \left(\dfrac{100}{1}\right)^{1/2} = 10$

$$\frac{Q_p}{Q_m} = \frac{A_pV_p}{A_mV_m} = \left(\frac{B_p}{B_m}\right)\left(\frac{Y_p}{Y_m}\right)\cdot\left(\frac{V_p}{V_m}\right) = \left(\frac{1000}{1}\right)\left(\frac{100}{1}\right)(10)$$

$$\frac{3000}{Q_m} = 10^6$$

$\therefore$ $Q_m = 0.003$ m^3/sec $= 3$ litre/sec **Ans.**

Substituting Q_m in Equation (A)

$$\frac{3000}{0.003} = \frac{10^5 \times (100)^{2/3}}{0.025} \times 10 \times \left(\frac{1}{1000}\right)^{1/2} n_m$$

$\therefore$ $n_m = 0.0367$ **Ans.**

EXAMPLE 14.12 A river model constructed with a horizontal scale of 1 : 5000 and a vertical scale of 1 : 256 has a discharge scale ratio 1 : 2 × 10^7. Obtain the scale ratio of velocity (V_m/V_p) and ratio of Manning's coefficients, i.e. $\left(\dfrac{n_p}{n_m}\right)$ for prototype and model.

Solution: $(L_r)_H = \dfrac{1}{5000}$, $(L_r)_V = \dfrac{1}{256}$, $\dfrac{Q_p}{Q_m} = \dfrac{1}{2 \times 107}$

$$\frac{Q_m}{Q_p} = \frac{A_m V_m}{A_p V_p} = \frac{(B_m Y_m)\ V_m}{(B_p Y_p)\ V_p}$$

$$\frac{Q_m}{Q_p} = \left(\frac{1}{256}\right)\left(\frac{1}{5000}\right)\left(\frac{V_m}{V_p}\right)$$

$$\frac{1}{2\times 10^7} = \left(\frac{1}{5000}\right)\left(\frac{1}{256}\right)\left(\frac{V_m}{V_p}\right)$$

$$\therefore \quad \frac{V_m}{V_p} = \frac{5000\times 256}{2\times 10^7} = 0.064 \qquad \textbf{Ans.}$$

$$\frac{V_m}{V_p} = \frac{\frac{1}{n_m}(R_m)^{2/3}(s_m)^{1/2}}{\frac{1}{n_p}(R_p)^{2/3}(s_p)^{1/2}} = \frac{n_p}{n_m}\left(\frac{1}{256}\right)^{2/3}\cdot\left(\frac{Y_m/L_m}{Y_p/L_p}\right)^{1/2} \quad \text{Assume } \frac{R_m}{R_p} = \frac{1}{256}$$

$$\frac{V_m}{V_p} = \frac{n_p}{n_m}\cdot\left(\frac{1}{256}\right)^{2/3}\left(\frac{Y_m}{Y_p}\right)^{1/2}\left(\frac{L_p}{L_m}\right)^{1/2}$$

$$0.064 = \frac{n_p}{n_m}\left(\frac{1}{256}\right)^{2/3}\left(\frac{1}{256}\right)^{1/2}\left(\frac{5000}{1}\right)^{1/2}$$

$$\therefore \quad \frac{n_p}{n_m} = 1.7124 \qquad \textbf{Ans.}$$

EXAMPLE 14.13 A spillway 7.2 m high and 150 m long discharges 2150 m^3/sec under a head of 4 m. If 1 : 16 model of spillway is to be constructed, find the model dimensions, head over the model and model discharge.

Solution:

$$\frac{L_p}{L_m} = \frac{H_p}{H_m} \Rightarrow \frac{16}{1} = \frac{4}{H_m} \Rightarrow H_m = \frac{4}{16}\text{ m} = 25\text{ cm} \qquad \textbf{Ans.}$$

$$\frac{L_p}{L_m} = \frac{H_{p\text{ height}}}{H_{m\text{ height}}} \Rightarrow \frac{16}{1} = \frac{7.2}{H_{m\text{ height}}} \quad \therefore H_{m\text{ height}} = \frac{7.2}{16} = 0.45\text{ m} = 45\text{ cm} \qquad \textbf{Ans.}$$

$$\frac{L_p}{L_m} = \frac{(\text{Length})_p}{(\text{Length})_m} \Rightarrow \frac{16}{1} = \frac{150}{\text{length}_m} \quad \therefore \text{Length}_m = \frac{150}{16} = 9.375\text{ m}$$

$$\frac{Q_p}{Q_m} = \frac{A_p\cdot V_p}{Q_m\cdot V_m} = \left(\frac{L_p}{L_m}\right)^2\frac{\sqrt{2gHp}}{\sqrt{2gHm}} = \left(\frac{L_p}{L_m}\right)\left(\frac{H_p}{H_m}\right)^{1/2} = \left(\frac{16}{1}\right)^2\left(\frac{16}{1}\right)^{1/2}$$

$$\therefore \quad Q_m = 2.0995\text{ m}^3/\text{sec} \qquad \textbf{Ans.}$$

14.9 CONCLUSION

The contents presented in this chapter gives the students a basic idea on dimensional analysis, similitude and analysis of model. Dimensions of different parameters used frequently in hydraulics and fluid mechanics are presented in Table 14.1. Dimensional homogeneity, two important methods of dimensional analysis, model laws, types of model, scale effect in model study have been discussed systematically with solved examples. Few problems as chapter-end exercises and a list of references like Langhaar[1], Duncan[2], Holt[3], Bridgeman[4], Sadov[5], National Bureau of Standard[6], Birkholf[7], Huntley[8] who did further works on dimensional analysis in the last century have included.

PROBLEMS

14.1 The torque T on a shaft of diameter revolves with a speed of N in a fluid of viscosity M and mass density ρ. Show by dimensional analysis, $T = (\rho d^5 N^2)\ \phi\left(\frac{v}{d^2 N}\right)$

14.2 A shallow river is 1500 m wide and maximum of depth of water is 5 m. The discharge is 3000 m^3/sec and the velocity of flow 51.5 m/sec. The horizontal and vertical scales of model are 1 : 800 and 1/40 respectively. If Manning's n of the river is 0.025, find the n for the model. The hydraulic radius may be assumed to be the depth as the river is wide.

(**Ans.** $n_m = 0.0605$)

14.3 For a laminar flow in a pipe, the pressure Δp is a function of the pipe length l, its diameter d, mean velocity V, and viscosity μ, using any method of dimensional analysis, obtain an expression Δp.

$$\left[\textbf{Ans. } \Delta p = \left(\frac{\mu V}{d}\right)\phi\left(\frac{l}{d}\right)\right]$$

14.4 A spillway model is to be made to a scale $\frac{1}{25}$ across a flume of 0.5 m wide. The prototype is 15 m high and the maximum head expected 2 m. Find the (a) height and head of the model. (b) If the flow over the model at 6 cm head is 0.02 m^3/sec, what flow per m length of the prototype is expected? (c) If the model shows a measured hydraulic jump of 3 cm, how high is the jump in the prototype?

[**Ans.** (a) 60 cm, 8 cm (b) 5 m^3/sec/m (c) 75 cm]

14.5 The critical depth y_c in a rectangular channel is a function of the discharge per unit width of and acceleration g due to gravity. Show by the Rayleigh method,

$$y_c = \left(\frac{g^2}{g}\right)^{1/3}$$

14.6 The velocity and discharge for $\frac{1}{50}$ scale model of a spillway are 0.6 m/sec and 0.18 m^3/sec., respectively. Calculate the corresponding velocity and discharge in the prototype.

(**Ans.** 4.24 m/sec, 3182 m^3/sec)

14.7 The efficiency η of a fan depends on density ρ, μ, angular velocity ω, and diameter D of the rotor discharge Q. Express η in terms of dimensionless parameters.

$$\left[\textbf{Ans.}\ \eta = \phi\left(\frac{\mu}{\rho w D^2}, \frac{Q}{D^2 \omega}\right)\right]$$

REFERENCES

1. Langhaar, H.L., *Dimensional Analysis and Theory of Models*, Joln Wiley & Sons, Inc., New York, 1951.
2. Duncan, W.J., *Physical Similarity and Dimensional Analysis*, Edward Arnold & Co., London, 1953.
3. Holt, M., "Dimensional Analysis," section 15, *Hand Book of Fluid Dynamics* edited by V.L. Streeter, McGraw-Hill Inc., New York, 1961.
4. Bridgeman, P.W., *Dimensional Analysis*, Yale University Press, New Heaven, Connecticut, 1963.
5. Sadov, L.I., *Similarity and Dimensional Methods in Mechanics*, Academic Press, New York, 1959.
6. National Bureau of Standards, *Technical News Bulletin*, Vol. 44, No. 12, December, 1960.
7. Birkhoff, G., *Hydrodynamics*, Dover Publications, Inc., New York, 1955.
8. Huntley, H.F., *Dimensional Analysis*, Dover Publications, Inc., New York, 1966.

Chapter 15

Compressible Flow

15.1 INTRODUCTION

In all the earlier chapters, analysis of flow has been made by assuming the fluid to be incompressible. In case of flow of gases, the density of fluid changes with pressure and temperature. Compressible flow is defined as that flow in which the density of fluid does not remain constant. Physical situations of compressible flow are flow of gases in the orifices and nozzles, flight of aircraft and projectiles at high altitude, water hammer problems, gas flow in turbines, gas flow in rockets, etc. Compressibility becomes predominant when velocity becomes equal to or more than the velocity of sound in the fluid medium. In compressible flow, thermodynamic behaviour of fluid is to be considered as the change of density of fluid is always accompanied by changes of pressure and temperature. Thermodynamic relations such as equation of state ($p = \rho RT$), isothermal process $\left(\frac{p_1}{\rho_1} = \frac{p_2}{\rho_2} = \frac{p}{\rho} = \text{constant}\right)$, adiabatic process ($p_1 v_1^{k1} = p_2 v_2^{k2} = pv^k$ = constant), entropy, enthalpy all are discussed in Chapter 2.

15.2 CONTINUITY EQUATION

Based on conservation of mass, the same continuity equation of incompressible flow in one dimension is applicable here also, i.e.

$$\rho AV = \text{Constant} \tag{15.1}$$

But here ρ is not constant. Hence Equation (15.1) may be written as:

$$\rho_1 A_1 V_1 = \rho_2 A_2 V_2 = \rho AV = \text{Constant} \tag{15.2}$$

Differentiating Equation (15.1):

$$\rho d(AV) + AV\, d\rho = 0$$

or

$$\rho V dA + \rho A dV + AV d\rho = 0$$

Dividing by ρAV,

$$\frac{dA}{A} + \frac{dV}{V} + \frac{d\rho}{\rho} = 0 \tag{15.3}$$

which is the differential form of the continuity equation in compressible flow.

In three-dimensional form, the continuity equation of compressible flow,

i.e. $$\frac{\partial \rho}{\partial t} + \frac{\partial(\rho w)}{\partial x} + \frac{\partial(\rho v)}{\partial y} + \frac{\partial(\rho w)}{\partial z} = 0 \text{ (5.4) is derived already.}$$

15.3 BERNOULLI'S EQUATION OR ENERGY EQUATION

With the same procedure like incompressible flow, Bernoulli's equation is already derived in Chapter 6, i.e.

$$VdV + gdZ + \frac{dp}{\rho} = 0 \tag{6.6}$$

In case of compressible flow, ρ is not constant and hence ρ cannot be taken outside the integration sign of Equation (15.4).

Integrating Equation (6.6),

$$\int VdV + \int gdz + \int \frac{dp}{\rho} = 0$$

or $$\frac{V^2}{2} + gz + \int \frac{dp}{\rho} = \text{Constant} \tag{15.4}$$

The value $\int \frac{dp}{\rho}$ in compressible flow will be worked for both isothermal and adiabatic process.

When it is the isothermal process:

$$\frac{p}{\rho} = \text{Constant} = C_1 \tag{2.5}$$

$\therefore$ $$\rho = \frac{p}{C_1}$$

Hence $$\int \frac{dp}{\rho} = \int \frac{dp}{p/C_1} = \int \frac{C_1 dp}{p} = C_1 \int \frac{dp}{p} = C_1 \log_e p$$

$\therefore$ $$C_1 = \frac{p}{\rho}, \int \frac{dp}{\rho} = \frac{p}{\rho} \log_e p \tag{A}$$

Substituting $\int \frac{dp}{\rho}$ from (A) in Equation (15.4):

$$\frac{V^2}{2} + gz + \frac{p}{\rho} \log_e p = \text{Constant}$$

Dividing by g:

$$\frac{V^2}{2g} + z + \frac{p}{\rho g} \log_e p = \text{Constant} \tag{15.5}$$

which is the Bernoulli's equation for compressible flow with isothermal process. When it is adiabatic process,

$$\frac{p}{\rho^k} = \text{Constant} = C \tag{2.6}$$

$$\rho^k = \frac{p}{C}$$

$$\therefore \qquad \rho = \left(\frac{p}{C}\right)^{1/k}$$

$$\therefore \qquad \int \frac{d_p}{\rho} = \int \frac{d_p}{(p/C)^{1/k}} = C^{1/k} \int \frac{1}{p^{1/k}} dp = C^{1/k} \int p^{-1/k} dp$$

$$\int \frac{dp}{\rho} = C^{1/k} \frac{p(-1/k+1)}{(-1/k+1)}$$

Simplifying:

$$\int \frac{dp}{\rho} = \left(\frac{k}{k-1}\right) \frac{p}{\rho}$$

Substituting: $\int \frac{dp}{\rho}$ in Equation (15.4)

$$\frac{V^2}{2} + gz + \left(\frac{k}{k-1}\right) \frac{p}{\rho} = \text{Constant}$$

Dividing by g:

$$\frac{V^2}{2g} + z + \left(\frac{k}{k-1}\right) \frac{p}{\rho g} = \text{Constant} \tag{15.6}$$

EXAMPLE 15.1 A gas is flowing through a horizontal pipe of cross-sectional area 30 cm^2. At a point the pressure is 30 N/cm^2 gauge and temperature is 20°C. At another section, the area of the cross-section is 15 cm^2 and pressure 25 N/cm^2 gauge. If mass flow rate of gas is 0.15 kg/sec, find the velocities of the gas at these two sections, assuming an isothermal change. Take $R = 287$ Nm/kg-k and atmospheric pressure is 10 N/cm^2.

Solution: For section 1:

Absolute temperature $T_1 = (20 + 273) = 293$ k

$A_1 = 30$ cm^2, $p_1 = 30$ N/cm^2 (gauge) + 10 N/cm^2 = 40 N/cm^2 (abs)

Section 2: $A_2 = 15$ cm^2, $p_2 = 25$ N/cm^2 + 10 N/cm^2 = 35 N/cm^2 (abs)

$R = 287$ N-m/kg.k

Mass flow rate = 0.15 kg/sec

Equation of state $\rho_1 R T_1 = p_1$ or $\rho_1 = \frac{p_1}{RT_1} = \frac{40 \times 10^4}{287 \times 293} = 4.7567$ kg/m^3

$$\rho_1 A_1 V_1 = 0.15$$

$$\therefore \qquad V_1 = \frac{0.15}{(30 \times 10^{-4}) \times 4.7567} = 10.511 \text{ m/sec}$$ **Ans.**

Similarly $$\rho_2 = \frac{p_2}{RT_2} = \frac{35 \times 10^4}{287 \times 293} = 4.162.\ \text{kg/m}^3$$

$\therefore$ $$V_2 = \frac{0.15}{(15 \times 10^{-4}) \times 4.162} = 24.027\ \text{m/sec}$$ **Ans.**

15.4 VELOCITY OF SOUND IN FLUID MEDIUM

If a pressure at the point of fluid is changed, the new pressure of the fluid transmitted to the rest of the fluid with velocity, which is called velocity of pressure wave, and it moves with a velocity of sound in the fluid medium. In order to find an expression of the pressure C or velocity of in the fluid medium, consider a long rigid tube of cross-section area A fitted with a piston at one end as shown in Figure 15.1.

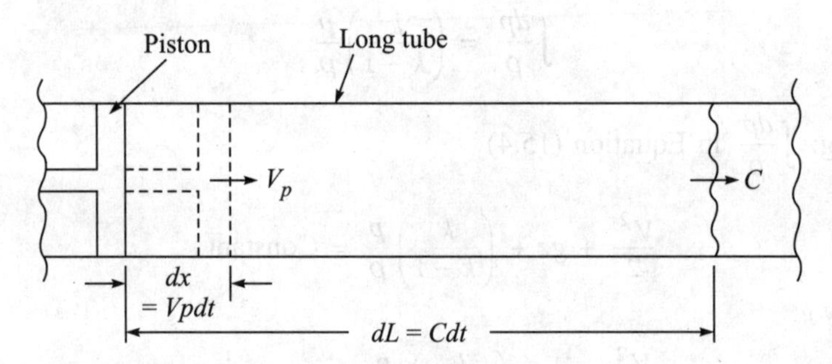

Figure 15.1 Pressure wave propagation.

If V_p is the velocity of the piston and C is the pressure wave velocity or velocity of sound is the fluid medium, in time dt, the piston moves a distance $dx = V_p dt$ and pressure wave moves in the same time dt a distance $dL = Cdt$. The pressure within the tube after compression ($p + dp$) and density ($\rho + d\rho$) continuity gives, mass of fluid in the tube before compression, and equal mass of fluid after compression, i.e.

$$\rho(AdL) = (\rho + d\rho)\, A(dL - dx)$$

$$\rho dL = (\rho + d\rho)\, (dL - dx)$$

or $$\rho Cdt = (\rho + d\rho)\, (Cdt - V\rho dt) = (\rho + d\rho)\, (C - V_\rho) dt$$

or $$\rho Cdt = \rho Cdt - \rho V_p dt + Cd\rho dt - V_p d\rho dt$$

or $$\rho C = \rho C - \rho V_p + Cd\rho - V_p d\rho$$

$\therefore$ $$Cd\rho - \rho V_p - d\rho V_p = 0$$

Since C is much greater than V_p, ($V_p d\rho$) term is neglected

$\therefore$ $$Cd\rho = \rho V_p$$

$\therefore$ $$C = \frac{\rho V_p}{d\rho} \tag{15.7}$$

When the piston moves with velocity V_p for time dt, the fluid which was initially at rest also moves with V_p and pressure in the fluid will increase from p to $p + dp$.

Applying impulse momentum equation,

$$(p + dp)A - pA = \text{Mass per sec} \times \text{change of velocity}$$

$$dpA = \frac{\rho AdL}{dt}(V_p - 0)$$

$$dp = \frac{\rho dL}{dt}V_p$$

$$\because \quad dL = Cdt,$$

$$dp = \frac{\rho Cdt}{dt}V_p$$

$$dp = \rho C V_p$$

$$\therefore \quad C = \frac{dp}{\rho Vp} \tag{15.8}$$

Now multiplying Equations (15.7) and (15.8):

$$C^2 = \frac{\rho V_p}{d\rho} \times \frac{dp}{\rho V_p} = \frac{dp}{d\rho}$$

$$\therefore \quad C = \sqrt{\frac{dp}{d\rho}} \tag{15.9}$$

which gives the velocity of pressure wave.

Equation (1.15) in Chapter 1, is

$$K = -\frac{dp}{d\forall/\forall} \tag{1.15}$$

$$\because \quad \forall \propto \rho \quad \therefore \ \forall\rho = \text{Constant}$$

Differentiating $\forall \ d\rho + \rho d\forall = 0$

$$\frac{d\forall}{\forall} = -\frac{d\rho}{\rho} \tag{15.10}$$

Equation (1.15) becomes:

$$K = \frac{dp}{\frac{d\rho}{\rho}} \tag{15.11}$$

or

$$\frac{K}{\rho} = \frac{dp}{d\rho}$$

$$\sqrt{\frac{K}{\rho}} = \sqrt{\frac{dp}{d\rho}} \tag{15.12}$$

Comparing Equations (15.9) and (15.12):

$$C = \sqrt{\frac{K}{\rho}} \tag{15.13}$$

This is another expression of C, which depends on the balk modulus of compression K.

15.4.1 Velocity of Sound for Isothermal and Adiabatic Processes

In the isothermal process, $\frac{p}{\rho}$ = Constant

or $$p\rho^{-1} = \text{Constant}$$

Differentiating the above expression:

$$\rho^{-1}dp + p(-1)\rho^{-2}d\rho = 0$$

or $$dp - p\rho^{-1}d\rho = 0$$

or $$dp - \frac{p}{\rho}d\rho = 0$$

or $$dp = \frac{p}{\rho}d\rho \quad \text{or} \quad \frac{dp}{d\rho} = \frac{p}{\rho}$$

Equation of state gives $\frac{p}{\rho} = RT$

$\therefore$ $$\frac{dp}{d\rho} = RT$$

Substituting this value of $\frac{dp}{d\rho}$ in Equation (15.9):

$$C = \sqrt{\frac{p}{\rho}} = \sqrt{RT} \tag{15.14}$$

When in the adiabatic process:

$$\frac{p}{\rho^k} = \text{Constant}$$

or $$p\rho^{-k} = \text{Constant}$$

Differentiating $\rho^{-k}dp + p(-k)\,\rho^{-k-1}\,d\rho = 0$

or $$dp + (-kp)\,\rho^{-1}d\rho = 0$$

or $$dp = \frac{kp}{\rho}d\rho = kRT$$

$\therefore$ $$\frac{dp}{d\rho} = kRT \quad \because \frac{p}{\rho} = RT$$

Substituting $\frac{dp}{d\rho}$ in Equation (15.9):

$$C = \sqrt{kRT} \tag{15.15}$$

For propagation sound wave in air, normally the process is assumed to be adiabatic.

15.5 MACH NUMBER

In Chapter 14, the definition of Mach number (M_a) is already given as:

$$M_a = \sqrt{\frac{\text{Inertia force}}{\text{Elastic force}}} = \sqrt{\frac{\rho A V^2}{KA}} = \sqrt{\frac{V^2}{K/\rho}} = \frac{V}{C}$$

$$\because \quad \sqrt{\frac{K}{\rho}} = C \text{ [Equation (15.13)]}$$

Hence Mach number $M_a = \dfrac{\text{Velocity of fluid or body in fluid}}{\text{Velocity of sound in fluid medium}}$

$$M_a = \frac{V}{C} \tag{15.16}$$

In compressible flow, this number is very important.

On the value of M_a, the flow is defined as:

(i) *Sub-sonic flow*, when $M_a < 1$, i.e. the velocity of flow or velocity of the moving body like aircraft is less than velocity of sound. The velocity V of the fluid or body is called sub-sonic velocity.

(ii) *Sonic flow*, when $M_a = 1$, i.e. when the velocity of fluid or moving body is equal to the velocity of sound. An aircraft is said to move with sonic speed if its velocity is equal to C.

(iii) *Supersonic flow*, when $M_a > 1$, i.e. when the velocity of fluid or moving body is greater than C. An aeroplane is said to supersonic, if its velocity is greater than the velocity of sound.

Sonic speed, i.e. velocity of sound in air, is approximately around 1100 km/hr.

EXAMPLE 15.2 Find the speed of sound waves in air at sea level when the pressure and temperature are 9.81 N/cm^2 (abs) and 20°C, respectively. Take R = 287 J/kg k, and k = 1.4.

Solution: $p = 9.81$ N/cm^2 = 9.81×10^4 N/m^2, $T = 273 + 20 = 293$ k, R = 287 J/kg k, ratio of specific heat k = 1.4.

For adiabatic process, velocity of sound $C = \sqrt{kRT}$ (i.e. Equation 15.15)

$$C = \sqrt{1.4 \times 287 \times 293} = 343.114 \text{ m/sec} = 1235.21 \text{ km/hr.}$$ **Ans.**

EXAMPLE 15.3 Calculate the Mach number at a point on a jet propelled aircraft at 900 km/hr at sea level where air temperature is 15°C, $k = 1.4$ and $R = 287$ J/kg k.

Solution: $C = \sqrt{kRT} = \sqrt{1.4 \times (287) \times (273 + 15)}$ m/sec = 340.174 m/sec

$= 1224.6266$ km/hr

$$\therefore \quad M_a = \frac{V}{C} = \frac{900}{1224.6266} = 0.7349 \simeq 0.735 \qquad \textbf{Ans.}$$

EXAMPLE 15.4 An aeroplane is flying at a height of 20 km where temperature is –40°C. The speed of the plane corresponds to the Mach number 1.8. Assuming $k = 1.4$, and $R = 287$ J/kg k, find the speed of the plane.

Solution: Mach number $M_a = \dfrac{V}{C}$

Now $C = \sqrt{kRT} = \sqrt{1.4 \times 287 \times (273 - 40)} = 305.9726$ m/sec

$= 1101.5$ km/hr.

$$\therefore \quad 1.8 = \frac{V}{1101.5} \quad \therefore V = 1982.704 \text{ km/hr} \qquad \textbf{Ans.}$$

15.6 PRESSURE WAVE PROPAGATION IN COMPRESSIBLE FLOW

When a disturbance is created in a compressible fluid, a pressure wave is generated. This disturbance is created when an object moves in a stationary fluid or when a fluid passes with some velocity around a stationary object. This pressure wave is transmitted radially with a velocity equal to C of the fluid medium. In order to study such propagation, consider a small projectile moving on a straight path with a velocity V in a stationary and compressible fluid. If $t = 0$, the object was A, then in time t, it moves a distance AB which is equal to Vt. Pressure wave generated at A by the object will grow into a surface of sphere of radius equal Ct as shown in Figure 15.2, along with the growth of other waves originated from the object at every $t/4$ interval of time as the projectile moves from A to B. Now, depending on the magnitude of the Mach number, i.e. the velocity of the object and velocity of sound, different pressure wave patterns are developed in three different cases as shown in Figure 15.2(a), (b), and (c).

(i) When the Mach number $M_a < 1$ or $V < C$:
In this case, as $V < C$, hence pressure or elastic waves travel ahead of the object in forward direction (i.e. towards B). Hence in Figure 15.2(a), the object at B is inside the spheres of radii Ct, $3/4Ct$, $\frac{1}{2}Ct$, $\frac{1}{4}Ct$. This is shown in Figure 15.2(a).

(ii) When the Mach number $M_a = 1$ or $V = C$:
In this case, as $V = C$, both the pressure wave and the object reach the point B at the same instant of time like t, $\frac{3}{4}t$, $\frac{1}{2}t$ and $\frac{1}{4}t$ as shown in Figure 15.2(b). The zone right of B is not disturbed and this zone is called the zone of silence and the left side of B is the zone of action.

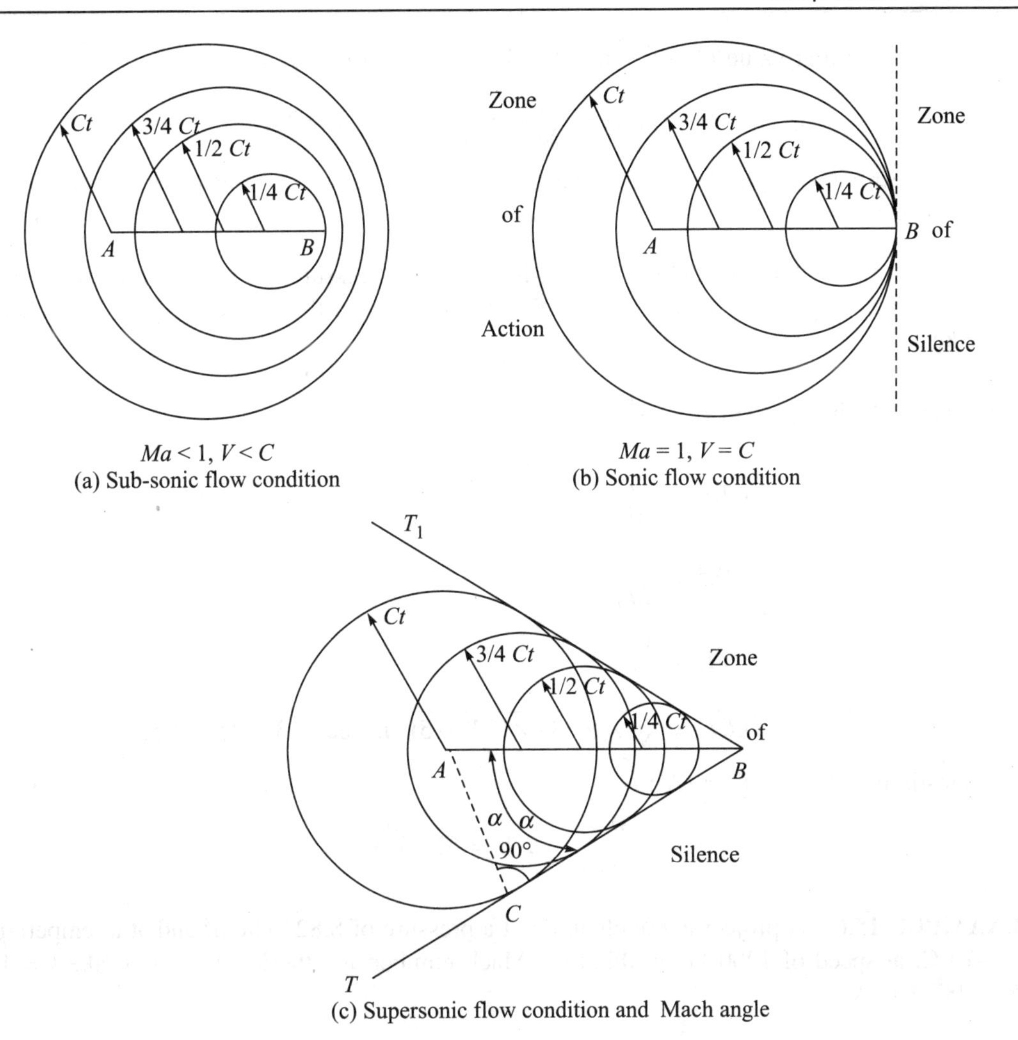

Figure 15.2 Pressure wave propagation in compressible flow.

(iii) Here $M_a > 1$ or $C > V$:

The object moves with a supersonic speed. Hence the distance $AB > Ct$ and thus B remains outside the sphere at any instant of time. If common tangents BT and BT_1 are drawn they become tangents to all the spheres drawn with radii Ct, $\frac{3}{4}Ct$, $\frac{1}{2}Ct$ and $\frac{1}{4}Ct$. A cone is formed at B by the two tangents. The semi-vertex angle of the cone is called the Mach angle,

i.e. $$\sin \alpha = \frac{AC}{AB} = \frac{Ct}{Vt} = \frac{C}{V} = \frac{1}{\dfrac{V}{C}} = \frac{1}{M_a} \qquad (15.17)$$

The limiting value of α is when $V = C$ (in case ii) i.e. $M_a = 1$

i.e. $$\sin \alpha = \frac{1}{M_a} = \frac{1}{1} = 1 \quad \therefore \alpha = 90°$$

The region inside the Mach cone is the zone of action and outside it, is the zone of silence.

EXAMPLE 15.5 A projectile is travelling in air having pressure and temperature of 8.829 N/cm^2 respectively, if the Mach angle is 30°, find the velocity of the projectile. Take $k = 1.4$, $R = 287$ J/kg k.

Solution: Mach angle $\sin \alpha = \dfrac{1}{M_a}$

i.e. $$\sin 30° = \frac{1}{M_a}$$

$$0.5 = \frac{1}{M_a} \quad \therefore M_a = 2$$

$$M_a = \frac{V}{C} \tag{A}$$

$$C = \sqrt{kRT} = \sqrt{1.4 \times 287 \times (273 - 5)} \text{ m/sec} = 328.15 \text{ m/sec}$$

Substituting M_a and C in (A):

$$2 = \frac{V}{328.15} \quad \therefore V = 656.3 \text{ m/sec}$$ **Ans.**

EXAMPLE 15.6 A projectile travels in air at a pressure of 8.829 N/cm^2 and at a temperature of –10°C, at speed of 1200 km/hr. Find the Mach number and the Mach angle (Take $k = 1.4$, $R = 287$ J/kg k)

Solution: $$C = \sqrt{kRT} = \sqrt{1.4 \times 287 \times (273 - 10)} \text{ m/sec} = 325.674453 \text{ m/sec}$$
$$= 1170.268 \text{ km/hr.}$$

$$\therefore \quad M_a = \frac{V}{C} = \frac{1200}{1170.268} = 1.0254$$ **Ans.**

Mach angle $$\sin \alpha = \frac{1}{M_a} = \frac{1}{1.0254} = 0.97522$$

$$\therefore \quad \alpha = 77.22°$$ **Ans.**

15.7 STAGNATION PRESSURE

Stagnation point is defined as the point in the fluid flow where the velocity is reduced to zero and the kinetic energy of flow is converted to pressure energy. The pressure at stagnation point is called stagnation pressure (See Figure 15.3). Consider a body placed in a flowing compressible fluid moving with free upstream velocity V_1. Streamlines of flow pattern around the stationary body are shown. The velocity of the middle streamline at S is zero which is the stagnation point. At S, K.E. of flowing fluid is converted into pressure energy and the pressure at S is called stagnation pressure p_s.

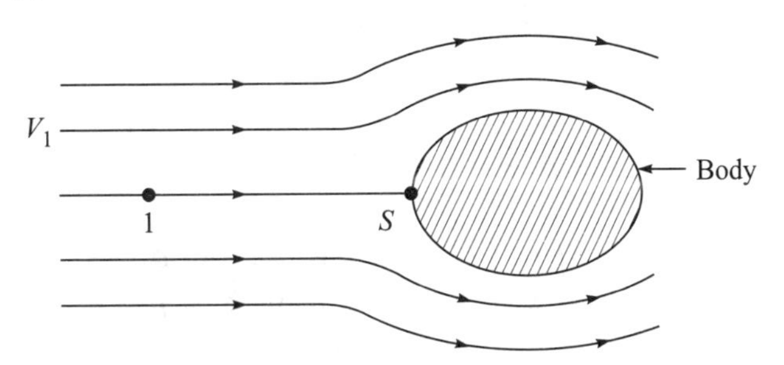

Figure 15.3 Stagnation point and stagnation pressure.

Writing Bernoulli's equation for adiabatic flow at point 1 and S:

$$\left(\frac{k}{k-1}\right)\frac{p_1}{\rho_1 g} + \frac{V_1^2}{2g} + Z_1 = \left(\frac{k}{k-1}\right)\frac{p_s}{\rho_s g} + \frac{V_s^2}{2g} + Z_2$$

∴ $V_s = 0$, assuming horizontal flow $Z_1 = Z_2$

$$\left(\frac{k}{k-1}\right)\frac{p_1}{\rho_1 g} + \frac{V_1^2}{2g} = \left(\frac{k}{k-1}\right)\frac{p_s}{\rho_s g}$$

or

$$\left(\frac{k}{k-1}\right)\left(\frac{p_1}{\rho_1} - \frac{p_s}{\rho_s}\right) = -\frac{V_1^2}{2}$$

or

$$\left(\frac{k}{k-1}\right)\frac{p_1}{\rho_1}\left(1 - \frac{p_s}{\rho_s} \times \frac{\rho_1}{p_1}\right) = -\frac{V_1^2}{2g}$$

or

$$\left(\frac{k}{k-1}\right)\frac{p_1}{\rho_1}\left(1 - \frac{p_s}{p_1} . \frac{\rho_1}{\rho_s}\right) = -\frac{V_1^2}{2} \quad \text{(A)}$$

For the adiabatic process: $\dfrac{p}{\rho^k} = \text{Constant} = \dfrac{p_1}{\rho_1^k} = \dfrac{p_s}{\rho_s^k}$

or

$$\left(\frac{\rho_1}{\rho_s}\right) = \left(\frac{p_1}{p_s}\right)^{1/k} \quad \text{(B)}$$

Substituting $\left(\dfrac{\rho_1}{\rho_s}\right)$ in (A):

$$\left(\frac{k}{k-1}\right)\frac{p_1}{\rho}, \left(1 - \frac{p_s}{p_1} \times \frac{p_1^{1/k}}{p_5^{1/k}}\right) = \frac{-V_1^2}{2}$$

or $$\left(\frac{k}{k-1}\right)\frac{p_1}{\rho_1}\left[1-\frac{p_s}{p_1}.\frac{p_s^{-1/k}}{p_1^{-1/k}}\right]=\frac{-V_1^2}{2}$$

or $$\left(\frac{k}{k-1}\right)\frac{p_1}{\rho_1}\left[1-\left(\frac{p_s}{p_1}\right)^{1-1/k}\right]=\frac{-V_1^2}{2}$$

or $$\left[1-\left(\frac{p_s}{p_1}\right)^{\frac{k-1}{k}}\right]=\frac{-V_1^2}{2}\times\left(\frac{k-1}{k}\right)\frac{\rho_1}{p_1}$$

or $$\left(\frac{p_s}{p_1}\right)^{\frac{k-1}{k}}=1+\frac{V_1^2}{2}\left(\frac{k-1}{k}\right)\frac{\rho_1}{p_1} \tag{C}$$

For the adiabatic process: $C=\sqrt{kRT}=\sqrt{k\frac{p}{\rho}}\ \because\ \frac{p}{\rho}=RT$

For point 1:

$$C_1=\sqrt{k\frac{p_1}{\rho_1}}\quad\therefore\ C_1^2=k\frac{p_1}{\rho_1}$$

Substituting $\frac{kp_1}{\rho_1}$ in (C):

$$\left(\frac{p_s}{p_1}\right)^{\frac{k-1}{k}}=1+\frac{V_1^2}{2}(k-1)\frac{1}{C_1^2}$$

or $$\left(\frac{p_s}{p_1}\right)^{\frac{k-1}{k}}=1+\frac{V_1^2}{2C_1^2}(k-1)$$

or $$\left(\frac{p_s}{p_1}\right)^{\frac{k-1}{k}}=1+\frac{M_1^2}{2}(k-1),\quad M_1=\text{Mach number at 1.}$$

or $$\left(\frac{p_s}{p_1}\right)^{\frac{k-1}{k}}=1+\left(\frac{k-1}{2}\right)M_1^2$$

$\therefore$ $$\frac{p_s}{p_1}=\left[1+\frac{(k-1)}{2}M_1^2\right]^{\frac{k}{k-1}} \tag{D}$$

$\therefore$ Stagnation pressure p_s:

$$p_s=p_1\left[1+\frac{k-1}{2}M_1^2\right]^{\left(\frac{k}{k-1}\right)} \tag{15.18}$$

In Equation (15.18), if $M_1 < 1$, the term $\left(\frac{k-1}{2} M_1^2\right)$ is less than 1 and there the right-hand side of the equation can be expanded by the Binomial theorem as:

$$p_s = p_1 \left[1 + \left(\frac{k-1}{k}\right)\left(\frac{k-1}{2} M_1^2\right) + \left(\frac{k}{k-1}\right)\left(\frac{k}{k-1} - 1\right)\left(\frac{k-1}{k} M_1^2\right)^2 2! + \left(\frac{k}{k-1}\right)\left(\frac{k}{k-1} - 1\right)\right.$$

$$\left.\left(\frac{k}{k-1} - 2\right)\left(\frac{k-1}{k} M_1^2\right)^3 \Big/ 3! + \ldots\right]$$

or
$$p_s = p_1 \left[1 + \frac{k}{2} M_1^2 + \frac{k}{8} M_1^4 + \frac{k(2-k)}{48} M_1^6 + \ldots\right]$$

or
$$p_s = p_1 + p_1 \left[\frac{k}{2} M_1^2 + \frac{k}{8} M_1^4 + \frac{k(2-k)}{48} M_1^6 + \ldots\right]$$

or
$$\frac{p_s - p_1}{p_1} = \frac{k}{2} M_1^2 \left[1 + \frac{1}{4} M_1^2 + \frac{(2-k)}{24} M_1^4 + \ldots\right]$$

or
$$\frac{p_s - p_1}{p_1} = \frac{k}{2} \frac{V_1^2}{C_1^2} \left[1 + \frac{1}{4} M_1^2 + \frac{(2-k)}{24} M_1^4 + \ldots\right]$$

$$= \frac{k}{2} \frac{V_1^2}{\left(k \frac{p_1}{\rho_1}\right)} \left[1 + \frac{1}{4} M_1^2 + \frac{(2-k)}{24} M_1^4 + \ldots\right] \quad \because C_1 = \sqrt{\frac{kp_1}{\rho_1}}$$

$$= \frac{1}{2} \cdot \frac{\rho_1 V_1^2}{p_1} \left[1 + \frac{1}{4} M_1^2 + \frac{(2-k)}{24} M_1^4 + \ldots\right]$$

$\therefore$
$$p_s - p_1 = \frac{1}{2} \rho_1 V_1^2 \left[1 + \frac{1}{4} M_1^2 + \frac{(2-k)}{24} M_1^4 + \ldots\right] \tag{15.19}$$

In Equation (15.19), if both V_1 and C_1 are small, M_1 is small, and three terms within big brackets on R.H.S. $\simeq 1$

$\therefore$ Equation (15.19) becomes:

$$p_s - p_1 = \frac{1}{2} \rho_1 V_1^2$$

$\therefore$
$$p_s = p_1 + \frac{1}{2} \rho_1 V_1^2 \tag{15.20}$$

If V_1 is high, M_1 is not small, hence Equation (15.20) becomes:

$$\frac{p_s - p_1}{\frac{1}{2} \rho_1 V_1^2} = 1 + \frac{1}{4} M_1^2 + \frac{(2-k)}{24} M_1^4 + \ldots \tag{15.21}$$

15.8 STAGNATION DENSITY

From Equation (B) of Section 15.7:

$$\frac{\rho_1}{\rho_s} = \left(\frac{p_1}{p_s}\right)^{1/k}$$

$$\therefore \qquad \rho_s = \rho_1\left(\frac{p_s}{p_1}\right)^{1/k}$$

Substituting $\left(\frac{p_s}{p_1}\right)$ from Equation (D) of Section 15.7:

$$\rho_s = \rho_1\left[\left(1+\frac{k-1}{2}M_1^2\right)^{\frac{k}{k-1}}\right]^{1/k}$$

$$\therefore \qquad \rho_s = \rho_1\left[1+\left(\frac{k-1}{2}\right)M_1^2\right]^{\frac{1}{k-1}} \qquad (15.22)$$

15.9 STAGNATION TEMPERATURE

Writing the equation of state at the stagnation point S, $\frac{p_s}{\rho_s} = R_s T_s$, where T_s is the stagnation pressure:

or $$T_s = \frac{1}{R_s}\cdot\frac{p_s}{\rho_s}$$

Substitute the values the of p_s and ρ_s from Equations (15.18) and (15.22):

$$T_s = \frac{1}{R_s}\cdot\frac{p_1\left[1+\left(\frac{k-1}{2}\right)M_1^2\right]^{k/k-1}}{\rho_1\left[1+\left(\frac{k-1}{2}\right)M_1^2\right]^{1/k-1}}$$

$$= \frac{1}{R_s}\cdot\frac{p_1}{\rho_1}\left[1+\left(\frac{k-1}{2}\right)M_1^2\right]^{\left(\frac{k}{k-1}\right)-\left(\frac{1}{k-1}\right)}$$

$$= \frac{1}{R_s}\cdot\frac{p_1}{\rho_1}\left[1+\left(\frac{k-1}{2}\right)M_1^2\right]^{\left(\frac{k-1}{k-1}\right)}$$

$$= \frac{1}{R_s}\cdot\frac{p_1}{\rho_1}\left[1+\left(\frac{k-1}{2}\right)M_1^2\right]$$

$$\therefore \qquad R_s = R, \quad \frac{p_1}{\rho_1} = RT_1$$

$$T_s = T_1\left[1+\left(\frac{k-1}{2}\right)M_1^2\right] \tag{15.23}$$

which gives the stagnation temperature T_s.

EXAMPLE 15.7 Find the Mach number when an aeroplane is flying at 900 km/hr through still air having pressure of 8 N/cm² and temperature of –15°C. Take k = 1.4 and R = 287 J/kg k. Calculate pressure, temperature and density of air at the stagnation point of the plane, i.e. at the nose of the plane.

Solution:

$$V = \frac{900\times 1000}{60\times 60} = 250 \text{ m/sec}$$

$$C = \sqrt{kRT} = \sqrt{1.4\times 287\times(273-15)} \text{ m/sec} = 321.97 \text{ m/sec}$$

$$\therefore \qquad M_a = \frac{V}{C} = \frac{250}{321.9} = 0.776 \qquad \textbf{Ans.}$$

$$p_s = p_1\left[1+\frac{k-1}{2}M_a^2\right]^{\frac{k}{k-1}} = 8\times 10^4\left[1+\left(\frac{1.4-1}{2}\right)(.776)^2\right]^{\frac{1.4}{1.4-1}} \text{ N/m}^2$$

$$= 8\times 10^4[1+0.2\times 0.602]^{\frac{1,4}{.4}} \text{ N/m}^2$$

$$= 119095.633 \text{ N/m}^2$$

$$= 11.90956 \text{ N/cm}^2 \qquad \textbf{Ans.}$$

$$T_s = T_1\left[1+\left(\frac{k-1}{2}\right)M_a^2\right] = (273-15)\left[1+\left(\frac{1.4-1}{2}\right)(.776)^2\right] \text{ °K}$$

$$= 258\times(1+0.2)(0.776)^2 \text{ °K}$$

$$= 289.072 \text{ °K}$$

$$\therefore \qquad t = (289.072-273)\text{°C} = 16.072\text{°C} \qquad \textbf{Ans.}$$

$$\rho_s = \frac{p_s}{RT_s} = \frac{11.90556\times 10^4}{287\times 289.072}$$

$$= 1.4355 \text{ kg/m}^3 \qquad \textbf{Ans.}$$

15.10 RELATIONSHIP OF AREA AND VELOCITY

In compressible flow, ρAV = Constant

Differentiating:

$$\rho d(AV) + AVd\rho = 0$$

or

$$\rho AdV + \rho VdA + AVd\rho = 0$$

Dividing by ρAV:

$$\frac{dV}{V} + \frac{dA}{A} + \frac{d\rho}{\rho} = 0 \qquad \text{(A)}$$

Euler the equation in compressible flow as $\frac{dp}{\rho} + VdV + gdz = 0$.

Assuming the horizontal flow Z term to be zero:

$$\therefore \quad \frac{dp}{\rho} + VdV = 0$$

or

$$\frac{dp}{d\rho} \cdot \frac{d\rho}{\rho} + VdV = 0$$

But Equation (15.9) gives $\frac{dp}{d\rho} = C^2$

$$\therefore \quad C^2 \frac{d\rho}{\rho} + VdV = 0$$

or

$$C^2 \frac{d\rho}{\rho} = -VdV = 0$$

or

$$\frac{d\rho}{\rho} = -\frac{VdV}{C^2} \qquad \text{(B)}$$

Substituting $\frac{d\rho}{\rho}$ from (B) in (A):

$$\frac{dV}{V} + \frac{dA}{A} - \frac{VdV}{C^2} = 0$$

$$\therefore \quad \frac{dA}{A} = \frac{VdV}{C^2} - \frac{dV}{V} = \frac{dV}{V}\left(\frac{V^2}{e^2} - 1\right)$$

$$\therefore \quad \frac{dA}{A} = \frac{dV}{V}(M_a^2 - 1) \qquad (15.24)$$

which gives the relationship between area A and velocity in terms of the Mach number.

If $M_a < 1$, the flow is sub-sonic, R.H.S. of Equation (15.24) is negative, i.e. with increase in area, the velocity decreases.

If $M_a = 1$, the flow is sonic, R.H.S. of Equation (15.24) is zero.

i.e. $\frac{dA}{A} = 0$ means that the area is constant.

If $M_a > 1$, R.H.S. is positive in supersonic flow. It means that with an increase in the area, the velocity also increases.

15.11 FLOW THROUGH ORIFICES OR NOZZLE

Figure 15.4 shows a large pressure tank from which the fluid flows through a small nozzle. Assuming that the pressure drop is large, the process is adiabatic.

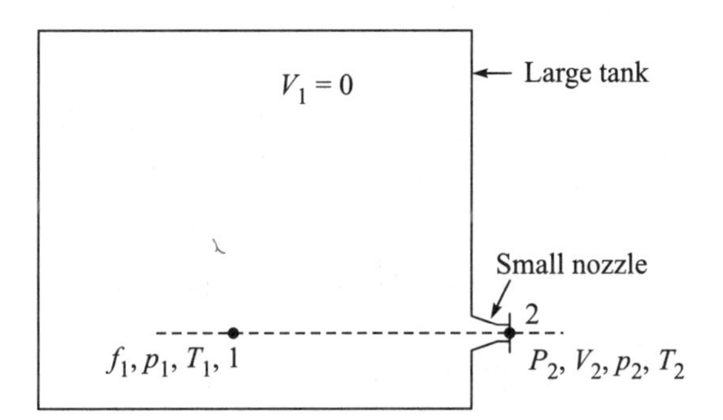

Figure 15.4 Pressure with a nozzle flow.

Writing Bernoulli's equation at points 1 and 2:

$$\left(\frac{k}{k-1}\right)\frac{p_1}{\rho_1 g} + \frac{V_1^2}{2g} + Z_1 = \left(\frac{k}{k-1}\right)\frac{p_2}{\rho_2 g} + \frac{V_1^2}{2g} + Z_2$$

But $V_1 = 0$, $Z_1 = Z_2$

$$\therefore \qquad \left(\frac{k}{k-1}\right)\frac{p_1}{\rho_1 g} = \left(\frac{k}{k-1}\right)\frac{p_2}{\rho_2 g} + \frac{V_2^2}{2g}$$

or
$$\left(\frac{k}{k-1}\right)\left[\frac{p_1}{\rho_1 g} - \frac{p_2}{\rho_2 g}\right] = \frac{V_2^2}{2g}$$

Dividing by g,

or
$$\left(\frac{k}{k-1}\right)\left[\frac{p_1}{\rho_1} - \frac{p_2}{\rho_2}\right] = \frac{V_2^2}{2}$$

$$\therefore \qquad V_2^2 = 2\left(\frac{k}{k-1}\right)\left[\frac{p_1}{\rho_1} - \frac{p_2}{\rho_2}\right]$$

$$V_2 = \sqrt{2\left(\frac{k}{k-1}\right)\left(\frac{p_1}{\rho_1} - \frac{p_1}{\rho_2}\right)}$$

$$\therefore \qquad V_2 = \sqrt{2\left(\frac{k}{k-1}\right)\frac{p_1}{\rho_1}\left(1 - \frac{p_2}{\rho_1}\cdot\frac{\rho_1}{p_2}\right)} \qquad \text{(A)}$$

In the adiabatic process: $\dfrac{p_1}{\rho_1^k} = \dfrac{p_2}{\rho_2^k}$

$$\frac{p_1}{p_2} = \left(\frac{\rho_1}{\rho_2}\right)^k$$

$$\therefore \qquad \frac{\rho_1}{\rho_2} = \left(\frac{p_1}{p_2}\right)^{1/k} \qquad \text{(B)}$$

Substituting $\left(\frac{\rho_1}{\rho_2}\right)$ in (A):

$$V_2 = \sqrt{2\left(\frac{k}{k-1}\right)\frac{p_1}{\rho_1}\left(1 - \frac{p_2}{p_1}\cdot\left(\frac{p_1}{p_2}\right)^{1/k}\right)}$$

or

$$V_2 = \sqrt{2\left(\frac{k}{k-1}\right)\frac{p_1}{\rho_1}\left[1 - \left(\frac{p_2}{p_1}\right)^{\frac{k-1}{k}}\right]} \qquad \text{(C)}$$

The mass rate of the flow through the nozzle = $\rho_2 A_2 V_2$

$$\therefore \qquad \text{Mass rate of flow} = \rho_2 A_2 \sqrt{\frac{2k}{k-1}\cdot\frac{p_1}{\rho_1}\left[1 - \left(\frac{p_2}{p_1}\right)^{\frac{k-1}{k}}\right]}$$

$$= A_2 \sqrt{\frac{2k}{k-1}\cdot\frac{p_1}{\rho_1} \times \rho_2 \left[1 - \left(\frac{p_2}{p_1}\right)^{\frac{k-1}{k}}\right]} \qquad \text{(D)}$$

From Equation *B*:

$$\rho_2 = \frac{\rho_1}{\left(\frac{p_1}{p_2}\right)^{1/k}}$$

$$\rho_2 = \rho_1 \left(\frac{p_2}{p_1}\right)^{1/k}$$

$$\rho_2^2 = \rho_1^2 \left(\frac{p_2}{p_1}\right)^{\frac{2}{k}}$$

$$\text{Mass rate of flow} = A_2 \sqrt{\frac{2k}{k-1}\cdot\frac{p_1}{\rho_1}\cdot\rho_1^2\left(\frac{p_2}{p_1}\right)^{2/k}\left[1 - \left(\frac{p_2}{p_1}\right)\frac{k-1}{k}\right]}$$

$$\text{Mass rate of flow } (m) = A_2 \sqrt{\frac{2k}{k-1}\cdot\rho_1 p_1\left[\left(\frac{p_2}{p_1}\right)^{2/k} - \left(\frac{p_2}{p_1}\right)^{\frac{k+1}{n}}\right]} \qquad (15.24)$$

which is the mass flow rate equation through the nozzle. The mass flow rate depends on $\left(\frac{p_2}{p_1}\right)$ for given values of p_1 and ρ_1 at 1.

15.11.1 Maximum Mass Flow Rate m for Values of $\left(\frac{p_2}{p_1}\right)$

The mass flow rate is maximum if the term $\left[\left(\frac{p_2}{p_1}\right)^{2/k} - \left(\frac{p_2}{p_1}\right)^{\frac{k+1}{n}}\right]$ is maximum as $\frac{2k}{k-1}\,\rho_1 p_1$ is constant.

Differentiating this term with respect to $\left(\frac{p_2}{p_1}\right)$:

$$\frac{d}{d(p_2/p_1)}\left[\left(\frac{p_2}{p_1}\right)^{2/k} - \left(\frac{p_2}{p_1}\right)^{\frac{k+1}{k}}\right] = 0$$

$$\frac{2}{k}\left(\frac{p_2}{p_2}\right)^{\frac{2}{k}-1} - \frac{k+1}{k}\left(\frac{p_2}{p_1}\right)^{\frac{k+1}{k}-1} = 0$$

or
$$\frac{2}{k}\left(\frac{p_2}{p_1}\right)^{\frac{2-k}{k}} = \frac{k+1}{k}\left(\frac{p_2}{p_1}\right)^{\frac{k+1-k}{k}}$$

or
$$\frac{2}{k}\left(\frac{p_2}{p_1}\right)^{\frac{2-k}{k}} = \frac{k-1}{k}\left(\frac{p_2}{p_1}\right)^{1/k}$$

or
$$\left(\frac{p_2}{p_1}\right)^{\frac{2-k}{k}} = \frac{k}{2}\cdot\frac{k+1}{k}\left(\frac{p_2}{p_1}\right)^{1/k}$$

$$\left(\frac{p_2}{p_1}\right)^{\frac{2-k}{k}} = \frac{k+1}{2}\left(\frac{p_2}{p_1}\right)^{1/k}$$

or
$$\left(\frac{p_2}{p_1}\right)^{\frac{2-k}{k}-1/k} = \frac{k+1}{2}$$

or
$$\left(\frac{p_2}{p_1}\right)^{\frac{2-k-1}{k}} = \frac{k+1}{2}$$

or
$$\left(\frac{p_2}{p_1}\right)^{\frac{1-k}{k}} = \frac{k+1}{2}$$

or
$$\left(\frac{p_2}{p_1}\right)^{\frac{-(k-1)}{k}} = \frac{k+1}{2}$$

or
$$\frac{1}{\left(\frac{p_2}{p}\right)^{\frac{k-1}{k}}} = \frac{1}{\left(\frac{2}{k+1}\right)}$$

$$\therefore \quad \left(\frac{p_2}{p}\right)^{\frac{k-1}{k}} = \frac{2}{k+1} \tag{15.25}$$

which is the condition for the maximum mass flow rate.

For $k = 1.4$:

$$\left(\frac{p_2}{p_1}\right)^{\frac{1.4-1}{1.4}} = \frac{2}{1.4+1}$$

$$\left(\frac{p_2}{p_1}\right)^{\frac{.4}{1.4}} = \frac{2}{2.4}$$

$$\left(\frac{p_2}{p_1}\right) = \left(\frac{2}{2.4}\right)^{\frac{1.4}{0.4}} = 0.52828 \tag{15.26}$$

15.11.2 Value V_2 in Nozzle for Maximum Flow Rate

Equation (15.25) is $\left(\frac{p_2}{p_1}\right)^{\frac{k-1}{k}} = \frac{2}{k+1}$

$$\therefore \quad \left(\frac{p_2}{p_1}\right) = \left(\frac{2}{k+1}\right)^{k/k-1} \tag{E}$$

Substituting $\left(\frac{p_2}{p_1}\right)$ from Equations (E) and (C)

$$V_2 = \sqrt{\left(\frac{2k}{k-1}\right)\left(\frac{p_1}{\rho_1}\right)\left[1-\left(\frac{p_2}{p_1}\right)^{\frac{k-1}{k}}\right]}$$

$$= \sqrt{\left(\frac{2k}{k-1}\right)\left(\frac{p_1}{\rho_1}\right)\left[1-\left(\frac{2}{k+1}\right)^{\frac{k}{k-1}\times\frac{k-1}{k}}\right]}$$

$$= \sqrt{\left(\frac{2k}{k-1}\right)\left(\frac{p_1}{\rho_1}\right)\left[1-\frac{2}{k+1}\right]}$$

$$= \sqrt{\left(\frac{2k}{k-1}\right)\left(\frac{p_1}{\rho_1}\right)\left(\frac{k+1-2}{k+1}\right)}$$

$$= \sqrt{\left(\frac{2k}{k-1}\right)\left(\frac{p_1}{\rho_1}\right)\left(\frac{k-1}{k+1}\right)}$$

$$= \sqrt{\left(\frac{2k}{k+1}\right)\left(\frac{p_1}{\rho_1}\right)}$$

$$\therefore \qquad V_2 = \sqrt{\left(\frac{2k}{k+1}\right)\left(\frac{p_1}{\rho_1}\right)} \qquad (15.27)$$

which gives the velocity in the nozzle for the maximum mass flow rate.

15.11.3 Maximum Flow Rate Through Nozzle

Equation (15.25) is $\left(\frac{p_2}{p_1}\right)^{\frac{k-1}{k}} = \frac{2}{k+1}$

$$\therefore \qquad \left(\frac{p_2}{p_1}\right) = \left(\frac{2}{k+1}\right)^{\frac{k}{k-1}}$$

Substituting $\left(\frac{p_2}{p_1}\right)$ in Equation (15.24) and simplifying similarly as above and taking $k = 1.4$, then simplifying, we get:

$$\text{Maximum mass flow rate} = 0.685\, A_z\sqrt{p_1\rho_1} \qquad (15.28)$$

15.11.4 Variation of Mass Flow Rate with $\left(\frac{p_2}{p_1}\right)$

Figure 15.5 shows the variation of mass rate of flow with pressure $\left(\frac{p_2}{p_1}\right)$ when the pressure ratio is less than 0.52828, i.e. given by Equation (15.26), the mass flow rate is constant if $\left(\frac{p_2}{p_1}\right)$ is less than 0.52828, and the mass flow rate decreases as shown by the dotted line.

15.11.5 Velocity V_2 of Nozzle for Maximum Flow Rate

Velocity V_2 at the outlet for the maximum flow rate is given by Equation (15.27), i.e.

$$V_2 = \sqrt{\left(\frac{2k}{k+1}\right)\frac{p_1}{\rho_1}} \qquad (15.27)$$

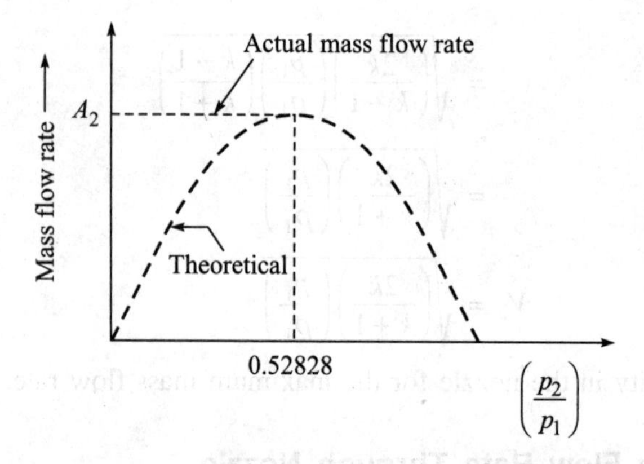

Figure 15.5 Variation of mass flow rate with $1\left(\frac{p_2}{p_1}\right)$.

Let the pressure ratio $\dfrac{p_2}{p_1} = n \quad \therefore p_1 = \dfrac{p_2}{n}$ (A)

for adiabatic flow $\dfrac{p_1}{\rho_1^k} = \dfrac{p_2}{\rho_2^k}$

$$\therefore \quad \frac{p_1}{p_2} = \left(\frac{\rho_1}{\rho_2}\right)^k$$

$$\therefore \quad \frac{\rho_1}{\rho_2} = \left(\frac{p_1}{p_2}\right)^{1/k} = n^{-1/k} \quad \therefore \rho_1 = \rho_2 n^{-1/k} \qquad \text{(B)}$$

Substituting p_1 and ρ_1 from Equations (A) and (B) in Equation (15.27):

$$V_2 = \sqrt{\frac{2k}{k+1} \cdot \frac{p_2}{n\rho_2 n^{-1/k}}}$$

$$= \sqrt{\frac{2k}{k+1} \cdot \frac{p_2}{\rho_2 n^{1-1/k}}}$$

$$= \sqrt{\frac{2k}{k+1} \cdot \frac{p_2}{\rho_2 n^{\frac{k-1}{k}}}}$$

But $n^{\frac{k-1}{k}} = \dfrac{2}{k+1}$ [Equation (15.25)]

$$= \sqrt{\frac{2k}{k+1} \cdot \frac{p_2}{\rho_2 \frac{2}{k+1}}}$$

$$= \sqrt{\frac{2k}{k+1} \cdot \frac{p_2}{\rho_2} \times \frac{k+1}{2}}$$

$$= \sqrt{\frac{k \cdot p_2}{\rho_2}}$$

$$= C_2 \quad \because \quad \sqrt{\frac{kp_2}{\rho_2}} = C_2$$

i.e. for maximum mass flow rate:

$$V_2 = C_2 \tag{15.28}$$

i.e. $\quad V_2$ = Sound Velocity.

EXAMPLE 15.8 Find the velocity of air flowing at the outlet of a nozzle fitted to a large tank containing air at a pressure of 294.3 N/cm^2 (abs) and at a temperature of 30°C. The pressure at the outlet of the nozzle is 137.34 N/cm^2 (abs). Take k = 1.4, R = 287 J/kg k.

Solution: Velocity V_2 of the nozzle is given by Equation ((C) in Section 15.11):

$$V_2 = \sqrt{\frac{2k}{k-1} \frac{p_1}{\rho_1}\left[1 - \left(\frac{p_2}{p_1}\right)^{\frac{k-1}{k}}\right]}$$

$$V_2 = \sqrt{\frac{2 \times 1.4}{1.4-1} \cdot \frac{294.3}{\rho_1}\left[1 - \left(\frac{137.34}{294.3}\right)\frac{1.4-1}{.4}\right]} \tag{A}$$

To find ρ_1, $\frac{p_1}{\rho_1} = RT_1 \Rightarrow \frac{294.3 \times 10^4}{\rho_1} = 287 \times (273 + 30)$

$$\therefore \quad \rho_1 = 0.029548 \text{ kg/m}^2$$

Substituting the value ρ_1 in (A)

$$V_2 = 242.98 \text{ m/sec.}$$ **Ans.**

15.12 VENTURIMETER TO MEASURE MASS RATE OF FLOW

In order to measure the compressible fluid, consider a venturimeter, inlet at 1 and throat at 2. For adiabatic flow, apply Bernoulli's equation at 1 and 2:

$$\left(\frac{k}{k-1}\right)\frac{p_1}{\rho g} + \frac{V_1^2}{2g} = \left(\frac{k}{k-1}\right)\frac{p_2}{\rho g} + \frac{V_2^2}{2g} \quad \text{(assuming horizontal)}$$

Multiplying by g, we get,

or

$$\frac{k}{k-1} \cdot \frac{p_1}{\rho_1} + \frac{V_1^2}{2} = \frac{k}{k-1} \cdot \frac{p_2}{\rho_2} + \frac{V_2^2}{2}$$

$$\left(\frac{k}{k-1}\right)\left[\frac{p_1}{\rho_1}-\frac{p_2}{\rho_2}\right]=\frac{V_2^2}{2}-\frac{V_1^2}{2}$$

$$\left(\frac{k}{k-1}\right)\frac{p_1}{\rho_1}\left[1-\frac{p_2}{p_1}\cdot\frac{\rho_1}{\rho_2}\right]=\frac{V^2}{2}-\frac{V_1^2}{2} \qquad \text{(i)}$$

For adiabatic flow,

$$\frac{p_1}{p_2}=\left(\frac{\rho_1}{\rho_2}\right)^k$$

or

$$\left(\frac{\rho_1}{\rho_2}\right)=\left(\frac{p_1}{p_2}\right)^{1/k}=\left(\frac{p_2}{p_1}\right)^{-1/k}$$

Substituting $\left(\frac{\rho_1}{\rho_2}\right)$ in (i):

$$\frac{k}{k-1}\frac{p_1}{\rho_1}\left[1\frac{p_2}{p_1}\cdot\left(\frac{p_2}{p_1}\right)^{-1/k}\right]=\frac{V_2^2}{2}-\frac{V_1^2}{2}$$

$$\frac{k}{k-1}\cdot\frac{p_1}{\rho_1}\left[1-\left(\frac{p_2}{p_1}\right)^{1-1/k}\right]=\frac{V_2^2}{2}-\frac{V_1^2}{2}$$

$$\frac{k}{k-1}\cdot\frac{p_1}{\rho_1}\left[1-\left(\frac{p_2}{p_1}\right)^{\frac{k-1}{k}}\right]=\frac{V_2^2}{2}-\frac{V_1^2}{2}$$

Applying continuity at 1 and 2:

$$\rho_1 A_1 V_1=\rho_2 A_2 V_2 \quad \therefore V_1=\frac{\rho_2 A_2 V_2}{\rho A_1}$$

Substitute V_1 above:

$$\frac{k}{k-1}\cdot\frac{p_1}{\rho_1}\left[1-\left(\frac{p_2}{p_1}\right)^{\frac{k-1}{k}}\right]=\frac{V_2^2}{2}-\frac{(\rho_2 A_2 V_2)^2}{2\rho^2 A_1^2}$$

$$\frac{k}{k-1}\cdot\frac{p_1}{\rho_1}\left[1-\left(\frac{p_2}{p_1}\right)^{\frac{k-1}{k}}\right]=\frac{V_2^2}{2}\left[1-\frac{\rho_2^2 A_2^2}{\rho_1^2 A_1^2}\right]$$

Again $\left(\frac{\rho_2}{\rho_1}\right)=\left(\frac{p_2}{p_1}\right)^{1/k}$:

$$\therefore \quad \frac{k}{k-1}\cdot\frac{p_1}{\rho_1}\left[1-\left(\frac{p_2}{p_1}\right)^{\frac{k-1}{k}}\right]=\frac{V_2^2}{2}\left[1-\left(\frac{p_2}{p_1}\right)^{\frac{2}{k}}\cdot\left(\frac{A_2}{A_1}\right)^2\right]$$

$$\therefore \quad V_2=\sqrt{\frac{\frac{2k}{k-1}\cdot\frac{p_1}{\rho_1}\left[1-\left(\frac{p_2}{p_1}\right)^{\frac{k-1}{k}}\right]}{1-\left(\frac{p_2}{p_1}\right)^{\frac{2}{k}}\cdot\left(\frac{A_2}{A_1}\right)^2}} \qquad \text{(A)}$$

$\therefore$ Mass rate of flow through:

$$m = \rho_2 A_2 V_2 = \rho_2 A_2 \sqrt{\frac{\frac{2k}{k-1} \cdot \frac{p_1}{\rho_1}\left[1-\left(\frac{p_2}{p_1}\right)^{\frac{k-1}{k}}\right]}{1-\left(\frac{p_2}{p_1}\right)^{\frac{2}{k}} \cdot \left(\frac{A_2}{A_1}\right)^2}} \tag{15.29}$$

EXAMPLE 15.9 Find the mass rate of flow of air through a venturimeter with inlet diameter 40 cm and throat diameter 20 cm. The pressure at the inlet of the venturimeter is 27.468 N/cm^2 (abs) and temperature of air at 20°C. The pressure at the throat is 25.506 N/cm^2 (abs) R = 287 J/kg k, k = 1.4.

Solution: $A_1 = \frac{\pi}{4}(0.4)^2 = 0.12663 \text{ m}^2$, $A_2 = \frac{\pi}{4}(.2)^2 = 0.003183 \text{ m}^2$

$p_1 = 27.468 \text{ N/cm}^2 = 27.468 \times 10^4 \text{ N/m}^2$, $p_2 = 25.506 \times 10^4 \text{ N/m}^2$

$T_1 = (273 + 20) = 293°\text{k}$, $R = 287$ J/kg k, $k = 1.4$

Density of air at the inlet is obtained as: $\rho_1 RT_1 = p_1 \quad \therefore \rho_1 = \frac{p_1}{RT_1} = \frac{27.468 \times 10^4}{287 \times 293}$

$$\rho_1 = 3.2664 \text{ kg/cm}^3$$

To find ρ_2,

$$\frac{p_1}{\rho_1^k} = \frac{p_2}{\rho_2^k}$$

$$\frac{27.468 \times 10^4}{3.2664^{1.4}} = \frac{25.506 \times 10^4}{\rho_2^{1.4}}$$

$$\therefore \quad \rho_2^{1.4} = \frac{3.2664^{1.4} \times 25.506 \times 10^4}{27.468 \times 10^4} = 4.86981$$

$$\therefore \quad \rho_2 = 3.0979 \text{ kg/cm}^2$$

Now using Equation (15.29):

$$\text{Mass flow rate } (m) = \rho_2 A_2 V_2 = \rho_2 A_2 \sqrt{\frac{\frac{2k}{k-1} \frac{p_1}{\rho_1}\left[1-\left(\frac{p_2}{p_1}\right)^{\frac{k-1}{k}}\right]}{1-\left(\frac{p_2}{p_1}\right)^{\frac{2}{k}} \left(\frac{A_2}{A_1}\right)^2}}$$

Substituting the values of ρ_2, A_2, k, ρ_1, p_z, p_1, A_2, A_1, m is calculated to be:

$$\therefore \quad m = 11.13 \text{ kg/sec}$$ **Ans.**

15.12 CONCLUSION

The equation of continuity and energy equation discussed in previous Chapters 5 and 6 for incompressible fluid cannot be applied for compressible fluid. Therefore, continuity and energy equations for compressible fluid have been derived. Velocity of sound, Mach number, pressure wave propagation, stagnation pressure, density and temperature, flow of fluid through orifices have been presented with their numerical formulations. Few numerical problems as examples are solved to show the application and use of the analytical formulations. Some problems as chapter-end exercises are provided with a list of reference books at the end.

PROBLEMS

15.1 In a horizontal, a gas is flowing with a velocity 350 m/sec at a point where pressure is 8 N/cm^2 (absolute) and temperature is 30°C. The pipe changes in diameter and at this section pressure 12 N/cm^2 (absolute). If R = 287 J/kgk, K = 1.4, find the velocity of the gas at this small section if the flow is adiabatic. (**Ans.** 218.65 m/sec)

15.2 An aeroplane moves with flight of Mach number Ma = 2 at an altitude 20 km where temperature –50°C, Determine the speed of the aeroplane if R = 287 J/kgk and K = 1.4. (**Ans.** 2155.2115 km/hr)

15.3 The atmospheric pressure at sea level is 10.05 N/cm^2 and temperature is 16°C. Find the pressure at a height 3.2 km assuming (i) Isothermal and (ii) Adiabatic change. Assume R = 297.14 J/kgk and K = 1.4. (**Ans.** (i) 9.734 N/cm^2 (ii) 9.376 N/cm^2)

15.4 A projectile travels in air at pressure of 6.4 N/cm^2 and at temperature –10°C. If the Mach angle is 30°, find the velocity of the projectile. Assume K = 1.4, R = 297 J/kgk. (**Ans.** 661.3785 m/sec)

15.5 A rocket moves in air of pressure 10 N/cm^2 at 16°C at a velocity 1640 km/hr. Find Mach number and Mach angle. Assume K = 1.4 and R = 287.13 J/kgk. (**Ans.** Ma = 1.3388, β = 48°20′)

15.6 An aircraft moving in still air attains a velocity of 800 km/hr. The atmospheric pressure and temperature on the aircraft are 6.867 N/cm^2 (abs) and –1°C respectively. Calculate pressure at nose of the aircraft and its Mach number. Take density of air 1.25 kg/m^3. (**Ans.** 10.732 N/cm^2, Ma = 0.82)

REFERENCES

1. Cambel, A.B., "Compressible Flow", section 8 of the *Handbook of Fluid Dynamics* edited by V.L. Streeter, McGraw-Hill, Inc., New York, 1961.
2. Hall, N.A., *Thermodynamics of Fluid Flow*, Prentice Hall Inc., Englewood Cliffs, N.J., 1950.
3. Shapiro, A.H., *The Dynamics and Thermodynamics of Compressible Fluid Flow*, The Ronald Press Company, New York, 1953.

4. Owezarek, J.A., *Fundamentals of Gas Dynamics*, International, Textbook Book Co., Scranton, Pa., 1964.
5. Liepmann, H.W. and A. Roshko, *Elements of Gas Dynamics*, John Wiley and Sons, Inc., New York, 1956.
6. Olson., R.M., *Engineering Fluid Mechanics*, International Textbook Company, Seranton, Pennsylvania, 1967.
7. Bansal, R.K., *A Textbook of Fluid Mechanics and Hydraulic Machines*, Laxmi Publications, New Delhi, 2000.
8. Ramamruthan, S., *Hydraulics, Fluid Mechanics and Fluid Machines*, Dhanpat Rai & Sons, Delhi, 1980.
9. Modi, P.N. and S.M. Seth, *Hydraulics and Fluid Mechanics including Hydraulic Machines*, Standard Book House, Delhi, 1975.

Chapter 16

Flow of Fluid Around Submerged Objects

16.1 INTRODUCTION

In some fields of engineering, problems involving fluid flow around submerged subjects or the movement of the objects in the stationary fluid or when both the object and fluid are in motion, are frequently encountered. The motion of sediment is water, fine sand particles in air, aeroplanes in air, automobiles on roads, submarines, ships in sea water, etc. are some examples of the flow of fluid in submerged objects. Also buildings, bridges submerged in moving air, and bridge piers submerged in moving water, are some other examples. In the analysis and design of such objects, the knowledge of determination of forces exerted by the fluid on the objects is of significance. These forces are very important in aerodynamics for the movement of aircraft, in marine engineering for the motion of submarines and ships and in structural engineering, specially in the designing of tall buildings.

16.2 FORCES ON THE BODY: DRAG AND LIFT FORCES

When the body is stationary and the fluid is in motion with a constant velocity V or the fluid is stationary and the body is on motion, the magnitude of forces experienced by the body remains the same. The force F exerted by the fluid on the body, in general is inclined to the direction of the motion as shown in Figure 16.1, in which the fluid flows past the stationary immersed body with free upstream velocity V of the fluid. This inclined force F thus has components in the direction of flow and perpendicular to flow. The component F_D in the direction of the flow is called drag force and component perpendicular to the flow is called the lift force (F_L). The mathematical expressions 'drag force' F_D and 'lift force' F_L may be obtained through dimensional analysis. In Chapter 14, it was shown in Example 14.5 that resistance R or F offered by the fluid on an aircraft depends on L, V, ρ_{air}, μ_{air}, and K of air,

i.e. $$F = \phi(L, V, \rho, \mu, K)$$

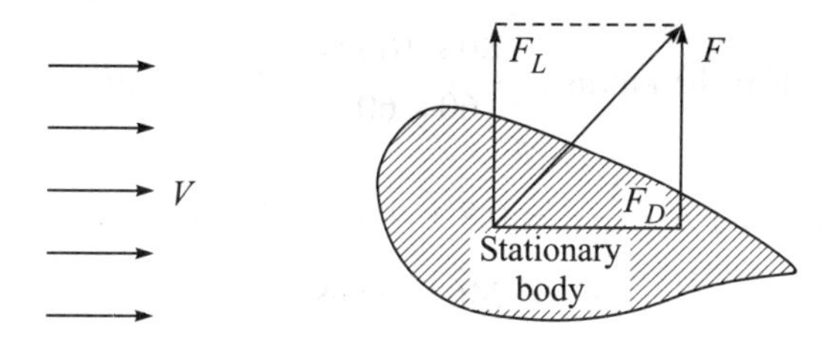

Figure 16.1 Forces on immersed stationary body.

By dimensional analysis, it was found that

$$\frac{F}{\rho L^2 V^2} = \phi\left[\left(\frac{\mu}{\rho LV}\right), \left(\frac{K}{\rho V^2}\right)\right]$$

or

$$F = \rho L^2 V^2 \, \phi\left[\left(\frac{\mu}{\rho LV}\right), \left(\frac{K}{\rho V^2}\right)\right] \tag{16.1}$$

This force F has two components, i.e. drag force F_D in the flow direction and lift force F_L, perpendicular to F_D.

$\therefore$

$$F_D = \frac{\rho L^2 V^2}{2} \times C_d$$

where C_d is a coefficient drag, which is the function of the Reynolds and Mach numbers

or

$$F_D = C_d \, A \, \frac{1}{2} \rho v^2$$

as $R_e = \dfrac{\rho LV}{\mu}$, $M_a = \dfrac{V}{\sqrt{k/\rho}}$, and thus writing $L^2 = A$ = projected area:

$$C_d = f\left(\frac{1}{R_e}, \frac{1}{M_a^2}\right)$$

$$F_D = C_d \, A \frac{\rho V^2}{2} \tag{16.2}$$

and

$$F_L = C_L \, A \frac{\rho V^2}{2} \tag{16.3}$$

where C_L is the coefficient of the lift and function of R_e, M_a.

EXAMPLE 16.1 A flat plate 2 m × 2 m moves at 40 km/hr in stationary air with a density of 1.25 kg/m^3. If the coefficients of drag and lift forces are 0.2 and 0.8, respectively find the lift force, drag force, resultant force and the power required to keep the plate in motion and the direction of the resultant force.

Solution: Area A of the plate $2 \times 2 = 4$ m^2 (Note for any other shape of the object, A is always a projected area.)

$$V = 40 \text{ km/hr} = \frac{40 \times 1000}{60 \times 60} = 11.111 \text{ m/sec}$$

Lift force $F_L = C_d\, A \cdot \dfrac{\rho V^2}{2} = 0.8 \times 4 \times \dfrac{1.25}{2} \times (11.111)^2$ N

$$F_L = 246.9135 \text{ N}$$ **Ans.**

$$F_D = C_d\, \rho A \frac{V^2}{2} = 0.2 \times 1.25 \times 4 \times \frac{(11.111)^2}{2}$$

$$F_D = 61.728 \text{ N}$$ **Ans.**

Resultant force $F = \sqrt{F_L^2 + F_D^2}$

$$F = 254.512 \text{ N}$$ **Ans.**

$$\text{Power} = F_D \times V = (61.728 \times 11.111) \text{ W}$$

$$\text{Power} = 685.86 \text{ W}$$

$$\text{Power} = 0.68586 \text{ kw}$$ **Ans.**

Direction of resultant force makes an angle θ with the flow direction

$$\therefore \quad \tan\theta = \frac{F_L}{F_D} = \frac{246.9135}{61.728} = 4$$

$$\therefore \quad \theta = 75.96°$$ **Ans.**

EXAMPLE 16.2 A truck having a projected area of 12 square metres travelling at a speed of 60 km/hr, has a total resistance 2943 N. Of this, 25% is due to rolling friction and 15% is due to surface friction. The rest is due to form drag. Calculate the coefficient of drag if ρ of air = 1.25 kg/m^3.

Solution: A (projected area) = 12 m^2

Total resistance F_R = 2943 N

Rolling friction $F_C = 2943 \times \dfrac{25}{100} = 735.75$ N

Surface friction resistance $F_S = 2943 \times \dfrac{15}{100} = 441.75$ N

$\therefore$ Form drag $F_D = F_R - F_C - F_S = (2943 - 735.75 - 441.75)$ W

$$F_D = 1765.8 \text{ N}$$

$$\therefore \quad F_D = C_d\, A \cdot \frac{\rho V^2}{2}, \quad \text{But } V = \frac{60 \times 1000}{60 \times 60} \text{ m/sec} = 16.6667 \text{ m/sec}$$

$$\therefore \quad 1765.8 = C_d \times 12 \times \frac{1.25 \times 16.6667^2}{2}$$

$$\therefore \quad C_d = 0.8475$$ **Ans.**

EXAMPLE 16.3 Find the diameter of a parachute with which a man of mass 80 kg descends to the ground from an aeroplane against the resistance of air with a velocity 25 m/sec. Take density = 1.25 kg/m^3 and C_D = 0.5.

Solution: Weight of man W = 80 kg = (80 × 9.81) N = 748.8 N

$$V = 25 \text{ m/sec}$$

$$C_d = 0.5$$

$$\rho = 1.25 \text{ kg/m}^2$$

Let D be the diameter of the parachute

$$\text{Projected Area } A = (\pi/4)D^2$$

Here $$W = F_D$$

$\therefore$ $$F_D = C_d\, A\frac{\rho V^2}{2} = 0.5 \times (\pi/4)D^2 \times \frac{1.25\times 25^2}{2}$$

$$748.8 = D^2\left(0.5\times\frac{\pi}{4}\right) \times \frac{1.25\times 25^2}{2}$$

$\therefore$ $$D = 2.21 \text{ m}$$ **Ans.**

EXAMPLE 16.4 A jet plane of weight 19.620 × 10^3 N has a wing area of 25 m^2. It flies at a speed of 200 km/hr. When the engine develops 588.6 kw, 70% of this power is used to overcome the drag resistance of the wing. Calculate the coefficient lift and drag for the wing. Take density as 1.25 kg/m^3.

Solution: Power used to overcome drag $= 588.6 \times \dfrac{70}{100} = 412.02$ kw

$$\text{This power } 412.02 = \frac{F_D\times V}{1000}$$

$$412.02 = \frac{F_D\times\dfrac{200\times 1000}{60\times 60}}{1000}$$

$\therefore$ $$F_D = 7416.36 \text{ W}$$

$$F_D = C_d\, A\frac{\rho V^2}{2}$$

$$7416 = C_d \times 25\frac{1.25\times\left(\dfrac{200\times 1000}{60\times 60}\right)^2}{2}$$

$\therefore$ $$C_d = 0.11397$$ **Ans.**

$$F_L = W = 19620 \text{ N}$$

$$\therefore \quad F_L = C_L A \frac{\rho V^2}{2} \Rightarrow 19620 = C_L \times 25 \times \frac{1.25 \times \left(\frac{200 \times 1000}{60 \times 60}\right)^2}{2}$$

$$\therefore \quad C_L = 0.4068$$ **Ans.**

16.3 ANALYTICAL EQUATIONS F_D AND F_L

Consider a stationary body held in real fluid flow at velocity *V*. Consider a small area *dA* of the body as shown in Figure 16.2. The force acting at any point of the elemental area is considered to have two components.

(i) *pdA* acts perpendicular to *dA*, which is the pressure force.
(ii) λdA acts along the tangential direction which is a shear force.

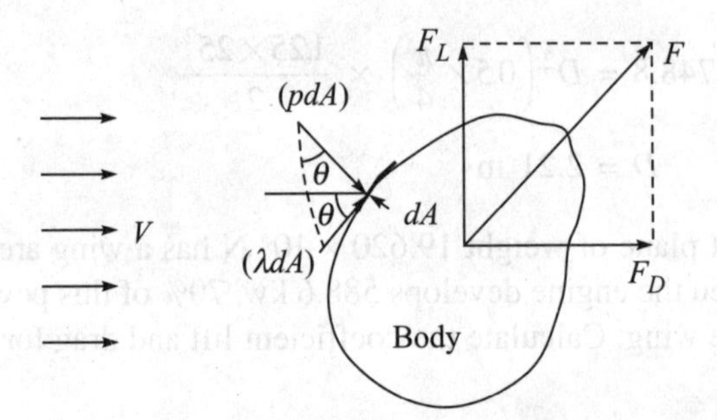

Figure 16.2 Components pressure and frictional force of F_D and F_L.

The drag on the body is therefore, given by the summation of components of these two forces acting over the whole surface area of the body in the direction *V*.

Total drag F_D = component of pressure force + component of shear force

$$F_D = \int_A pdA \sin\theta + \int_A \lambda dA \cos\theta \qquad (16.4)$$

The lift on the body is similarly given by summation of components. These two forces acting over the whole surface area of the body in the perpendicular direction of *V*

$\therefore$ F_L = Component of pressure force + component of shear force

$$F_L = \int_A pdA \cos\theta + \int_A \lambda dA \sin\theta \qquad (16.5)$$

The relative magnitude of two components of total drag, i.e. friction drag and pressure depends on the shape and position of the body immersed in the fluid flow.

In Figure 16.3(a), a thin plate is held parallel to the flow. In such a situation, pressure drag $\left(\int_A pdA \sin\theta\right)$ is practically zero and the total drag in this case is equal to the friction drag.

Here
$$F_D = \int_A \lambda dA \cos\theta \qquad (16.4a)$$

In Figure 16.3(b), the thin plate is held perpendicular to the flow direction. In this case, the friction drag is practically zero. Hence the total drag is only due to pressure force.

Here $$F_D = \int_A pdA \sin\theta \tag{16.4b}$$

Thus Equation (16.4a) is called pressure or form drag, and Equation (16.4b) is known as friction or shear drag.

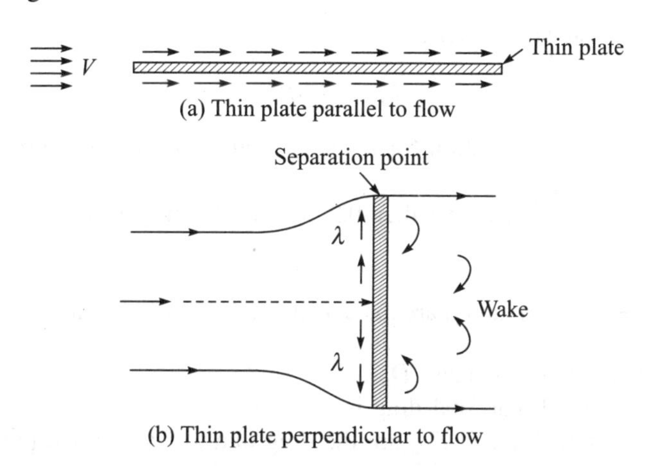

Figure 16.3 Position of thin plate.

16.4 STREAMLINED BODY AND BLUFF BODY

A streamlined body is defined as that body whose surface coincides with streamlines when the body is held in the flow. In such a body, the separation of flow takes place at the trailing edge of the body as shown in Figure 16.4(a). Although the boundary layer is formed, separation takes place at the rearmost point and thus the wake formation zone becomes very small, i.e. the pressure drag is small or negligible. Thus the total drag on the streamlined body is due to friction only. The shape of the aircraft and its wings are always streamlined to decrease the drag force.

On the other hand, a body is said to be a bluff body if its surface does not coincide with streamlines. Separation takes place much ahead of the trailing edge forming a bigger wake region.

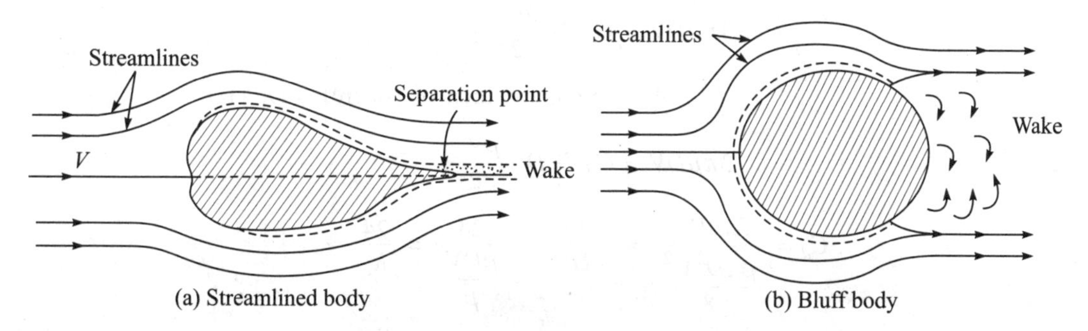

Figure 16.4 Streamlined and bluff bodies placed in the flow.

Pressure drag is much higher than friction drag. Thus in a bluff body, the pressure drag plays an important role in the total drag force. Figure 16.4(b) shows a bluff body in which the wake region is quite large.

16.5 DRAG ON A SPHERE

The Stokes law (Equation 11.37) gives:

$$F_D = 3\pi\mu VD \tag{11.37}$$

G.G. Stokes derived this equation for a sphere immersed in a flowing fluid for which the Reynolds number $\left(R_e = \dfrac{\rho DV}{\mu}\right)$ is less than 0.2, so that the inertia force may be assumed to be negligible. In the $R_e = \dfrac{\rho VD}{\mu}$, ρ and μ are the density and viscosity of the fluid, which flows with velocity V over the sphere of diameter D.

Stokes further observed that total drag is:

$F_D = F_{D_f} + F_{D_p}$ where F_{D_f} and F_{D_p} are the drag due to friction and pressure, respectively in the following proportion of total drag:

$$\therefore \qquad F_{D_f} = \frac{2}{3}F_D = \frac{2}{3} \times 3\pi\mu VD = 2\pi\mu VD$$

and

$$F_{D_p} = \frac{1}{3}F_D = \frac{1}{3} \times 3\pi\mu VD = \pi\mu VD$$

From the equation of Stokes law [Equation (11.37)], the coefficient of drag for the sphere may be obtained as follows:

General equation of drag is:

$$F_D = C_d\, A\, \frac{\rho V^2}{2} \tag{16.2}$$

Equating (11.37) and (16.2):

$$3\pi\mu DV = C_d\, A \cdot \frac{\rho V^2}{2}$$

$$A = \text{(projected area of sphere)} = (\pi/4)D^2$$

$$\therefore \qquad 3\pi\mu DV = C_d\, \frac{\pi}{4} D^2 \cdot \frac{\rho V^2}{2}$$

$$\therefore \qquad C_d = \frac{3\pi\mu DV}{\dfrac{\rho\pi D^2 V^2}{8}} = \frac{24\,\mu}{\rho DV} = \frac{24}{\dfrac{\rho DV}{\mu}} = \frac{24}{R_e}$$

$$\therefore \qquad C_d = \frac{24}{R_e} \tag{16.6}$$

When the Reynolds number increases, the inertia force also increases, $0.2 \leq R_e \leq 5$. In 1927, C.W. Oseen, a Swedish physicist, made an improvement in Stokes C_d given by Equation (16.6), i.e.

$$C_D = \frac{24}{R_e}\left[1 + \frac{3}{16R_e}\right] \tag{16.7}$$

It appears from Stokes Equation (16.6), Oseen Equation (16.7) and further investigation of C_D by increasing the Reynolds number that the value of C_d goes on decreasing as the Reynolds number increases. A plot C_d against increasing Reynolds for sphere and a circular disc is shown in Figure 16.5.

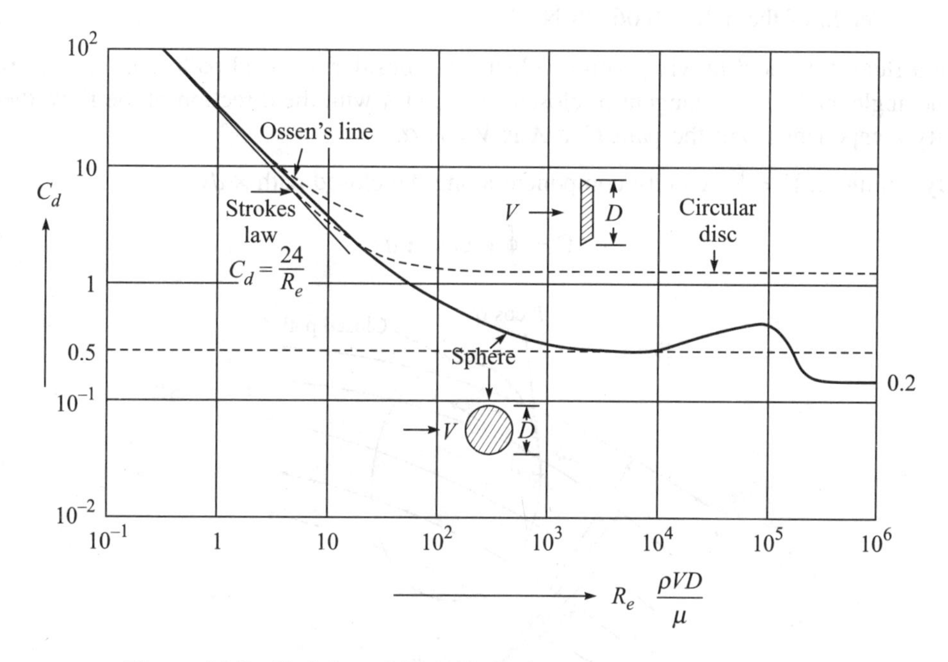

Figure 16.5 Variation of C_D with R_e for sphere and circular disc.

Figure 16.5 shows the variation C_d for a sphere in R_e values 0.2 to 10^6. Also it gives the variation of a circular disc (a body of maximum bluffness). It has variation of a low R_e but when $R_e > 10^2$, there is no variation.

EXAMPLE 16.5 Calculate the weight of ball of diameter 5 cm which is just supported in a vertical air stream which is flowing at a velocity of 10 m/sec. The density of air is 1.25 kg/m³ and Kinematic viscosity = 1.5 stores.

Solution: Diameter of ball = 0.05 m $V = 10$ m/sec $\rho = 1.25$ kg/m³

$$v = 1.5 \text{ stokes} = 1.5 \times 10^{-4} \text{ m}^2/\text{sec}$$

$$\therefore \quad R_e = \frac{VD}{\mu} = \frac{10 \times 0.05}{1.5 \times 10^{-4}} = 3333.33$$

Thus value of $C_D \simeq 0.5$ as $10^3 \angle R_e \angle 10^5$

When the fall is supported by the vertical air stream, weight W of the ball to F_D:

$$\therefore \quad F_D = 0.5\, A \cdot \frac{\rho V^2}{2} \quad A = \text{projected of the ball} = \frac{\pi}{4} D^2$$

$$A = \pi/4\ (0.05)^2 = 0.0019635 \text{ m}^2$$

$$\therefore \quad F_D = \left(0.5 \times 0.001935 \times \frac{1.25 \times 10^2}{2}\right) \text{N}$$

$$= 0.06136 \text{ N}$$

$\therefore$ Weight of the ball = 0.06136 N **Ans.**

In a fluid flow field moving with a velocity V, consider a closed path C in Figure 16.6. If α is the angle made by the tangent of closed path C at A with the direction of the flow, then the velocity component along the path C at A is $V \cos \alpha$.

By definition $\Gamma = \oint$ Velocity component along the closed path $\times\, ds$

$$\Gamma = \oint V \cos \alpha \, ds \tag{5.29}$$

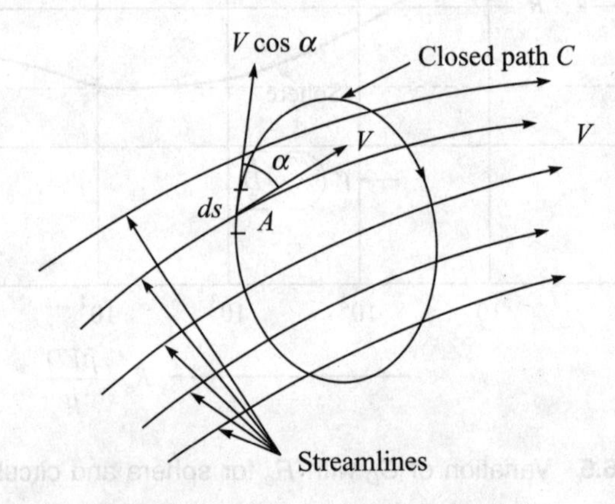

Figure 16.6 Circulation in a fluid flow.

If this circulation is considered along a free vortex, (shown in Figure 16.7),

Here $\cos \alpha = 1$ since $\alpha = 0$

$$\therefore \quad \Gamma = \oint V ds$$

$$= 2\pi R_1 V_1 = \text{Constant}$$

$$\therefore \quad \Gamma = 2\pi R_1 V_1 = 2\pi R_2 V_2 = 2\pi R_3 V_3 = \text{Constant} = 2\pi R V$$

$$\therefore \quad V = \frac{\Gamma}{2\pi R} \tag{16.8}$$

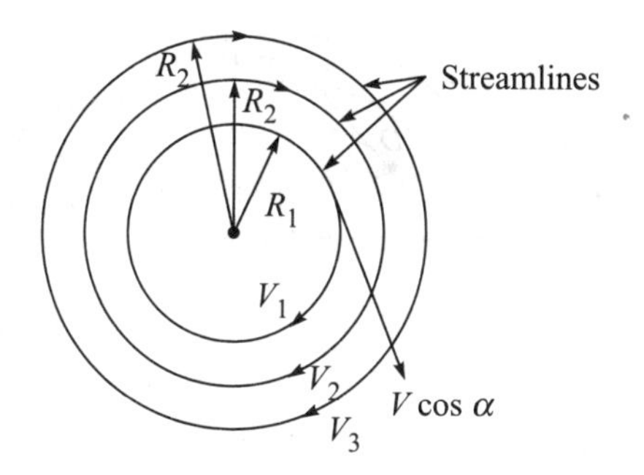

Figure 16.7 Circulation along a free vortex.

Now consider a stationary cylinder placed in an ideal fluid flow field (Figure 16.8a). If V is free stream velocity, R is the radius of the cylinder, θ is the angle made on any point D of the cylinder surface in the direction of flows, velocity component at D is 2 V sin θ. This is obtained from an analysis of classical hydrodynamics. As the flow pattern is symmetrical about the horizontal axis, the pressure distribution on lower and upper halves of the cylinder is identical and hence no lift is produced.

16.6 LIFT AND CIRCULATION

When a symmetrical body is placed in a flow of fluid with its axis parallel to the flow, the resultant force acting on the body is in the direction of the flow. There will be no force component on the body perpendicular to the flow. But the component of force perpendicular to the flow direction is called lift. Hence in such a case, the lift force is zero. If a cylinder which is symmetrical, is placed in the fluid and axis of the cylinder is parallel to the flow direction, the lift force is zero. However, if the cylinder is rotated, i.e. circulation imparted to the cylinder, the axis of the cylinder no longer remains parallel to the flow direction. In such a situation of rotating cylinder, the lift force will act on the cylinder.

16.6.1 Circulation

Circulation (Γ) is defined as the flow along a closed path. It is already discussed in Chapter 5.

If a constant circulation Γ is imparted to the same cylinder, the flow pattern around the cylinder will consist of streamlines which are a series of concentric circles as shown in Figure 16.8(b). From mathematical definition, peripheral velocity on the surface of the cylinder is given $V_P = \dfrac{\Gamma}{2\pi R}$. If the two flow patterns, Figures 16.8(a) and 16.8(b) are superimposed one above the other, the composite flow pattern will be obtained as shown in Figure 16.8(c). It is seen that the resulting flow pattern is unsymmetrical at the top and at the bottom of the cylinder about the horizontal axis. The resulting velocity on the surface of the cylinder is:

$$V_r = 2V \sin \theta + \frac{\Gamma}{2\pi R} \tag{16.9}$$

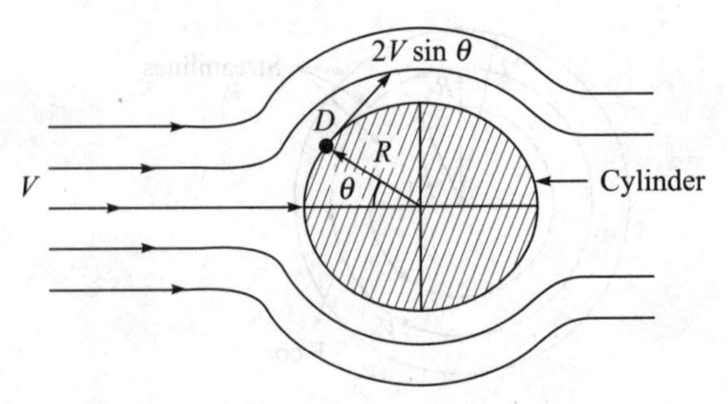

(a) Ideal flow around a stationary cylinder without any lift

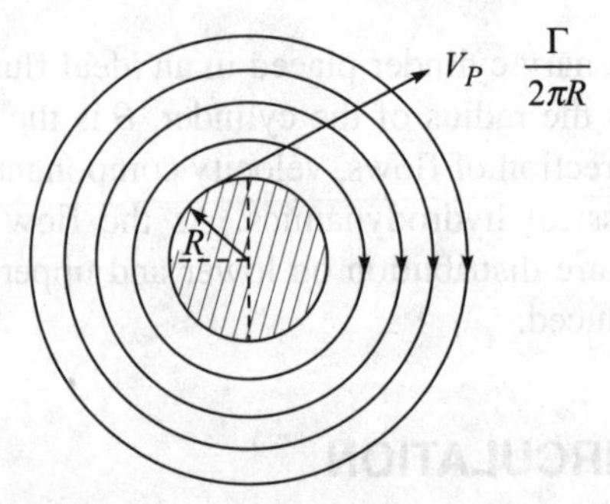

(b) Constant circulation imparted

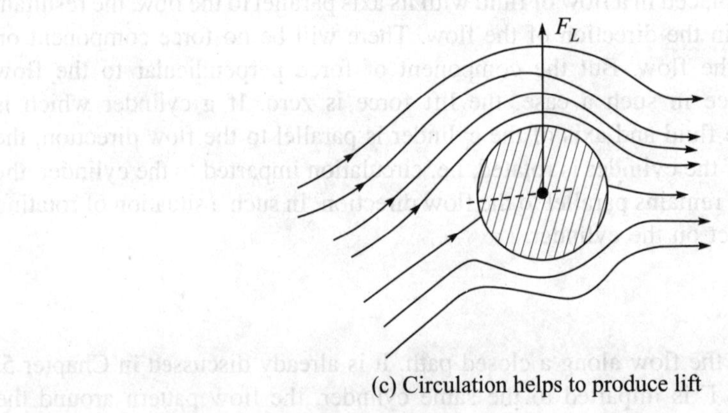

(c) Circulation helps to produce lift

Figure 16.8 Circulation and lift in different conditions.

As the circulation is clockwise, superimposition causes the velocity on the upper half to be more than the lower half as in the upper half both the velocity of flow and circulation in the same direction while in the lower half both the velocity of flow and circulation are from the opposite direction. Due to an increase of velocity in the upper half and a decrease in the lower half, by Bernoulli's equation, pressure at the upper half is less than that at the lower half and a resulting force acts perpendicular to the direction of motion, i.e. a lift force is produced.

The position of the stagnation point on the surface of the cylinder may be determined by considering V_r on the surface equal to zero Equation (16.9) becomes $0 = 2V \sin \theta + \dfrac{\Gamma}{2\pi R}$

or $$2V \sin \theta = -\frac{\Gamma}{2\pi R} \quad \therefore \sin \theta = -\frac{\Gamma}{4\pi RV} \tag{16.10}$$

The negative sign or value of $\sin \theta$ indicates that the value of θ is equal to $(180° + \theta)$. When $\sin \theta > -1$, there will be two stagnation points on the lower surface of the cylinder. But when $\sin \theta = -1$, both the stagnation points coincide at the lower extremity of the vertical diameter which is the limiting condition for the location of stagnation points on the surface of the cylinder. Now substituting $\sin \theta = -1$, and $\Gamma = 2\pi R$ in above Equation (16.10),

$$-1 = -\frac{2\pi RV_R}{4\pi RV}$$

$\therefore$ $$\frac{V_R}{V} = 2 \tag{16.11}$$

When $\sin \theta < -1$, stagnation will not lie on the surface of the cylinder and will move out into the fluid flow.

The magnitude of lift exerted on the cylinder due to the composite flow pattern may be obtained by integrating over the whole surface area of cylinder normal to the direction of flow.

Applying Bernoulli's equation at any point of the unaffected flow and at any point on the surface of the cylinder, the pressure at any point on the cylinder is obtained as:

$$\frac{p}{\rho g} + \frac{V_r^2}{2g} + z_1 = \frac{p_0}{\rho g} + \frac{V^2}{2g} + z_2$$

$z_1 = z_2$, and p_0 is the pressure at some distance of the cylinder where the velocity is V.

or $$\frac{p}{\rho} + \frac{V_r^2}{2} = \frac{p_0}{\rho} + \frac{V^2}{2}$$

or $$p + \frac{\rho V_r^2}{2} = p_0 + \frac{\rho V^2}{2}$$

$\therefore$ $$p = p_0 + \frac{\rho V^2}{2} - \frac{\rho Vr^2}{2}$$

or $$p = p_0 + \frac{\rho}{2}\left[V^2 - \left(2V \sin \theta + \frac{\Gamma}{2\pi R}\right)^2\right]$$

The lift dF_L acting on an elementary surface area of cylinder ($LRd\theta$) is given by:

$$d_{FL} = -(LRd\theta)p \sin \theta$$

where L is the length of the cylinder. The component of pressure p, i.e. $p \sin \theta$ acts downwards and hence it is a negative sign for lift. The total lift F_L is:

$$F_L = \int_0^{2\pi} -(LRd\theta)p \sin \theta$$

$$= \int_0^{\pi} -LR\left[p_0 + \frac{\rho}{2}\left\{V^2 - \left(2V\sin\theta + \frac{\Gamma}{2\pi R}\right)^2\right\}\right]\sin d\theta$$

Simplification of the above reduces it to a simple equation:

$$F_L = \rho V L \Gamma \tag{16.12}$$

Equation (16.12) is popularly known as the Kutta–Joukowski equation in honour of the German scientist M.W. Kutta (1902) and the Russians scientist, N.E. Joukowski (1905) who derived this equation independently almost at the same time. This is the equation which relates lift force F_L with circulation Γ.

Now
$$F_L = C_L A \frac{\rho}{2} V^2 = \rho V L \Gamma$$

But
$$A = 2RL \text{ (projected area)}$$

$$\therefore \quad C_L\,(2RL)\,\frac{\rho}{2} V^2 = \rho V L \Gamma$$

$$\therefore \quad C_L = \frac{\Gamma}{RV} \tag{16.13}$$

16.7 MAGNUS EFFECT

The German physicist, H.G. Magnus initially studied in 1853 the phenomenon of lift produced by circulation around a cylinder of circular cross-section. Thus this phenomenon of lift with circulation is known as the Magnus effect. The phenomenon was mentioned by Newton in 1652 and investigated experimentally by Magnus. In real fluid, this effect may be produced by a ping-pong ball, for example, by making it spin as it travels through air. Spin is the circulation introduced on the ball. Bottom spin, as shown in Figure 16.9, gives an upward force causing the lift of the ball.

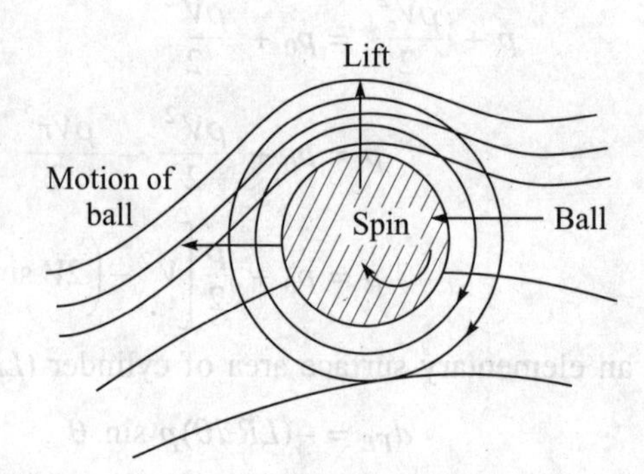

Figure 16.9 Magnus effect on a ping-pong ball.

Thus the Magnus effect shows that in both cases of ideal or real fluid, circulation is necessary for producing lift.

16.8 KARMAN VORTEX STREET

When fluid flows past a cylinder and when the Reynolds number exceeds 30, two vortices are formed in the wake region first and washed downstream. Another two vortices are formed, which are also washed downstream. In this manner, a series of a pair of vortices are formed showing a street or trail of vortices. Von Karman was the first to study and analyse the stability of the vortices in the wake region behind a cylinder and this formation of vortices is called the Karman vortex street or trail (See Figure 16.10). These vortices are unstable unless $\frac{a}{L} = \frac{1}{\pi}$ sin $h^{-1} = 0.281$, the frequency f with which vortices are shed from the cylinder is given by the following empirical formula:

$$f = 0.198 \frac{V}{D}\left[1 - \frac{19.7}{R_e}\right]$$

or

$$\frac{fD}{V} = 0.198\left[1 - \frac{19.7}{R_e}\right] \tag{16.14}$$

$\frac{fD}{V}$ is non-dimensional and is called the Strouhal number, after V. Strouhal, a Czech physicist.

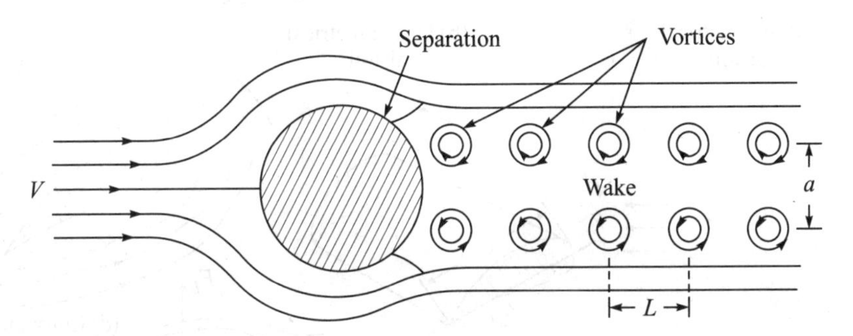

Figure 16.10 Karman vortex street or trail.

Alternate shedding of the vortices produces periodic transverse forces on the cylinder which tends to cause traverse oscillations. If the natural frequency of vibrations of the cylinder is in resonance with the frequency of vortex shedding, severe deflection or damage can result. This consideration is vital in the design of elastic strictures like suspension bridges, tall chimneys, etc. which are exposed to wind.

Similarly 'singing' of telephone or transmission line wire in high winds is caused when the frequency of vortex shedding is close to the natural frequency of the wires.

16.9 LIFT ON AN AIRFOIL WITH CIRCULATION

An airfoil is a streamlined body which may be either symmetrical or unsymmetrical as shown in Figure 16.11.

Let C be the chord length and α be the angle of attack (angle between the flow direction and chord length) of the airfoil and L be the span. As the section is streamlined, drag is always very small, negative pressure is created on the upper part of the airfoil and as such lift force on the airfoil is created.

The uniform flow of an ideal fluid flowing past an airfoil of span L (or infinite) is considered in Figure 16.11(b) and (c). Next a constant circulation imparted to the same airfoil in Figure 16.11(d) is considered when flows in (b) and (d) are superimposed then the resulting flow pattern is shown in Figure 16.11(e). Joukowski showed that the pattern of flow around a circular cylinder could be used to deduce the flow pattern around a body of any shape, provided there is a circulation around it. By properly adjusting the circulation, it is possible to obtain the flow pattern such that a streamline at the trailing edge of the airfoil is tangential to it as shown in Figure 16.11(e). The circulation Γ required to arrive at such a situation has been found analytically as:

$$\Gamma = \pi CV \sin \alpha \tag{16.15}$$

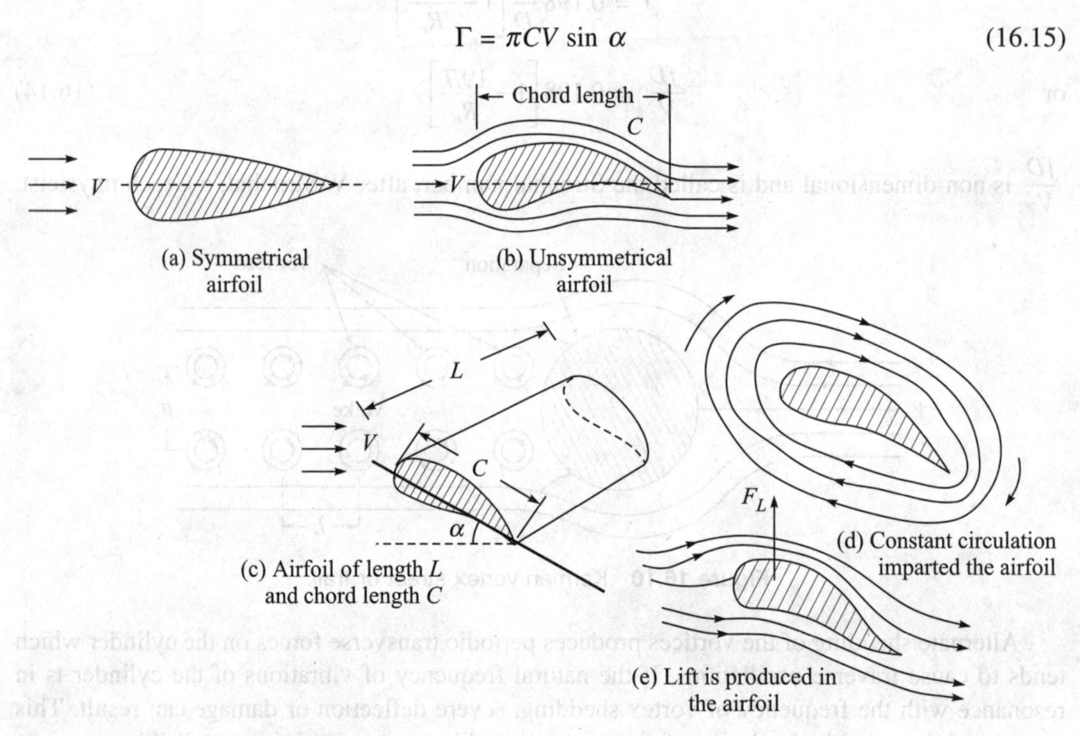

Figure 16.11 Types of airfoil sections, circulation and lift.

C and α are already defined above. The Kutta–Joukowski equation $F_L = \rho VL\Gamma$ is introduced in Equation (16.15) which gives:

$$F_L = \pi CL\rho V^2 \sin \alpha \tag{16.16}$$

$$F_L = C_L A \frac{\rho V^2}{2} = \pi CL \rho V^2 \sin \alpha$$

$$\therefore \quad C_L = \frac{2\pi CL \rho V^2 \sin \alpha}{A \rho V^2}$$

But $$A = \text{(projected area of airfoil)} = C \cdot L$$

$$\therefore \quad C_L = \frac{2\pi(CL)(\rho V^2)}{(CL)(\rho V^2)} \sin \alpha$$

$$\therefore \quad C_L = 2\pi \sin \alpha \tag{16.17}$$

It is seen from Equation (16.17), that the angle of attack α has a significant effect on the C_L and on the lift force F_L. Thus while designing an airfoil, α should be properly selected. For convenience, another angle of attack α' is considered. α' is the angle which is made by the inclination of the tangent to the lower boundary of the airfoil profile with horizontal as shown in Figure 16.12 and α' is slightly smaller than α. The variation of C_L and C_D with angle α' for a Joukowski airfoil is shown in Figure 16.12.

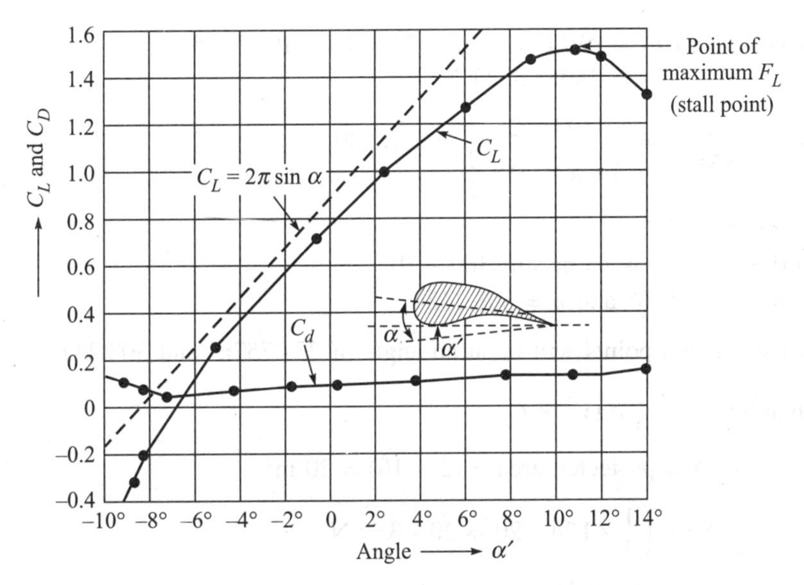

Figure 16.12 Lift and drag coefficients of a typical Joukowski airfoil of infinite span.

The dotted line on Figure 16.12 represents the theoretical value of $C_L = 2\pi \sin \alpha$ for an airfoil in an irrotational flow. The solid lines represent the actual values of C_L and C_D obtained experimentally.

The plot shows that the theoretical values of $C_L = 2\pi \sin \alpha$ are very close to the experimental values with α' upto small angle nearer to 7°. When α' reaches 8°, a remarkable difference or deviation C_L begins to occur. If we further increase α' upto 10° or little more around 11°, C_L attains a maximum value of approximately 1.5. Further with an increase of α', the value of C_L decreases from its maximum value. The point of maximum C_L is called stall point and the angle at which it becomes maximum is called stalling angle which is between 10° and 12° as evident from the plot. The drag coefficient C_D for an airfoil with α or α' remains more or less the same, i.e. there is a little variation with α or α'. The lifting efficiency is defined as the ratio of (C_L/C_D). The average of C_D for α or α' is below 0.2 and above 0.1 as seen from the plot. The maximum lifting efficiency at about 11°, is $\dfrac{1.5}{0.1} \simeq 15$.

EXAMPLE 16.6 A cylinder whose axis is perpendicular to the stream of air having a velocity of 20 m/sec, rotates at 300 rpm. The cylinder is 2 m in diameter and 10 m long. Find: (i) circulation, (ii) theoretical lift force, (iii) position of stagnation points, (iv) actual lift and drag and direction of resultant force. Take density of air 1.24 kg/m^3. For actual drag and lift, take C_L = 3.4 and C_d = 0.65. If a single stagnation point is there, find the speed of rotation of the cylinder.

Solution: $V = 20$ m/sec, $N = 300$ rpm, $\therefore V_p = \dfrac{\pi DN}{60} = \dfrac{\pi \times 2 \times 300}{60} = 31.416$ m/sec

(i) $\therefore$ Circulation $\Gamma = 2\pi R V_P = 2 \times \pi \times \dfrac{2}{2} \times 31.416.$

$$\Gamma = 197.3925 \text{ m}^2\text{/sec}$$ **Ans.**

(ii) Theoretical lift $F_L = \rho V L \Gamma = 1.24 \times 20 \times 10 \times 197.3925$

$$F_L = 48953.3523 \text{ N}$$ **Ans.**

(iii) $\text{Sin } \theta = \dfrac{\Gamma}{4\pi VR} = \dfrac{197.3925}{4 \times \pi \times 20 \times \dfrac{2}{2}} = 0.78539$

$\therefore \quad \theta = 51.7576°$

$\therefore \quad \sin\theta = \sin(180 + \theta) \text{ or } \sin(360 - \theta)$ **Ans.**

$\theta = 231.7576°$ and $\theta = 308.2424°$

Thus stagnation points will be at an angle of 231.7576° and 308.2424°

(iv) Actual lift $F_L = \dfrac{1}{2}\rho A V^2 \times C_L$

A = projected area = (2 × 10) = 20 m^2

$$\therefore \quad F_L = \left(\frac{1}{2} \times 1.24 \times 20^2 \times 20 \times 3.4\right) \text{N}$$

$F_L = 16864$ N. **Ans.**

Actual drag $F_D = \left(\dfrac{1}{2}\rho A V^2 \times C_d\right) = \left(\dfrac{1}{2} \times 1.24 \times (20) \times 20^2 \times .65\right)$ N

$F_D = 3224$ N. **Ans.**

(v) Resultant $F = \sqrt{F_L^2 + F_D^2} = 17169.41$ N

Direction F with horizontal,

$$\tan\theta = \frac{F_L}{F_D} = \frac{16864}{3224} = 5.23076$$

$\therefore \quad \theta = 79.177°$ **Ans.**

(vi) Speed of rotation of the cylinder for single stagnation point:

$$\Gamma = 4\pi VR = \left(4 \times \pi \times 20 \times \frac{2}{2}\right) \text{m}^2\text{/sec}$$

$$\Gamma = 251.3274 \text{ m}^2\text{/sec}$$

$$V_P = \frac{\Gamma}{2\pi R} = \frac{251.3274}{2\pi \times \frac{2}{2}} = 40 \text{ m/sec}$$

$$V_P = \frac{\pi DN}{60} \Rightarrow = 40 = \frac{\pi \times 2 \times N}{60}$$

$$\therefore \quad N = \frac{60 \times 40}{2\pi} = 381.97 \text{ rpm} \qquad \textbf{Ans.}$$

16.10 CONCLUSION

The contents presented in this chapter have shown that drag and lift forces, circulation and lift, depending on shape of the body have lot of influence on the resulting drag and lift. Details of research to modify the shape of aircraft, missiles, submarine, ships, tall building etc. till have been continuing to have better results by reducing the drag force. The knowledge of such situations has been the topic in with aerodynamics. This chapter deals only with the preliminary and basic knowledge and an idea to the beginners.

Few examples are solved and few problems are left for beginners to attempt at the end. A list of references for the readers given at the end. A substantial works on hydrodymics in the field have presented by Lamb[1], Von Karman[2], Sutton[3], Abott and Doenhoff[4] Cochran[5], Cooper and Lutzky[6] for further more works.

PROBLEMS

16.1 A flat plate 2 m × 2 m moves with 60 km/hr in stationary air of density 1.25 kg/m^3. If the coefficient drag C_D and lift C_L are 0.2 and 0.65, respectively, determine lift force F_L, drag force F_D, resultant force F and power required to keep the plate moving.

(**Ans.** F_L = 451.388 N, F_D = 138.888 N, F = 47.272 N, P = 2.3456 kW)

16.2 A circular disc of 3 m in diameter is held normal to a 26.4 m/sec wind of density 1.25 kg/m^3. What force is required to hold it at rest? Assume Cd = 1.1.

(**Ans.** 3385.26 N)

16.3 A circular cylinder of infinite length and of diameter 0.03 m in which the air of density 1.26 kg/m^3 is flowing past with a velocity 0.06 m/sec. If the total drag coefficient is equal to 1.4 and shear drag coefficient is 0.185, find the total drag, shear drag and pressure drag per unit length of the cylinder.

(*Hints:* A = (1 × 0.03) m^2, Pressure drag = Total drag – Shear drag
Ans. $F_D = 9.349 \times 10^{-5}$ N, $F_{D_S} = 1.236 \times 10^{5}$ N, $F_{D_P} = 8.113 \times 10^{-5}$ N)

16.4 A semi-tubular cylinder of 7.5 cm radius with concave side upstream is submerged in flow water with velocity 0.6 m/sec. If cylinder is 7.2 m long, calculate drag if C_D = 2.3.

(**Ans.** 223.56 N)

16.5 A cylinder 1.2 m in diameter is rotated about its axis in air having a velocity of 35.555 m/sec. A lift of 5886 N per metre length of cylinder is developed on the body. Assuming the ideal fluid theory, find the rotational speed and location of stagnation point. Take ρ of air as 1.26 kg/m^3.

(**Ans.** Γ = 131.38 m^2/sec), N = 566 rpm, θ = –30° and –30° and 210°)

16.6 An areoplane weighing 117720 N has a wing span of 12 m and a wing area of 20 sq m. It flies at a velocity of 360 km/hr in air at a steady level. Find C_L, total drag on the wind if C_D = 0.06, power required to keep the areoplane at this velocity, theoretical value of circulation around the wind, ρ of air 1.25 kg/m^3.

(**Ans.** C_L = 0.942, 7499.745 N, 750 kw, 78.5 m^2/sec)

REFERENCES

1. Lamb, H., *Hydrodynamics*, Cambridge University Press, London, 1932.
2. Von Karman, T., *Aerodynamics*, Cornell University Press, Ithaca, New York, 14853, 1954.
3. Sutton, O.G., *The Science of Flight*, Penguin Book, Inc., Baltimore, 1955.
4. Abott, I.H. and A.E. Von Doenhoff, *Theory of Wing Section*, McGraw-Hill, New York, 1949.
5. Cochran, W.G., “Flow due to Rotating Disk” *Proceedings Cambridge Philosophical Society*, Vol. 30, 1934.
6. Cooper, R.D. and M. Lutzky, “Explanatory Investigation of the Turbulent wakes behind Bluff Bodies”, *Research and Development Report 963*, David Taylor Model Basin, October, 1955.
7. Modi, P.N. and S.M. Seth, *Hydraulic and Fluid Mechanics including Hydraulic Machines*, Standard Book House, Delhi, 1975.

Chapter 17

Impact of Jets

17.1 INTRODUCTION

The jet of water issuing out from a nozzle fitted on the end of a pipe has some velocity or kinetic energy. The impact of this jet of water takes place when it strikes a stationary or moving flat or curved plate. This impact exerts a dynamic force on the plate due to a change of its momentum. The magnitude of this dynamic force in different conditions of plates or vanes which the jet impacts on will be considered. This basic knowledge of the impact of a jet is a pre-requisite for understanding the working principles of turbomachines.

17.2 DYNAMIC FORCES OF THE JET IN DIFFERENT SITUATIONS

When the Plate is Stationary and Vertical

In Figure 17.1, let W be the weight of the water striking the plate with a velocity V. If a is the area of the jet, the mass of liquid strikes the plate,

$$= \frac{w}{g} aV(V - 0)$$

$$= \rho a V^2$$

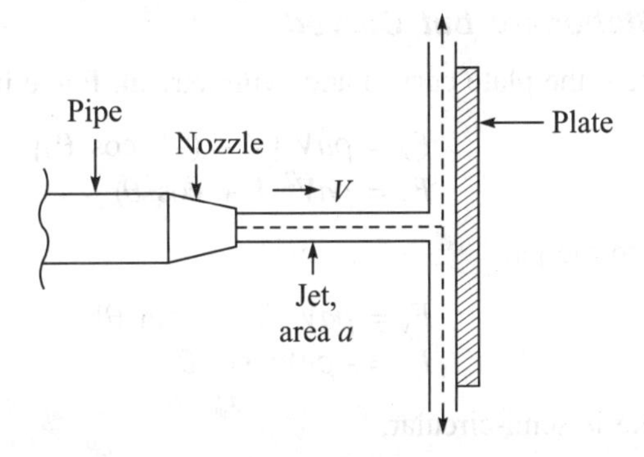

Figure 17.1 Impact on a stationary vertical plate.

The dynamic force exerted on the plate by the jet in the flow direction $= P = \dfrac{\text{Mass}}{\text{sec}}$ $(V - O)$

$$P = \rho a V^2 \tag{17.1}$$

When the Plate is Stationary and Inclined

Consider Figure 17.2 where the jet strikes an inclined plate. Velocity component normal to the plate $= V \sin \theta$

$$\therefore \quad P = \rho a V(V \sin \theta - 0)$$

$$P = \rho a V^2 \sin \theta \tag{17.2}$$

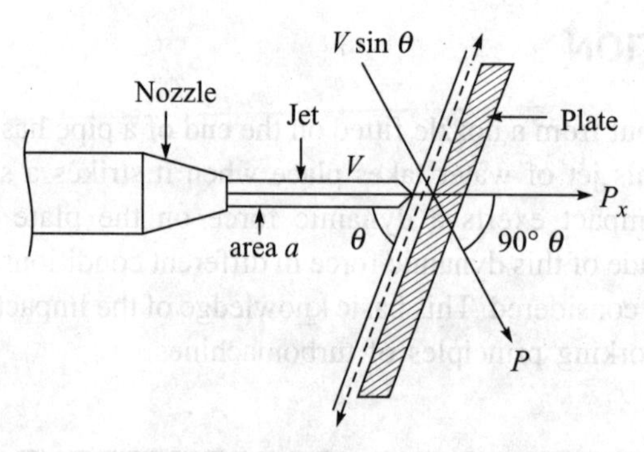

Figure 17.2 Impact on an inclined plate.

which is the force normal to the plate. Force or component in the direction of jet (i.e. x)

$$P_x = P \cos (90° - \theta) = \rho a V^2 \sin \theta \, [\cos (90° - \theta)]$$

$$P_x = \rho a V^2 \sin^2 \theta \tag{17.3}$$

The force component perpendicular to the plate $P_y = \rho a V^2 \sin \theta \sin (90° - \theta)$

$$P_y = \rho a V^2 \sin \theta \cos \theta \tag{17.4}$$

When the Plate is Stationary but Curved

Consider Figure 17.3 where the plate curved and symmetrical. Force in the direction of the jet:

$$F_x = \rho a V \, [V - (-V \cos \theta)]$$

$$F_x = \rho a V^2 (1 + \cos \theta) \tag{17.5}$$

Force perpendicular to the jet:

$$F_y = \rho a V \, [0 - V \sin \theta]$$

$$F_y = -\rho a V^2 \sin \theta \tag{17.6}$$

When $\theta = 0$, the plate is semi-circular,

$$F_x = 2 \rho a V^2 \tag{17.7}$$

$$F_y = 0$$

$$\therefore \quad F = \sqrt{(2\rho aV^2)^2 + o^2} = 2\rho aV^2 \tag{17.8}$$

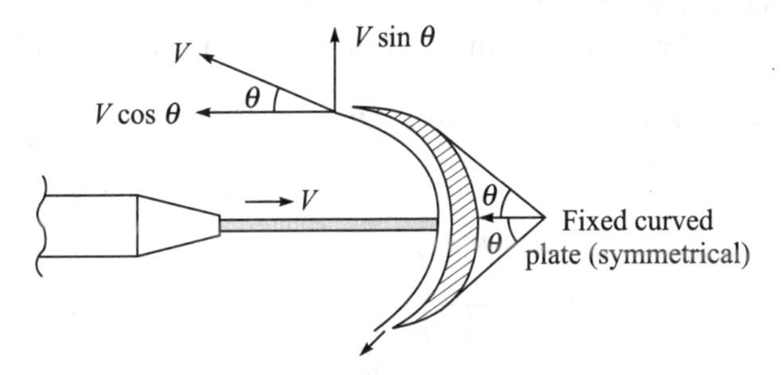

Figure 17.3 Impact on a stationary curved plate.

When the Symmetrical Plate is Curved and Jet Strikes at One End Tangentially (Figure 17.4).

Hence

$$F_x = \rho aV\,[V \cos\,\theta - (-V \cos\,\theta)]$$

$$F_x = 2\rho aV^2 \cos\,\theta \tag{17.9}$$

$$F_y = \rho aV\,[V \sin\,\theta - V \sin\,\theta]$$

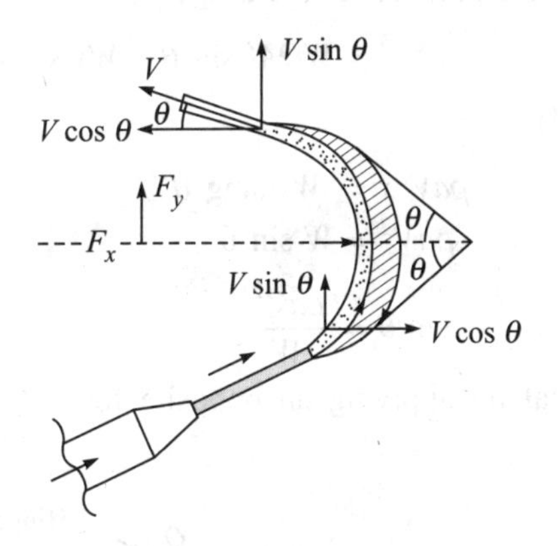

Figure 17.4 Impact tangentially on a curved plate.

$\therefore \quad F_y = 0$ if outlet edge of the plate makes an angle Q_1

$$\therefore \quad F_x = \rho AV\,(V \sin\,\theta + V \cos\,\phi)$$

$$F_y = \rho AV\,(V \sin\,\theta - V \cos\,\phi) \tag{17.9a}$$

If the plate is semi-circular, $\theta = 0$

$$F = F_x = 2\rho aV^2 \times 1$$

$\therefore \quad F = 2\rho aV^2$ which is the same as Equation (17.8).

When the Plate is Hinged at the Top

The jet strikes in the middle of hinged plate shown in Figure 17.5. The plate is deflected by angle θ about the hinge due to the force of the jet. Two forces are acting about the hinge, the force of jet and the weight of the plate. The moment of force of jet about the hinge

$$= F_n \times OB = \rho a V^2 \sin(90° - \theta) \times OB = \rho a V^2 \cos\theta \cdot \frac{oA}{\cos\theta} = \rho a V^2 x \qquad \text{(I)}$$

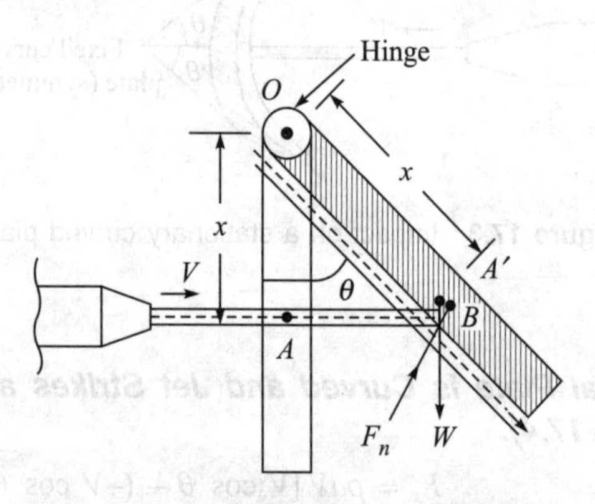

Figure 17.5 Impact on a hinged plate.

Moment of weight W of the plate about the hinge

$$= WOA' \sin\theta = Wx \sin\theta \qquad \text{(II)}$$

For equilibrium (I) = (II)

$$\therefore \quad \rho a V^2 x = Wx \text{ sing } \theta$$

$$\therefore \quad \rho a V^2 = W \sin\theta$$

$$\therefore \quad \sin\theta = \frac{\rho a V^2}{W} \qquad (17.10)$$

The plate is kept vertical by applying an opposite force Q at the bottom as shown in Figure 17.6.

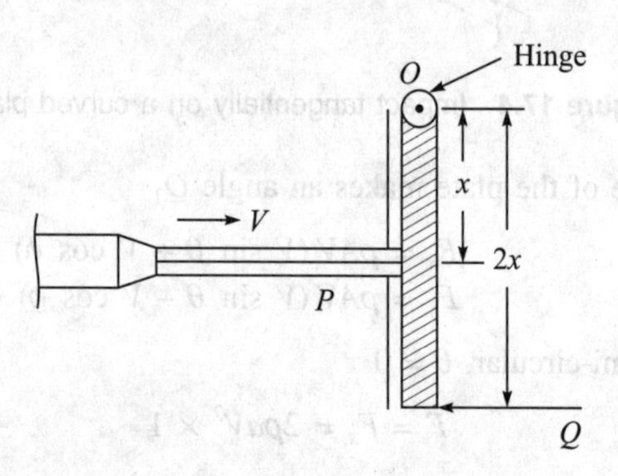

Figure 17.6 Opposite force to keep the plate vertical.

Taking moment of P, applied force Q, and weight W about hinge point O i.e. hinge.

$$\rho a V^2 x = Q \times 2x + W \times O$$

$$\therefore \quad Q = \frac{\rho a V^2}{2} \tag{17.11}$$

EXAMPLE 17.1 A jet of water from a nozzle of 4 cm diameter impinges normally with a velocity of 25 m/sec on a stationary plate. Find the force exerted by the jet on the plate.

Solution: In the above problem, we have to use Equation (17.1).
If P is the force,

$$\therefore \quad P = \rho a V^2$$

$$\therefore \quad P = [1000 \times \pi/4 (0.04)^2 \times 25^2] \text{ N}$$

$$P = 785.398 \text{ N}$$ **Ans.**

EXAMPLE 17.2 A jet of 5 cm diameter has a direct impact on a fixed plate and exerts a force of 1226.25 N over it. Find the discharge of the jet.

Solution: Force exerted by the plate $P = \rho a V^2$

$$\rho a V^2 = 1226.25$$

$$1000 \times (\pi/4)(0.05)^2\, V^2 = 1226.25$$

$$V^2 = 624.524$$

$$\therefore \quad V = 24.9904 \text{ m/sec}$$

$\therefore$ Discharge of the jet $Q = aV$

$$Q = \pi/4 (0.05)^2 \times 24.9904$$

$$Q = 0.049068 \text{ m}^3\text{/sec}$$ **Ans.**

EXAMPLE 17.3 A jet of diameter 15 cm moving at a velocity of 25 m/sec strikes a plate which is stationary. If the plate makes an angle of 30° to the jet, find the force exerted by the jet normal to the plate.

Solution: For normal to the plate, using Equation (17.2):

i.e.

$$P = \rho a V^2 \sin \theta$$

$$P = [1000 \times \pi/4 (0.15)^2 \times 25^2 \sin 30] \text{ N}$$

$$P = 5522.33 \text{ N}$$ **Ans.**

EXAMPLE 17.4 A jet of water of 4 cm diameter moving with a velocity of 35 m/sec strikes a fixed symmetrical curved van at the centre. If the jet is deflected by an angle of 125°, find the force exerted by the jet in its direction. If the plate is semi-circular, find the force exerted by the jet in its flow direction.

Solution: Here we have to use the equation of force in the x direction, i.e.

$$F_x = \rho a V^2 (1 + \cos \theta) \quad (17.5)$$

$$180° - \theta = 125 \;\therefore\; \theta = 55°$$

$$\therefore \quad F_x = 1000 \times \pi/4 (0.04)^2 \times 35^2 (1 + \cos 55°)$$

$$F_x = 2422.33 \text{ N}$$ **Ans.**

If the plate is semi-circular:

$$F_x = 2\rho aV^2$$
$$F_x = [2 \times 1000 \times (\pi/4)(0.04)^2 \times 35^2]\ \text{N}$$
$$= 3078.76\ \text{N}$$ **Ans.**

EXAMPLE 17.5 A jet of water with a diameter of 7 cm, moving with a velocity 25 m/sec, strikes a curved fixed plate tangentially at one end at an angle of 20° to the horizontal. The jet leaves the plate and angle 15° to the horizontal. Find the force exerted by the jet on the plate in the horizontal and vertical direction.

Solution: $F_x = \rho aV^2(\sin\theta + \cos\phi) = 1000 \times \pi/4(0.07)^2 \times 25^2(\sin 20° + \cos 15°)$

$F_x = 1825.173\ \text{N}$ **Ans.**

$F_y = \rho aV^2(\sin\theta - \sin\phi) = 1000 \times \pi/4(0.07)^2 \times 25^2(\sin 20° - \sin 15°)$

$F_y = 200.12\ \text{N}$ **Ans.**

EXAMPLE 17.6 A square plate of 32 cm weighing 225 N is hinged at the upper edge. A horizontal jet 3 cm in diameter, having a velocity of 16 m/sec, impinges at the middle of the plate placed vertically. What force is required at the bottom edge of the plate to keep the plate vertical? If the plate is allowed to swing freely without any force at the bottom, find the inclination to the vertical which the plate assumes under the force of the jet.

Solution: When the bottom force Q is applied to keep the plate vertical,

$$\text{Force of plate } P = \rho AV^2 = [1000 \times \pi/4(0.03)^2 \times 16^2]\ \text{N}$$
$$P = 180.955\ \text{N}$$

If Q is the horizontal force at the bottom to keep the plate vertical, taking the moment of P and Q about the upper hinge, we get:

$$Q \times 32 = P \times 16$$
$$Q \times 32 = 180.955 \times 16$$
$$\therefore \quad Q = 90.4778\ \text{N.}$$ **Ans.**

When Q is removed, the plate makes an angle θ with the vertical.

$$\therefore \quad \sin\theta = \frac{\rho AV^2}{W} \quad \text{(i.e. Equation 17.9)}$$
$$\sin\theta = \frac{180.955}{225}$$
$$\therefore \quad \theta = 53.53°$$ **Ans.**

17.3 DIRECT IMPACT OF JET ON A SERIES OF FLAT VANES MOUNTED ON THE PERIPHERY OF A LARGE WHEEL

Velocity of jet = V. Jet striking on the flat plates makes the whole to rotate. After striking the plate, velocity of jet V reduces to v. (See Figure 17.7).

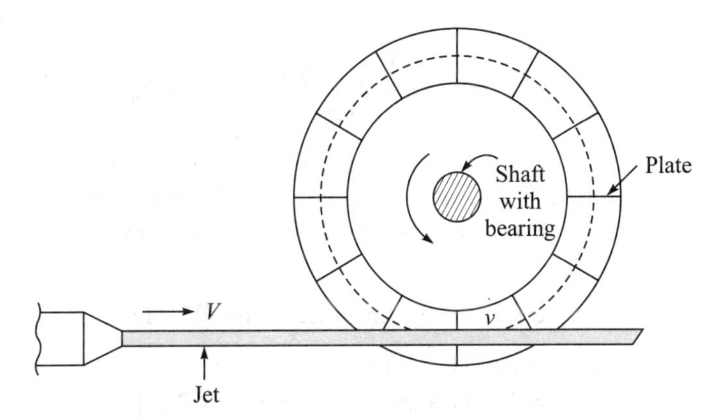

Figure 17.7 Impact of jet on series of flat vanes.

∴ Force exerted by the jet = mass × change of velocity

$$= \rho a V(V - v) \text{ N} \tag{17.12}$$

Work done by this force per second

$$= \rho a V(V - v)\, v \text{ Nm/sec}$$

Energy supplied by the jet/sec $= \rho a v \dfrac{V^2}{2}$

∴ Hydraulic efficiency $= \eta_H = \dfrac{\text{Workdone}}{\text{K.E. supplied}}$

∴
$$\eta_H = \frac{\rho a V(V - v)}{\rho a V \dfrac{V^2}{2}} = \frac{2(V - v)v}{V^2}$$

Condition for maximum η:

$$\eta = \frac{2(V - v)v}{V^2} \tag{17.13}$$

Differentiating: η_{max}, $\dfrac{d\eta}{dv} = \dfrac{2}{V^2}[V - 2v] = 0$, for maximum

∴
$$V - 2v = 0$$

∴
$$v = \frac{V}{2} \tag{17.14}$$

Substituting Equation (17.14), i.e. $v = \dfrac{V}{2}$, in Equation (17.13)

$$\eta_{max} = \frac{2\left(V - \dfrac{V}{2}\right)\dfrac{V}{2}}{V^2}$$

$$\eta_{max} = \frac{V^2 - V^2/2}{V^2}$$

or
$$\eta_{max} = \frac{V^2\left(1-\frac{1}{2}\right)}{V^2} = \frac{1}{2} = 50\%$$

EXAMPLE 17.7 A jet of diameter 40 cm impacts on a series of flat plates or vanes mounted on the periphery of a large wheel. The velocity of the jet is 5 m/sec. The velocity of vanes due to the impact of the jet is 2.5 m/sec. Calculate the force exerted by the jet on the vanes and hydraulic efficiency.

Solution: Force exerted by the jet is as given in Equation (17.11), i.e.

$$\text{Force} = \rho aV(V - v)$$
$$= [1000 \times \pi/4(0.4)^2 \times 5\,(5 - 2.5)]\ \text{N}$$

Since $d = 0.4$ m, $V = 5$ m/sec, $v = 2.5$ m/sec

$$= 1570.8\ \text{N}$$ **Ans.**

$$\text{Hydraulic efficiency } \eta_H = \frac{2(V-v)}{V^2} = \frac{2(5-2.5)}{5^2} = 0.5 = 50\%$$ **Ans.**

17.4 JET IMPACTS ON UNSYMMETRICAL MOVING CURVED VANES TANGENTIALLY

Figure 17.8 shows a moving unsymmetrical vane with velocity u in x-direction, on which the jet strikes tangentially at the inlet.

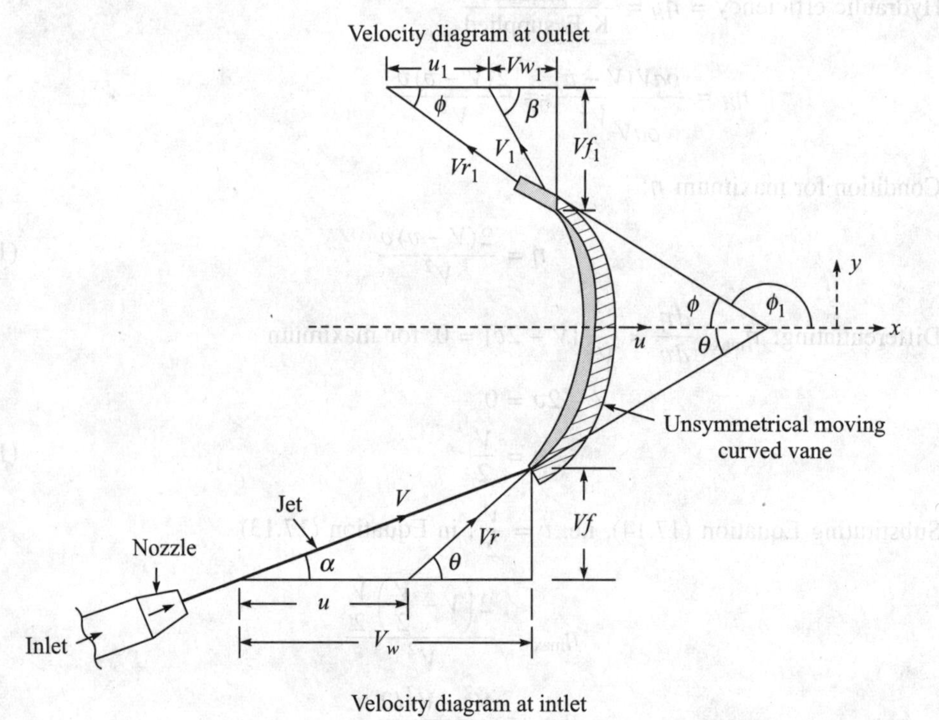

Figure 17.8 Jet strikes tangentially at one tip of moving unsymmetrical curved vane.

Let V be the velocity of the jet at the inlet, Vr is the relative velocity of the jet at the inlet due to its own velocity V and the vane velocity u. Vw is the velocity component of V in the x-direction and Vf is the velocity component of V perpendicular to the x-direction. Vw is called velocity of whirl and Vf is called velocity of flow at the inlet.

α at the inlet is the angle at which the jet strikes the vane and it is angle between the directions V and u. θ is the angle between the directions Vr and u, θ is the angle between directions of Vr and u and θ denotes the angle of the vane at the inlet. Similarly, β is the angle between Vr_1 and direction of motion of the vane at the outlet. Similarly, ϕ at the outlet is the angle between Vr_1 (relative velocity at the outlet) and the direction vane moment. ϕ thus denotes the angle of the vane at the outlet. If the blade is smooth, no loss of energy occurs on the blade and it is thus assumed that $Vr = Vr_1$, i.e. the jet leaves without shock and has the same vane velocity at the inlet and the outlet, i.e, $u = u_1$.

Now the mass of water striking the blade/sec = ρaVr, a is the area of the jet.

Force exerted by jet in the direction of motion = $F_x = \rho aVr$ [Initial velocity in the x-direction – final velocity in the x-direction]

$$F_x = \rho aVr\,[(Vw - u) - (-u_1 + Vw_1)] \because u = u_1$$

$$F_x = \rho aVr\,(Vw + Vw_1) \tag{A}$$

In Equation (A), $\beta = 90°$, $Vw_1 = 0$

$$\therefore \quad F_x = \rho aVr\, Vw$$

If β is an obtuse angle, the expression in Equation (A) becomes:

$$F_x = \rho aVr\,(Vw - Vw_1)$$

Thus the general equation of F_x is:

$$F_x = \rho aVr\,(Vw \pm Vw_1) \tag{17.15}$$

Work done by the jet on the vane per sec

$$= F_x\, u = \rho aVr\,(Vw \pm Vw_1)u \tag{17.15a}$$

Work done by the jet on the vane per sec per unit mass

$$= \rho aVr\,(Vw \pm Vw_1)u/\rho aVr$$

$$= (Vw \pm Vw_1)\,u \tag{17.15b}$$

EXAMPLE 17.8 A jet of water with a velocity of 40 m/sec strikes a curve vane which moves with a velocity of 20 m/sec. The jet makes an angle of 30° with the direction of the motion of the vane at the inlet and leaves at 90° to the direction of motion of the vane at the outlet. Determine vane angles at the inlet and the outlet if water enters and leaves the vane without shock.

Solution: $V = 40$ m/sec, $u = 20$ m/sec, $\alpha = 30°$, $\beta = 90°$ here $u = u_1 = 20$ m/sec as water enters and leaves without shock.

From the inlet diagram, $\tan\theta = \dfrac{Vf}{Vw - u} = \dfrac{V \sin 30°}{V \cos 30° - 20}$

$$\tan\theta = \frac{40 \sin 30°}{40 \cos 30° - 20}$$

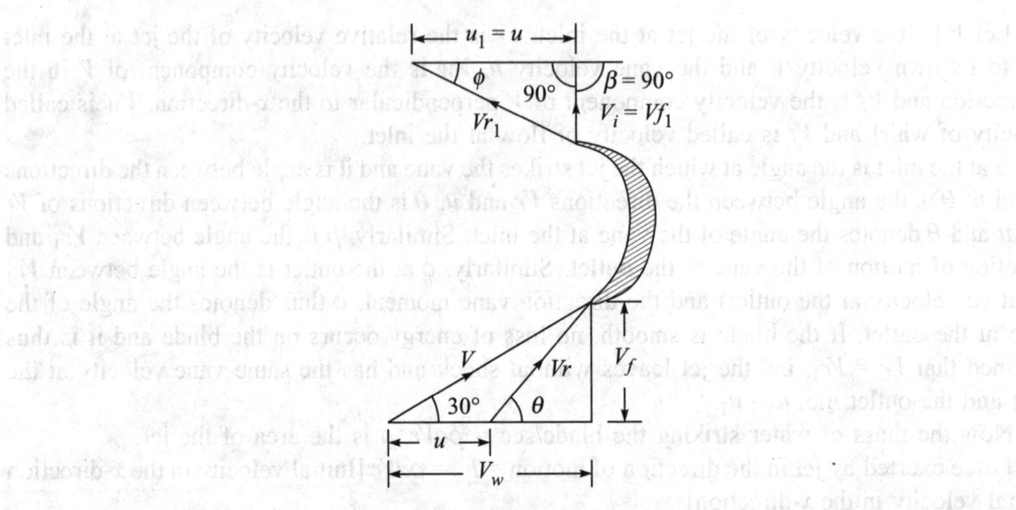

Figure 17.9 Visual of Example 17.8 to determine vane angles.

$\therefore$ $$\theta = 53.8° \qquad \textbf{Ans.}$$

$$\frac{Vf}{Vr} = \sin\theta = 53.8°$$

$\therefore$ $$Vr = \frac{Vf}{\sin 53.8°} = \frac{40 \sin 30°}{\sin 53.8°}$$

$$Vr = 24.784 \text{ m/sec}$$

Since the water or jet leaves without shock, $Vr = Vr_1 = 24.784$ m/sec.

$\therefore$ $$\cos\phi = \frac{u_1}{Vr_1} = \frac{u}{Vr} = \frac{20}{24.784} = 00.80696$$

$\therefore$ $$\phi = 36.2° \qquad \textbf{Ans.}$$

17.5 FORCE EXERTED BY A JET ON A SERIES OF RADIAL CURVED VANES

Figure 17.10 is a part of the wheel of radial flow hydraulic turbines. The wheel rotates with ω shown in the figure due to the impact of the jet. Let R and R_1 be the radii at the inlet and outlet ends of the vanes, respectively. Since the entire wheel rotates with an angular velocity of ω, the tangential velocity at the inlet and at the outlet tips of the vanes will be different. Let u and u_1 be the tangential velocity at the inlet and outlet tips of the vane.

$\therefore$ $$u = \omega R = \frac{2\pi N}{60} R$$

and $$u_1 = \omega R_1 = \frac{2\pi N}{60} R_1$$

where N is the speed of the wheel in rpm.

Let V be the velocity of the jet at the inlet.

Vr is the relative velocity of jet and vane at the inlet.

α is the guide vane blade angle at the inlet.

θ is the angle made by the relative velocity Vr with the direction of u and is called the vane angle at the outlet.

Vw and Vf are the components of V in the direction of u and perpendicular to u, respectively. Vw is called the velocity of the whirl and V_F is called the velocity of the flow at the inlet. The above V, Vr, α, θ, Vw, Vf are shown at the inlet with the inlet velocity diagram in Figure 17.10.

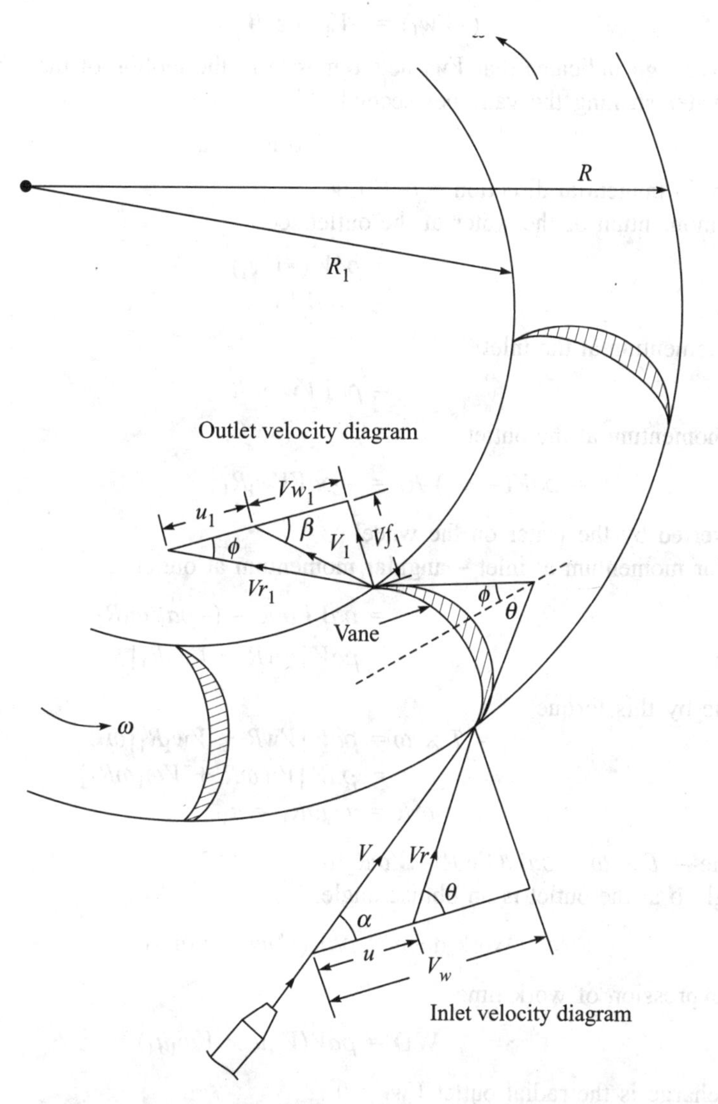

Figure 17.10 Impact of jet on series of radial curved vanes with angular velocity.

Similarly, the outlet velocity diagram is drawn at the outlet of the vane where V_1 is the velocity of the jet at the outlet, β is the angle of V_1 with u_1. Vr_1 is the relative velocity of the jet and vane at outlet, ϕ is the angle made by the relative velocity Vr_1, with direction u_1 at the outlet and is called the vane angle at the outlet. Vw_1 and Vf_1 are components of V_1 at the outlet and are called the velocity of the whirl and flow at the outlet, respectively. Velocity diagram at the outlet with V_1, Vr_1, θ, ϕ, Vw_1 and Vf_1 is drawn in Figure 17.10.

From the inlet diagram:

$$Vw = V \cos \alpha$$

Similarly at the outlet diagram:

$$(-Vw_1) = -V_1 \cos \beta$$

The negative sign indicates that Vw_1 acts opposite to the motion of the wheel.

Mass of water striking the vane per second,

$$= \rho aV, \text{ } a \text{ is the area of the jet.}$$

Momentum in tangential direction $= \rho aV\, Vw$

Similarly momentum of the water at the outlet/sec

$$= \rho aV\,(-Vw_1)$$
$$= -\rho aVVw_1$$

Angular momentum at the inlet

$$= \rho aVVw \times R$$

Angular momentum at the outlet

$$= \rho aV(-Vw_1)\, R_1 = -\, \rho aVVw_1R_1$$

Torque exerted by the water on the wheel

$T =$ angular momentum at inlet – angular momentum at outlet

$$= \rho aVV\omega R \, - \, (-\rho aV\omega_1 R_1)$$
$$= \rho aV\,[VwR \, + \, Vw_1R_1]$$

Work done by this torque

$$= T \times \omega = \rho aV\,[VwR \, + \, Vw_1R_1]\,\omega$$
$$= \rho aV\,[Vw\omega R \, + \, V\omega_1\omega R_1]$$

$$\therefore \quad \omega R = u, \; \omega R_1 = u_1$$

$\therefore$ Work done $= T \times \omega = \rho aV/(VwR - V\omega R_1)\,\omega$

If the angle β at the outlet is an obtuse angle:

$$\text{Work done} = \rho aV(V\omega u \, - \, V\omega_1 u_1)$$

$\therefore$ General expression of work time

$$\text{WD} = \rho aV(V_w u \pm V\omega_1 u_1) \tag{17.16}$$

If the discharge is the radial outlet $Vw_1 = 0$

$$\therefore \quad \text{WD} = \rho AV(V_w u) \tag{17.17}$$

$$\text{Efficiency} = \frac{\text{WD/sec}}{\text{KE/sec}} = \frac{\rho a V(Vwu \pm Vw_1 u_1)}{\frac{1}{2}(\text{mass/sec}) \times V^2}$$

$$= \frac{\rho a V(V_w u \pm V_{w_1} u_1)}{\frac{1}{2}\rho a V(V)^2} = \frac{2(V_w u \pm V_{w_1} u_1)}{V^2} \quad (17.18)$$

EXAMPLE 17.9 A jet of water having a velocity of 30 m/sec, strikes a series of radial curved vanes mounted on a wheel which is rotating at 300 rpm. The jet makes an angle of 30° with the tangent to the wheel at the inlet and leaves the wheel with a velocity of 4 m/sec at an angle of 120° to the tangent to the wheel at the outlet. Water is flowing from the outlet in the radial direction. The outer and inner radii of the wheel are 0.6 m and 0.3 m, respectively (See Figure 17.11). Determine: (i) the vane angles at the inlet and outlet, (ii) work done/sec per kg of water, (iii) efficiency of the wheel.

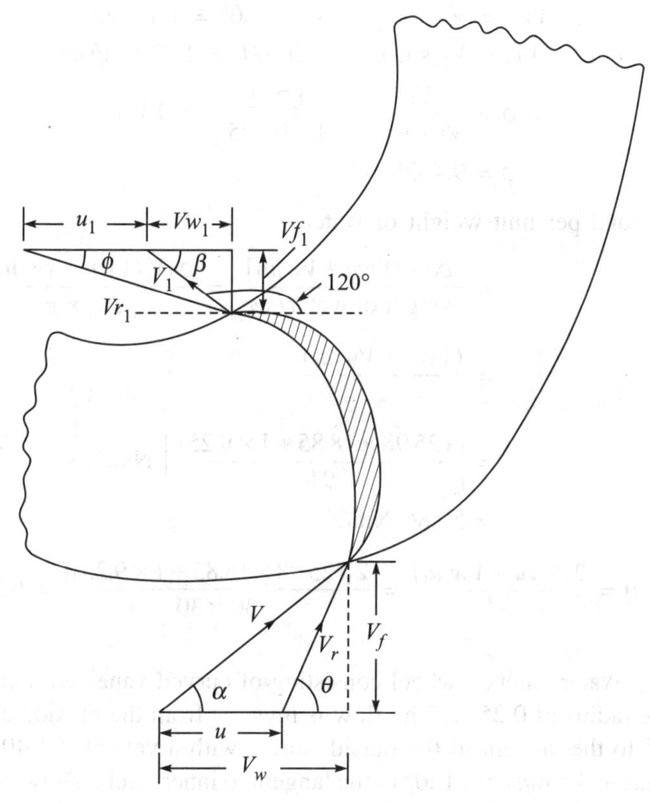

Figure 17.11 Visual for Example 17.9.

Solution: Velocity of jet $= V = 30$ m/sec

$N = 300$ rpm

$$\text{Angular speed } \omega = \frac{2\pi N}{60} = 31.416 \text{ rad/sec}$$

Angle of jet or guide vanes angle, i.e. $\alpha = 30°$

V_1 (velocity of jet at outlet) = 4 m/sec

$$\beta = 180° - 120 = 60°$$

$$R = 0.6 \text{ m}, R_1 = 0.3 \text{ m}$$

$$u = \omega R = 31.416 \times 0.6 = 189.85 \text{ m/sec}$$

$$u_1 = \omega R_1 = 31.416 \times 0.3 = 9.425 \text{ m/sec}$$

$$Vw = V \cos \alpha = 30 \cos 30° = 25.98 \text{ m/sec}$$

$$Vf = V \sin \alpha = 30 \sin 30° = 15 \text{ m/sec}$$

$$\therefore \quad \tan \theta = \frac{Vf}{Vw - u} = \frac{15}{25.98 - 18.85} = 2.10378$$

$$\therefore \quad \theta = 64.567°$$ **Ans.**

From the outlet velocity diagram:

$$Vw_1 = V_1 \cos \beta = 4 \cos 60° = 1 \text{ m/sec}$$

$$Vf_1 = V_1 \sin \beta = 4 \sin 60° = 1.732 \text{ m/sec}$$

$$\therefore \quad \tan \phi = \frac{Vf_1}{Vw_1 + u_1} = \frac{1.732}{1 + 9.425} = 0.166139$$

$$\phi = 9.433°$$ **Ans.**

(ii) WD per second per unit weight of water

$$= \frac{\rho a V_1 [Vwu + Vw_1 u_1]}{\text{weight of water/sec}} = \frac{\rho a V_1 (Vwu + vw_1 u_1)}{\rho a V_1 \times g}$$

$$= \frac{(Vwu + Vw_1 u_1)}{g}$$

$$= \left[\frac{(25.98 \times 18.85 + 1 \times 9.25)}{9.81}\right] \text{Nm/N}$$

$$= 50.86 \text{ Nm/N}$$ **Ans.**

(iii) Efficiency $\eta = \dfrac{2[Vwu + Vw_1 u_1]}{V^2} = \dfrac{2[(25.98 \times 18.85 + 1 \times 9.25)]}{30 \times 30} = 0.5544 = 55.44\%$

Ans.

EXAMPLE 17.10 Water enters a wheel consisting of curved vanes with an outside radius of 0.5 m and an inside radius of 0.25 m. The flow is inwards from the outside circumference. The supply jet is at 30° to the tangent to the outside circle with a velocity of 40 m/sec. The water leaves the wheel vane at 3.5 m/sec at 120° to the tangent to inner circle. Draw the velocity triangle for the inlet and the outlet and find suitable blade angles if the wheel runs at 360 rpm. (See Figure 17.12).

Solution: $R = 0.5$ m, $R_1 = 0.25$ m

$N = 360$ rpm

$$\omega = \frac{2\pi N}{60} = \frac{2\pi \times 360}{60} = 37.7 \text{ rad/sec}$$

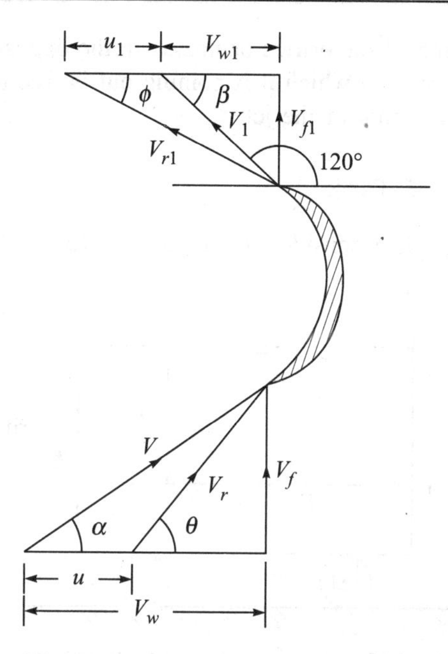

Figure 17.12 Visual for Example 17.10: Velocity diagrams at inlet and outlet.

$$\therefore \quad u = \omega R = 37.7 \times 0.5 = 18.85 \text{ m/sec}$$

$$u_1 = \omega R_1 = 37.7 \times 0.25 = 9.425 \text{ m/sec}$$

$$\beta = 180° - 120° = 60°$$

$$\alpha = 30°, \; V = 40 \text{ m/sec}, \; V_1 = 3.5 \text{ m/sec}$$

$$Vw = V \cos \alpha = 40 \cos 30°$$

$$\therefore \quad Vw = 34.641 \text{ m/sec}$$

$$\therefore \quad Vf = V \sin \alpha = 40 \sin 30° = 20 \text{ m/sec}$$

$$\tan \theta = \frac{Vf}{Vw - u} = \frac{20}{34.641 - 18.85} = 1.2665$$

$$\theta = 51.70$$ **Ans.**

$$Vw_1 = V_1 \cos \beta = 3.5 \cos 60° = 1.75 \text{ m/sec}$$

$$Vf_1 = V_1 \sin \beta = 3.5 \sin 60° = 3.031 \text{ m/sec}$$

$$\tan \phi = \frac{Vf_1}{Vw_1 + u_1} = \frac{3.031}{1.75 + 9.425} = 0.27123$$

$$\therefore \quad \phi = 15.175°$$ **Ans.**

17.6 JET PROPULSION: ACTION AND REACTION OF THE JET

This is the principle used to utilise the action and reaction of the jet to propel vessels like tanks and ships. A jet of fluid issuing out of an orifice or nozzle exerts a force when the impact of the jet takes place on the plate. The force exerted by the jet is called the action of the jet. Newton's law states that every action has an equal and opposite reaction. Hence, a jet coming out of an orifice or nozzle exerts a force on the orifice or nozzle in the opposite direction which is called reaction of the jet. If the body from which it is coming out is free to move, the body will start moving in opposite direction to that of the jet.

17.6.1 Jet Propulsion of Tank

A tank with and orifice at the side is considered in Figure 17.13. Let the velocity of jet be V under a constant head H.

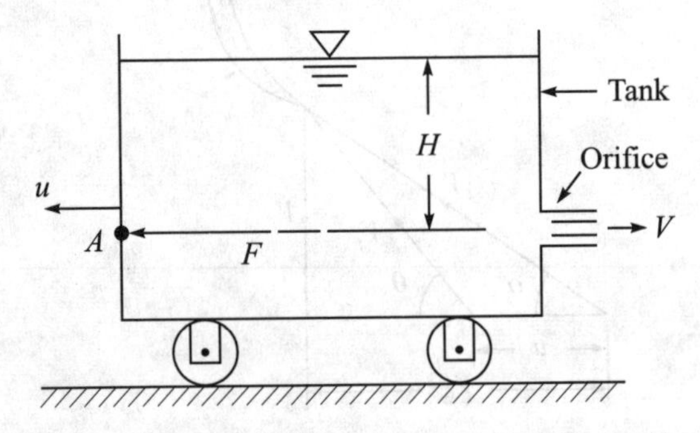

Figure 17.13 Jet propulsion of tank with an orifice.

C_v is the coefficient of velocity and a is the area of the tank.

(i) Initially the tank is at rest, i.e. $u = 0$

Mass of water coming out of the orifice $= \rho \times$ volume per second

$$= \rho \text{ (area} \times \text{velocity)}$$
$$= \rho a V$$
$$= Cv \rho a \sqrt{2gH}$$
$$= Cv \sqrt{2g}\, \rho a \sqrt{H}$$

$\therefore$ Force acting by the jet on the tank = Mass per second × (final velocity – initial velocity)

Force $\quad F = \rho a V(V - 0)$

Force $\quad F = \rho A v^2 \qquad (17.19)$

This $\quad F = \rho a V^2$ acts at a point A of the tank.

(ii) When the tank has an initial velocity u opposite to V, then the relative velocity of the jet with respect to the tank is:

$$Vr = V - (-u) = V + u$$

Mass of water coming out of the orifice $= \rho a Vr = \rho a (V + u)$

Force exerted on the tank $= F = [\rho a(V + u)][(V + u) - u]$

$$F = \rho a(V + u)\, V \qquad (17.20)$$

Work done by this force $= F \times u = \rho a(V + u) V u$

$$\text{Efficiency } \eta = \frac{\text{Work done}}{\text{KE of the jet}}$$

$$\eta = \frac{\rho a(V+u)\,Vu}{\frac{1}{2}\rho a(V+u)(V+u)^2}$$

$$\eta = \frac{2Vu}{(V+u)^2} \tag{17.21}$$

The condition for maximum efficiency is obtained by differentiating η with respect to u and equating with zero for maximum:

$$\frac{d\eta}{du} = \frac{d}{du}\left[\frac{2Vu}{(V+u)^2}\right] = 0$$

or $\quad \frac{d}{du}[(2Vu)(V + u)^{-2}] = 0$

Simplifying: $\quad u = V \quad$ (17.22)

Substituting this $u = V$ in (Equation 17.20), from maximum efficiency:

$$\eta = \frac{2V^2}{(2V)^2} = \frac{2V^2}{4V^2} = \frac{1}{2} = 50\% \tag{17.22a}$$

EXAMPLE 17.11 The head of water from the centre of the orifice fitted on a tank is 6 m. The diameter of the orifice is 15 cm. The tank is fitted with frictionless wheels at the bottom and it is moving with a velocity of 4 m/sec due to the reaction of the jet coming out of the orifice. Determine: (i) Propelling force on tank, (ii) Work done per sec per kg of water, (iii) Efficiency of the wheel.

Solution: Assume that $C_v \simeq 0.97$

$\therefore \quad V = C_v\sqrt{2gH} = 0.97\sqrt{2\times 9.81\times 6} = 10.524$ m/sec

$u = 4$ m/sec

$\therefore$ (i) Propelling force $= \rho a(V + u)V = \rho[(\pi/4)(0.15)^2](10.54 + 4) \times 10.524$

$= (1000 \times 0.0176714 \times 14.524 \times 10.524)$ N

$= 2701.083$ N **Ans.**

(ii) Work done = force $\times\, u = (2701.083 \times 4)$ N-m/sec

$= 10804.33467$ N-m/sec **Ans.**

(iii) Efficiency $\eta = \frac{2Vu}{(V+u)^2} = \frac{2\times 10.524\times 4}{(10.524+4)^2} = 0.39911 = 39.911\%$

17.6.2 Jet Propulsion of Ships

The principle of jet propulsion is used in ships by discharging water through a jet fitted at the back or stern of the ship. The ship is provided with a centrifugal pump which draws water from the region surrounding the ship, and the water is discharged at the back through an orifice. Two possible systems are adopted. These are discussed below.

When the Inlet of Water is Normal to the Direction of Motion of the Ship.

Consider Figure 17.14.

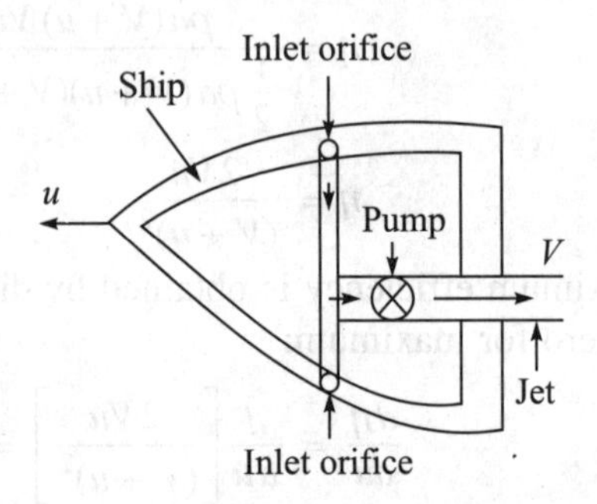

Figure 17.14 Inlet orifices are at a right angle.

Let u be the velocity of the ship.

V is the absolute velocity of the jet

Vr is the relative of the jet = $V + u$.

Mass of water discharged by the orifice $= \rho a Vr$

$$= \rho a (V + u)$$

where a is the area of the orifice.

Propulsive force exerted on the ship F = mass × change of velocity

$$F = \rho a (V + u)[(V + u) - u]$$

$$F = \rho a (V + u)V \tag{17.20}$$

The same equation is used here as that for tanks.

Work done $= F \times u = \rho a (V + u)Vu$

Efficiency η, condition for maximum efficiency ($V = u$) and expression of maximum efficiency (50%) are same as in the case of the tank.

When the Inlet Orifice Faces the Flow Direction of the Ship.

Consider Figure 17.15. Here water enters at the front or bow of the ship as shown in Figure 17.15. In this case also, the expressions of propelling force, and work done per second are the same as in the first case. But expressions for energy supplied and efficiency will be different. Here water enters through the inlet orifice at the bow and enters with a velocity equal to the ship velocity u.

$$\text{Hence kinetic energy supplied} = \frac{1}{2}\ (\text{mass of water per sec}) \times [(V + u)^2 - u^2]$$

$$\therefore \quad \text{KE} = \frac{1}{2}\rho a (V + u)[(V + u)^2 - u^2] \tag{17.23}$$

$$\text{Efficiency} = \frac{\text{Work Done}}{\text{KE supplied}} = \frac{\rho a (V + u)\, Vu}{\frac{1}{2}\rho a (V + u)[(V + u)^2 - u^2]} = \frac{2\,Vu}{V^2 + 2\,Vu}$$

$$\therefore \quad \eta = \frac{2u}{V + 2u} \tag{17.24}$$

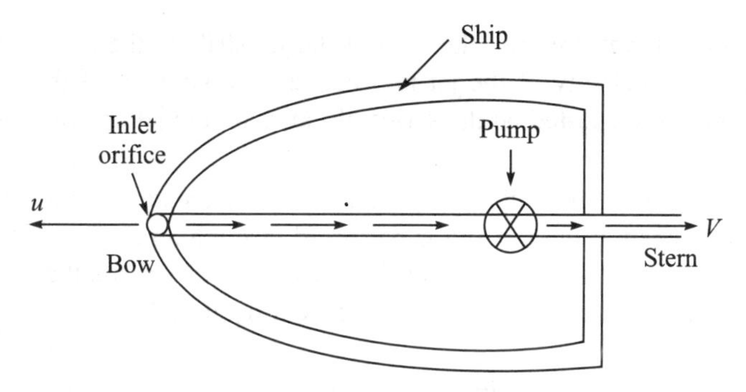

Figure 17.15 Inlet orifice faces the flow direction.

EXAMPLE 17.12 The water in a jet-propelled ship is drawn through inlet openings facing the direction of motion of the ship. It is moving in sea water with a speed of 36 km/hr. The absolute velocity of jet of the water discharged at the back is 25 m/sec and the area of the jet is 0.04 m^2. Find the propelling force and efficiency of propulsion.

Solution: $u = \dfrac{36\text{ Km}}{\text{hr}} = \left(\dfrac{36 \times 1000}{60 \times 60}\right)\text{m/sec} = 10\text{ m/sec}$

$V = 25$ m/sec

$a = 0.04$ m^2

$\therefore$ Propelling force $= [\rho a(V + u) \times V] = [1000 \times 0.04(25 + 10) \times 20]$ N

$= 28000$ N **Ans.**

Efficiency $= \dfrac{2u}{V + 2u} = \dfrac{2 \times 10}{25 + 2 \times 10} = 0.4444 = 44.44\%$ **Ans.**

17.7 CONCLUSION

In order to understand the basic concept of working principles of turbomachines, the knowledge of impact of jet is essential. Keeping this view in mind, dynamic forces, work done by the jet of water on different plates, vanes under their static and moving conditions are derived. The theory and solved examples in this chapter will help the beginners to understand the theory of different turbomachines. The technique of adding or extracting energy from fluid by impacting the jet on the blades or vanes of the rotating wheel has been presented in this chapter. It is felt that content of this chapter is the prerequisite of turbomachines.

PROBLEMS

17.1 A jet of water with diameter 10 cm strikes a stationary flat plate with a velocity of 30 m/sec normally. Find the force exerted by the jet. (**Ans.** 7068.6 N)

17.2 A jet of water of diameter 5 cm with a velocity of 20 m/sec strikes a fixed plate in such a way that the angle between the jet and the plate is 60°. Find the force exerted by the jet on the plate: (i) in the direction normal to the plate, (ii) in the direction of the jet. (**Ans.** 683.13 N, 589 N)

17.3 In an undershoot water wheel, the area of the jet striking the series of radial flat wheel is 0.125 m^2 and velocity of the jet is 5 m/sec. The velocity of the vanes is 2.5 m/sec. Calculate the force exerted on the series of vanes by the jet and the hydraulic efficiency.

(**Ans.** 15622.53 N, 50%)

17.4 A plate is acted upon at its centre by a jet of 2 cm diameter with a velocity of 20 m/sec. The plate is hinged at the top and is deflected through an angle of 15°. Find the weight of the plate. If the plate is not allowed to swing, what will be the force required at the edge of the plate to keep the plate in a vertical position?

(**Ans.** 485.5 N, 62.8 N)

17.5 A jet of water having a velocity of 30 m/sec impinges on a series of vanes moving with a velocity 15 m/sec. The jet makes an angle of 30° to the direction of motion of the vanes when entering, and leaves at an angle of 120° to the direction of motion of vanes. Draw the triangle of velocities at the inlet and outlet, and find: (i) The angles of the vane at the tips so that water enters and leaves the vane without shock, (ii) Work done per kg of water entering the vanes, (iii) Hydraulic efficiency.

(**Ans.** ϕ = 15°40′, 44.16 kg m/sec/kg/water, 96.27%)

17.6 A jet of water having a velocity of 30 m/sec strikes a series of radial vanes mounted on a wheel which is rotating at 200 rpm. The jet makes an angle of 20° with the tangent to the wheel at the inlet and leaves the wheel with a velocity of 5 m/sec at an angle of 130° to the tangent to the wheel at the outlet. Water is flowing from the outward with a radial direction. The outer and inner radii of the wheel are 0.5 m, and 0.25 m respectively. Determine (i) Vane angles at the inlet and the outlet, (ii) Work done per unit weight of water, (iii) Efficiency of the wheel.

(**Ans.** θ = 30°4′, ϕ = 24°24′, η = 69.3%)

17.7 The head of water from the centre of the orifice which is fitted on one side of the tank is maintained at 6 m of water. C_v of the orifice is 0.98, and diameter of the orifice is 15 cm. Find the force exerted by the jet on the tank. If the tank is fitted with a frictionless pulley or wheels and allowed to move with a velocity of 4 m/sec, find the propelling force on the tank and the efficiency of the propulsion. (**Ans.** 45995.5 N, 34%)

REFERENCES

1. Rouse, H. and J.W. Howe, *Basic Mechanics of Fluid*, John Wiley & Sons, Inc., New York, 1953.
2. Shepherd, D.G., *Principles of Turbomachinery*, Macmillan Company, New York, 1956.
3. Subramanya, K., *Theory and Applications of Fluid Mechanics including Hydraulic Machines*, Tata McGraw-Hill, New Delhi, 1992.
4. Bansal, R.K., *Fluid Mechanics and Hydraulic Machines*, Laxmi Publications, New Delhi, 1983.
5. Lal, J., *Hydraulic Machines*, Metropolitan Book Co., New Delhi, 1959.

Chapter 18

Turbomachines: Hydraulic Turbines

18.1 INTRODUCTION

A turbomachine is a rotating machine which adds energy to or extracts energy from a fluid by virtue of a rotating system of blades within the machine. The rotating system is called a runner in a hydraulic turbine, a rotor in a gas or steam turbine, an impeller in a pump and rotor in a compressor. Examples of turbomachines are water, gas or steam turbines, centrifugal axial-flow and mixed flow pumps, compressors, fans and blowers, etc. This chapter contains a comprehensive discussion on hydraulic turbines.

18.2 HYDRAULIC TURBINES

One of the most remarkable developments pertaining to the exploitation of water resources throughout the world is hydroelectric power production. The greatest hydroelectric power developments have taken place in countries like Brazil (93%), Canada (62%) and Norway (99%). In India, the total hydro-potential has been assessed at 84044 MW at 60% Load Factor (LF) against which the present developed capacity stands at 20976 MW, i.e. about 40% of the existing potential. Hydroelectric power has been developed by hydraulic turbines which are hydraulic machines that convert hydraulic energy or hydro-potential into mechanical energy. This mechanical energy developed by turbines is used for running electric generators directly coupled to the shaft of turbines. Thus, by the use of hydraulic turbines, hydropower is finally converted into hydroelectric power for the use of mankind. The generation of hydroelectric power is relatively cheaper than power generated by coal, oil, etc. Greater efforts have been made for the use of hydraulic turbines or development of hydroelectric and multi-power projects in India and other countries.

The idea of utilising hydraulic energy to develop mechanical energy had first been mooted by J.V. Poncelet, who used and undershot wheel made of wood. With the advancement of science and civilisation, modern hydraulic turbines were developed by G. Coriolis and J.B. Francis for reaction turbines by L.A. Pelton for impulse turbines, and by V. Kaplan for propeller type turbine.

In general, a water turbine consists of a wheel called runner (or rotor) having a number of specially designed vanes or blades or buckets. When water with potential or kinetic energy strikes the runner, it works on the runner and causes it to rotate. The mechanical energy so developed is supplied to the generator coupled to the runner shaft, which converts the supplied energy into electrical energy.

18.3 COMMON TERMS ASSOCIATED WITH HYDROPOWER PLANTS

Hydraulic turbines constitute the main element of hydro-power plants. It is essential to know some common terms, their names, use and functions. Figure 18.1 represents a general view of the plan and sectional views of common hydroelectric plants with high head of water. The source of water may be a natural or artificial lake or reservoir formed across a river by a dam at a higher altitude to create the necessary head of water. The water surface in the reservoir is called *head race level*. Water from a reservoir is carried to the power house or turbine by a tunnel or a high pressure conduit. For a high head water power plant, a *surge tank* is provided just upstream of the power house. The main function of the surge tank is to relieve the conduit from high pressure developed when the valves are closed to shut down the turbine. *Penstocks* are usually pipes designed to withstand high pressure with a valve which connects a surge tank and power house (i.e. turbines). For a lower head water power plant, a *forebay* is provided in place of a surge tank.

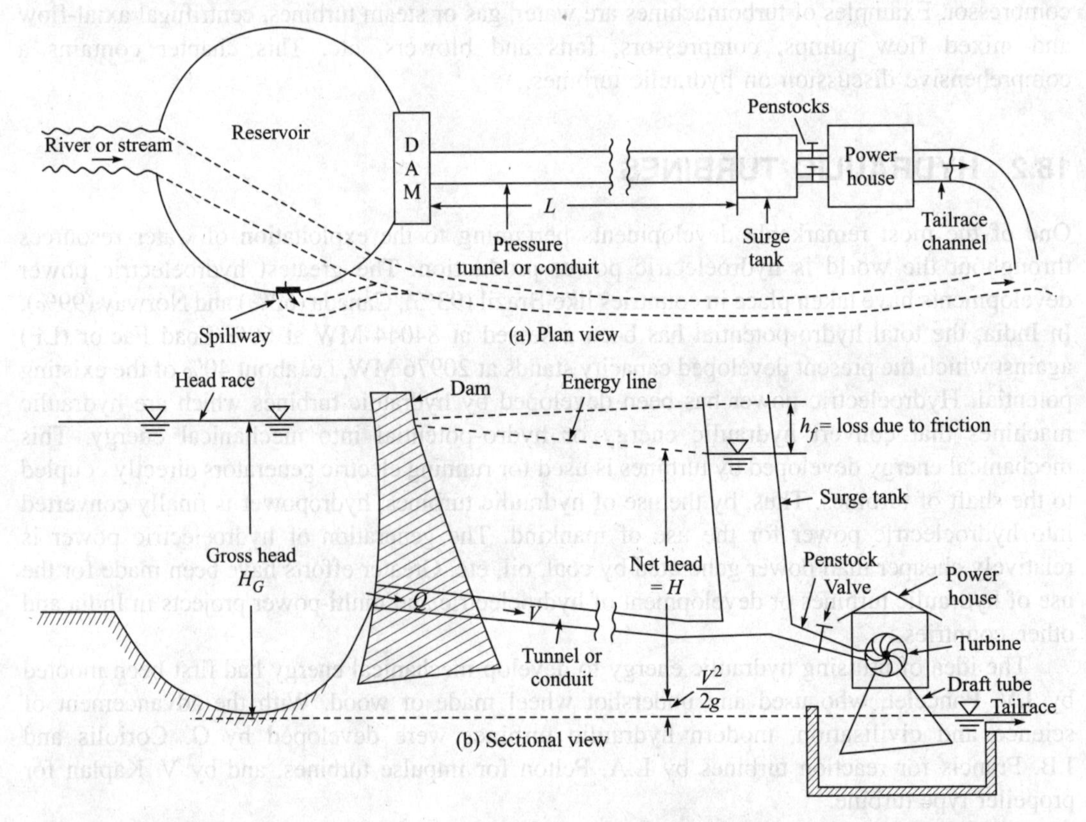

Figure 18.1 Elements of common hydroelectric plants with plan and sectional views.

A forebay is another small storage reservoir connecting upstream the main reservoir and downstream the power house. Water from penstocks discharges on blades or vanes or buckets of the turbines and then goes to the trail race channel through the draft tube, which is usually water-tight.

18.3.1 Heads and Efficiencies of Turbines

(i) *Gross head* (H_G): Gross head H_G shown in Figure 18.1(b) is the difference between the head race and the tail race.

(ii) *Net or effective head* (H) $= H_G - h_f$ (18.1)

where h_f is the head lost due to friction, i.e. $h_f = \dfrac{fLV^2}{2gD}$, D is the diameter of the conduit.

(iii) *Efficiencies*

(a) *Hydraulic efficiency* (η_h): It is the ratio of the power given by the water to the runner to the power supplied by water at the inlet of the turbine.

i.e.
$$\eta_h = \frac{\text{Power delivered to runner}}{\text{Power supplied at inlet}} \qquad (18.2)$$

(b) *Mechanical efficiency* (η_m): Some mechanical losses occur when the power delivered by water to the runner is transmitted to the shaft of the turbine and hence

$$\eta_m = \frac{\text{Power at the shaft of turbine}}{\text{Power delivered by water to the runner}} \qquad (18.3)$$

(c) *Volumetric efficiency* ($\eta_\forall$): The volume of water striking the runner of the turbine is slightly less than the volume supplied to the turbine. A small amount of water escapes to the tail race without striking the runner.

Hence
$$\eta_\forall = \frac{\text{Volume of water actually striking the runner}}{\text{Volume of water supplied to the turbines}}$$

(d) *Overall efficiency* (η_o): $= \dfrac{\text{Shaft power}}{\text{Water power}} = \eta_m \times \eta_h$

$$\therefore \qquad \eta_o = \eta_m \cdot \eta_h \qquad (18.4)$$

18.4 CLASSIFICATION OF TURBINES

Hydraulic turbines are classified based on the basis of the following considerations:

(i) On the basis of hydraulic action or type of energy at the inlet as:
   (a) *Impulse turbine* (Pelton wheel or turbine)
   (b) *Reaction turbine* (Francis turbine)

(ii) On the bssis of the direction of flow through the runner:
   (a) *Tangential flow turbine* (Pelton wheel)
   (b) *Radial flow turbine* (Francis turbine, Thomsen and Girard turbines)

(c) *Axial flow turbine* (Kaplan turbine)
(d) *Mixed flow turbine* (Modern Francis turbine)

(iii) On the basis of the head of water (H):

(a) *High head turbine* (Pelton wheel, $H > 250$ m)
(b) *Medium head turbine* (Modern Francis turbine, 60 m to 250 m)
(c) *Low head turbine* (Kaplan turbine, $H < 60$ m)

(iv) On the basis of the specific speed N_s of the turbine $\left(\dfrac{N\sqrt{P}}{H^{5/4}}\right)$:

(a) *Low specific speed turbine* (Pelton wheel, N_s : 10 to 35)
(b) *Medium specific speed turbine* (Francis turbine, N_s : 60 to 400)
(c) *High specific speed turbine* (Kaplan turbine, N_s : 300 to 1000)

18.5 PELTON WHEEL OR TURBINE

A pelton wheel is the most commonly used impulse or tangential flow turbine. It is named in honour of Lester A. Pelton (1829–1908), an American engineer who contributed immensely towards the development of this turbine. It is quite suitable to be used in high head hydroelectric plants.

Figure 18.2 shows the sketch of a Pelton wheel. The runner is a circular disc with a number of evenly spaced vanes or buckets. The buckets have the shape of double semi-ellipsoidal cups shown in Figures 18.3(a) and 18.3(b). Each bucket is divided into two symmetrical parts by a sharp-edged ridge called splitter.

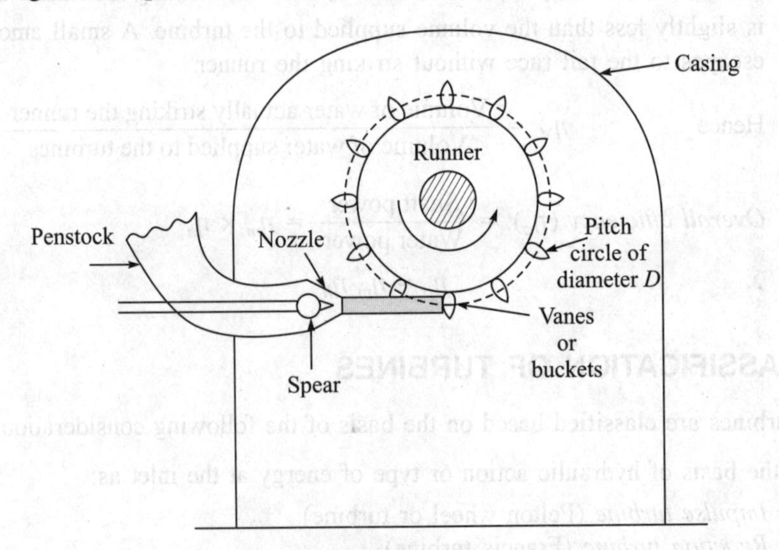

Figure 18.2 Sketch of Pelton turbine.

The jet of water impinges on the splitter dividing the jet into two equal portions and leaves the bucket at its outer edge, as shown in Figure 18.3(b). The buckets are so shaped to ensure that the jet leaves with the angle varying from 10° to 20°. Normally, it is kept at 15°. The jet is deflected through an angle of 160° to 170°. The nozzle is fitted with a spear to control the flow through the nozzle (Figure 18.2).

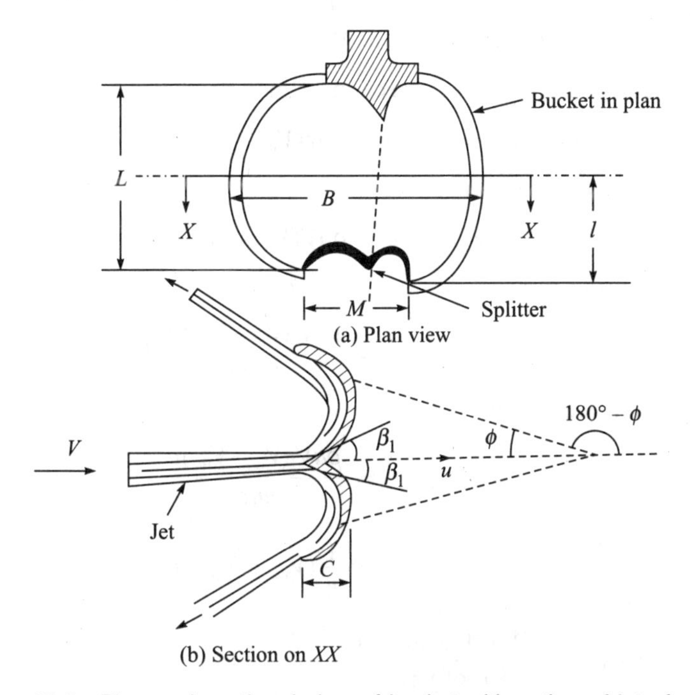

Figure 18.3 Plan and sectional view of bucket with action of jet of water.

The casing shown in Figure 18.2 has no hydraulic function. It is used to prevent the splashing of water.

18.5.1 Velocity Triangles or Inlet, Outlet Diagram

These are shown in Figure 18.4. The diagram at the inlet is a straight line, where $V = Vw$, $V_f = 0$.
$Vr = V - u$, $\alpha = 0$, $\theta = 0$.

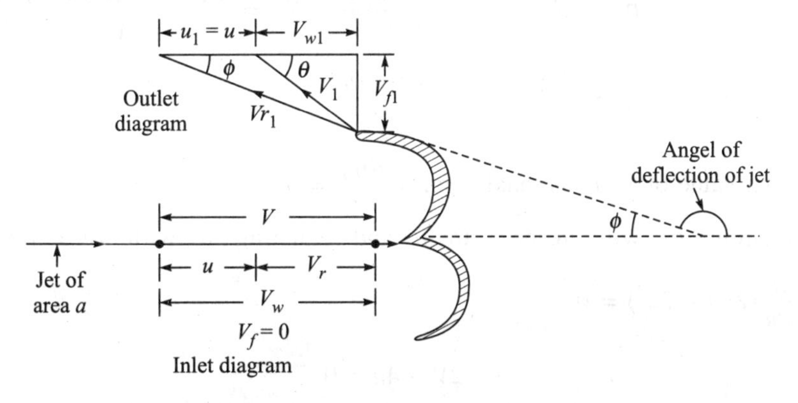

Figure 18.4 Velocity diagrams at inlet and outlet.

At the outlet, $u_1 = u$

Assuming a smooth flow on the bucket, no loss occurs

$$\therefore \qquad Vr_1 = Vr$$

$$Vw_1 = Vr_1 \cos\phi - u_1 = V_r \cos\phi - u = (V - u)\cos\phi - u$$

The force exerted by the jet in the direction of motion

$$F_x = \rho av[V_w - (-V_{w1})] = \rho av(V_w + V_{w1}) \tag{18.5}$$

Work done by the jet on the runner/sec

$$= F_x \times u = \rho aV(V_w + V_{w1})u \text{ N-m/sec} \tag{18.6}$$

Power obtained by the runner

$$= \frac{\rho aV(V_w + V_{w_1})u}{1000} \text{ kW} \tag{18.7}$$

Work done per unit weight of water

$$= \frac{\rho aV(V_w + V_{w_1})u}{\rho gaV}$$

$$= \frac{1}{g}(V_w + V_{w1})u \tag{18.8}$$

$$\text{Hydraulic efficiency } \eta_h = \frac{\text{Work done/sec}}{\text{KE of jet/sec}}$$

$$\eta_h = \frac{\rho aV(V_w + V_{w_1})u}{\frac{1}{2}(\rho aV)V^2} = \frac{2(V_w + V_{w_1})u}{V^2} \tag{18.9}$$

Substituting $V_w = V$ and $V_{w_1} = (V - u)\cos\phi - u$ in Equation (18.9):

$$\eta_h = \frac{2[V + (V - u)\cos\phi - u]u}{V^2}$$

or

$$\eta_h = \frac{2[(V - u) + (V - u)\cos\phi]u}{V^2} = \frac{2(V - u)(1 + \cos\phi)u}{V^2}$$

$$\therefore \qquad \eta_h = \frac{2(V - u)(1 + \cos\phi)u}{V^2} \tag{18.10}$$

For a given value of V, η_h is maximum, if $\dfrac{d\eta_h}{du} = 0$.

Differentiating η_h with respect to u and equating it with zero for maximum,

$$\frac{(1 + \cos\phi)}{V^2}\frac{d}{du}(2Vu - 2u^2) = 0.$$

$$2V - 4u = 0$$

$\therefore$ $$2V = 4u$$

$\therefore$ $$u = \frac{V}{2} \qquad (18.11)$$

It shows that the hydraulic efficiency is the maximum when the velocity of wheel u is half the velocity of jet V.

Substituting $u = \frac{V}{2}$ in Equation (18.10):

$$\eta_{h_{\max}} = \frac{2\left(V - \frac{V}{2}\right)(1+\cos\phi) \times \frac{V}{2}}{V^2}$$

$$= \frac{2 \times \frac{V}{2} \times \frac{V}{2}(1+\cos\phi)}{V^2}$$

$$= \frac{V^2(1+\cos\phi)}{2V^2}$$

$\therefore$ $$\eta_{h_{\max}} = \frac{(1+\cos\phi)}{2} \qquad (18.12)$$

18.5.2 Design or Working Proportions of a Pelton Wheel

(i) If the turbine is working under a net head of H, then $\sqrt{2gH}$ is called the ideal or spouting velocity of the wheel. But the actual velocity is slightly less than $\sqrt{2gH}$ due to the friction loss of the nozzle. Hence velocity V of the jet at the inlet is:

$$V = C_V\sqrt{2gH} \qquad (18.13)$$

The value of the coefficient of velocity C_V ranges from 0.97 to 0.99.

(ii) In Equation (18.11), the velocity of the wheel at the pitch circle is $\frac{1}{2}V$ for maximum hydraulic efficiency η_h. But in actual practice, $\eta_{h\max}$ occurs when $u \simeq 0.46$ V. If u is expressed in terms of the ratio $\phi = \frac{u}{\sqrt{2gH}}$, which is called speed ratio, the actual speed ratio is:

$$\phi = \frac{u}{0.46V} = \frac{u}{0.46 \times 0.98\sqrt{2gH}}$$

$$u = 0.46\ V = 0.45\sqrt{2gH} \qquad (18.14)$$

In practice, ϕ ranges from 0.43 to 0.47.

(iii) For the design or working of a Pelton wheel, the angle through the jet is deflected after the impact on the bucket is taken as 165°, if not mentioned, i.e. ϕ at the outlet velocity triangle is equal to 15°.

(iv) Least diameter d of the jet is obtained as follows:

$$Q = aV = \frac{\pi}{4} d^2 \cdot C_V \sqrt{2gH} = \frac{\pi}{4} d^2 C_V \sqrt{2g}\sqrt{H}$$

or

$$d^2 = \left(\frac{4Q}{\pi C_V \sqrt{2g}\sqrt{H}} \right)$$

$\therefore$

$$d = \left(\frac{4}{\pi \times 0.98 \times \sqrt{2g}} \right)^{1/2} \left(\frac{Q}{\sqrt{H}} \right)^{1/2}$$

or

$$d = 0.5416 \left(\frac{Q}{\sqrt{H}} \right)^{1/2} \quad (18.15)$$

(v) If D is the mean pitch diameter of the pitch circle (shown in Figure 18.2) and N is the rotation of the wheel in rpm., then:

$$u = \frac{\pi DN}{60}$$

$\therefore$

$$D = \frac{60u}{\pi N} = \frac{60 \times \phi \sqrt{2gH}}{\pi N} \quad (18.16)$$

(vi) The ratio $\left(\frac{D}{d} \right)$ is called the jet ratio, which is an important parameter in the design of a Pelton wheel. For maximum efficiency, the value of $\left(\frac{D}{d} \right)$ is taken as 11 to 14. The normal value of $\left(\frac{D}{d} \right)$ is taken as 12 if not mentioned otherwise.

(vii) Dimensions of buckets usually adopted are: (Figure 18.3a) to (18.3b).

$$\left.\begin{aligned} B &= (4 \text{ to } 5)\, d \\ L &= (2.4 \text{ to } 3.2)\, d \\ M &= (1.1 \text{ to } 1.25)\, d \\ l &= (1.2 \text{ to } 1.9)\, d \\ \phi &= 10° \text{ to } 15° \\ \beta_1 &= 5° \text{ to } 8° \\ c &= (0.81 \text{ to } 1.05)\, d \end{aligned}\right\} \quad (18.17)$$

(viii) In order to maintain the volumetric efficiency nearer to unity, the number of buckets (b_n) for the Pelton wheel should be provided in such a way that the jet is always intercepted by buckets. The number of buckets normally provided is more than 15. Taygun developed the following empirical formula for number of buckets (b_n):

$$b_n = \left(\frac{1}{2}\frac{D}{d} + 15\right) = \left(0.5\frac{D}{d} + 15\right) \tag{18.18}$$

But $\frac{D}{d}$, i.e. the jet ratio is usually around 12.

$$\therefore \quad b_n = (0.5 \times 12 + 15) = 21$$

EXAMPLE 18.1 A Pelton wheel has a mean bucket speed of 35 m/sec. The discharge rate of the jet is 1 m^3/sec under a head of 270 m. The bucket deflects the jet through an angle of 170°. Calculate the power delivered to the runner and hydraulic efficiency of the turbine. Assume $C_V = 0.98$. (See Figure 18.5).

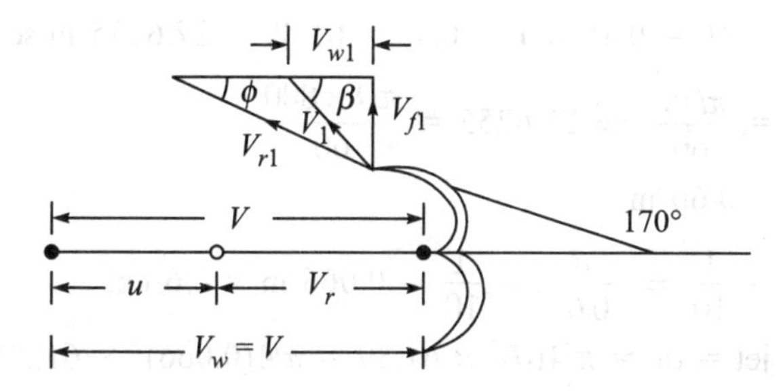

Figure 18.5 Visual of Example 18.1.

Solution: $u = u_1 = 35$ m/sec, $Q = 1$ m^3/sec, $\phi = 180° - 170° = 10°$, $H = 270$ m.

$C_V = 0.98 \quad \therefore$ Velocity of the jet $V = 0.98\sqrt{2gH}$

$$\therefore \quad V = 0.98\sqrt{2 \times 9.81 \times 270} = 71.3275 \text{ m/sec.}$$

$$V_r = V - u = 71.3275 - 35 = 36.3275 \text{ m/sec}$$

From the inlet velocity diagram:

$$V_w = V = 71.3275 \text{ m/sec}$$

Assuming that there is no loss, $V_r = V_{r_1} = 36.3275$ m/sec

$$V_{w_1} = V_{r_1} \cos \phi - u_1 = V_r \cos 10° - u$$

$$V_{w_1} = 36.3275 \cos 10° - 35 = 0.7756 \text{ m/sec}$$

Work done by the jet/sec on the runner $= \rho(av)\,(V_w + V_{w_1})u$

$$= 1000 \times Q(V_w + V_{w_1})u \quad \because aV = Q$$

$$= [1000 \times 1 \times (71.3275 + 0.7756) \times 35] \text{ N-m/sec}$$

$$= 2523608.628 \text{ N-m/sec}$$

Power given to the turbine $= \dfrac{2523608.628}{1000} = 2523.608$ kW **Ans.**

Hydraulic efficiency $\eta_h = \frac{2(V_w + V_{w_1})u}{V^2} = \frac{2(71.3275 + 0.7756) \times 35}{71.3275^2} = 0.9922 = 99.22\%$ **Ans.**

EXAMPLE 18.2 Design a Pelton with the following data. Shaft power = 735.75 kW, $H = 200$ m, $N = 800$ rpm,

$$\eta_0 = 0.86, \ \frac{D}{d} = 10, \ C_V = 0.98, \ \phi = 0.45$$

Determine D, d, and number of jets.

Solution: Shaft Power = 735.75 kW, H = 200 m, N = 800 rpm.

Overall efficiency $\eta_o = 0.86 = 86\%$ $\frac{d}{D} = \frac{1}{10}$, $C_V = 0.98$, $\phi = 0.45$

$$V = C_V\sqrt{2gH} = 0.98\sqrt{2 \times 9.81 \times 200} = 61.39 \text{ m/sec}$$

$$u = u_1 = \phi V = 0.45 \times V = 0.45 \times 61.39 = 27.6255 \text{ m/sec}$$

$$u = \frac{\pi DN}{60} \Rightarrow 27.6255 = \frac{\pi D \times 800}{60}$$

$\therefore$ $D = 0.66$ m **Ans.**

Given that $\frac{d}{D} = \frac{1}{10} \Rightarrow \frac{d}{0.66} = \frac{1}{10} = 0.066$ m = 6.6 cm **Ans.**

Discharge of one jet = $av = \pi/4(d)^2 \times 61.39 = \pi/4(0.066)^2 \times 61.39 = 0.21$ m³/sec

$$\eta_o = \frac{\text{Shaft Power}}{\text{Wheel Power}} = \frac{735.75}{\frac{\rho g Q H}{1000}}$$

or $$0.86 = \frac{735.75 \times 1000}{1000 \times 9.81 \times Q \times 200}$$

$$Q = 0.436 \text{ m}^3\text{/sec}$$

$\therefore$ $$\text{No. of jets} = \frac{Q}{\text{Discharge of one jet}} = \frac{0.436}{0.2} = 2.076 \simeq 2 \text{ Nos.}$$ **Ans.**

EXAMPLE 18.3 A Pelton wheel has a mean bucket diameter 0.8 m and is running at 1000 rpm. The net head of the turbine is 400 m. If the side clearance angle (i.e. ϕ) is 15°, and discharge through the nozzle is 0.15 m³/sec, find the power available at nozzle and hydraulic efficiency.

Solution: $D = 0.8$ m, $N = 1000$ rpm,

$\therefore$ $$u = u_1 = \frac{\pi DN}{60} = \frac{\pi \times 0.8 \times 1000}{60} = 41.88 \text{ m/sec}$$

Net head available = 400 m, $\phi = 15°$, $Q = 0.15$ m³/sec

Assuming value of $C_V = 0.98$

$$V = 0.98 \times \sqrt{2 \times 9.81 \times 400} = 86.817 \text{ m/sec}$$

$$\text{Power available at the nozzle} = \frac{\rho g QH}{1000} = \left(\frac{1000 \times 9.81 \times 0.15 \times 400}{1000}\right) \text{kW}$$

$$= 588.6 \text{ kW}$$ **Ans.**

$$\text{Hydraulic efficiency } \eta_h = \frac{2(V - u)(1 + \cos\phi)u}{V^2}$$

$$= \frac{2(86.817 - 41.88)(1 + \cos 15^\circ) \times 41.88}{86.817^2}$$

$$= 0.98174$$

$$= 98.174\%$$ **Ans.**

EXAMPLE 18.4 The penstock supplies to the Pelton wheel with a gross head of 450 m. The loss of head due to friction in the pipe is one-third of the gross head. Flow through the nozzle is 1.8 m^3/sec. Angle of deflection is 168°. Determine the power given by the water to the runner and η_h. Take speed ratio $\phi = 0.46$, $C_V = 0.98$.

Solution: $H_g = 450$ m, $h_f = \frac{450}{3} = 150$ m, $H = 450 - 150 = 300$ m.

$Q = 1.8$ m^3/sec, $\phi = 180° - 168° = 12°$, $\phi = 0.46$, $C_V = 0.98$

$\therefore \quad V = C_V\sqrt{2gH} = 0.98\sqrt{2 \times 9.81 \times 300} = 75.185$ m/sec

$u = \phi V = 0.46V = 0.46 \times 75.185 = 34.585$ m/sec $= u_1$

$V_r = V - u = 75.185 - 34.585 = 40.6$ m/sec

$V_w = V = 75.185$ m/sec

At the outlet velocity diagram: (See Figure 18.6)

$$V_{r_1} = V_r = 40.6 \text{ m/sec}$$

$$V_{r_1} \cos\phi = u_1 + V_{w_1}$$

or $\quad 40.6 \cos 12° = 34.585 + V_{w_1}$

$\therefore \quad V_{w_1} = 5.1278$ m/sec

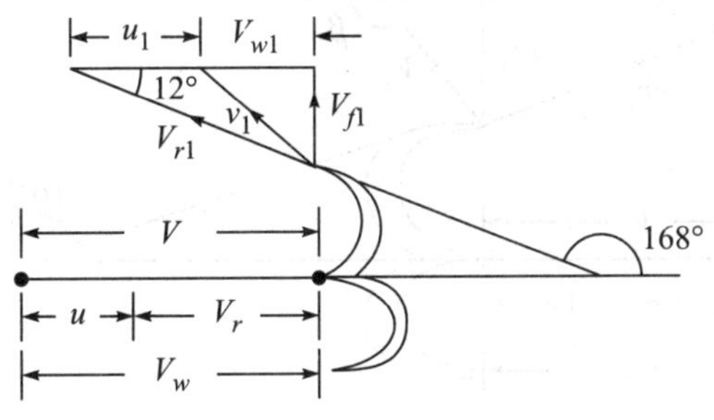

Figure 18.6 Visual of Example 18.4.

Work done by the jet/sec on the runner

$$= \rho aV(V_w + V_{w_1})u$$

$$= 1000 \times 1.8\ (75.185 + 5.1278) \times 34.585$$

$$= 4999712.738 \text{ N-m/sec}$$

Power given by the water to the runner in kW

$$= 4999.712738 \text{ kW}$$

$$\simeq 5000 \text{ kW} \qquad \textbf{Ans.}$$

$$\eta_h = \frac{2(V_w + V_{w_1})u}{V^2} = \frac{2(75.185 + 5.1278) \times 34.585}{75.185^2}$$

$$= 0.98274$$

$$= 98.274\% \qquad \textbf{Ans.}$$

EXAMPLE 18.5 A jet of 0.13 m diameter impinges on a bucket of a Pelton wheel and the jet is deflected through an angle of 165° by the buckets. The head of water is 400 m. The bucket is not smooth and hence there is a 20% reduction of relative velocity. See Figure 18.6(a). If $C_V = 0.97$, $\phi = 0.45$, find the force exerted by the jet in the tangential direction, and the power developed.

Solution: Given that $d = 0.13$ m, $\phi = 15°$, $H = 400$ m, $C_V = 0.97$, $\phi = 0.45$

$$\therefore \quad a = \pi/4\ (0.13)^2 = 0.013273 \text{ m}^2$$

Relative velocity at the outlet $= V_{r_1} = 0.80 \times V_r$

$$V_{r_1} = 0.8(V - u)$$

$$u = u_1 = 0.45\sqrt{2 \times 9.81 \times 400} = 39.865 \text{ m/sec}$$

$$V = 0.97\sqrt{2 \times 9.81 \times 400} = 88.589 \text{ m/sec} = V_w$$

$$\therefore \quad V_{r_1} = 0.8(88.589 - 39.865) = 38.98 \text{ m/sec}$$

$V_{r_1} \cos \phi = 38.98 \cos 15° = 37.65$ m/sec $< u_1$, hence velocity diagram at the outlet is shown in Figure 18.6(a), i.e. β is an obtuse angle.

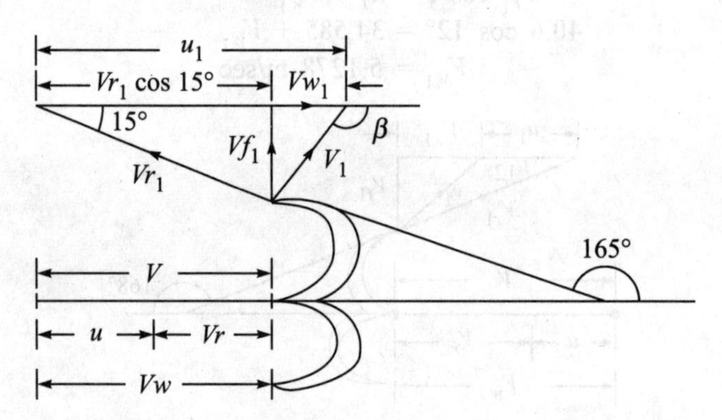

Figure 18.6(a) Visual of Example 18.5.

From the outlet velocity diagram, (Figure 18.6(a))

$$V_{w_1} = u_1 - V_r \cos 15° = 39.865 - 37.65 = 2.215 \text{ m/sec}$$

Force exerted by the jet in the tangential direction:

$$F_x = \rho a V(V_w - V_{w1}) \quad \text{(Here } V_{w1} \text{ is } -V_c \text{ in the same direction of } V_w)$$

$$F_x = 1000 \times 0.013273 \times 88.589(88.589 - 2.215)$$

$$F_x = 101562.16 \text{ N}$$ **Ans.**

$$\therefore \quad \text{Power} = \frac{F_x \times u}{1000} \text{ kW} = \frac{101562.16 \times 39.865}{1000}$$

$$\text{Power} = 4048.775 \text{ kW}$$ **Ans.**

18.6 REACTION TURBINES

In a reaction turbine only a part of the total head of water is converted into velocity head before it reaches the runner. Thus, the pressure of water changes gradually as it passes through the runner. The most commonly used reaction turbines are the Francis turbine and the Kaplan turbine developed by James B. Francis (1815–1892), an American engineer, and V. Kaplan (1876–1934), an Austrian engineer, respectively. Reaction turbines may be radial flow or axial flow turbines. Again radial flow may be an inward flow reaction turbine or an outward flow reaction turbine, depending on whether the water at the inlet flows inward or outward.

18.6.1 Main Parts of Radial Flow Turbine

(i) *Casing*: It is a spiral shape casing in which water from the penstocks enters. The area of the cross-section goes on decreasing (Figure 18.7) gradually to maintain the approximately equal velocity and pressure. It is made of steel and concrete, and is a water-tight casing.

(ii) *Guide wheel and guide blades or vanes*: A guide wheel is stationary and circular in shape and fitted with guide blades to guide the water to the runner vanes. Guide blades help the water strike on the runner blades without shock at the inlet (Figure 18.7).

(iii) *Runner and runner vanes*: A runner is also a circular wheel in which a series of radial curved vanes called runner vanes are fixed. The surface of these vanes is smooth and they are so shaped to ensure that water enters and leaves the runner without shock. They are made of iron or stainless steel. They are keyed to the shaft.

(iv) *Draft tube*: Water passing through a runner flows to the tail race through a pipe to the tail race. This pipe, shown in Figure 18.8, is called draft tube. It is usually diverging in the cross-section and is fully airtight to increase the extra head of water, i.e. upto the tailrace level. The pressure at the exit of the runner of a reaction turbine is generally less than atmospheric pressure. Thus pressure within the draft tube is always negative. Thus, the pipe is gradually increasing in area to connect the runner exit to the tailrace.

There are different types of draft tubes like straight divergent, Moody spreading tube Simple elbow type and Elbow type with circular at the inlet and rectangular at the outlet. (See Figure 18.9).

The most common type is the straight divergent type. It has observed that for this type central cone angle should not be more than 8°. If it is more than 8°, water discharged to the tail

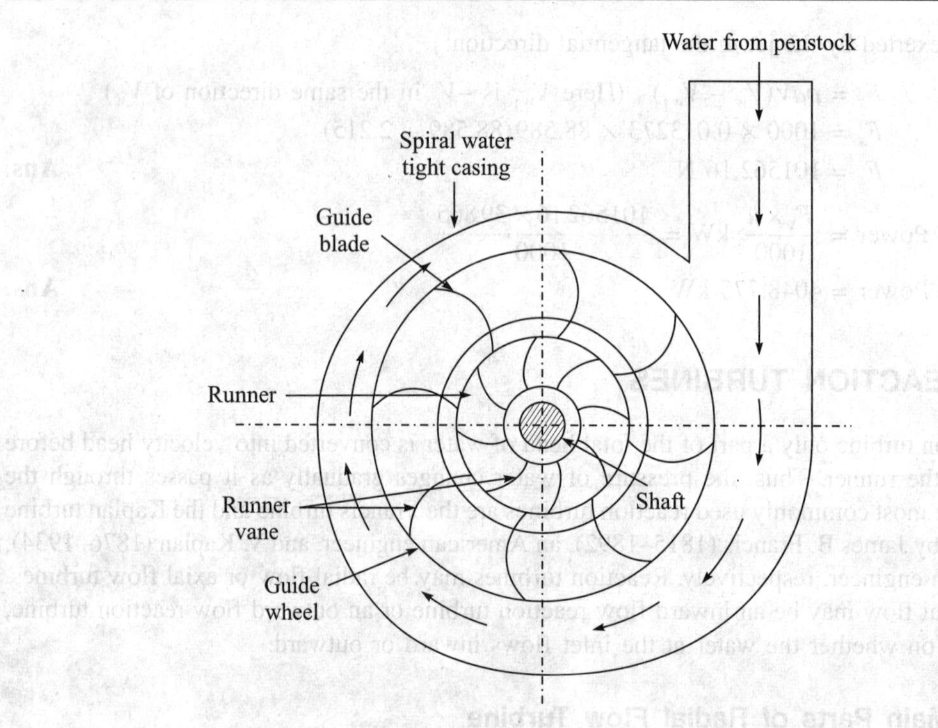

Figure 18.7 Radial flow reaction turbine showing main parts.

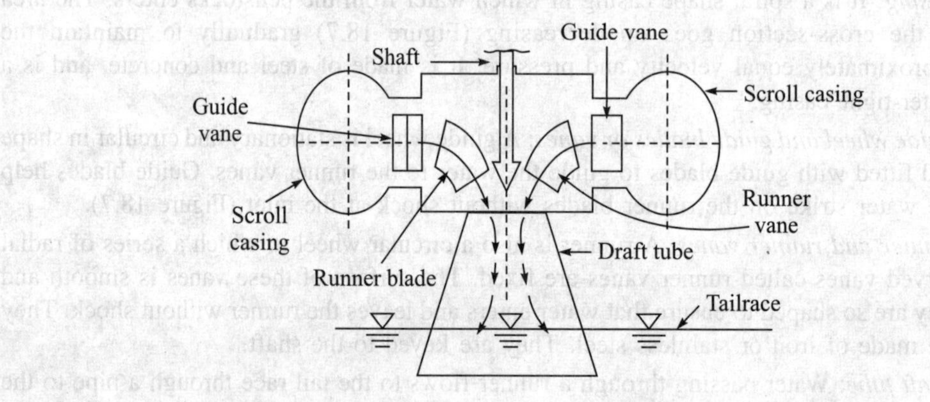

Figure 18.8 Vertical sectional view showing draft tube and other parts.

race does not remain in contact with its runner surface with the result of formation of eddies to reduce the efficiency of the draft tube.

18.7 INWARD FLOW REACTION TURBINE

Figure 18.10 shows the inward flow reaction turbine in which water from the casing enters the stationary guide wheel with guiding vane.

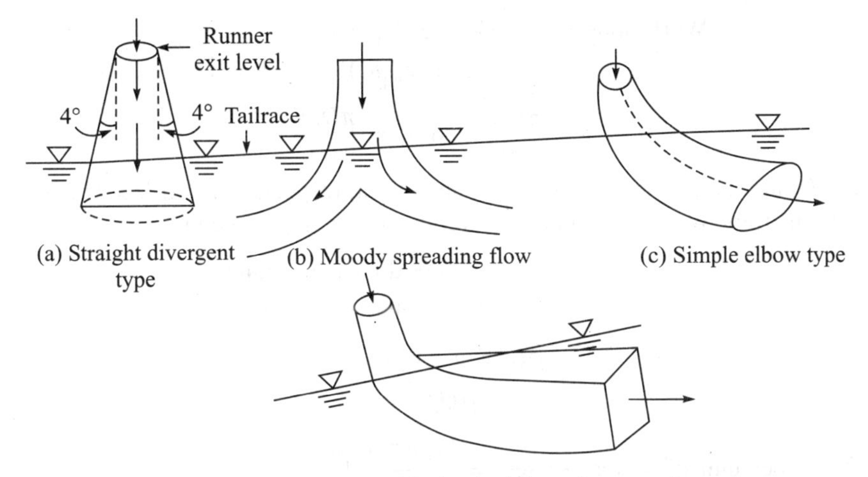

Figure 18.9 Different types of draft tubes.

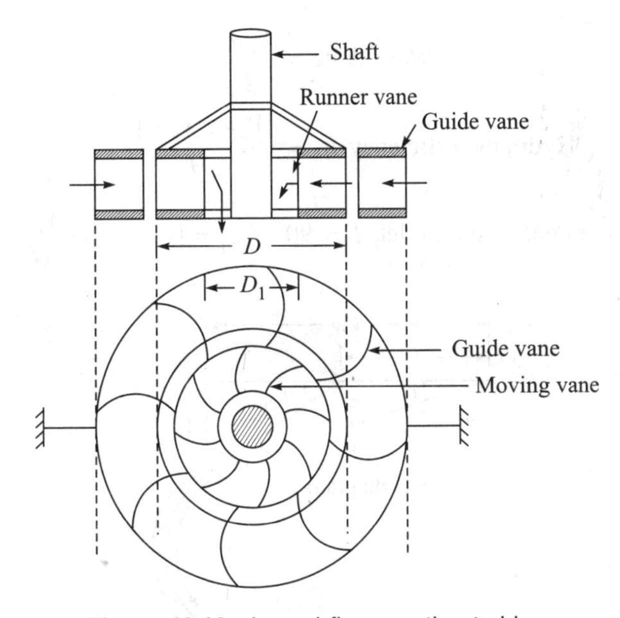

Figure 18.10 Inward flow reaction turbine.

The guide vanes guide the water to enter the runner vanes. Water flows in an inward radial direction and discharges at the inner diameter D_1 of the runner. The outer diameter D is at the inlet and the inner diameter D_1 is at the outlet.

18.7.1 Velocity Triangles and Work Done by the Runner

We have already worked the general expression of work done by the water on the curve vane as follows:

$$\text{Work done} = \rho a V(V_w u \pm V_{w_1} u_1)$$
$$= \rho Q(V_w u \pm V_{w_1} u_1) \quad \because aV = Q$$

Here
$$u = \frac{\pi DN}{60} \text{ and } u_1 = \frac{\pi D_1 N}{60}$$

where D and D_1 are the diameters at the inlet and outlet, respectively and $D > D_1$, N is the speed of the turbine in rpm. Work done per unit weight of water per second:

$$= \frac{\text{Work done per second}}{\text{Weight of water striking per second}}$$

$$= \frac{\rho Q[V_w u \pm V_{w_1} u_1]}{\rho g Q}$$

$$\therefore \quad \text{Work done per unit of water per sec.} = \frac{[V_w u \pm V_{w_1} u_1]}{g} \tag{18.19}$$

when β is the acute angle, it is a positive sign, when β is obtuse angle, it is a negative sign. When $\beta = 90°$, $V_{w_1} = 0$.

$$\therefore \quad \text{Work done} = \frac{V_w u}{g} \tag{18.19a}$$

$$\text{Hydraulic efficiency } \eta_h = \frac{[V_w u \pm V_{w_1} u]}{gH} \tag{18.20}$$

If the discharge is radial at the outlet, $\beta = 90°$, $V_{w_1} = 0 \therefore \eta_h = \frac{V_w u}{gH}$ (18.21)

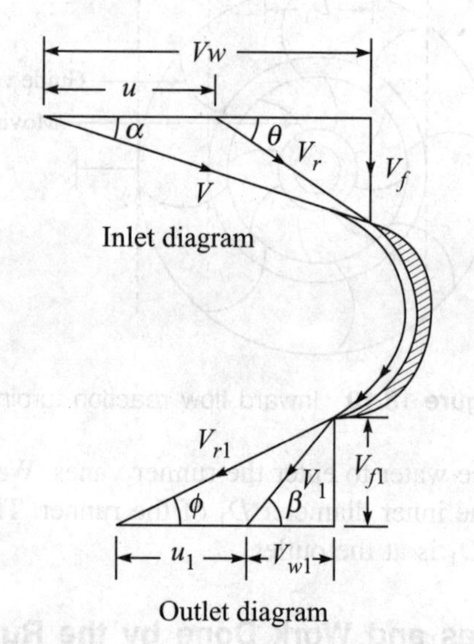

Figure 18.11 Inlet and outlet velocity triangles for inward flow reaction turbine.

18.7.2 Some Terms and Conditions Used in Reaction Turbines

(i) Speed ratio $= \dfrac{u}{\sqrt{2gH}}$ (18.22)

(ii) Flow ratio $= \dfrac{V_f}{\sqrt{2gH}}$ (18.23)

(iii) Discharge equation: $Q = \pi DBV_f = \pi D_1B_1V_f$ (18.24)
B and B_1 are widths of the runner at the inlet and outlet.

(iv) If t is the thickness of the vanes and this t is considered and if there are n number of vanes, then

$$Q = (\pi D - nt)BV_f \quad (18.25)$$

(v) If p is the pressure at the inlet, then $H = \dfrac{p}{\rho g} + \dfrac{V^2}{2g}$ (18.26)

(vi) When the flow is radial at the outlet, $\beta = 90°$, $V_{w_1} = 0$ and when the flow is radial at the inlet, $\alpha = 90°$, $V_w = 0$.

(vii) Assuming that there is no loss of energy when water flows through the vanes, $H - \dfrac{V_1^2}{2g}$

$$= \frac{1}{g}(V_wu \pm V_{w_1}u_1) \quad (18.27)$$

EXAMPLE 18.6 An inward flow reaction turbine has external and internal diameters as 1.2 m and 0.6 m, respectively. The velocity of flow through the runner is constant and is equal to 1.8 m/sec. Determine: (i) The discharge through the runner.

(ii) Width at the outlet if width at the inlet is 20 cm.

Solution: External diameter $D = 1.2$ m, Internal diameter $D_1 = 0.6$ m, $V_{f_1} = V_f = 1.8$ m/sec, $B = 20$ cm.

Let the width at outlet be B_1:

$$Q = \pi DBV_f = (\pi \times 1.2 \times 0.2 \times 1.8) \text{ m}^3\text{/sec}$$

Writing discharge:

$$Q = 1.357168 \text{ m}^3\text{/sec}$$ **Ans.**

Again equating discharge at the inlet and outlet:

$$\pi DBV_f = \pi D_1B_1V_f$$

But $$V_f = V_{f_1}$$

$\therefore$ $$DB = D_1B_1$$

$\therefore$ $$B_1 = \frac{DB}{D_1} = \frac{1.2 \times 0.2}{0.6} = 0.4 \text{ m}$$

$$= 40 \text{ cm}$$ **Ans.**

EXAMPLE 18.7 A reaction turbine works at 500 rpm under a head of 100 m. The diameter of the turbine at the inlet is 100 cm and the flow area is 0.35 m². The angle made by the absolute and relative velocities at the inlet are 15° and 60°, respectively with the tangential velocity direction. Determine: (a) Q (b) Power developed (c) Efficiency. Assume that the velocity of whirl at the outlet is *zero*.

Solution: The 'data' are: N = 500 rpm, H = 100 m, D = 1 m flow area = πDB = 35 m², $\alpha = 15°$, $\theta = 60°$, $V_{w_1} = 0$.

Velocity diagrams at the inlet and outlet are drawn:

$$u = \frac{\pi DN}{60} = \frac{\pi \times 1 \times 500}{60} = 26.18 \text{ m/sec}$$

From the inlet diagram,

$$\tan 15° = \frac{V_f}{V_w} \quad \therefore \quad \frac{V_f}{V_w} = 0.268$$

and

$$V_f = 0.268\ V_w$$

$$\tan 60° = \frac{V_f}{V_w - u} = \frac{V_f}{V_w - 26.18}$$

or

$$1.732 = \frac{0.268\,V_w}{V_w - 26.18}$$

or

$$1.732\ V_w - 45.345 = 0.268\ V_w$$

or

$$1.464\ V_w = 45.345$$

$$\therefore \quad V_w = 30.973 \text{ m/sec}$$

$$\because \quad V_f = 0.268\ V_w$$

$$\therefore \quad V_f = 0.268 \times 30.973 = 8.3 \text{ m/sec}$$

$$\therefore \quad Q = (\pi DB)V_f = (.35 \times 8.3) \text{ m}^3\text{/sec} \quad \because \pi DB = 0.35 \text{ m}^2$$

$$\therefore \quad Q = 2.9053 \text{ m}^3\text{/sec}$$ **Ans.**

$$\text{Power developed} = \frac{\text{Work done}}{1000} \text{ kW} = \frac{\rho Q(V_w u)}{1000} \quad \because V_{w_1} = 0$$

$$= \frac{1000 \times 2.9053 \times 30.973 \times 26.18}{1000} \text{ kW}$$

$$= 2355.83 \text{ kW}$$ **Ans.**

$$\text{Efficiency } (\eta_G) = \frac{V_w u}{gH} \quad \because V_{w_1} = 0$$

$$= \frac{30.973 \times 26.18}{9.81 \times 100} = 0.826578$$

$$= 82.6578\%$$ **Ans.**

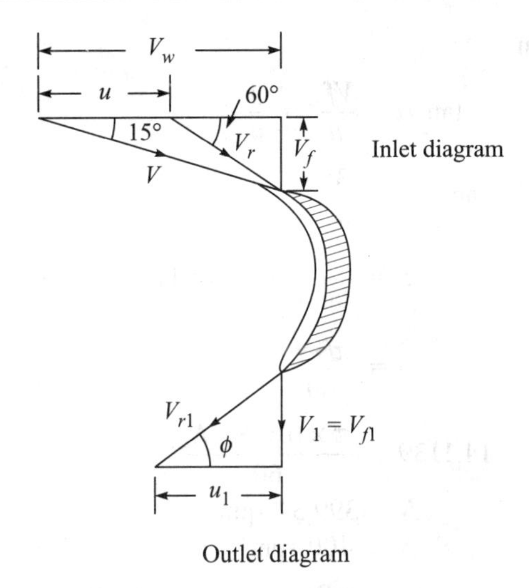

Figure 18.12 Visual of Example 18.7.

EXAMPLE 18.8 An inward flow turbine has an external diameter of runner equal to 67.5 cm and corresponding width or breadth of 15 cm. The effective head is 21 m. If the velocity of flow at the inlet is 3 m/sec, and the guide vane angle 12° runner vane is radial at the inlet, find the speed of the turbine, outlet vane angle and power developed. Assume that the turbine discharges radially at outlet also.

Solution: Given that D = 67.5 cm, B = 15 cm = 0.15 m, D_1 = 50 cm = 0.5 m

B_1 = 22.5 cm = 0.225 m, Vane angle radial at inlet, hence $V_w = u \quad \because \beta = 90°, V_r = V_f$.

Discharge is a radial at the outlet, hence $V_{w_1} = 0$, $V_1 = V_{f_1}$, $\alpha = 12°$.

Velocity diagrams are drawn (as shown in Figure 18.13).

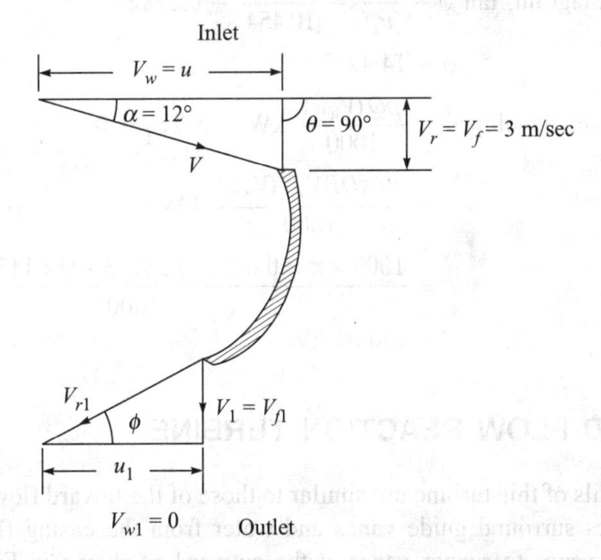

Figure 18.13 Visual of Example 18.8.

From the inlet diagram

$$\tan \alpha = \frac{Vf}{u} = \frac{3}{u}$$

$\therefore$ $$\tan 12° = \frac{3}{u}$$

$\therefore$ $$u = \frac{3}{\tan 12°} = 14.1139 \text{ m/sec}$$

But $$u = \frac{\pi DN}{60}$$

or $$14.1139 = \frac{\pi \times 0.675 \times N}{60}$$

$\therefore$ $$N = 399.34 \text{ rpm}$$

$$N \simeq 400 \text{ rpm}$$ **Ans.**

$$\frac{u}{u_1} = \frac{\pi DN}{60} \times \frac{60}{\pi D_1 N} = \frac{D}{D_1} = \frac{0.675}{0.5}$$

or $$\frac{14.1139}{u_1} = \frac{0.675}{0.5}$$

$\therefore$ $$u_1 = 10.454 \text{ m/sec}$$

Equating Q: $$Q = \pi D V_f B = \pi D_1 V_{f_1} B_1$$

$\therefore$ $$V_{f_1} = \frac{DB}{D_1 B_1} V_f = \frac{0.675 \times 0.15}{0.5 \times .225} \times 3$$

$\therefore$ $$V_{f_1} = 2.7 \text{ m/sec}$$

From the outlet diagram, $\tan \phi = \dfrac{V_{f_1}}{u_1} = \dfrac{2.7}{10.454} = 0.25827$

$\therefore$ $$\phi = 14.481°$$ **Ans.**

$$\text{Power developed} = \frac{\rho Q (V_w u)}{1000} \text{ kW} \quad \because V_{w_1} = 0$$

$$= \frac{\rho (\pi DBV_f)(V_w u)}{1000} \text{ kW}$$

$$= \frac{1000 \times \pi \times 0.675 \times 0.15 \times 3 \times (14.1139 \times 14.1139)}{1000} \text{ kW}$$

$$= 190.09 \text{ kW}$$ **Ans.**

18.8 OUTWARD FLOW REACTION TURBINE

The construction details of this turbine are similar to those of the inward flow turbine. But in this turbine, moving vanes surround guide vanes and water from the casing flows to the outward direction from guide vanes to runner vanes at the outward as shown in Figure 18.14.

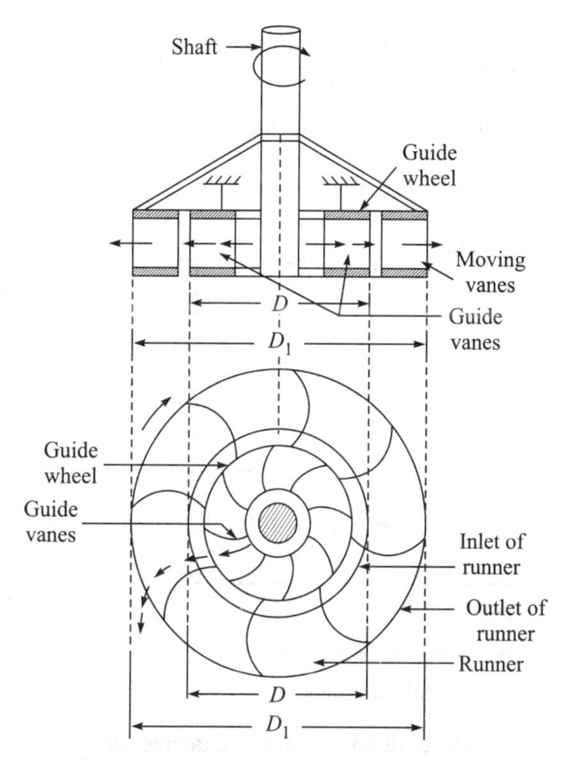

Figure 18.14 Outward flow reaction turbine.

The stationary guide wheel with its guide vanes directs the water to enter the runner. Here inlet is at the inner diameter D of the runner and the outlet is at the outer diameter D_1. The velocity diagram of the outward flow reaction turbine will be drawn in the same manner as that for the inward flow turbine. The work is done by the water on the runner per second, the power developed and hydraulic efficiency will be obtained from the velocity triangles. In this turbine, as the inlet diameter D of the runner is less than the outlet diameter D_1,

so
$$u < u_1$$

EXAMPLE 18.9 An outward flow reaction turbine has internal and external diameters of runner as 0.5 m and 1 m, respectively. The guide blade angle is 15° and velocity of flow through the runner is constant and equal to 3 m/sec. If the speed of the turbine is 250 rpm, the head on turbine is 10 m, and discharge at outlet is radial, determine: (i) Runner vane angles at the inlet and outlet, (ii) Work done by the water on the runner per second per unit weight striking per second and hydraulic efficiency.

Solution: The given data are: $D = 0.5$ m, $D_1 = 1$ m, $\alpha = 15°$, $V_f = V_{f_1} = 3$ m/sec

$N = 250$ rpm, $H = 10$ m, Discharge at the outlet is radial, i.e. $Vw_1 = 0$, $V_{f_1} = V_1$

tangential velocity at the inlet $u = \dfrac{\pi DN}{60} = \dfrac{\pi \times 250 \times 0.5}{60}$

$$u = 6.545 \text{ m/sec}$$

tangential velocity at the outlet $u_1 = \dfrac{\pi \times 1 \times 250}{60} = 13.090$ m/sec

Inlet and outlet velocity diagrams are drawn. (See Figure 18.15).

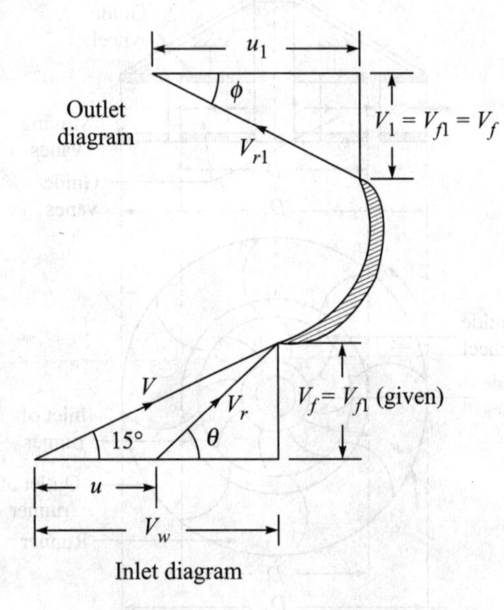

Figure 18.15 Visual of Example 18.9.

From the inlet diagram:

$$\tan 15° = \frac{Vf}{Vw} = \frac{3}{Vw}$$

$$\therefore \quad V_w = \frac{3}{\tan 15°} = 11.196 \text{ m/sec}$$

Runner vane angle at the inlet is θ.

$$\tan \theta = \frac{V_f}{V_w - u} = \frac{3}{11.196 - 6.545} = 0.645$$

$$\therefore \quad \theta = 32.823°$$ **Ans.**

Runner vane angle at the outlet is ϕ.

$$\tan \phi = \frac{V_{f_1}}{u_1} = \frac{3}{13.09} = 0.22918$$

$$\therefore \quad \phi = 12.9°$$ **Ans.**

Work done by water per second per unit weight striking per second

$$= \frac{V_w u}{g} \quad \because V_{w1} = 0$$

$$= \frac{11.196 \times 6.545}{9.81} = 7.4697 \text{ N-m/sec}$$ **Ans.**

$$\eta_h = \frac{V_w u}{gH} = \frac{7.4697}{10} = 0.74697 = 74.697\%$$ **Ans.**

EXAMPLE 18.10 The peripherial velocity at the inlet of and outward flow reaction turbine is 12 metre/sec. The internal diagram is 0.8 times the external diameter. The vanes are radial at entrance and vane angle at outlet is 20°. The velocity of flow through the runner at outlet is 4 m/sec. If the final discharge is radial and the turbine is situated 1 m below the tail water level, determine: (i) Guide vane angle, (ii) Absolute velocity of water leaving the guides, (iii) The head on the turbine and (iv) Hydraulic efficiency.

Solution: The given data are: u = 12 m/sec, $D = 0.8\ D_1$ i.e. $\left(\frac{D_1}{D}\right) = \frac{1}{0.8}$

Vanes are radial at the inlet, i.e. $\theta = 90°$ (See Figure 18.16)

and $\phi = 20°$, $V_f = 4$ m/sec

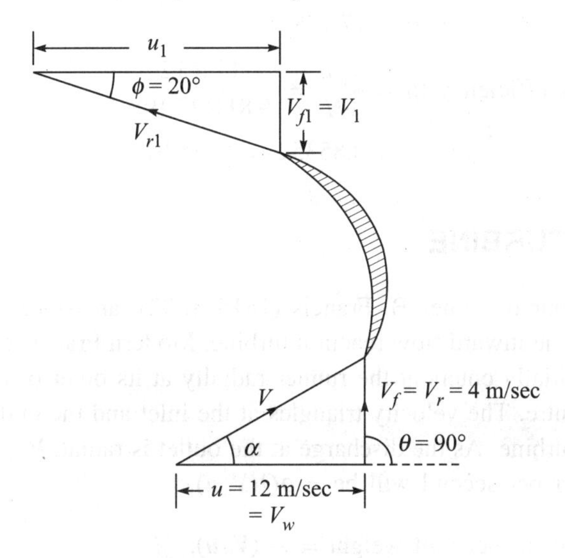

Figure 18.16 Visual of Example 18.10.

Final discharge is radial, $\beta = 90°$ i.e. $V_1 = V_{f1}$, $V_{w1} = 0$
Pressure head at the outlet = 1 m
From the inlet diagram, $u = V_w = 12$ m/sec

$$\tan\alpha = \frac{V_f}{V_w} = \frac{4}{12}$$

$\therefore$ $$\alpha = 18°26'$$ **Ans.**

$$V = \sqrt{V_w^2 + V_f^2} = \sqrt{12^2 + 4^2} = 12.6491 \text{ m/sec}$$ **Ans.**

If N is the speed of turbine in rpm:

$$u = \frac{\pi DN}{60} \quad \therefore N = \frac{60u}{\pi D}$$

and $$u_1 = \frac{\pi DN}{60} \quad \therefore N = \frac{60u_1}{\pi D_1}$$

Equating: $$N, \frac{60u}{\pi D} = \frac{60u_1}{\pi D_1}$$

$\therefore$ $$u_1 = \left(\frac{D_1}{D}\right) u = \left(\frac{1}{0.8}\right) \times 12 = 15 \text{ m/sec}$$

$$V_1 = V_{f1} = u_1 \tan 20° = 15 \tan 20° = 5.4595 \text{ m/sec}$$

Head on turbine H = Energy head at outlet + Work done per unit weight per second + losses

$$\text{Neglecting losses, } H = \left(\frac{V_1^2}{2g} + 1\right) + \left(\frac{V_w u}{g}\right) = \left(\frac{5.4595^2}{2 \times 9.81} + 1\right) + \left(\frac{12 \times 12}{9.81}\right)$$

$$H = 17.198 \text{ m}$$

$\therefore$ $$\text{Hydraulic efficiency } \eta_h = \frac{V_w u}{gH} = \frac{12 \times 12}{9.81 \times 17.198}$$

$$= 0.85352 = 85.352\%$$ **Ans.**

18.9 FRANCIS TURBINE

This is named in honour of James B. Francis (1815–1892), an American engineer who first designed it in 1849 as the inward flow reaction turbine. Modern Francis turbine is a mixed flow turbine. Here water initially enters at the runner radially at its outer or outward periphery and leaves axially at its centre. The velocity triangles at the inlet and the outlet are the same as the inward flow reaction turbine. As the discharge at the outlet is radial, $V_{w_1} = 0$, hence work done by water on the runner per second will be = $\rho Q(V_w u)$.

and work done per second per unit weight = $\frac{1}{g}(V_w u)$.

$$\text{Hydraulic efficiency } \eta_h = \frac{V_w u}{gH}$$

The ratio of width of wheel to its diameter = $\frac{B}{D}$ (0.1 to 0.4)

$$\text{Flow ratio} = \frac{V_f}{\sqrt{2gH}} \text{ [0.15 to 0.3]}$$

$$\text{Speed ratio} = \frac{u}{\sqrt{2gH}} \text{ (0.6 to 0.9)}$$

EXAMPLE 18.11 A Francis turbine with an overall efficiency of 70% is required to produce 147.15 kW. It is working under a head of 8 m. The peripherial velocity = $0.3\sqrt{2gH}$ and radial velocity of flow at the inlet is $0.96\sqrt{2gH}$. The wheel runs at 200 rpm and the hydraulic losses in the turbine are 20% of the available energy. Assuming radial discharge, determine: (i) Guide blade angle, (ii) The wheel vane angle at the inlet, (iii) Diameter of the wheel at inlet, and (iv) Width of the wheel at the inlet.

Solution: The given data are: η_o = 75% = 0.75, Power produced = 147.15 kW,

$$H = 8 \text{ m}, \; u = 0.3\sqrt{2gH} = 0.3\sqrt{2 \times 9.81 \times 8} = 3.7585 \text{ m/sec}$$

$V_f = 0.96\sqrt{2gH}$ = 12.0272 m/sec, N = 200 rpm, Hy. Losses = 20% of the available energy, radial discharge i.e. $V_{f_1} = V_1$ and $V_{W_1} = 0$.

Now
$$\eta_h = \frac{\text{Total head} - Hy \text{ Loss}}{\text{Head at inlet}} = \frac{H - 0.2H}{H} = \frac{0.8H}{H} = 0.8$$

But
$$\eta_h = \frac{V_w u}{gH} \quad \therefore V_w = \frac{\eta_h gH}{u} = \frac{0.8 \times 9.81 \times 8}{3.75} = 16.7424 \text{ m/sec}$$

From the inlet diagram, (Figure 18.17)

$$\tan \alpha = \frac{Vf}{Vw} = \frac{12.0272}{16.7424} = 0.71836$$

$$\therefore \quad \alpha = 35°43' \qquad \textbf{Ans.}$$

$$\tan \theta = \frac{Vf}{V_w - u} = \frac{12.0272}{16.7424 - 3.75}$$

$$\tan \theta = 0.92571$$

$$\therefore \quad \theta = 42°47' \qquad \textbf{Ans.}$$

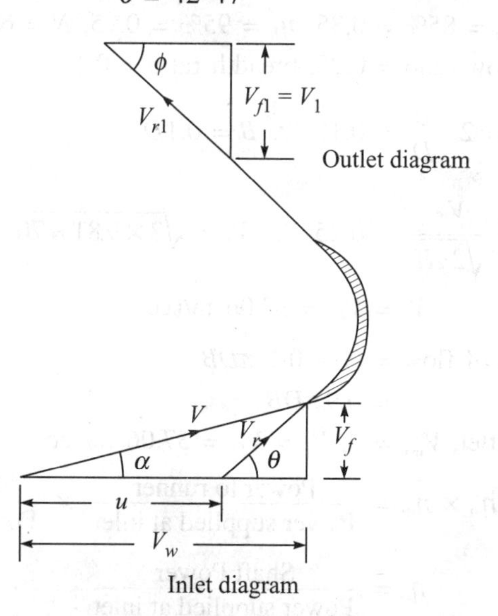

Figure 18.17 Visual of Example 18.11.

We know $\eta = \dfrac{\pi DN}{60}$ $\quad \therefore D = \dfrac{60u}{\pi N} = \dfrac{60 \times 3.75}{\pi \times 200} = 0.358$ m **Ans.**

$$\text{We know} \quad \eta_o = \eta_h \times \eta_m = \eta_h \times \frac{\text{Power at the shaft}}{\text{Power delivered by water to the runner}}$$

$$0.75 = 0.8 \times \frac{147.15}{\dfrac{W}{g}\left(\dfrac{V_w u}{1000}\right)}$$

$$0.75 = \frac{0.8 \times 147.15 \times 1000}{1000 \times Q \times \left(\dfrac{16.7424 \times 3.75}{1000}\right)}$$

$$\because \quad \frac{W}{g} = \frac{\rho g Q}{g}$$

Solving for $\quad Q = 2.54846 \text{ m}^3/\text{sec}$

$$\therefore \quad Q = \pi DBVf$$

$$2.54846 = \pi \times 0.358 \times B \times 12.0272$$

$$\therefore \quad B = 0.1884 \text{ m} = 18.84 \text{ cm}$$ **Ans.**

EXAMPLE 18.12 The following data are given for a Francis turbine. Net $H = 70$ m, $N = 600$ rpm, shaft power = 367.875 kW, $\eta_o = 85\%$, $\eta_h = 95\%$, flow ratio = 0.25, breadth ratio = 0.1, outer diameter of the runner = 2 × inner diameter of runner. The thickness of vanes occupies 10% of the circumferential area of the runner. The velocity of flow is constant at the inlet and the outlet. Determine the guide blade angle, runner vane angle at the inlet and outlet, diameters of the runner at the inlet and the outlet, and width of the wheel at inlet.

Solution: The data given are: $H = 70$ m, $N = 600$ rpm, shaft power = 367.875 kW,

$$\eta_o = 85\% = 0.85, \ \eta_h = 95\% = 0.95, \ N = 600 \text{ rpm}$$

Flow ratio = 0.25, breadth ratio = 0.1

$$\frac{D}{D_1} = 2, \ \frac{B}{D} = 0.1, \quad \therefore B = 0.1D$$

$$\frac{V_f}{\sqrt{2gH}} = 0.25 \quad \therefore V_f = \sqrt{2 \times 9.81 \times 70} = 37.06 \text{ m/sec}$$

$$\therefore \quad V_f = V_{f_1} = 37.06 \text{ m/sec}$$

$$\text{Actual area of flow} = (1 - 0.1)\pi DB$$

$$= 0.9\pi DB$$

Discharge radial at outlet, $V_{w1} = 0$, $V_1 = V_{f_1} = 37.06$ m/sec

$$\text{Overall efficiency } \eta_o = \eta_h \times \eta_m = \frac{\text{Power to runner}}{\text{Power supplied at inlet}} \times \frac{\text{Shaft Power}}{\text{Power to runner}}$$

$$\therefore \quad \eta_o = \frac{\text{Shaft Power}}{\text{Power supplied at inlet}}$$

$$0.85 = \frac{367.875}{\text{Power supplied at inlet}}$$

$\therefore$ Power supplied at inlet = 0.85 × 367.875 = 312.693 kW

or
$$\frac{\rho g Q H}{1000} = 312.693$$

$$\frac{1000 \times 9.81 \times Q \times 70}{1000} = 312.693$$

$$Q = 0.4553 \text{ m}^3\text{/sec}$$

$$Q = \text{Actual area of flow} \times \text{Velocity of flow}$$

$$0.4553 = 0.9\ \pi DB\ V_f$$

$$0.4553 = 0.9\pi D \times 0.1D \times 37.06 = 3.3354\ D^2$$

$\therefore$
$$D = 0.3694 \text{ m}$$

But
$$\frac{B}{D} = 0.1 \quad \therefore B = .1 \times 0.3694 = 0.03694 \text{ m} = 3.694 \text{ cm}$$

Now
$$u = \frac{\pi DN}{60} = \frac{\pi \times 0.3694 \times 600}{60} = 11.605 \text{ m/sec}$$

Now
$$\eta_o = \frac{V_w u}{gH} = \frac{V_w \times 11.605}{9.81 \times 70}$$

or
$$0.93 = \frac{V_w \times 11.605}{9.81 \times 70}$$

$\therefore$
$$V_w = 55.03 \text{ m/sec}$$

From the inlet velocity diagram: (Figure 18.18)

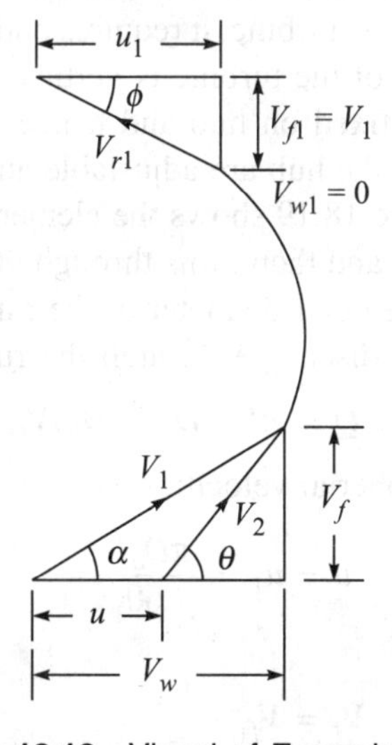

Figure 18.18 Visual of Example 18.12.

$$\tan \alpha = \frac{Vf}{Vw} = \frac{37.06}{55.03} = 0.6734$$

$\therefore$ $\alpha = 33°57.5'$ **Ans.**

$$\tan \theta = \frac{Vf}{V_w - u} = \frac{37.06}{55.03 - 11.605} = 0.8534$$

$\therefore$ $\theta = 40°29'$ **Ans.**

From the outlet velocity diagram,

$$\tan \phi = V_{f1}/u_1 = \frac{37.06}{u_1} \quad \because V_f = V_{f1}$$

$$u_1 = \frac{\pi D_1 N}{60} = \frac{\pi D \times N}{2 \times 60} = \frac{\pi \times 0.3694 \times 600}{2 \times 60} = 5.802 \text{ m/sec}$$

$\therefore$ $$\tan \phi = \frac{37.06}{5.802} = 6.38745$$

$\therefore$ $\phi = 81°6'$ **Ans.**

Diameter of runner at inlet $D = 0.3694$ m **Ans.**

Diameter of runner at outlet $D_1 = 0.1847$ m

Width at the inlet $= B = 0.1D = 0.03694$ m

$= 3.694$ cm **Ans.**

18.10 KAPLAN TURBINE

A Kaplan turbine is a type of propeller type turbine developed by the Austrian engineer, V. Kaplan (1870–1934). It is an axial flow turbine that is relatively suitable for low heads and therefore, the amount of discharge is high enough to produce larger power.

Since it is also one type of reaction turbine, it requires entirely a closed conduit from the head race to the tail race. Here the shaft of the turbine is vertical. The lower portion is larger and is called hub or boss. The vanes are fixed on hub and hence the hub acts as runner. In the case of the Kaplan turbine, the vanes on the hub are adjustable and if the vanes are not adjustable, it is called a 'propeller turbine'. Figure 18.19 shows the elements of a Kaplan turbine. Water from the scroll casing enters guide vanes and then turns through 90° as shown in Figure 18.19(a) and flows axially the runner. If D_0 is the outer diameter of the runner, D_h is the diameter of the hub, V_f is the velocity of flow, then the discharge through the runner is:

$$Q = \pi/4 \ (D_0^2 - D_h^2) V_f \tag{18.28}$$

In the Kaplan turbine, the peripherial velocities are:

$$u = u_1 = \frac{\pi D_0 N}{60} \tag{18.29}$$

Velocities of flow are:

$$V_f = V_{f1} \tag{18.30}$$

Area of flow at the inlet = area of flow at the outlet = $\pi/4(D_0 - D_h)^2$ (18.31)

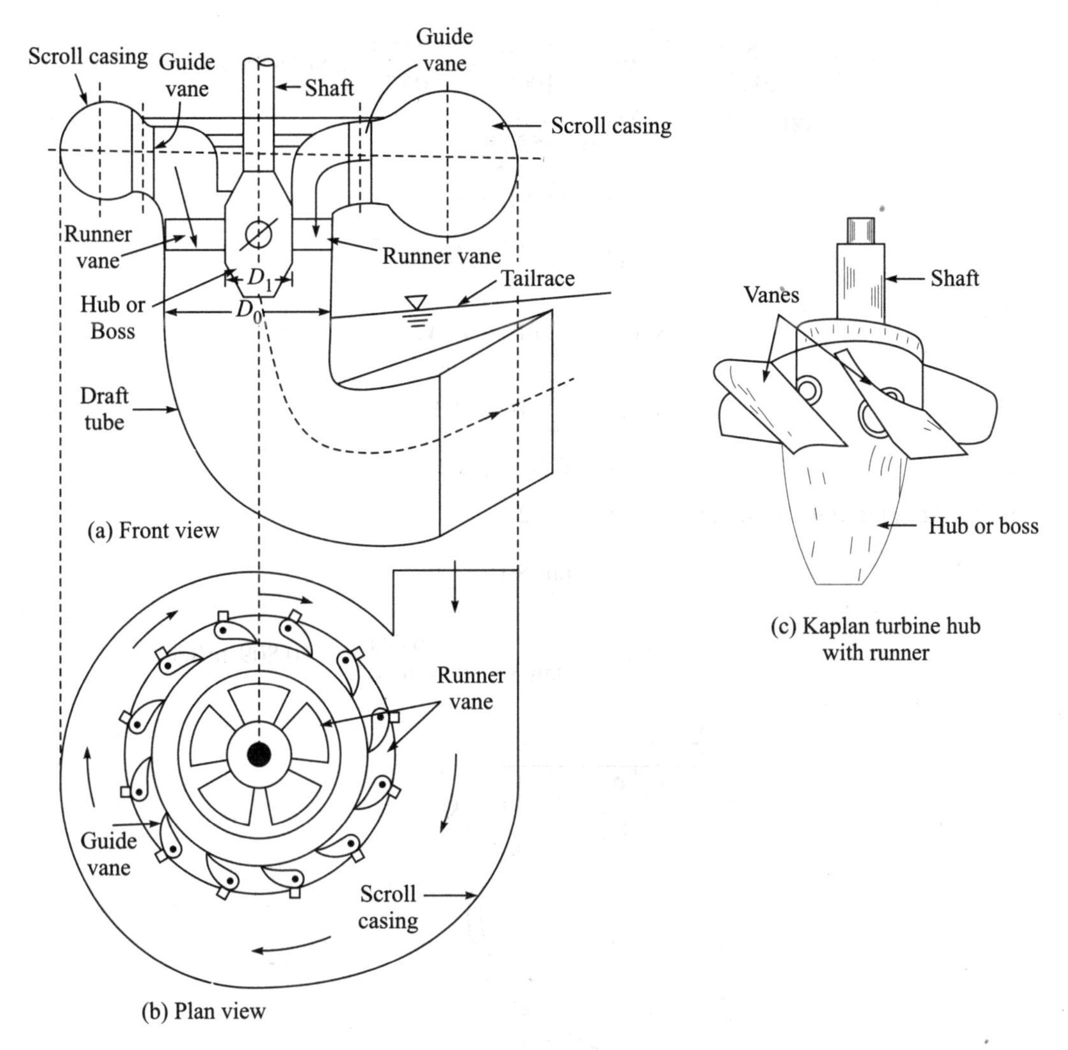

Figure 18.19 Kaplan turbine arrangements.

EXAMPLE 18.13 A Kaplan turbine working under a head of 15 m develops 7357.5 kW shaft power. The outer diameter of the runner is 4 m and the hub diameter is 2 m. The guide blade angle at the extreme edge of the runner is 30°. The hydraulic and overall efficiency are 90% and 85%, respectively. If the velocity of the whirl (V_{w_1}) at the outlet is zero, determine runner vane angles at the inlet and outlet at the extreme edge of the runner and speed of turbine.

Solution: The given data are: H = 15 m, Shaft Power (SP) = 7357.5 kW,

$$D_0 = 4 \text{ m}, D_h = 2 \text{ m}, \eta_h = 90\% = 0.9,$$

$$\eta_0 = 85\% = 0.85, V_{w_1} = 0, \alpha = 30°$$

We know

$$\eta_0 = \frac{\text{Shaft Power (S.P.)}}{\text{Water Power Supplied at inlet (W.P.)}}$$

$$\therefore \quad 0.85 = \frac{7357.5}{WP} \quad \therefore\ WP = \frac{\rho g Q H}{1000} = \frac{7357.5}{0.85} = 8655.88 \text{ kW}$$

or

$$\frac{1000 \times 9.81 \times Q \times 15}{1000} = 8655.88$$

$$Q = 58.8235 \text{ m}^3\text{/sec}$$

Again

$$Q = \frac{\pi}{4}(D_0^2 - D_h^2)Vf$$

$$58.823 = \frac{\pi}{4}(4^2 - 2^2)Vf$$

$$58.823 = \frac{\pi}{4}(16 - 4)Vf$$

$$\therefore \quad V_f = 6.241 \text{ m/sec}$$

From the inlet velocity diagram (Figure 18.20),

$$\tan \alpha = \tan 30° = \frac{Vf}{Vw}$$

$$\therefore \quad V_w = \frac{Vf}{\tan 30°} = \frac{6.241}{\tan 30°} = 10.809 \text{ m/sec}$$

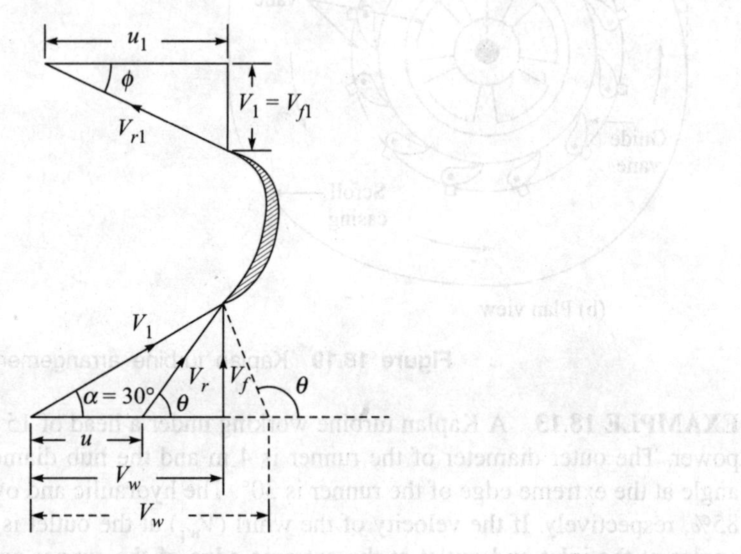

Figure 18.20 Visual of Example 18.13.

Hydraulic efficiency $\eta_h = \dfrac{V_w u}{gH}$

$$0.9 = \frac{10.809 \times u}{9.81 \times 15}$$

$$u = 12.252 \text{ m/sec}$$

If N is the speed of the turbine, $u = u_1 = \dfrac{\pi D_0 N}{60}$

or $$12.252 = \frac{\pi \times 4 \times N}{60}$$

$$N = 58.5 \text{ rpm}$$ **Ans.**

Runner angle θ is from the inlet diagram,

$$\tan \theta = \frac{V_f}{V_w - u} = \frac{12.252}{10.809 - 12.252} = \frac{12.252}{-1.443} = -8.49$$

$\therefore$ $$\theta = (180° - 8328°)$$

$$\theta = 96.71°$$ **Ans.**

From the outlet velocity diagram:

$$\tan \phi = \frac{V_{f_1}}{u_1} = \frac{V_f}{u} = \frac{6.241}{12.252} = 0.5102$$

$\therefore$ $$\phi = 27.03°$$ **Ans.**

18.11 THEORY OF DRAFT TUBE ON PRESSURE AND EFFICIENCY

Let H_s be the vertical height of the draft tube above the tailrace and y be the distance of the tailrace from the bottom of the draft tube as shown in Figure 18.21.

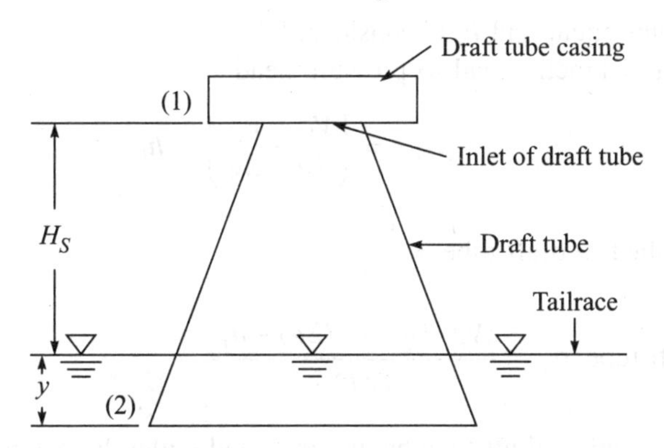

Figure 18.21 Draft tube.

Taking the bottom line as the datum, applying Bernoulli's equation at (1) and at (2):

$$\frac{p_1}{w} + \frac{V_1^2}{2g} + (y + H_s) = \frac{p_2}{w} + \frac{V_2^2}{2g} + 0 + h_f \tag{A}$$

where p_1 and V_1 are the pressure and velocity at the inlet of the draft tube p_2 and V_2 are the pressure and velocity at the outlet of the draft tube.

Note that $\left(\dfrac{p_2}{w}\right)$ is the atmospheric pressure $+y$

$$\therefore \qquad \left(\frac{p_2}{w}\right) = \left(\frac{p_a}{w} + y\right)$$

h_f is the loss of head due to friction in the draft tube.

Substituting $\left(\frac{p_2}{w} + y\right)$ in Equation (A) for $\frac{p_2}{w}$,

$$\frac{p_1}{w} + \frac{V_1^2}{2g} + y + H_s = \frac{p_a}{w} + y + \frac{V_2^2}{2g} + h_f$$

$$\frac{p_1}{w} + \frac{V_1^2}{2g} + H_s = \frac{p_a}{w} + \frac{V_2^2}{2g} + h_f$$

$$\frac{p_1}{w} = \frac{p_a}{w} + \frac{V_2^2}{2g} + h_f - \frac{V_1^2}{2g} - H_s = \left(\frac{p_a}{w} - H_s\right) - \left(\frac{V_1^2}{2g} - \frac{V_2^2}{2g} - h_f\right) \tag{18.32}$$

Equation (18.32) shows that the pressure at the inlet of the draft tube is less than atmospheric, i.e. the pressure in draft tube is negative.

If the draft tube is cylindrical, i.e. area at (1) is equal to area at (2), $V_1 = V_2$ and head loss due to friction h_f neglected as the length of tube is small, Equation (18.32) is

$$\frac{p_1}{w} = \frac{p_a}{w} - H_s \tag{18.33}$$

i.e. the pressure in draft tube is less than the atmospheric pressure by an amount of static head H_s.

If V_1 and V_2 are not equal and h_f is considered,

Actual conversion of kinetic head to pressure head

$$= \left(\frac{V_1^2}{2g} - \frac{V_2^2}{2g}\right) - h_f$$

Kinetic head supplied to draft tube $= \dfrac{V_1^2}{2g}$

$$\therefore \quad \text{Efficiency of draft tube } \eta_d = \frac{(V_1^2/2g - V_2^2/2g) - h_f}{V_1^2/2g} \tag{18.34}$$

EXAMPLE 18.14 A conical draft tube having inlet and outlet diameters of 0.8 m and 1.2 m, respectively, discharges water at the outlet with a velocity of 3 m/sec. The total length of the draft tube is 8 m and 2 m if it is immersed in water. If the atmospheric pressure 10.3 m of water and loss of head due to friction is 0.25 times the velocity head at the outlet of the tube, find the pressure head at the inlet and efficiency of the draft tube.

Solution: The given data are: Diameter at inlet = D_i = 0.8 m,

Diameter at outlet = D_o = 1.2 m, Velocity at outlet = V_2 = 3 m/sec

Total length = $H_s + y$ = 8 m, y = 2 m $\therefore H_s = 8 - 2 = 6$ m

$$\frac{p_a}{w} = 10.3 \text{ m of water, } h_f = 0.25\frac{V_2^2}{2g},$$

$\therefore \quad Q = A_2V_2 = \frac{\pi}{4} d_0^2 \; V_2 = \pi/4(1.2)^2 \times 3 = 3.393 \text{ m}^3/\text{sec}$

$\therefore$ Velocity at inlet $V_1 = \frac{Q}{A_1} = \frac{3.393}{\pi/4(.8)^2} = 6.75 \text{ m/sec}$

Writing Equation (18.32), $\frac{p_1}{w} = \frac{p_a}{w} - H_s - \left(\frac{V_1^2}{2g} - \frac{V_2^2}{2g}\right) - h_f$

$$\frac{p_1}{w} = 10.3 - 6 - \left(\frac{6.75^2}{2 \times 9.81} - \frac{3^2}{2 \times 9.81}\right) - 0.25 \frac{3^2}{2 \times 9.81}$$

$$\frac{p_1}{w} = 4.3 - 1.8635 - 0.1146$$

$$\frac{p_1}{w} = 2.3219 \text{ m of water}$$ **Ans.**

$$\eta_\alpha = \frac{(V_1^2/2g - V^2/2g) - h_f}{\frac{V_1^2}{2g}} = \frac{\left(\frac{6.75^2}{2 \times 9.81} - \frac{3^2}{2 \times 9.81}\right) - 0.25 \frac{32}{2 \times 9.81}}{\frac{6.75^2}{2 \times 9.81}}$$

$$\eta_\alpha = \frac{2.3222 - 0.4587 - 0.1146}{2.3222} = \frac{1.74878}{2.3222} = 0.7530$$

$\therefore \quad \eta_\alpha = 75.3\%$ **Ans.**

18.12 PERFORMANCE OF TURBINES

In order to predict the behaviour of turbines working under different conditions to facilitate comparison between the performance of turbines of the same type but having different output and speed and working under different heads, it is convenient to express the test result in terms of certain unit quantities such as unit speed, unit power and unit discharge under unit head.

Unit speed (N_u): It is defined as the speed of the turbine working under unit head (1 m)

We know $\quad u \propto V$ and $V \propto \sqrt{H}$

$\therefore \quad u \propto \sqrt{H}$ and $u = \frac{\pi DN}{60}$, for a given turbine,

D is constant $\therefore$ $u \propto N$ or $N \propto u$

or $\quad N \propto \sqrt{H}$

$\therefore \quad N = C\sqrt{H}$ where C is a constant.

If $\quad H = 1$ m, $N = N_u \;\therefore\; N_u = C\sqrt{H} \;\therefore\; C = N_u$

$$\therefore \quad N = N_u\sqrt{H}$$

$$\therefore \quad N_u = \frac{N}{\sqrt{H}} \tag{18.35}$$

Unit discharge (Q_u) is defined as the discharge passing through a turbine, which is working under a unit head (1 m)

$$Q \propto V \quad \text{and} \quad V \propto \sqrt{H}$$

$$\therefore \quad Q \propto \sqrt{H}$$

$$\therefore \quad Q = C\sqrt{H} \quad \text{where } C_1 \text{ is constant}$$

When $H = 1$ m $Q_u = C_1\sqrt{1}$ $\therefore$ $C_1 = Q_u$

$$\therefore \quad Q = Q_u\sqrt{H}$$

$$\therefore \quad Q_u = \frac{Q}{\sqrt{H}} \tag{18.36}$$

Unit Power (P_u) is defined as the power developed by a turbine working under a unit head.

$$\eta_0 = \frac{\text{Power developed}}{\text{Water Power}} = \frac{P}{\frac{wQH}{1000}}$$

$$\therefore \quad P = \eta_0 \frac{wQH}{1000}$$

$$\therefore \quad P \propto Q \times H \text{ but } Q \propto \sqrt{H}$$

$$\therefore \quad P \propto H^{3/2}$$

$$\therefore \quad P = C_2 H^{3/2}$$

when $\quad H = 1$ m, $P = P_u = C_2(1)^{3/2}$

$$\therefore \quad C = P_u$$

$$\therefore \quad P = P_u H^{3/2}$$

$$P_u = \frac{P}{H^{3/2}} \tag{18.37}$$

Performance or behaviour of turbines working under different heads can be known easily from the values of the above unit quantities.

If $H_1, H_2, H_3\ldots$ are heads under which the turbine works, $N_1, N_2, N_3\ldots$ are corresponding speeds and $Q_1, Q_2, Q_3,\ldots, P_1, P_2, P_3,\ldots$ are the discharges and powers respectively.

Then,

$$N_u = \frac{N_1}{\sqrt{H_1}} = \frac{N_2}{\sqrt{H_2}} = \frac{N_3}{\sqrt{H_3}},\ Q_u = \frac{Q_1}{\sqrt{H_1}} = \frac{Q_2}{\sqrt{H_2}} = \frac{Q_3}{\sqrt{H_3}},\ P_u = \frac{P_1}{H_1^{3/2}} = \frac{P_2}{H_2^{3/2}} = \frac{P_3}{H_3^{3/2}} \tag{18.38}$$

Thus Equation (18.38) is used to predict the head, discharge, power of the turbine when speed, discharge and power under a head are known.

EXAMPLE 18.15 A turbine develops 8000 kW of water running under a head of 25 m at 90 rpm. If the head of the turbine is reduced to 15 m, determine the speed and power developed by the turbine.

Solution: The given data are: $P_1 = 8000$ kW, $N_1 = 90$ rpm,

$$H_1 = 25 \text{ m}, H_2 = 15 \text{ m}$$

Equation (18.38) gives:

$$\frac{N_1}{\sqrt{H_1}} = \frac{N_2}{\sqrt{H_2}}$$

or
$$\frac{90}{\sqrt{25}} = \frac{N_2}{\sqrt{15}} \quad \therefore N_2 = 69.7 \text{ rpm.} \qquad \textbf{Ans.}$$

and
$$\frac{p_1}{H_1^{3/2}} = \frac{p_2}{H_2^{3/2}}$$

or
$$\frac{8000}{25^{3/2}} = \frac{p_2}{15^{3/2}} \quad \therefore P_2 = 3718 \text{ kW} \qquad \textbf{Ans.}$$

18.13 SPECIFIC SPEED (N_S)

It is defined as the speed of a turbine which develops a unit power when working under a unit head. It is used to compare the different types of turbines which are identical in shape, geometrical dimensions, blade angles, gate opening, etc.

An expression of specific speed N_s of turbine, we may start with overall efficiency η_o.

i.e.
$$\eta_o = \frac{\text{Shaft Power } P}{\text{Water Power}} = \frac{\rho}{\dfrac{wQH}{1000}}$$

$\therefore$
$$P = \left(\eta_o \frac{w}{1000}\right) QH$$

$\therefore$
$$P \propto QH \qquad \text{(A)}$$

But
$$u \propto V \text{ and } V \propto \sqrt{H} \quad \therefore u \propto \sqrt{H} \qquad \text{(B)}$$

again
$$u = \frac{\pi DN}{60} \quad \text{or } u \propto DN \qquad \text{(C)}$$

From Equations (B) and (C), $\sqrt{H} \propto DN$

or
$$D \propto \frac{\sqrt{H}}{N} \qquad \text{(D)}$$

$$Q = \text{area} \times \text{velocity}$$

But
$$\text{area} \propto D^2$$

and
$$\text{Velocity} \propto \sqrt{H}$$

$$\therefore \qquad Q \propto D^2 \times \sqrt{H}$$

or
$$Q \propto \left(\frac{\sqrt{H}}{N}\right)^2 \sqrt{H}$$

or
$$Q \propto \frac{H^{3/2}}{N^2}$$

From Equation (A),
$$P \propto \frac{H^{3/2}}{N^2} \times H$$

$$P \propto \frac{H^{5/2}}{N^2}$$

$$\therefore \qquad P = C\frac{H^{5/2}}{N^2} \qquad \text{(E)}$$

$\therefore$ When Equation $P = 1$ kW, $H = 1$ m, then by definition $N = N_S$

$$\therefore \qquad 1 = C\frac{1}{N_S^2} \qquad \therefore C = N_S^2$$

Substituting Equation C in Equaiton (E):

$$P = N_S^2 \frac{H^{5/2}}{N^2}$$

$$\therefore \qquad N_S = \frac{N\sqrt{P}}{H^{5/4}} \qquad (18.39)$$

which give the expression of specific speed. It plays an important role in selecting the type of turbine. The performance of turbine can be predicted through specific speed. Normally, specific speed N_S of Pelton wheel varies from 8 to 50, for Francis turbine it varies from 51 to 255, and for Kaplan it varies from 256 to 860.

EXAMPLE 18.16 A turbine operates under a head of 30 m at 300 rpm. The discharge is 10 m^3/sec. If the overall efficiency is 90%, determine the specific speed, power generated and types of turbine.

Solution: The given data are: $H = 30$ m, $N = 300$ rpm, $Q = 10$ m^3/sec

$$\eta_0 = 90\%, = 0.9$$

We know
$$\eta_0 = \frac{P}{\frac{wQH}{1000}}$$

$$\therefore \qquad P = \eta_0 \frac{\rho g Q H}{1000} = \frac{0.9 \times 1000 \times 9.81 \times 10 \times 30}{1000} \text{ kW}$$

$$P = 2648.7 \text{ kW} \qquad \textbf{Ans.}$$

$$N_S = \frac{N\sqrt{P}}{H^{5/4}} = \frac{300\sqrt{2648.7}}{30^{5/4}} = 219.9 \quad \textbf{Ans.}$$

Since the Francis turbine range of specific speed is 51 to 255, type of turbine is Francis. **Ans.**

18.14 GOVERNOR TO TURBINE

The modern hydraulic turbines are directly coupled to the electric generator. The generators are allowed to run at a constant speed irrespective of variation in the load. The constant speed N (rpm) of the generator is given by the expression $f = \frac{pN}{60}$, where f is the frequency of the power generator in cycles per second and p is the number of pairs of poles usually

$$f = 50, \quad \therefore N = \frac{3000}{P} \tag{18.40}$$

The speed of the generator can be maintained constant only if the speed of the turbine runner is constant which is equal to N given by Equation (18.40). It is known as synchronous speed of the turbine runner for which it is designed. In order to maintain this synchronous speed between the turbine and generator shaft under all conditions of working, it is required to regulate the quantity of water flowing through the runner in accordance with the variation of load. Such an operation of regulation of speed is known as governing of turbine and is usually done automatically by means of an oil pressure governor shown in Figure 18.22. The main components of this oil pressure governor are:

(1) Servometer or relay cylinder
(2) Relay or control or distribution valve.
(3) Pendulum or actuator or centrifugal governor
(4) Oil sump
(5) Gear pump
(6) Spear or needle
(7) Pipe connecting oil sump with control valve.

18.14.1 Working Principle

When the load on the generator drops, the turbine speed tends to increase. Since the pendulum is driven by the turbine shaft, balls move up due to an increase in speed resulting in upward movement of the sleeves. As the sleeves moves up, the control valve moves downwards, closes valve V_1 and opens valve V_2. The oil pumped from the oil pump to V_1, will flow through V_2 to the servometer and exert a pressure on face A of the piston of the relay cylinder. The piston with spear will move towards the right decreasing the area of the flow at the outlet of the nozzle, which consequently decreases the speed of the turbine and brings it back to normal speed. When the load on the generator increases, the speed of the shaft and turbine increases, the fly ball comes down, servometer piston moves right, flow area increases and thus the turbine speed also increases to synchronise with the generator speed.

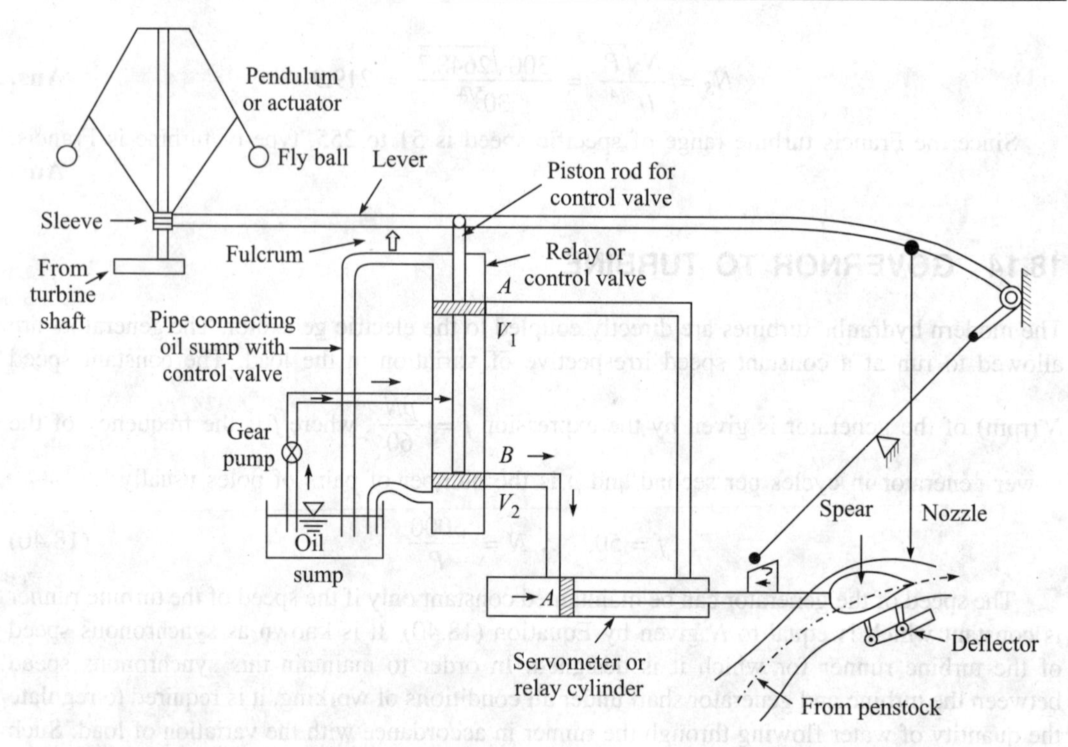

Figure 18.22 Oil pressure governor.

18.15 CHARACTERISTIC CURVES OF TURBINES

The turbines are normally designed to work at particular values of Q, H, N, P and η_o. But often they are required to work under varying conditions other than the designed conditions. Therefore, it is essential to determine the exact behaviour of turbines under varying conditions by carrying out tests either on actual turbines or on models. The results of these tests are usually represented graphically. The resulting curves are known as characteristic curves. These curves are plotted in terms of unit quantities like N_u, Q_u, P_u, etc. Following are the three types of characteristic curves:

(1) Constant head or main characteristic curves
(2) Constant speed or operating characteristic curves
(3) Constant efficiency curves.

Main or Constant Head Characteristic Curves

These curves are obtained by maintaining constant head and constant gate opening on the turbine. Keeping H and the gate opening constant, the speed is varied by changing the load. For each value of speed, corresponding values P, Q and η_0 are determined. Then unit quantities Q_u, N_u, P_u are calculated. Again for a constant head, gate openings are changed in the step one-fourth from full gate to three-fourth gate opening, half gate opening and one-fourth gate opening and for each gate opening, the corresponding value of Q_u, P_u, N_u and η_0 are determined and Q_u, P_u and η_0 are plotted against N_u for four different gate openings. The main characteristics curves are shown in Figure 18.23 for the Pelton turbine.

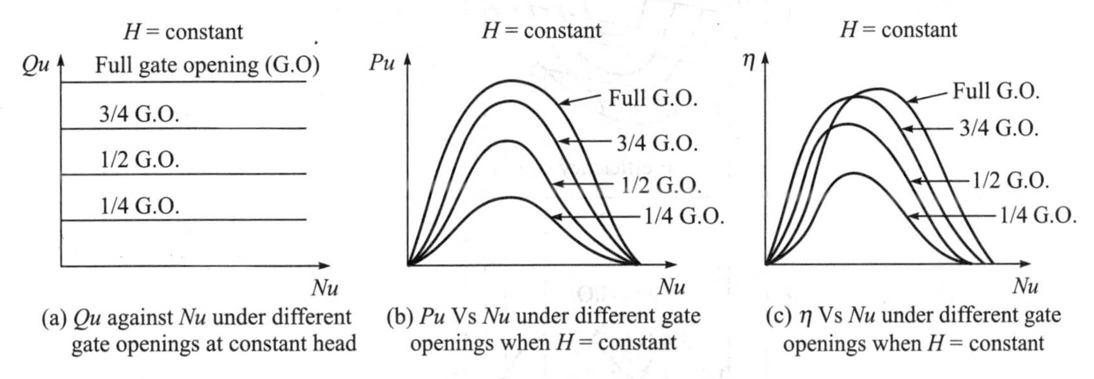

(a) *Qu* against *Nu* under different gate openings at constant head
(b) *Pu* Vs *Nu* under different gate openings when *H* = constant
(c) *η* Vs *Nu* under different gate openings when *H* = constant

Figure 18.23 Main characteristic or constant head characteristic curves for Pelton turbine.

For other reaction turbines like Francis and Kaplan turbines, almost more or less similar curves are obtained.

Constant Speed Characteristic Curves

These constant speed or operating characteristic curves are plotted keeping speed N and H constant. Variations Q with respect to P and η_0 are plotted as shown in Figure 18.24.

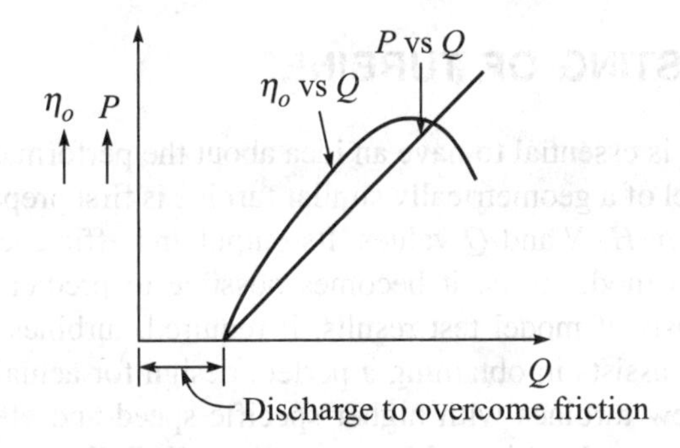

Figure 18.24 Constant speed or operating characteristic curves.

The power and efficiency curves show that a certain amount of discharge is required initially to overcome friction and then it starts producing P and efficiency η_o starts declining at some stage of increasing discharge.

Constant Efficiency Curves or Iso-efficiency Curves

Keeping H constant at different Gate Opening (G.O.), these curves are obtained by plotting N Vs Q and N Vs η_o as shown in Figure 18.25.

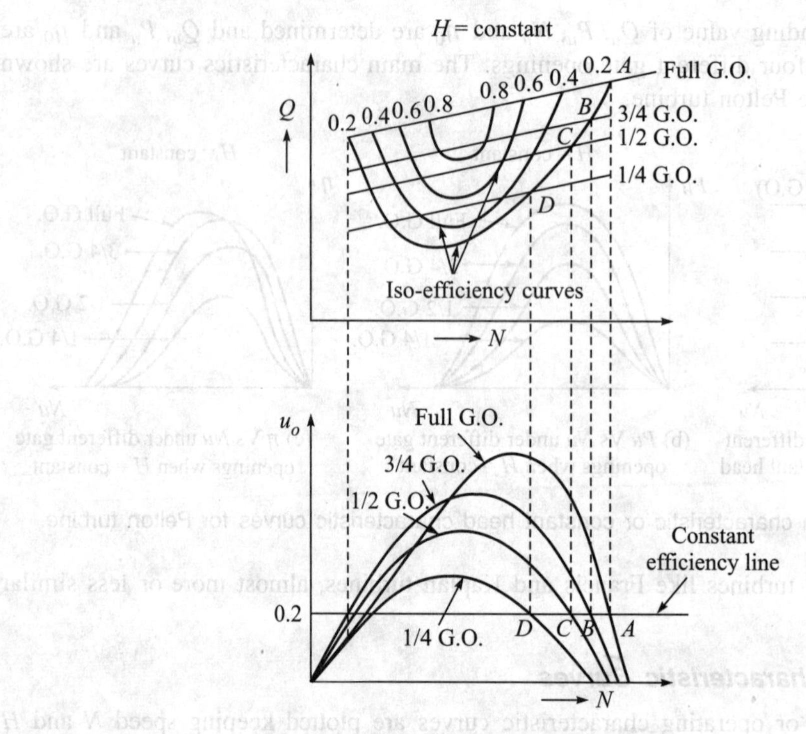

Figure 18.25 Constant or Iso-efficiency curves.

18.16 MODEL TESTING OF TURBINES

Model testing of turbines is essential to have an idea about the performance of the actual turbine in advance. A small model of a geometrically similar turbine is first prepared. This model turbine is then tested under known H, N and Q values. Its output and efficiency are determined. From the results obtained from model tests, it becomes possible to predict the performance of the actual turbine. On the basis of model test results, if required, turbines may be modified. Thus model testing of turbines assists in obtaining a perfect design for actual turbines. Also, it helps in the development of new turbines with higher specific speed and efficiency.

The various variables involved in turbines are Q, H, N, D, P, ρ and μ of the fluid. In the problems of turbomachines, generally shaft work gH is used as one of the variables instead of H. These variables with the help of dimensional analysis may be grouped into the following dimensionless numbers or parameters:

$$\left(\frac{Q}{ND^3}\right), \left(\frac{gH}{N^2D^2}\right), \left(\frac{P}{\rho g\, HND^3}\right), \left(\frac{\mu}{\rho ND^2}\right).$$

(i) The parameter $\left(\frac{Q}{ND^3}\right)$ is known as the discharge number or flow number which may also be expressed as $\left(\frac{Q}{D^2\sqrt{gH}}\right)$.

(ii) The parameter $\left(\dfrac{gH}{N^2 D^2}\right)$ is known as the head number.

(iii) The parameter $\left(\dfrac{P}{\rho g\, HND^3}\right)$ is known as power number which may also be expressed as $\left(\dfrac{P}{\rho g H^{3/2} D^2}\right)$.

(iv) The parameter $\left(\dfrac{\mu}{\rho ND^2}\right)$ represents the Reynolds number.

Now dividing the square root of power number by (head number)$^{3/4}$, we get

$$\frac{P^{1/2}}{\rho^{1/2}\, g^{1/2}\, H^{1/2}\, N^{1/2}\, D^{3/2}} \times \frac{N^{3/2}\, D^{3/2}}{g^{3/4}\, H^{3/4}} = \frac{N\sqrt{P}}{\rho^{1/2}\, g^{5/4}\, H^{5/4}} = n_s \qquad \text{(A)}$$

This n_s is known as the dimensionless specific speed or shape number of the turbine. (A) may be expressed as

$$n_s = \left(\frac{1}{\rho^{1/2}\, g^{5/4}}\right)\left(\frac{N\sqrt{P}}{H^{5/4}}\right)$$

$$n_s = \frac{1}{\rho^{1/2}\, g^{3/4}} N_s \qquad (18.40)$$

$$\because \quad \frac{N\sqrt{P}}{H^{5/4}} = N_s \text{ (sp. speed)}$$

which is the relationship between dimensionless specific speed n_s and specific speed N_s.

In order to ensure complete similarity between the model (m) and prototype (p) turbines, the following conditions need to be satisfied.

$$\left(\frac{1}{\rho^{1/2}\, g^{5/4}} N_S\right)_m = \left(\frac{1}{\rho^{1/2}\, g^{5/4}} N_S\right)_p \qquad (18.41)$$

$$\left.\begin{aligned} \left(\frac{Q}{ND^3}\right)_m &= \left(\frac{Q}{ND^3}\right)_p \\ \text{or} \quad \left(\frac{Q}{D^2\sqrt{gH}}\right)_m &= \left(\frac{Q}{D^2\sqrt{gH}}\right)_p \end{aligned}\right\} \qquad (18.42)$$

$$\left(\frac{gH}{N^2 D^2}\right)_m = \left(\frac{gH}{N_2 D_2}\right)_p \qquad (18.43)$$

$$\left(\frac{P}{\rho g H^{3/2} D^2}\right)_m = \left(\frac{P}{\rho g H^{3/2} D^2}\right)_p$$

or

$$\left(\frac{P}{\rho g HND^3}\right)_m = \left(\frac{P}{\rho g HND^3}\right)_p \tag{18.44}$$

In most cases, values of ρ and g for both the model and the prototype are the same.

18.17 CAVITATION IN TURBINES

Pressure in any part of the turbine may sometimes reach vapour pressure. When such a stage is reached, water boils and small bubbles of vapour form in large numbers. These vapour-filled cavities are carried along by the flow and as they are carried to a region of high pressure, bubbles suddenly collapse. Due to this sudden collapse of bubbles, surrounding liquid rushes in to the zone to fill it up. The liquid moving in from all directions collides at the centre of cavity thus giving rise to very high local pressure. The solid surface in the vicinity is also subjected to these high pressure. Alternate formation and collapse of vapour bubbles may cause severe damage to the surface which eventually fails by fatigue and the surface becomes badly scored and pitted. This phenomenon is called cavitation.

In a reaction turbine, pressure at the runner exit or inlet to draft tube may be reduced considerably. Thus there is every likelihood of occurring cavitation in this zone. Due to this cavitation, metal at the runner exit and inlet of draft tube may be eaten gradually, which results in lowering the efficiency of the turbine. Therefore, the components of the turbine are so designed to eliminate cavitation as far as possible.

D. Thoma of Germany developed a dimensionless parameter known as Thoma's cavitation factor σ, which is expressed as:

$$\sigma = \frac{H_a - H_v - H_s}{H} \tag{18.45}$$

where H_a, H_v, H_s and H are the atmospheric, vapour, suction and working pressure head, respectively.

A critical Thoma cavitation factor σ_c is calculated by empirical relationship as:

For a Francis turbine $\sigma_c = 0.625\left(\frac{N_S}{380.78}\right)^2$

$$\sigma_c = 431 \times 10^{-8}\, N_s^2 \tag{18.46}$$

For propeller turbines $$\sigma_c = 0.28 + \left[\frac{1}{7.5}\left(\frac{N_S}{380.78}\right)^3\right] \tag{18.47}$$

where N_s is (rpm, kW, m) units.

In order to avoid cavitation, the value of σ should not fall below the value given by σ_c. Practically the pressure of water at any point of the turbine should not fall below the vapour pressure, i.e. 2.6 m of water.

18.18 CONCLUSION

A comprehensive presentation of hydraulic turbines has been made in this chapter. Details of classification and types of turbine like Pelton wheel, reaction turbines both inward and outward flow, Francis and Kaplan have been discussed. Draft tube theory, performance of turbines, specific speed, governor, characteristics of turbine, model testing, unit speed, power and discharge, cavitation all are discussed. Numerical examples of each type have been solved and few problems with answers are given at the end for the students as chapter-end exercises.

The works by Daily[1], Baumeister[2], Shephard[3], Wislicenus[4], Beitler[5], Kovalev[6] on turbomachines are worth to be mentioned.

PROBLEMS

18.1 Design a Pelton wheel for a head of 80 m and speed 300 rpm. It develops 103 kW shaft power. Take $C_V = 0.98$, speed ratio = 0.45 and $\eta_0 = 0.80$. Determine the wheel diameter D, jet diameter d and no. of jets z. (**Ans.** $D = 1.135$ m, $d = 7.26$ cm, $z = 23$)

18.2 A Pelton wheel has a mean bucket speed of 10 m/sec with a jet of water flowing at the rate of 0.7 m^3/sec under a head of 30 m. The bucket deflects the jet by an angle of 160°. Calculate the power given by the water to the runner and hydraulic efficiency of the turbine. Take $C_V = 0.98$. (**Ans.** 186.97 kW, 94.54%)

18.3 The penstock supplies water from a reservoir to the Pelton wheel with a gross head of 450 m. One-third of the gross of the nozzle is 1.8 m^3/sec. The angle of deflection is 162°. Determine the power given by the water to the runner and η_h of the Pelton wheel. Take speed ratio = 0.45, $C_V = 0.99$.

18.4 A hydraulic turbine under a head of 27 m develops 10,000 kW running at 120 rpm. What is the specific speed of the turbine? What type of turbine is this? Also find the normal speed and output if the head on the turbine is reduced to 20 m.

(**Ans.** 195, 103 rpm, 6375 kW)

18.5 Estimate the main dimensions and blade angles of the inward flow reaction turbine working under a head of 68 m at 750 rpm with an output of 331.2 kW. The flow ratio is 0.15, $\eta_h = 95\%$, $\eta_0 = 85\%$, the ratio of the wheel width to the diameter at entry = 0.1, inner diameter = $\frac{1}{2}$ outer diameter. Assume that 5% of the circumferential area of the runner is blocked by the thickness of the blades. The turbine discharges radially at output. The velocity of flow through runner is constant.

(**Ans.** 0.6 m, 0.06 m, $\theta = 60°49'$, $\alpha = 11°38'$, $\phi = 24°57'$, 0.3 m, 0.12 m)

18.6 A inward flow reaction turbine discharging radially at the outlet having an overall efficiency of 80% is required to develop 147.2 kW. The head is 9 m. The velocity of the periphery of the wheel is $0.9\sqrt{2gH}$ and the radial velocity of the flow is $0.8\sqrt{2gH}$. The wheel rotates with 200 rpm and hydraulic efficiency is 85%. Determine: (I) Angle of guide blade at inlet, (II) Wheel vane angle at inlet, (III) Diameter of the wheel, (IV) Width of the wheel at the inlet.

(**Ans.** $\alpha = 32°28'$, $\theta = 144°57'$, $d = 1.14$ m, $b = 0.15$ m)

18.7 Show that the hydraulic efficiency for an inward flow reaction turbine discharging radially at out is given by:

$$\eta_h = \frac{1}{1 + \left[\dfrac{\frac{1}{2}\tan^2\alpha}{1 - \dfrac{\tan\alpha}{\tan\theta}}\right]}$$

18.8 An outward flow reaction turbine utilises 5.2 m^3/sec of water. The internal and external diameters of the runner are 1.5 m and 2.5 m, respectively. The width of the runner is 30 cm at the inlet as well as at the outlet. If the head on the turbine is 50 m and speed of the runner is 190 rpm, find the runner vane angles at the inlet and outlet, and guide blade angle. Neglect the thickness of the vanes and assume that the turbine discharges radially at outlet. (**Ans.** 11°41′, 5°5′, 6°25′)

18.9 A Kaplan turbine working under a head of 18 m develops 18400 kW at an overall efficiency of 85%. The boss diameter is 0.3 times the runner diameter. If the velocity of flow is 9.05 m/sec, calculate the discharge and diameter of the runner and the hub.

(**Ans.** $Q = 122.59$ m^3/sec, $D_0 = 4.35$ m, $D_h = 1.3$ m)

18.10 Water enters an inward flow turbine at an angle 20° to the wheel tangent to the outer rim and leaves the turbines radially. If the speed of the wheel is 310 rpm and the velocity of flow is constant at 3.25 m/sec, find the necessary angles of the blade at the inlet and outlet, if the inner and outer diameter of the turbine are 30 cms and 60 cms, respectively. If the width of the wheel at the inlet is 15 cm, find power developed. Neglect thickness of the blades. (**Ans.** 104°, 34°43′, 79.93 kW)

REFERENCES

1. Daily, J.W., "Hydraulic Machinery", Chapter 13 of *Engineering Hydraulics*, edited by H. Rouse, John Wiley & Sons, Inc., New York, 1950.
2. Baumeister, T., *Turbomachinery*, Section 19 of *Handbook of Fluid Dynamics*, edited by V.L. Streeter, McGraw-Hill Inc., 1961.
3. Shepherd, D.G., *Principles of Turbomachinery*, Macmillan, New York, 1956.
4. Wislicenus, G.F., *Fluid Mechanics of Turbomachinery*, McGraw-Hill, Inc., 1953.
5. Beilter, S.R. and E.J. Lindahl, *Hydraulic Machinery*, Ronald Press, 1954.
6. Kovalev, N.N., *Hydroturbines Design and Construction*, Israd Program for Scientific Translation, Jerusalem, 1949.
7. Subramanya, K., *Theory and Applications of Fluid Mechanics including Hydraulic Turbines*, Tata McGraw-Hill, New Delhi, 1992.

8. Lal, J., *Hydraulic Machines*, Metroplitan Book Co., New Delhi, 1992.
9. Romamruthunm, S., *Hydraulics, Fluid Mechanics and Fluid Machines*, Dhanpat Rai and Sons., Delhi, 1980.
10. Bansal, R.K., *Fluid Mechanics and Hydraulic Machines*, Laxmi Publications New Delhi, 1983.
11. Mott, R.L., *Applied Fluid Mechanics*, Prentice-Hall, New Jersey. 1999.

Chapter 19

Centrifugal Pumps

19.1 INTRODUCTION

A pump is defined as a mechanical device which when interposed in a pipe or a conduit, converts the mechanical energy supplied to it from some external source into hydraulic energy and transfers the same to the liquid flowing through the pipe. On the basis of the mode of action of conversion of mechanical energy into hydraulic energy, pumps are classified as: (i) Rotadynamic pumps, and (ii) Positive displacement pumps. A centrifugal pump is a rotadynamic pump which has a rotating element called an 'impeller'. As the liquid passes through the impeller, its angular momentum changes due to which the pressure energy of liquid is converted into potential energy as the liquid is lifted from a lower level to a higher level. Thus, the basic principle of a centrifugal pump on which the pump acts, is that when the mass of liquid is made to rotate by the impeller, the liquid is thrown away from the centre of rotation, and the centrifugal head is impressed, which helps the liquid to rise to a higher level. Since the liquid is delivered at a higher level by the centrifugal head or force impressed on it, this pump is commonly called a centrifugal pump. Thus the action of a centrifugal pump is just the reverse of that of a radially inward flow reaction turbine.

19.2 COMPONENTS OF A CENTRIFUGAL PUMP

Figure 19.1 show the sketch of a centrifugal pump which lifts water from the sump level to a delivery tank along with its components. The main components are:

(1) *Impeller:* It is a rotating wheel or rotor fitted with a series of backward curved vanes or blades. It is mounted on a shaft which is connected to the shaft of an electric motor.

(2) *Casing:* It is an air-tight passage surrounding the impeller which is similar to the casing of a reaction turbine as shown in Figure 19.1.

A centrifugal pump is designed in such a way that the kinetic energy of water discharge at the outlet of the impeller is converted into pressure energy before water leaves the casing and enters the delivery pipe. This casing may be of three types:

(a) Volute casing (Figure 19.1)
(b) Vortex casing
(c) Casing with guide blades.

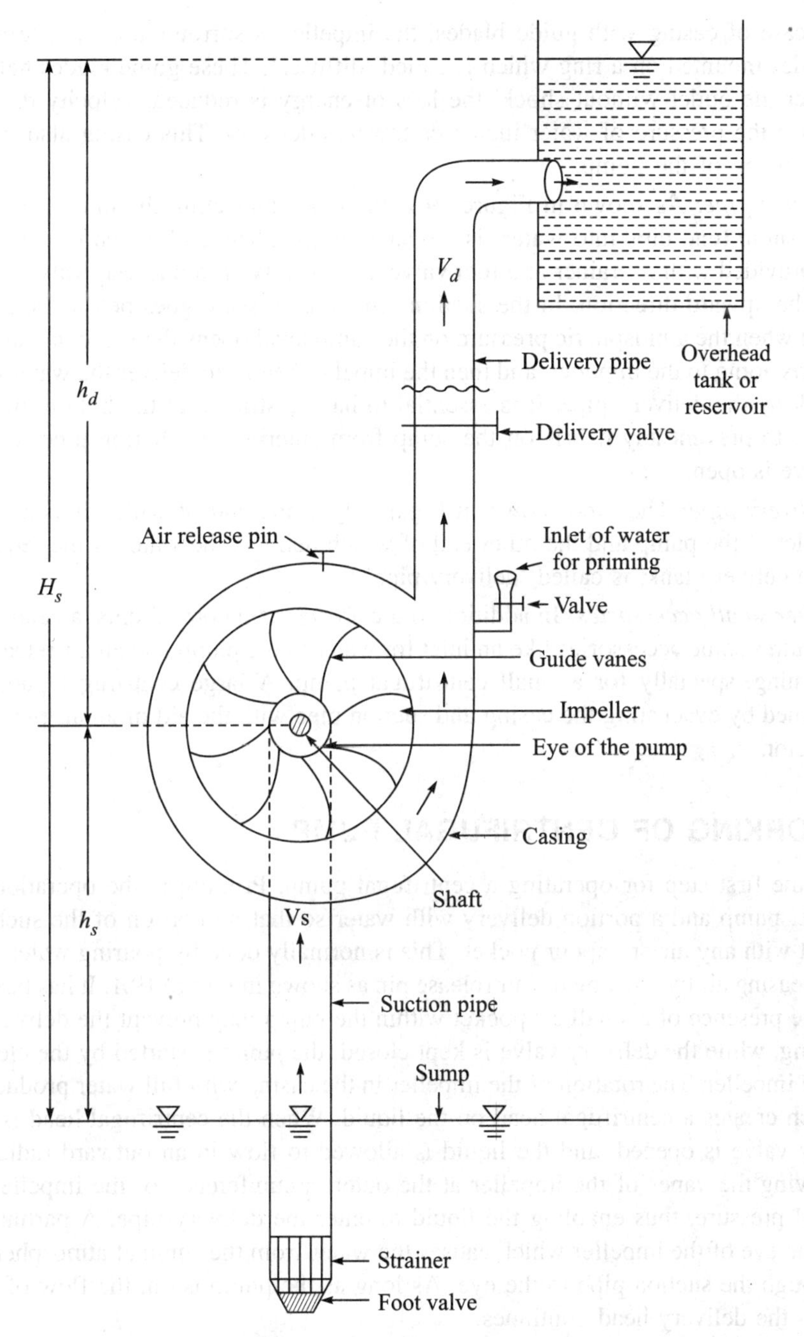

Figure 19.1 Centrifugal pump showing components.

Vortex casing is used to increase the efficiency of the turbine when a circular chamber is introduced between the volute casing and the impeller, loss of energy due to formation of eddies between the impeller and volute casing is reduced and efficiency of the pump can be increased.

In case of casing with guide blades, the impeller is surrounding by a series of guide blades mounted on a ring which is called 'diffuser'. These guide blades help the water enter the outlet without shock, the loss of energy is reduced, velocity decreases, and hence the pressure of water increases towards delivery. This casing also increases the efficiency of the pump.

(3) *Suction pipe:* As shown in Figure 19.1, the pipe connecting the inlet of the pump and the sump (end dips into water) is the suction pipe. The end which dips into the sump is provided with a strainer and foot valve. A foot valve is a one-way valve opening only in the upward direction. In the suction pipe, the pressure goes below the atmospheric, and when the atmospheric pressure on the sump level opens the valve upward, the water starts going to the impeller, and then the impeller begins to deliver the water to the upper tank by the delivery pipe. It is essential to have a strainer at the bottom of the suction pipe to prevent any debris on the sump from entering the suction pipe when the foot valve is open.

(4) *Delivery pipe:* The pipe shown in Figure 19.1, one end of which is connected to the outlet of the pump and the other end of which delivers the water at the required height to a delivery tank, is called 'delivery pipe'.

(5) *Other small accessories:* In addition to the above main components, a centrifugal pump requires some accessories like an inlet for water to the pump and an air release valve for priming, specially for a small centrifugal pump. A large centrifugal pump has been primed by evacuating the casing and suction pipe with the aid of an air pump or stream ejector.

19.3 WORKING OF CENTRIFUGAL PUMP

Priming is the first step for operating a centrifugal pump. Priming is the operation to fill the suction pipe, pump and a portion delivery with water so that no portion of the suction pipe or pump is left with any air or vapour pocket. This is normally done by pouring water through the inlet and releasing air by opening the air release pin as shown in Figure 19.1. It has been observed that even the presence of a small air pocket within the pump may prevent the delivery of water. After priming, while the delivery valve is kept closed, the pump is started by the electric motor to rotate the impeller. The rotation of the impeller in the casing with full water produces a forced vortex which creates a centrifugal head on the liquid. When the centrifugal head is impressed, the delivery valve is opened, and the liquid is allowed to flow in an outward radial direction, thereby leaving the vanes of the impeller at the outer circumference of the impeller with high velocity and pressure, thus enabling the liquid to enter the delivery pipe. A partial vacuum is created at the eye of the impeller which causes the water from the sump at atmospheric pressure to rush through the suction pipe to the eye. As long as the pump is on, the flow of water from the sump to the delivery head continues.

19.4 WORK DONE BY THE PUMP

In order to find an expression for the work done by the pump on water, velocity triangles at the inlet and outlet are drawn in Figure 19.2. The water enters the impeller radially at the inlet for best efficiency, i.e. $\alpha = 90°$ at the inlet and $V_w = 0$. Let N be the rotation of the impeller in rpm. D and D_1 are the diameters of the impeller at the inlet and the outlet, respectively.

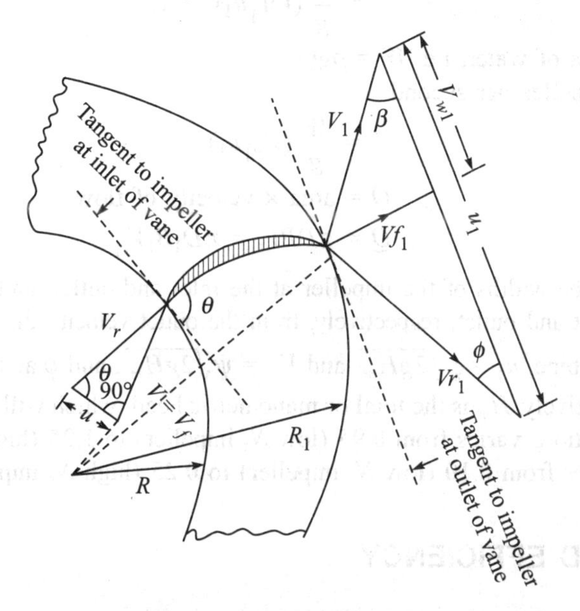

Figure 19.2 Velocity triangles at inlet and outlet of the impeller.

If u and u_1 are the tangential velocities of the impeller at the inlet and outlet:

$$u = \frac{\pi DN}{60}, \text{ and } u_1 = \frac{\pi D_1 N}{60}$$

Let V be the absolute velocity and V_r the relative velocity at the inlet. $\alpha = 90°$ is the angle made by V at the inlet within the direction of motion of the vane, θ is the angle made by V_r in the same direction. The corresponding values at the outlet are V_1, V_{r1}, β and ϕ, as shown in the outlet velocity triangles.

It has already been mentioned that a centrifugal pump is the reverse of a radially inward flow reaction turbine. In case of the turbine, the work done by water on the runner per second per unit weight of water striking per second is already given as:

$$= \frac{1}{g}[V_w u - V_{w1} u_1]$$

Since the centrifugal pump is the reverse of the turbine, the work done by the impeller on the water per second per unit weight of water striking per second

$$= -\left[\frac{1}{g}(V_w u - V_{w_1} u_1\right]$$

$$= \frac{1}{g}[V_{w1} u_1 - V_w u]$$

Since $V_w = 0$ in a centrifugal pump,

∴ work done by the impeller per unit weight per second

$$= \frac{1}{g}(Vw_1 u_1) \quad (19.1)$$

If W is the weight of water, i.e. $W = \rho g Q$,
work done by the impeller per second

$$= \frac{W}{g}(V_{w_1} u_1) \quad (19.2)$$

and $$Q = \text{area} \times \text{velocity of flow}$$

∴ $$Q = \pi DBV_f = \pi D_1 B_1 V_{f_1} \quad (19.3)$$

where B and B_1 are the widths of the impeller at the inlet and outlet, and V_f, V_{f_1} are velocities of the flow at the inlet and outlet, respectively, from the outlet velocity diagram, $V_{w_1} = (u_1 - V_{f_1} \cot \phi)$ and like the turbine, $u_1 = \phi\sqrt{2gH_m}$ and $V_{f_1} = \psi\sqrt{2gH_m}$, and ϕ and ψ are the speed ratio and flow ratio, respectively, H_m is the total or manometric head, which will be defined in the next section. The speed ratio ϕ varies from 0.95 (low N_s impeller) to 1.25 (high N_s impeller), while the flow ratio ψ varies from 0.10 (low N_s impeller) to 0.25 (high N_s impeller).

19.5 HEAD AND EFFICIENCY

19.5.1 Heads

(i) *Suction head (h_s)*: As shown in Figure 19.1, it is the vertical height from the sump level to the centre line of the pump.

(ii) *Delivery head (h_d)*: As shown in Figure 19.1, the vertical distance between the centre line of the pump to the water surface in the delivery tank.

(iii) *Static head (H_s)*: It is the sum of the suction head and delivery head (i.e. $H_s = h_s + h_d$). It is also shown in Figure 19.1.

(iv) *Manometric head (H_m)*: It is the head against which the pump has to work. If no loss were to occur, $H_m = \frac{V_{w_1} u_1}{g}$, but loss will occur in the impeller and casing.

∴ $$H_m = \frac{V_{w_1} u_1}{g} - \text{loss of head in impeller and casing} \quad (19.4)$$

Also H_m = Total head at outlet of the pump – Total head at inlet of the pump.

$$H_m = \left(\frac{p_0}{w} + \frac{V_0^2}{2g} + Z_0\right) - \left(\frac{p_i}{w} + \frac{V_1^2}{2g} + Z_i\right) \quad (19.5)$$

where $\frac{p_0}{w} = h_d$, $\frac{V_0^2}{2g}$ is the velocity head at the outlet $= \frac{V_d^2}{2g}$, Z_0 is the vertical height of the outlet from the datum and $\frac{p_i}{w}$, $\frac{V_d^2}{2g}$ and Z_i are the corresponding values at the inlet, i.e. $$\frac{p_i}{w} = h_s, \ \frac{V_d^2}{2g} \text{ and } Z_s$$

And again $$H_m = (h_s + h_d) + h_{f_s} + h_{f_d} + \frac{V_d^2}{2g} \quad (19.6)$$

h_{f_s} is the frictional loss in the suction pipe, and h_{f_d} is the friction loss in the delivery pipe.

19.5.2 Efficiencies

In this pump, the power is first transmitted from the shaft of the electric motor to the shaft of the pump and then to the impeller. From the impeller, it is transmitted to the water. Thus, in these processes of shifting or transmitting of power from the electric motor to the pump shaft, then to the impeller and to the water, the power decreases. Therefore, the following efficiencies are involved.

(i) Manometric efficiency $(\eta_{\text{mano}}) = \dfrac{\text{Manometic head}}{\text{Head imparted by impeller to water}}$

$$\eta_{\text{mano}} = \frac{H_m}{\dfrac{V_{w_1} u_1}{g}} = \frac{g H_m}{V_{w_1} u_1} \quad (19.7)$$

The power given to the water at the outlet of the pump $= \dfrac{W H_m}{1000}$ kW

Power at the impeller $= \dfrac{\text{Work done by impeller per sec}}{1000}$ kW

$$= \frac{W}{g} \cdot \frac{V_{w_1} u_1}{1000} \text{ kw}$$

$$\therefore \quad \mu_{\text{mano}} = \frac{W H_m / 1000}{\dfrac{W}{g} \cdot \dfrac{V_{w_1} u_1}{1000}} = \frac{g H_m}{V_{w_1} u_1}$$

(ii) Mechanical efficiency $(\eta_m) = \dfrac{\text{Power at impeller}}{\text{Power at the shaft}}$

$$\eta_m = \frac{\dfrac{W}{g} \cdot \left(\dfrac{V_{w_1} u_1}{1000} \right) \text{in kW}}{\text{Shaft Power (S.P.) in kW}} \quad (19.8)$$

(iii) Overall efficiency $(\eta_0) = \dfrac{\text{Weight of Water lifted} \times H_m}{\text{Power supplied by motor to shaft}}$

$$\eta_0 = \frac{(W H_m / 1000) \text{in kW}}{\text{S.P. in kW}} \quad (19.9)$$

and also $$\eta_0 = \eta_{\text{mano}} \times \eta_m \quad (19.10)$$

EXAMPLE 19.1 The internal and external diameters of the impeller of a centrifugal pump are 30 cm and 60 cm, respectively. The pump runs at 1000 rpm. The vane angles at the inlet and outlet are 20° and 30°, respectively. The water enters the impeller radially and the velocity of the flow is constant. Determine the work done by the impeller per unit weight of water.

Solution: The given data are:

$D = 30$ cm $= 0.3$ m, $D_1 = 60$ cm $= 0.6$ m, $N = 1000$ rpm

Vane angle at the inlet $\theta = 20°$, vane angle at the outlet $\phi = 30°$

Water enters radially, i.e. $\alpha = 90°$, hence $V_w = 0$, Velocity of flow is constant, i.e. $V_f = V_{f1}$

We know $u = \dfrac{\pi DN}{60} = \dfrac{\pi \times 0.3 \times 1000}{60}$ 15.7 m/sec

$$u_1 = \frac{\pi D_1 N}{60} = \frac{\pi \times 0.6 \times 1000}{60} = 31.4 \text{ m/sec}$$

From the inlet velocity diagram: (Figure 19.3)

$$\tan 20° = \frac{V_f}{u} = \frac{V_f}{15.7}$$

$$\therefore \quad V_f = 5.714 \text{ m/sec} = V$$

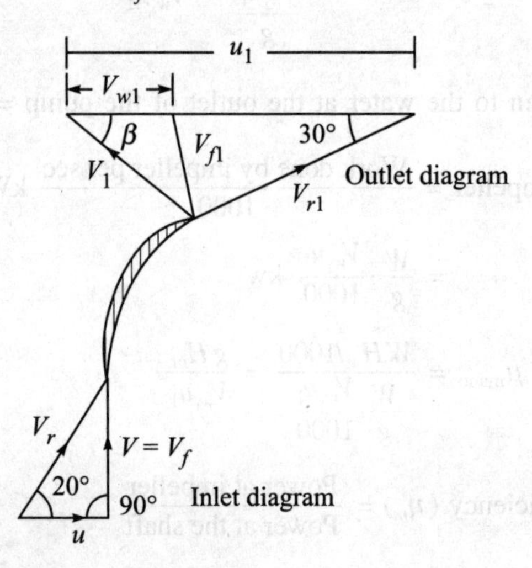

Figure 19.3 Visual of Example 19.1.

From the outlet velocity diagram: (Figure 19.3)

$$\tan 30° = \frac{V_{f_1}}{u_1 - V_{w_1}} = \frac{5.714}{31.4 - V_{w_1}} \quad \because V_{f1} = V_f$$

$$-V_{w_1} \tan 30° = 31.4 \tan 30° = 5.714$$

$$V_{w_1} \tan 30° = 31.4 \tan 30° - 5.714$$

$$\therefore \quad V_{w_1} = \frac{31.4 \tan 30° - 5.714}{\tan 30°} = 21.5 \text{ m/sec}$$

$\therefore$ Work done by impeller per unit weight per second is:

$$= \frac{1}{g} V_{w_1} u_1 = \left(\frac{1}{9.81} \times 21.5 \times 31.4\right) \text{Nm/N}$$

$$= 68.827 \text{ Nm/N}$$ **Ans.**

EXAMPLE 19.2 The outer diameter of a centrifugal pump is equal to two times the inner diameter. The pump runs at 1200 rpm and works against the total head of 75 m. The velocity of flow through the impeller is constant and equal to 3 m/sec. The vanes are set back at an angle of 30° at the outlet. If the outer diameter of the impeller is 60 cm and width at the outlet 5 cm, determine: (i) Vane angle at the inlet, (ii) Work done by the impeller per second and (iii) Manometric efficiency.

Solution: The given data are: $D_1 = 60 \text{ cm} = 0.6 \text{ m} \therefore D = \frac{D_1}{2} = 0.3 \text{ m}$

$$N = 1200 \text{ rpm}, H_m = 75 \text{ m}, V_f = V_{f_1} = 3 \text{ m/sec}$$

$$\phi = 30°, B_1 = 5 \text{ cm} = 0.05 \text{ m}$$

The tangential velocity $u = \frac{\pi DN}{60} = \frac{\pi \times 0.3 \times 1200}{60} = 18.85$ m/sec;

$$u_1 = \frac{\pi D_1 N}{60} = \frac{\pi \times 0.6 \times 1200}{60} = 37.7 \text{ m/sec}$$

Discharge $Q = \pi D_1 B_1 V_{f_1} = (\pi \times 0.6 \times 0.05 \times 3) \text{ m}^2\text{/sec} = 0.2827 \text{ m}^2\text{/sec}$

From the inlet diagram, vane angle θ is (Figure 19.4)

$$\tan \theta = \frac{V_f}{u} = \frac{3}{18.85}$$

(i) $\theta = 9° \, 2.5'$ **Ans.**

From the outlet velocity diagram, (Figure 19.4)

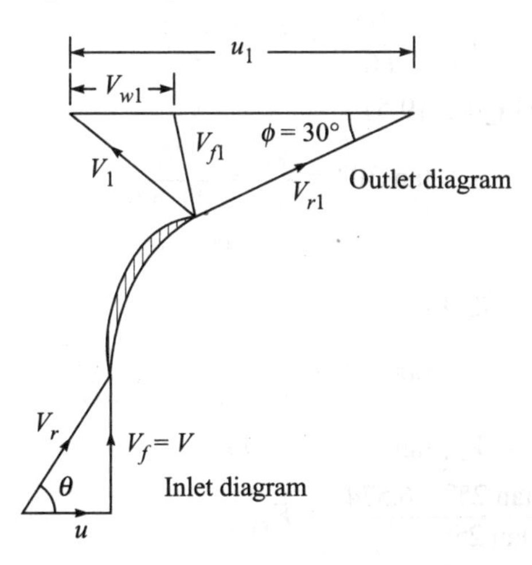

Figure 19.4 Visual of Example 19.2.

$$\tan 30° = \frac{V_{f_1}}{u_1 - V_{w_1}} = \frac{3}{37.7 - V_{w_1}}$$

or $$37.7 \tan 30° - V_{w1} + \tan 30° = 3$$

or $$\frac{37.7 \tan 30° - 3}{\tan 30°} = V_{w1}$$

∴ $$V_{w1} = 32.5 \text{ m/sec}$$

(ii) Work done by the impeller on water per sec

$$= \frac{W}{g} V_{w1} u_1 = \frac{\rho g Q}{g} V_{w1} u_1$$

$$= (1000 \times 0.2827 \times 32.5 \times 37.7) \text{ Nm/sec}$$

$$= 346419.18 \text{ Nm/sec}$$

$$= 346.419 \text{ Kw/sec}$$ **Ans.**

(iii) Manometric efficiency $\eta_{mano} = \frac{gH_m}{V_{w_1} u_1} = \frac{9.81 \times 75}{32.5 \times 37.7} = 0.60048 = 60.048\%$ **Ans.**

EXAMPLE 19.3 The impeller of the centrifugal pump is 30 cm at the outlet of 15 cm at the inlet. Impeller vane angles are 30° and 25° at the inner and outer peripheries, respectively and the speed is 1450 rpm. The velocity of the flow through the impeller is constant. Find the work done per unit weight per second.

Solution: The given data are: $D_1 = 30 \text{ cm} = 0.3 \text{ m}$, $D = 15 \text{ cm} = 0.15 \text{ m}$

$$\theta = 30°, \quad \phi = 25°, \quad N = 1450 \text{ rpm}, \; V_f = V_{f1}$$

Find $$u = \frac{\pi D_1 N}{60} = \frac{\pi \times 0.15 \times 1450}{60} = 11.388 \text{ m/sec}$$

$$u_1 = \frac{\pi D N}{60} = \frac{\pi \times 0.3 \times 1450}{60} = 22.776 \text{ m/sec}$$

From the inlet diagram: (Figure 19.5)

$$\tan 30° = \frac{V_f}{u} = \frac{V_f}{11.388}$$

∴ $$V_f = 6.5748 \text{ m/sec} = V_{f1}$$

From the outlet diagram: (Figure 19.5)

$$\tan 25° = \frac{V_f}{u_1 - V_{w_1}} = \frac{6.574}{22.776 \; V_{w_1}}$$

∴ $$22.776 \tan 25° - V_{w1} \tan 25° = 6.574$$

∴ $$\frac{22.776 \tan 25° - 6.574}{\tan 25°} = V_{w1}$$

∴ $$V_{w1} = 8.678 \text{ m/sec}$$

$\therefore$ Work done by unit weight per sec $= \dfrac{V_{w_1}u_1}{g}$

$$= \left(\frac{8.678 \times 22.776}{9.81}\right) \text{ N–m/N}$$

$$= 20.1478 \text{ Nm/N}$$ **Ans.**

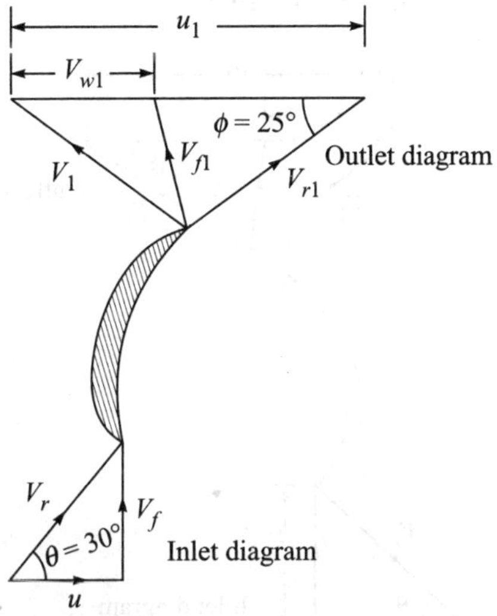

Figure 19.5 Visual of Example 19.3.

EXAMPLE 19.4 A centrifugal pump discharges water at 0.11 m^3/sec at 1450 rpm against a head of 23 m. If the diameter of the impeller at the outlet is 0.25 m and its width 0.05 m, find the vane angle at outer periphery. Take $\eta_{\text{mano}} = 0.75$ (i.e. 75%).

Solution: The given data are: $Q = 0.11$ m^3/sec, $N = 1450$ rpm

$$H_m = 23 \text{ m} \quad D_1 = 0.25 \text{ m}, \quad B_1 = 0.05 \text{ m}$$

$$\eta_{\text{mano}} = 0.75$$

Now $$u_1 = \frac{\pi D_1 N}{60} = \frac{\pi \times 0.25 \times 1450}{60} = 18.98 \text{ m/sec}$$

$$\eta_{\text{mano}} = \frac{H_m}{\dfrac{(V_{w_1}u_1)}{g}} = \frac{gH_m}{V_{w_1}u_1}$$

or $$0.75 = \frac{9.81 \times 23}{V_{w_1} \times 18.98}$$

$\therefore$ $$V_{w_1} = \frac{9.81 \times 23}{0.75 \times 18.98} = 15.85 \text{ m/sec}$$

Again $$Q = \pi D_1 B_1 V_{f_1}$$

$$0.11 = \pi \times 0.25 \times 0.05 \times V_{f_1}$$

$$\therefore \quad V_{f_1} = 2.801 \text{ m/sec}$$

From the outlet velocity diagram: (Figure 19.6)

$$\tan \phi = \frac{V_{f_1}}{u_1 - V_{w_1}} = \frac{2.801}{18.98 - 15.85}$$

$$\tan \phi = 41° \; 49.5'$$ **Ans.**

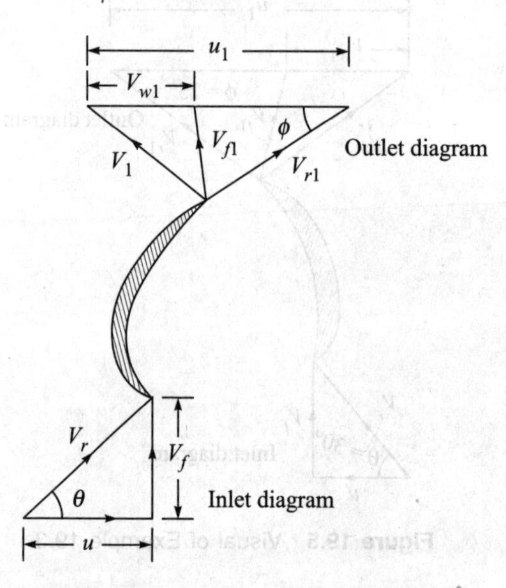

Figure 19.6 Visual of Example 19.4.

EXAMPLE 19.5 A centrifugal pump runs at 1000 rpm and delivers water against a head of 15 m. The impeller diameter and width at the outlet are 0.3 m and 0.05 m, respectively. The vanes are curved back at an angle of 30° with the periphery at the outlet. $\eta_{mano} = 0.92$, find Q.

Solution: The given data are: $N = 1000$ rpm, $H_m = 15$ m

$$D_1 = 0.3 \text{ m}, \quad B_1 = 0.05$$

$$\phi = 30°, \quad \eta_{mano} = 0.92$$

$$\text{Find } u_1 = \frac{\pi D_1 N}{60} = \frac{\pi \times 0.3 \times 1000}{60} = 15.707 \text{ m/sec}$$

$$\eta_{mano} = \frac{H_m}{\dfrac{V_{w_1} u_1}{g}} = \frac{g H_m}{V_{w_1} u_1}$$

$$0.92 = \frac{9.81 \times 15}{V_{w_1} \times 15.707}$$

$$\therefore \quad V_{w_1} = 10.1824 \text{ m/sec}$$

From the outlet diagram: (Figure 19.7)

$$\tan 30° = \frac{V_{f_1}}{u_1 - V_{w_1}} = \frac{V_{f_1}}{15.707\ 10.1824}$$

$$\tan 30° = \frac{V_{f_1}}{5.5246}$$

$\therefore$ $$V_{f1} = 3.1896 \text{ m/sec}$$

$\therefore$ $$Q = \pi D_1 B_1 V_{f1} = (\pi \times 0.3 \times 0.05 \times 3.1896) \text{ m}^3\text{sec}$$

$$Q = 0.1503 \text{ m}^3\text{/sec}$$ **Ans.**

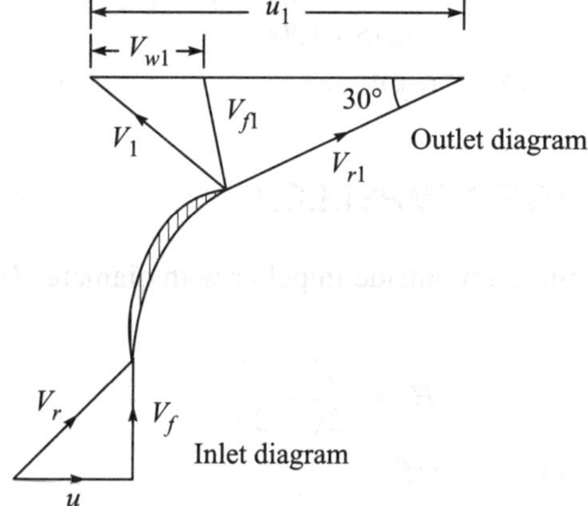

Figure 19.7 Visual of Example 19.5.

EXAMPLE 19.6 A centrifugal pump delivers 0.03 m^3/sec of water to a height 18.25 m through a 10 cm pipe diameter which is 90 m long. If the overall efficiency of the pump is 75%, find the power required to drive the pump. Take friction factor $f = 0.04$.

Solution: The given data are: $Q = 0.03$ m^3/sec, $h_s + h_d = 18.25$ m

$$(l_s + l_d) = 90 \text{ m}, \quad d_p = 10 \text{ cm} = 0.1 \text{ m}$$

$$\eta_0 = 75\%, \quad f = 0.04$$

$$\text{Velocity of water in the pipe} = V_s = V_d = \frac{Q}{D_4 (d_p)^2}$$

$$V = V_s = V_d = \frac{0.03}{D_4 (0.1)^2} = 3.8197 \text{ m/sec}$$

$$\text{Loss of head due to friction} = (h_{f_s} + h_{fd}) = \frac{fLV^2}{2gd_p} = \frac{0.04 \times 90 \times 3.8197^2}{2 \times 9.81 \times 0.1}$$

$$= 26.77 \text{ m}$$

Again $$\frac{V_d^2}{2g} = \frac{3.8187^2}{2 \times 9.81} = 0.7436 \text{ m.}$$

Now Manometric $H_m = [(h_s + h_d) + (h_{f_s} + h_{f_s}) + V_d^2/2g]$

$$= (18.25 + 26.77 + 0.7436) \text{ m}$$

$$= 45.7636 \text{ m}$$

$$\text{Overall efficiency } \eta_0 = \frac{(WH_m/1000) \text{ in kW}}{\text{Power reqd to drive the pump i.e. S.P.}}$$

or

$$0.75 = \frac{(\rho g Q/1000)45.7636}{\text{S.P.}}$$

$\therefore$

$$\text{S.P.} = \frac{1000 \times 9.81 \times 0.03 \times 45.7636}{0.75 \times 1000}$$

$$\text{S.P.} = 17.9576 \text{ kW.}$$

Ans.

19.6 LEAST DIAMETER IMPELLER

It is possible to find the minimum outside impeller with diameter D_1 to enable the pump to start at its normal speed.

We know that

$$H_m = \frac{u_1^2}{2g} - \frac{u^2}{2g}$$

But $u_1 = \omega R_1$ and $u = \omega R$

$$\therefore \quad H_m = \frac{(\omega R_1)^2}{2g} + \frac{(\omega R)^2}{2g}$$

$$\because \quad R_1 = \frac{D_1}{2}, \text{ and } R = \frac{D}{2}$$

$$\therefore \quad H_m = \frac{\omega^2}{8g}(D_1^2 - D^2)$$

In normal design, $D_1 = 2D$

$$\therefore \quad H_m = \frac{\omega^2}{8g}\left[D_1^2 - \left(\frac{D_1}{2}\right)\right]$$

$$H_m = \frac{\omega^2}{8g}\left(D_1^2 - \frac{D_1^2}{4}\right)$$

$$H_m = \frac{\omega^2}{32} \cdot \frac{3D_1^2}{g} = \frac{3}{32}\left(\frac{\omega^2 D_1^2}{g}\right)$$

$$\therefore \quad D_1 = \sqrt{\frac{32 \times 9.81 \times H_m}{3\omega^2}} = \sqrt{\frac{32 \times 9.81 \times H_m}{3}} \cdot \frac{1}{\omega}$$

$$\therefore \quad D_1 = \frac{10.2293}{\omega}\sqrt{H_m}$$

If the rotational speed is N rpm, $\omega = \dfrac{2\pi N}{60}$

$$D_1 = \frac{10.229}{\dfrac{2\pi N}{60}}\sqrt{H_m}$$

$$D_1 = 97.683\,\frac{\sqrt{H_m}}{N} \text{ metre} \qquad (19.11)$$

If the manometric efficiency is 75%,

$$H_m = 0.75\ \text{H}$$

where H is the theoretical lift.

$$\therefore \qquad D_1 = \frac{84.596}{N}\sqrt{H} \qquad (19.12)$$

Thus Equations (19.11) and (19.12) give the least diameter without efficiency and with the need for considering manometric efficiency.

19.7 MINIMUM STARTING SPEED

When the pump is started, no flow of water takes place until the pressure difference in the impeller is large enough to overcome the gross manometric head. When the pump rotates, the pressure head developed due to centrifugal force is $\left(\dfrac{u_1^2}{2g} - \dfrac{u^2}{2g}\right)$. Since $u_1 = \omega R_1$ and $u = \omega R$, the pressure head developed $= \left(\dfrac{\omega^2 R_1^2}{2g} - \dfrac{\omega^2 R^2}{2g}\right)$. The flow will commence if $\left(\dfrac{\omega^2 R_1^2}{2g} - \dfrac{\omega^2 R^2}{2g}\right) \geq H_m$

Thus for maximum speed, $\left(\dfrac{\omega^2 R_1^2}{2g} - \dfrac{\omega^2 R^2}{2g}\right) = H_m$

or

$$\left(\frac{u_1^2}{2g} - u^2/2g\right) = H_m \qquad (19.13)$$

But

$$\eta_{\text{mano}} = \frac{gH_m}{V_{w_1}u_1}, \qquad \because H_m = \eta_{\text{mano}} \cdot \frac{V_{w_1}u_1}{g}$$

Hence for minimum speed $\left(\dfrac{\omega^2 R_1^2}{g} - \dfrac{\omega^2 R^2}{g}\right) = \eta_{\text{mano}} \times \dfrac{V_{w_1}u_1}{g}$

But

$$\omega R_1 = u_1 = \frac{\pi D_1 N}{60}, \qquad \omega R = u = \frac{\pi DN}{60}$$

$$\therefore \qquad \frac{1}{2g}\left(\frac{\pi D_1 N}{60}\right)^2 - \frac{1}{2g}\left(\frac{\pi DN}{60}\right) = \eta_{\text{mano}} \cdot \frac{V_{w_1}(\pi D_1 N/60)}{g}$$

Dividing by $\left(\dfrac{\pi N}{g \times 60}\right)$,

$$\frac{\pi D_1^2 N}{120} - \frac{\pi D^2 N}{120} = \eta_{\text{mano}} \cdot V_{w_1} D_1$$

$$\frac{\pi N}{120}(D_1^2 - D^2) = \eta_{\text{mano}} \times V_{w_1} D_1$$

$$N = \frac{120 \times \eta_{\text{mano}} \times V_{w_1} D_1}{\pi (D_1^2 - D^2)} \tag{19.14}$$

which is the minimum starting speed of the centrifugal pump.

EXAMPLE 19.7 Find the minimum speed at which a centrifugal pump will start functioning against a head of 7.5 m of diameter of the impeller at the outlet and inlet area, which are 100 cm and 50 cm, respectively.

Solution: H_m = 7.5 m, D_1 = 100 cm = 1 m, D = 50 cm = 0.5 m

Since $D_1 = 2D$, $u_1 = 2u$

∴ For minimum speed, use Equation (19.13):

$$\frac{u_1^2}{2g} - \frac{u^2}{2g} = H_m$$

or

$$\frac{u_1^2}{2g} - \frac{(u_1/2)^2}{2g} = H_m$$

∴

$$3/4\left(\frac{u_1^2}{2g}\right) = H_m$$

$$\frac{3}{4}\left(\frac{u_1^2}{2g}\right) = 7.5$$

$$u_1^2 = \frac{4 \times 7.5 \times 2 \times 9.81}{3} = 196.2$$

∴

$$u_1 = 14.00714 \text{ m/sec}$$

or

$$\frac{\pi D_1 N}{60} = 14.00714$$

∴

$$N = \frac{60 \times 14.00714}{\pi \times 4} = 267.516 \text{ rpm}$$ **Ans.**

EXAMPLE 19.8 The diameters of the impeller of a centrifugal pump at the inlet and outlet are 30 cm and 60 cm, respectively. The velocity of flow at the outlet is 2.5 m/sec and the vanes are set back at an angle of 45° at the outlet. Determine the minimum starting speed of the pump if the manometric efficiency is 75%.

Solution: The given data are: D = 30 cm = 0.3 m, D_1 = 60 cm = 0.6 m

$$V_{f_1} = 2.5 \text{ m/sec}, \quad \phi = 45°, \quad \eta_{mano} = 75\% = 0.75$$

Take the minimum starting speed as N.

Then N (by Equation 19.14) is:

$$N = \frac{120 \times \eta_{mano} \times V_{w_1} \times D_1}{\pi(D_1^2 - D^2)} \tag{19.14}$$

Now, we know the relation of $\tan \phi = \dfrac{V_{f_1}}{u_1 - V_{w_1}}$

$$u_1 - V_{w_1} = \frac{V_{f_1}}{\tan 45°} = \frac{2.5}{\tan 45°} = 2.5$$

$\therefore$ $$V_{w_1} = u_1 - 2.5$$

and $$u_1 = \frac{\pi D_1 N}{60} = \frac{\pi \times 0.60 N}{60} = 0.031416\, N$$

$\therefore$ $$V_{w_1} = (0.031416\, N - 2.5)$$

Using the above equation:

$$N = \frac{120 \times 0.75 \times (0.031416\, N - 2.5 \times 0.6}{\pi\,(0.6^2 - 0.3^2)}$$

Solving for N = 159.3 rpm **Ans.**

19.8 MULTI-STAGE CENTRIFUGAL PUMPS

A centrifugal pump with a single impeller cannot develop a delivery head if it is quite high. For the requirement of a high head centrifugal pump, a number impellers are connected in series or on the shaft as shown in Figure 19.8 (two-stage). Here the discharge from the first impeller is guided to the inlet of the second impeller. The discharge from the second impeller is guided to the third impeller and this may be continued to the nth impeller. Finally, the discharge from the last or nth impeller is directed to the delivery pipe. If the head impressed on each impeller is H_m, the total head developed i.e. $H_{total} = n\, H_m$. The same Q passes through each impeller which can be delivered in a higher head. Thus, this series connection can develop high delivery head.

But when the impellers or pumps are connected in parallel as shown in Figure 19.9, high discharge is delivered for the given head H_m. Here impellers or pumps are mounted on separate shafts (shown in Figure 19.4). The delivery pipe of each pump is connected to a common delivery pipe, which collects all discharges from different pumps in parallel.

If Q_1, Q_2, Q_3, ..., Q_n are the discharges of different pumps, the total discharge is:

$$Q_{total} = Q_1 + Q_2 + Q_3 +, \ldots, + Q_n.$$

Thus this arrangement of pumps in parallel can produce a high discharge with a small head H_m.

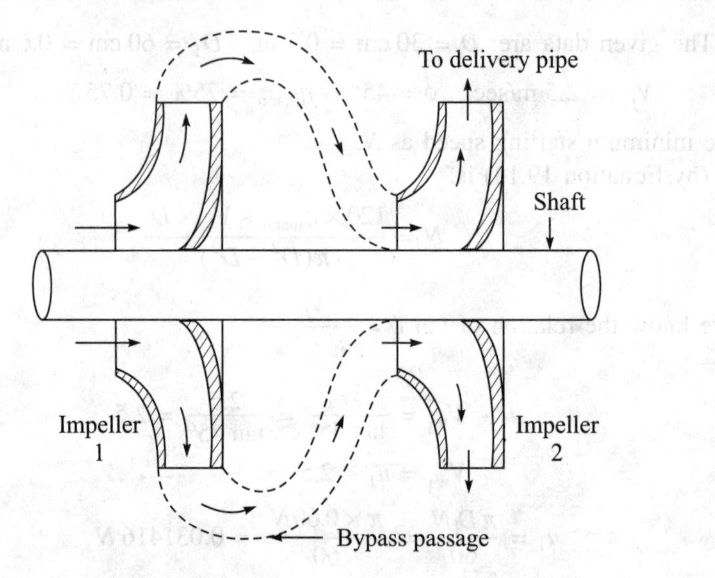

Figure 19.8 Impeller in series (2-stage).

EXAMPLE 19.9 A two-stage centrifugal pump with an impeller in series has an outer diameter of 40 cm and a width of 2.5 cm. The discharge of the pump is 0.06 m^3/sec at a speed of 1000 rpm. The vane angle at the outlet is 30°. Assuming $\eta_{mano} = 80\%$, calculate the manometric head developed by the pump.

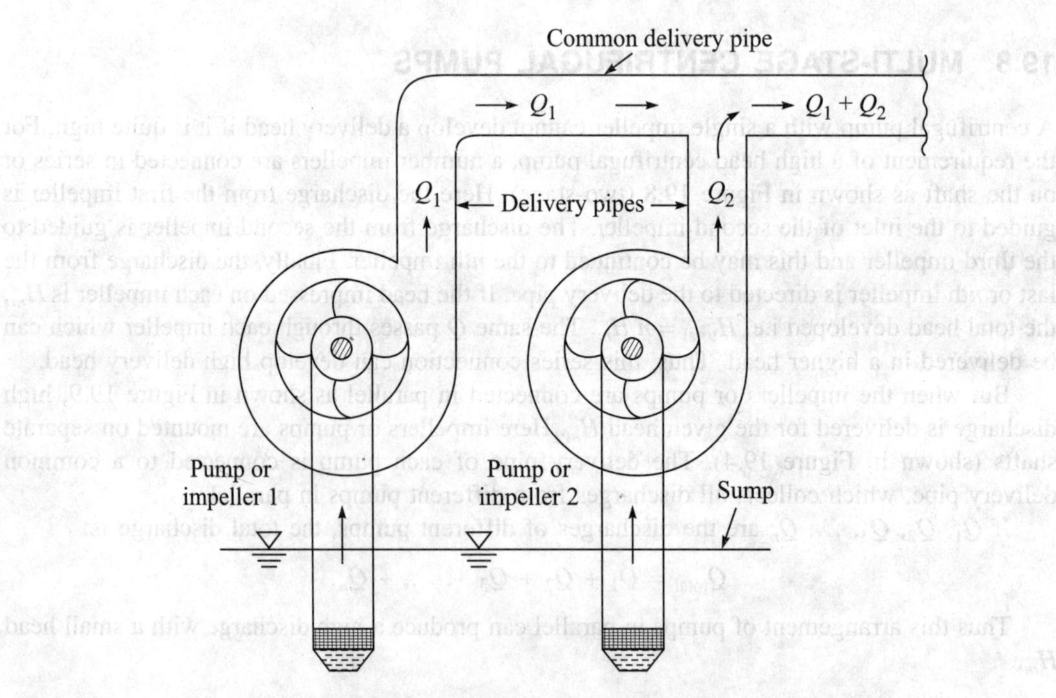

Figure 19.9 Pumps (or impellers) in parallel to higher discharge.

Solution: The given data are: two-stage pump with two impeller in series

$$D_1 = 40 \text{ cm} = 0.4 \text{ m}, \quad B_1 = 2.5 \text{ cm} = 0.025 \text{ m}$$

$$Q = 0.06 \text{ m}^3\text{/sec}, \; N = 1000 \text{ rpm}, \quad \phi = 30°$$

$$\phi_{\text{mano}} = 80\%$$

Now

$$u_1 = \frac{\pi D_1 N}{60} = \frac{\pi \times .4 \times 1000}{60} = 20.943 \text{ m/sec}$$

$$Q = \pi D_1 B_1 V_{f_1} \quad \Rightarrow \quad 0.06 = \pi \times 0.4 \times 0.025 \times V_f$$

$\therefore$

$$V_f = 1.9098 \text{ m/sec}$$

From the outlet diagram: Figure (19.10)

$$\tan 30° = \frac{u_1 - V_{w_1}}{V_{f_1}} \quad \Rightarrow \quad \tan 30° = \frac{20.943 - V_{w_1}}{1.9098}$$

or

$$1.10265 = 20.943 - V_{w_1}$$

$\therefore$

$$V_{w_1} = 19.84 \text{ m/sec}$$

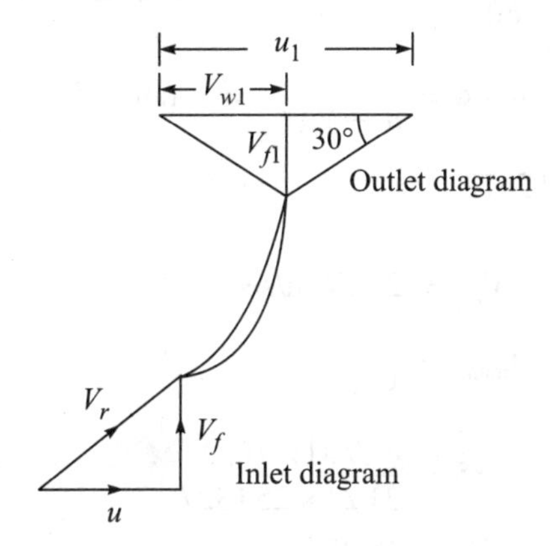

Figure 19.10 Visual of Example 19.9.

The head generated by one impeller $= \dfrac{V_{w_1} u_1}{g} = \dfrac{19.84 \times 20.943}{9.81}$

$$= 42.355 \text{ m}$$

If H_m is the manometric head developed by one impeller,

$$H_m = \eta_{\max} \times \left(\frac{V_{w_1} u}{g}\right)$$

$$= 0.80 \times 42.355$$

$$= 33.8845 \text{ m}$$

Since it is a two-stage pump,
Total head developed = 2 × 33.8845
= 67.769 m **Ans.**

EXAMPLE 19.10 A three-stage centrifugal pump has an impeller of 40 cm in diameter and 2.5 cm wide the outlet. The vanes are curved back at the outlet at 30° and reduce the circumferential area by 15%. The manometric efficiency is 85% and the overall efficiency is 75%. Determine the head generated by the pump when it is running at 1200 rpm and discharging water at 0.06 m³/sec. Also find the shaft power.

Solution: The given data are: $n = 3$, $D_1 = 40$ m, $B_1 = 2.5$ cm $= 0.025$ m

$\phi = 30°$, Reduction of circumferential area $= 0.15$, $\eta_{mano} = 85\% = 0.85$

$\eta_0 = 75\% = 0.75$, $Q = 0.06\ \text{m}^3/\text{sec}$

$N = 1200$ rpm

Now the area at outlet $(\pi D_1 B_1)(1 - .15) = \pi \times .4 \times 0.25 \times .85 = 0.0267\ \text{m}^2$

$$\therefore \quad V_{f1} = \frac{Q}{\text{outlet area}} = \frac{0.06}{0.0267} = 2.2472\ \text{m/sec}$$

$$u_1 = \frac{\pi D_1 N}{60} = \frac{\pi \times .4 \times 1200}{60} = 25.1327\ \text{m/sec}$$

$$\tan\phi = \frac{V_{f_1}}{u_1 - V_{w_1}} \Rightarrow \tan 30° = \frac{2.2472}{25.1327 - V_{w_1}}$$

or
$$\frac{25.1327 \tan 30° - 2.2472}{\tan 30°} = V_{w1}$$

$$\therefore \quad V_{w1} = 21.24\ \text{m/sec}$$

$$\eta_{mano} = \frac{gH_m}{V_{w_1} u_1}$$

$$0.85 = \frac{9.81 \times H_m}{21.24 \times 25.1327}$$

$$H_m = 46.253\ \text{m}$$

$\therefore$ Total head generated $= n\,H_m = (3 \times 46.253)$ m

$= 138.76$ m **Ans.**

$$\text{Again } \eta_0 = \frac{\text{Power output of pump}}{\text{Power output at pump}} = \frac{\rho g Q \times 138.76}{1000 \times \text{S.P.}}$$

$$0.75 = \frac{1000 \times 9.81 \times 0.06 \times 138.76}{1000 \times \text{S.P.}}$$

$\therefore$ S.P $= 108.8988$ kW **Ans.**

19.9 SPECIFIC SPEED OF CENTRIFUGAL PUMP (N_S)

The specific speed N_s of a centrifugal pump is defined as the speed of a geometrically similar pump which delivers 1 cubic metre of liquid per second under a head of 1 metre.

To derive an expression of N_s,
We know $Q = \pi DBV_f$

$\therefore \quad Q \propto DBV_f$

But $\quad B \propto D$

$\therefore \quad Q \propto D^2V_f \qquad$ (A)

We also know that $u = \dfrac{\pi DN}{60} \quad \therefore \; u \propto DN$

Both u and V_f are related to $\sqrt{H_m}$

$\therefore \quad u \propto V_f \propto \sqrt{H_m}$

$\therefore \quad \sqrt{H_m} \propto DN$

$\therefore \quad D \propto \dfrac{\sqrt{H_m}}{N}$

Equation (A) becomes $\quad Q \propto \dfrac{H_m}{N^2}\, V_f$

or $\quad Q \propto \dfrac{H_m}{N^2}\sqrt{H_m} \qquad \therefore \; V_f \propto \sqrt{H_m}$

or $\quad Q \propto \dfrac{H_m^{3/2}}{N^2}$

$\therefore \quad Q = C\,\dfrac{H_m^{3/2}}{N^2}$

When $Q = 1\ \text{m}^3/\text{sec}$ and $H = 1$ m, $N = N_s$

$$1 = \frac{C}{N_s^2}$$

$\therefore \quad C = N_s^2$

$\therefore \quad Q = N_s^2\,\dfrac{H_m^{3/2}}{N^2}$

$\therefore \quad N_s^2 = \dfrac{N^2 Q}{H^{3/2}}$

$\therefore \quad N_s = \dfrac{N\sqrt{Q}}{H^{3/4}} \qquad$ (19.15)

Equation 19.15 gives the expression of specific speed N_s.

EXAMPLE 19.11 Water is to be pumped out of a deep well under a total head of 156 cm. There are a number of identical pumps of design speed of 1000 rpm and specific speed of 20 with a rated capacity of 150 litres/sec. Find the number of pumps required.

Soluton: The given data are: Total head = 156 m

Revolution of identical pumps $N = 1000$ rpm

$$N_s = 20$$

$$Q = 150 \text{ lits/sec} = 0.15 \text{ m}^3\text{/sec}$$

$$N_s = \frac{N\sqrt{Q}}{H^{3/4}}$$

$$20 = \frac{1000\sqrt{0.15}}{H_m^{3/4}}$$

or

$$H_m^{3/4} = \frac{1000\sqrt{0.15}}{20}$$

$\therefore$

$$H_m = 52 \text{ m}$$

$$\text{The number of pumps required} = \frac{\text{Total head}}{\text{Head developed by one pump}}$$

$$= \frac{156}{52} = 3$$

Ans.

19.10 MODEL TESTING OF PUMPS

As in turbines, tests on model are necessary to predict the performance of prototypes. Such model tests are usually done with the help of some terms, which give a constant value when there is a complete similarity between actual pumps (prototype) and models. Complete similarity between models and their prototypes exists if the following conditions are satisfied:

(a) Specific speed of model = Specific speed of prototype

$$(N_s)_m = (N_s)_p$$

$$\left(\frac{N\sqrt{Q}}{H_m^{3/4}}\right)_m = \left(\frac{N\sqrt{Q}}{H_m^{3/4}}\right)_p \tag{19.16}$$

(b)

$$u = \frac{\pi DN}{60} \quad \therefore u \propto DN$$

and

$$u \propto \sqrt{H}$$

$\therefore$

$$\sqrt{H} \propto DN$$

$\therefore$

$$\frac{\sqrt{H}}{DN} = \text{const.}$$

$\therefore$

$$\left(\frac{\sqrt{H}}{DN}\right)_m = \left(\frac{\sqrt{H}}{DN}\right)_p \tag{19.17}$$

(c)

$$Q = \pi DBV_f$$

$\therefore$

$$Q \propto DBV_f$$

$$D \propto B$$

$\therefore$ $$Q \propto D^2 V_f$$

But $$V_f \propto u \propto DN$$

$\therefore$ $$Q \propto D^2 \cdot DN$$

$\therefore$ $$Q \propto D^3 N$$

$\therefore$ $$\frac{Q}{D^3 N} = \text{constant}$$

$\therefore$ $$\left(\frac{Q}{D^3 N}\right)_m = \left(\frac{Q}{D^3 N}\right)_p \tag{19.18}$$

(d) Power of the pump $P = \dfrac{\rho g Q H_m}{1000}$

$\therefore$ $$P \propto Q H_m$$

But $$Q \propto D^3 N$$

and $$\sqrt{H} \propto DN$$

or $$H \propto D^2 N^2$$

$\therefore$ $$P (D^3 N)\,(D^2 N^2)$$

or $$P \propto D^5 N^3$$

$\therefore$ $$\frac{P}{D^5 N^3} = \text{constant}$$

$\therefore$ $$\left(\frac{P}{D^5 N^3}\right)_m = \left(\frac{P}{D^5 N^3}\right)_p \tag{19.19}$$

EXAMPLE 19.12 The diameter of a centrifugal pump which discharges 0.035 m^3/sec of water against a total head of 25 m, is 0.5 m. The pump is running at 1200 rpm. Find the head, discharge and ratio of power of a geometrically similar pump of diameter 0.3 m when it is running at 2000 rpm.

Solution: The given data are: First pump data

$$Q_1 = 0.035 \text{m}^3/\text{sec}, \quad H_1 = 25 \text{ m}$$
$$D_1 = 0.5 \text{ m}, \quad N_1 = 1200 \text{ rpm}$$

The data for a geometrically similar pump are:

$$D_2 = 0.3 \text{ m}, \quad N_2 = 2000 \text{ rpm}$$

Since the pumps are geometrically similar,

$$\frac{\sqrt{H_1}}{D_1 N_1} = \frac{\sqrt{H_2}}{D_2 N_2} \tag{19.17}$$

$$\frac{\sqrt{25}}{0.5 \times 1200} = \frac{\sqrt{H_2}}{0.3 \times 2000}$$

$$0.5 \times 1200\ \sqrt{H_2} = 0.3 \times 2000 \times \sqrt{25}$$

$$\sqrt{H_2} = \frac{0.3 \times 2000 \times \sqrt{25}}{0.5 \times 1200} = 5$$

$$H_2 = 25 \text{ m}$$ **Ans.**

Again
$$\left(\frac{Q}{D^3 N}\right)_1 = \left(\frac{Q}{D^3 N}\right)_2 \quad (19.18)$$

$$\left(\frac{0.035}{0.5^3 \times 1200}\right) = \frac{Q^2}{0.3^3 \times 2000}$$

∴
$$Q_2 = 0.0126 \text{ m}^3/\text{sec}$$ **Ans.**

Again
$$\left(\frac{P}{D^5 N^3}\right)_1 = \left(\frac{P}{D^5 N^3}\right)_2 \quad (19.19)$$

$$\frac{P_1}{P_2} = \frac{D_1^5 N_1^3}{D_2^5 N_2^3} = \left(\frac{D_1}{D_2}\right)^5 \left(\frac{N_1}{N_2}\right)^3$$ **Ans.**

$$\frac{P}{P_2} = \left(\frac{0.5}{0.3}\right)^5 \left(\frac{1200}{2000}\right)^3$$

$$\frac{P_1}{P_2} = 2.7777$$ **Ans.**

19.11 PERFORMANCE OF PUMPS: CHARACTERISTIC CURVES

A pump is designed for a particular speed, flow rate and head. But in practice, it has to run for some other speed, Q and H. Due to this changed state, the behaviour of the pump may be different. When Q changes, velocities like V_w, u, etc. will change and eventually the efficiency may be low as the loss of energy has taken place in a changed state. In order to predict the performance and behaviour of the pump under different varying conditions of Q, H, P, tests are performed and the results are plotted. These plotted curves are called 'characteristic curves'. The following are three characteristic curves for a centrifugal pump.

(i) Main and Operating Characteristic Curves
(ii) Constant Efficiency Characteristic Curves
(iii) Constant Head and Discharge Curves

Main and Operating Characteristic Curves

Here a speed of the pump is selected, (say, 1000 rpm), and for this speed Q is varied by adjusting the delivery valve. For different values of Q, the corresponding values of H_m, shaft power (S.P.), η_0 are calculated. Again, another speed is selected, say 800 rpm. Again for different values of Q, H_m, S.P. and efficiency are calculated. The same operation is repeated for different speeds. Then $Q\ V_r\ H_m$, $Q\ V_r$ S.P., $Q\ V_s$ and η_0 are plotted for different speeds as shown in

Figure 19.11(a) which represents the main characteristics of the pump. These curves indicate the performance of a pump at different speeds.

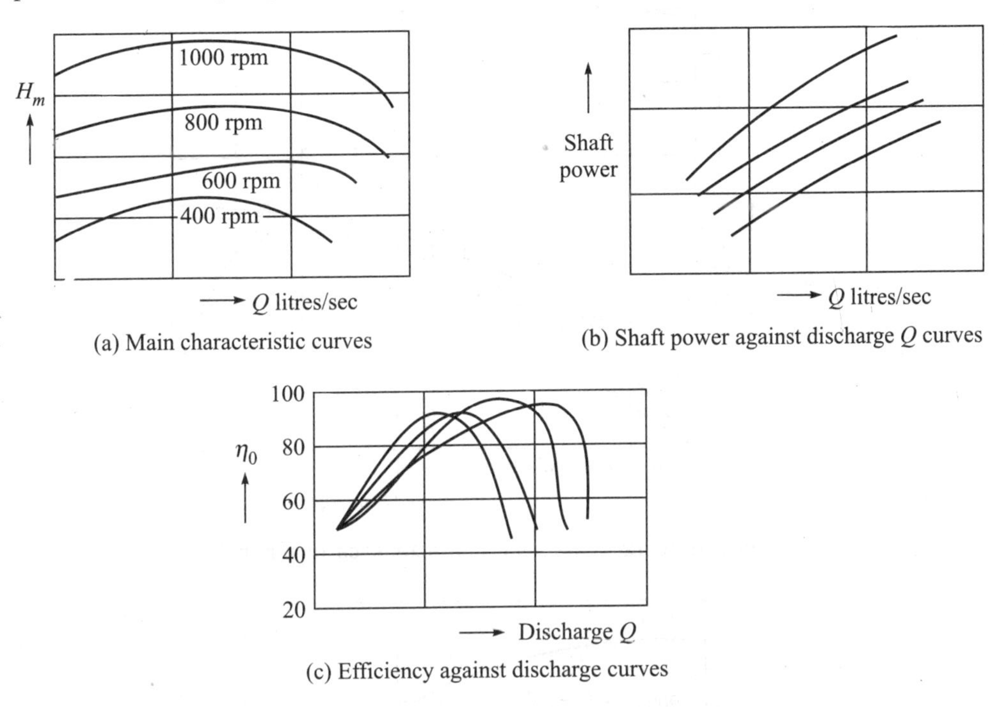

Figure 19.11 Main characteristic of centrifugal pump.

During operation, the pump is normally required to run at a designed speed, which is constant. A particular set of main operating characteristics which correspond to this designed speed are used in operation and are therefore, known as operating characteristics. A set of such characteristics of the pump are shown in Figure 19.11(b). The normal or designed head and discharge correspond to maximum efficiency.

These curves show whether the pump is running at the desired speed and discharge, and whether the motor is overloaded in the operating condition, i.e. they help monitor and predict the operating condition of the pump.

Constant Efficiency Characteristic Curves

Constant efficiency curves are drawn from $Q\ V_s\ H_m$ and $Q\ V_s\ \eta$ curves in Figure 19.12. These curves also show the efficiency of the pump on the basis of the Q, H_m and N values.

Constant Head and Constant Discharge Curves

These curves are shown in Figure 19.13. They are useful in determining the performance of the variable speed pump for which the speed constantly varies. If head H_m is kept constant, then as the speed N varies, the rate Q will also vary. Hence N, V_s and Q plots can be prepared which can be used to determine the required speeds to discharge variation speed at the constant head. As $Q\ \alpha\ N$, $N\ V_s\ Q$ plots a straight line as shown in the figure. Similarly if Q is constant, then H varies

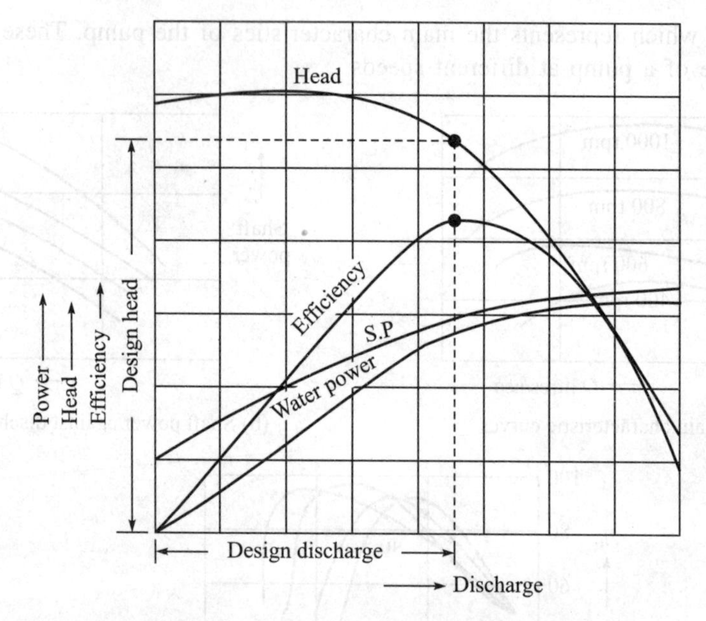

Figure 19.12 Operating characteristics of pump.

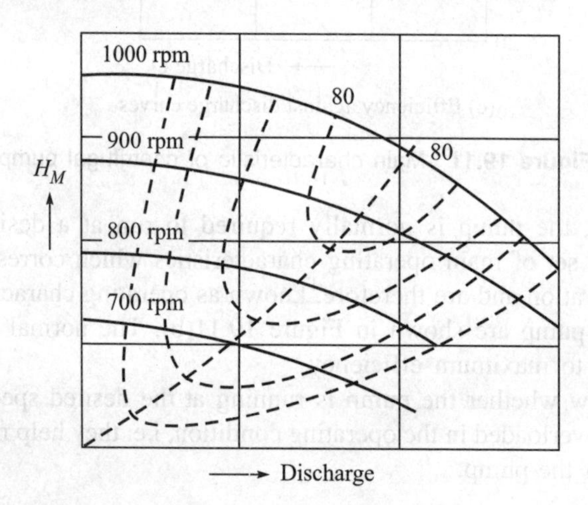

Figure 19.13 Constant efficiency curves.

with N. As $H_m \; \alpha \; N^2$, hence $N \; V_s H_m$ plot is parabolic. Similarly $P \; V_s N^3$ may be plotted. (Figure 19.14).

19.12 CAVITATION IN CENTRIFUGAL PUMP: LIMITATION TO SUCTION LIFT

If $\frac{p_s}{\omega}$ is the absolute pressure head inlet, $\frac{p_s}{\omega} = \frac{p_a}{\omega} - \left[\frac{V_s^2}{2g} + h_s + h_{f_s} \right]$ (19.20)

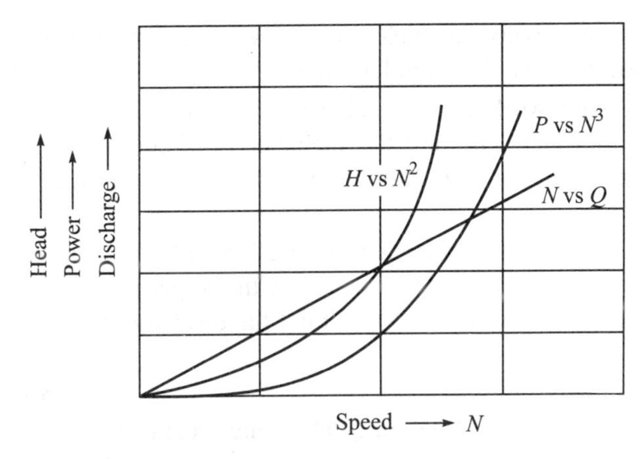

Figure 19.14 *N* vs *Q*, *N* vs *H*, and *N* vs *P* curves.

where p_a is the atmospheric pressure, h_s is suction lift, $\left(\frac{V_s^2}{2g} + h_{f_s}\right)$ is the loss suction pipe. If p_v is the vapour. Absolute pressure at the inlet cannot be less than value used to show any case to avoid cavitation to be $\left(\frac{p_a}{\omega} - \frac{p_v}{\omega}\right)$.

$$\therefore \qquad \text{suction lift } h_s = \left(\frac{p_a}{w} - \frac{p_v}{w}\right) - \frac{V_s^2}{2g} - h_{f_s} \tag{19.21}$$

Such lift in no case can be greater than that given by Equation (19.21), otherwise vaporization of liquid takes place which may ultimately lead to cavitation.

It has already been discussed cavitation in turbines and its effect on turbine blades. Similar is the case for a pump if cavitation is allowed at the inlet. Here also the Thoma cavitation member is applied. The details have already been discussed in the case of turbines.

19.13 CONCLUSION

Centrifugal pump is most commonly used to lift water for irrigation, domestic and municipal water supply, sluice gate and some other miscellaneous purposes. Therefore, its working principles, workdone, heads, efficiency, least diameter, minimum speed for starting, its use in multistage, its performance etc. are essential. The chapter deals with all these points with numerical solved examples and figures. The contents offered in this chapter is enough for beginners. Few problems with answers are given for readers to solve and a list of some reference textbooks are presented at the end. The presentation by Addison[1] and Lewitt[5] are original.

PROBLEMS

19.1 An impeller of a centrifugal pump has its internal and external diameters as 15 cm and 30 cm, respectively. The pump runs at 1000 rpm. The inlet and outlet vane angles are

15° and 25°, respectively. The velocity of flow is constant water, which enters radially. Find the work done by the impeller per unit weight of water. (**Ans.** 17.992 N_m/N)

19.2 A centrifugal pump working under a head of 25 m and speed 1450 rpm has to discharge 0.118 m^3/sec of water. The impeller diameter is 25 cm and the width at the outlet is 5 cm. If the manometric efficiency is 70%, find the vane angle at outer periphery.
(**Ans.** 45°58′)

19.3 A centrifugal pump works under a head of 20 m with a design speed of 1200 rpm. The vanes are curved back at an angle of 25° with the periphery. The impeller diameter is 35 cm and the outlet width is 5.5 cm. Find the discharge if the manometric efficiency is 80%.
(**Ans.** 0.19845 m^3/sec)

19.4 A centrifugal pump has the following data. Outer diameter = 2 times the inner diameter, N = 1000 rpm, H = 40 m, $V_f = V_{f1}$ = 2.5 m/sec, ϕ = 40°, diameter of impeller = 50 cm and width at outlet = 5 cm. Determine: (i) θ (ii) Work done by the impeller per second (iii) η_{mano}
(**Ans.** (i) 10°45, (ii) 119.228 Nm/sec, (iii) 64.6%

19.5 A centrifugal pump works at 800 rpm against a head of 22 m. The velocity of flow at the outlet is 2 m/sec outlet vane angle is 45° and discharge is 0.225 m^3/sec. Determine: (i) Diameter of the impeller, (ii) Width of the impeller at the outlet. Neglect losses.
(**Ans.** (i) 37.5 cm, (ii) 9.5 cm)

19.6 Prove that in general, for a centrifugal pump running at N rpm giving a discharge Q, the manometric head H_m can be expressed in the form of:

$$H_m = A(N^2) + B(NQ) + C(Q)^2$$

where A, B and C are constants.

19.7 A centrifugal pump runs at 1500 rpm and discharges water at 0.12 m^3/sec against a head of 25 m. If the diameter of the impeller is 25 cm and its width is 8 cm, find the vane angle at outer periphery. η_{mano} = 0.75.
(**Ans.** 32° 30′)

19.8 Show that the rise of pressure head in the impeller of a centrifugal force with usual rotation is given by $\frac{1}{2g}[V_r^2 + u_1^2 - V_{f1}^2 \text{ cosec } \phi]$.

19.9 Find the minimum speed at which the centrifugal pump will start functioning against a head of 10 m if the diameter of the impeller at the inlet and outlet are 45 cm and 90 cm, respectively.
(**Ans.** 343.2 rpm)

19.10 A large centrifugal pump has to be predicted from that of a scale model one fourth the diameter. The model absorbs 16.1865 kW when pumping under a head of 7.25 m at its speed of 450 m. The prototype is required to pump against a head of 20 m. What will be the working speed and power required to drive it and what will be the ratio of quantities discharged by the larger pump and the model?
(**Ans.** 186.25 rpm, 1186 kW, 26.67)

REFERENCES

1. Addison, H., *Centrifugal and Rotadynamic Pumps*, Chapman & Hall., London, 1948.
2. Modi, P.N. and S.M. Seth, *Hydraulics and Fluid Mechanics including Hydraulic Machines*, 3rd. ed., Standard Book House, New Delhi, 1977.
3. Bansal, R.K., *Fluid Mechanics and Hydraulic Machines*, Laxmi Publications, New Delhi, 1983.
4. Ramamrutham, S., *Hydraulics, Fluid Mechanics and Fluid Machines*, Dhanpat Rai and Sons, Delhi, 1980.
5. Lewitt, E.H., *Hydraulics and the Mechanics*, Pitman, 1955.
6. Subramanya, K., *Theory and Application of Fluid Mechanics including Hydraulic Machines*, Tata MeGraw-Hill, New Delhi, 1992.

Chapter 20

Reciprocating Pumps

20.1 INTRODUCTION

The reciprocating pumps are the positive displacement pumps in which liquid is sucked and then it is pushed or displaced due to the thrust exerted on it by a moving member, which results in lifting the liquid to a required height. The pumps have usually one or more chambers which are alternately fitted with liquid to be pumped and then emptied again. Thus, a reciprocating action is being continued within the pump chamber and therefore, this pump is called a reciprocating pump.

20.2 COMPONENTS OF RECIPROCATING PUMPS

Figure 20.1 shows the diagrammatic view of a single acting reciprocating pump. The pump consists of the following components:

(1) Cylinder
(2) Piston or plunger
(3) Suction pipe with a one-way suction valve
(4) Delivery pipe with a one-way control valve which admits water from the cylinder
(5) Piston rod
(6) Connecting rod
(7) Crank

20.3 WORKING PRINCIPLES

The piston moves forwards and backwards in the close fitting cylinder. This movement of the piston is obtained by connecting the piston rod to a crank by means of a connection rod. The crank is rotated clockwise by means of an electric motor. The suction and delivery pipes with one-way suction and delivery valves are connected to the cylinder. The suction valve allows water from the suction pipe to the cylinder and the delivery valve allows this amount of water in the cylinder to the delivery pipe. The piston with the arrangement of piston rod and connecting rod moves to and fro in the cylinder when the crank starts rotating. As the crank rotates from

A to B (Figure 20.1) (i.e. from 0° to 180°), the piston moves towards the right end of the cylinder and a partial vacuum is created within the cylinder. The pressure in the sump is atmospheric which is more than the pressure within the cylinder. Thus liquid from the sump to the cylinder is forced to enter through the suction pipe by opening the suction valve. When the crank rotates from B to A (i.e. from 180° to 360°), the piston from its extreme right position starts moving towards the extreme left position. This movement of the piston increases the pressure, the suction valve closes and the delivery valve opens. The liquid is forced into the delivery pipe to discharge at a required height.

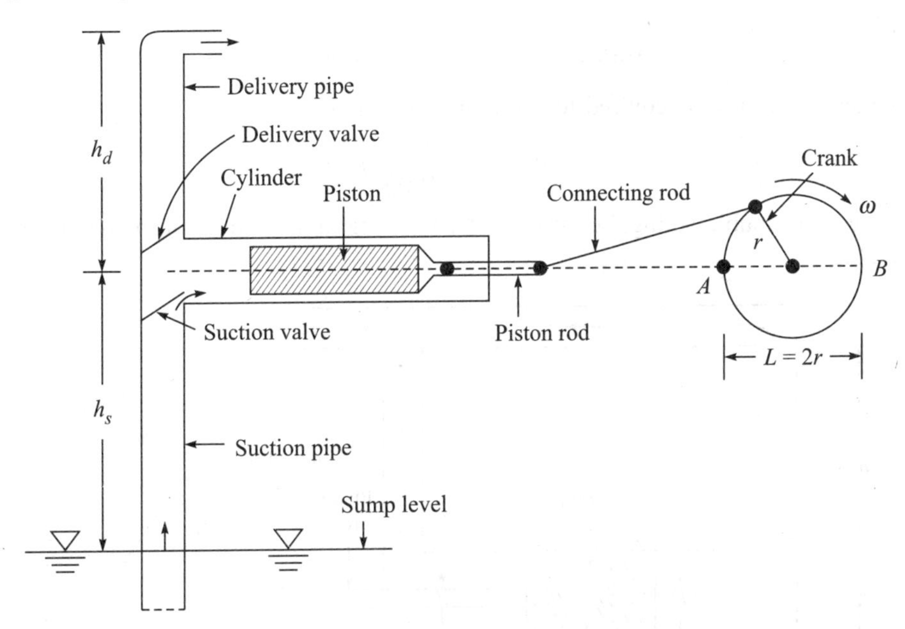

Figure 20.1 Components of reciprocating pump.

20.4 DISCHARGE, WORK DONE AND POWER REQUIRED

Consider the pump to be a single acting pump, i.e. water is acting on one side of the piston as shown in Figure 20.1.

Let A be cross-sectional area of the piston or cylinder.

If D is the diameter of the piston, $A = \frac{\pi}{4}D^2$.

If r is the radius of the crank, stroke length is $L = 2r$.

Let N be the rotation of the crank in rpm.

Let h_s and h_d be the suction head and delivery head, respectively.

Now volume of water delivered in one rotation = $A \times L$

If N is in rpm, the number of revolutions per second = $\frac{N}{60}$

$\therefore$ Discharge of the pump per second, $Q = \dfrac{ALN}{60}$ (20.1)

$\therefore$ Weight of water delivered per second, $W = \dfrac{wALN}{60}$

or $$W = PgALN/60 \tag{20.2}$$

Work done by the pump per second = weight per second × total height lifted

or $$\text{Work done} = W(h_s + h_d)$$

$\therefore$ $$\text{Work done} = \frac{\rho gALN}{60}(h_s + h_d) \tag{20.3}$$

Thus theoretical power required to drive the pump is given by:

$$P = \frac{\rho gALN}{60 \times 1000}(h_s + h_d) \text{ kW} \tag{20.4}$$

If the pump is double-acting, i.e. when water is acting on both sides of the piston, as shown in Figure 20.2:

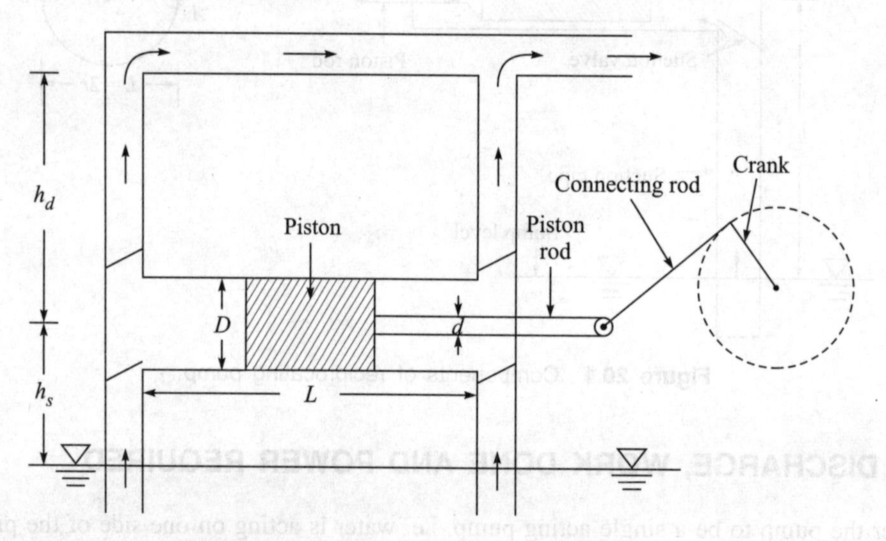

Figure 20.2 Double-acting reciprocating pumps.

Area on left side of piston is $A = \dfrac{\pi}{4}D^2$

If the diameter of piston rod is d,
Area on right side of the piston is = $(\pi/4)(D^2 - d^2)$
Volume of water deliver in one revolution

$$= (\pi/4)D^2L + \pi/4(D^2 - d^2)L$$

$\therefore$ Discharge per second is

$$Q = \left[\frac{\pi}{4}D^2L + \frac{\pi}{4}(D^2 - d^2)L\right]\frac{N}{60} \tag{20.5}$$

If d is very small as compared to D, it may be neglected and Q becomes

$$Q = \left[\frac{\pi}{4}D^2 + \frac{\pi}{4}D^2\right]\frac{LN}{60}$$

$$Q = 2\left[\frac{\pi}{4}D^2\right] LN/60$$

$$Q = \frac{2ALN}{60} \qquad (20.5a)$$

which is two times that takes place in the discharge of single-acting pump.

Work done per second (neglecting piston rod area) = weight of water × total height

$$\text{Work done} = \frac{LwALN}{60}(h_s + h_d) = \frac{2\rho gALN}{60}(h_s + h_d) \qquad (20.6)$$

$$\text{Power required} = \frac{2\rho gALN(h_s + h_d)}{60 \times 1000}\,\text{kW} \qquad (20.7)$$

20.5 SLIP IN RECIPROCATING PUMP

The difference between the theoretical discharge and actual discharge is known as 'slip of the pump'.

The discharge equations given by Equations. (20.1) and (20.5) give the theoretical discharge. There is some leakage of water in the system and hence the actual discharge is always less than the theoretical discharge.

$$\therefore \qquad \text{slip} = Q_{th} - Q_{ac}$$

But the slip is always expressed in terms of percentage.

$$\therefore \quad \text{Percentage slip} = \frac{Q_{th} - Q_{ac}}{Q_{th}} \times 100$$

$$= \left(1 - \frac{Q_{ac}}{Q_{th}}\right) \times 100$$

$$= (1 - c_d) \times 100 \qquad \because \frac{Q_{ac}}{Q_{th}} = c_d \text{ (coefficient of discharge)}$$

$$\therefore \quad \text{Percentage slip} = (1 - c_d) \times 100 \qquad (20.8)$$

In normal case, c_d is less than unity, therefore, the slip is positive. But sometimes Q_{th} is less than Q_{ac} and c_d is more than unity, and therefore the slip becomes negative. This happens in the case of pumps having long suction pipes and low delivery heads, especially when the pump runs at high speed. In such a pump, inertia pressure in the suction pipe becomes larger in comparison to the pressure outside the delivery valve, which causes the delivery valve to open before the suction is completed. Some liquid thus enters the delivery pipe before the delivery stroke is commenced, which results in making Q_{ac} more than Q_{th}. In such a case,

$$\text{Percentage slip} = -(1 - c_d) \times 100$$

20.6 CLASSIFICATION OF RECIPROCATING PUMP

Reciprocating pumps can be classified according to the liquid being in contact with one side or both sides of the piston and according to the number of pistons or cylinders provided.

According to the liquid contact on sides of the piston, pumps are classified as:

(a) Single-acting pump
(b) Double-acting pump

These two types of pumps are already explained and shown in Figures 20.1 and 20.2.

According to the number of cylinders used or provided, pumps are classified as:

(a) Single cylinder pump
(b) Double cylinder pump
(c) Triple cylinder pump
(d) Duplex double acting pump
(e) Quintuplex pump

A single cylinder pump may be single-acting or double-acting (Figures 20.1 and 20.2) delivering water with a single cylinder. Double cylinder and triple cylinder pumps are shown in Figure 20.3.

A double cylinder pump has two single acting cylinders, each equipped with one suction and one delivery pipe with appropriate valves and with separate pistons for each of the cylinders as shown in Figure 20.3(a). Both the pistons are simultaneously driven from cranks set at 180°.

A triple acting pump has three single acting cylinders, each equipped with one suction and one delivery pipe with appropriate valves and separate piston as shown in Figure 20.3(b). All the three pistons are simultaneously driven from cranks set at 120°.

A duplex double acting pump is formed by combining either double acting single cylinder pumps or two double acting double cylinder pumps. The two cranks are set at 90°.

A quintuplex or five-throw pump has five single acting cylinders driver from a crank set at 72°. In general, a reciprocating pump having more than one cylinder is known as a multi-cylinder pump.

EXAMPLE 20.1 A single-acting reciprocating pump running at 30 rpm, delivers 0.012 m³/sec of water. The diameter of piston is 25 cm and stroke length is 50 cm. Determine:

(i) Theoretical discharge (Q_{th})
(ii) Coefficient of discharge (c_d)
(iii) Slip and percentage of slip

Solution: The given data are: $N = 30$ rpm, $Q_{ac} = 0.012$ m³/sec

$$D = 25 \text{ cm} = 0.25 \text{ m}, \; L = 50 \text{ cm} = 0.5 \text{ m}$$

Now $$Q_{th} = \frac{A \times L \times N}{60} = \frac{\pi/4(0.25)^2 \times .5 \times 30}{60} = 0.01227 \text{ m}^3/\text{sec} \quad \textbf{Ans.}$$

$$\text{Slip} = Q_{th} - Q_{ac} = 0.01227 - 0.012 = 0.00027 \text{ m}^3/\text{sec} \quad \textbf{Ans.}$$

$$c_d = \frac{Q_{ac}}{Q_{th}} = \frac{0.012}{0.01227} = 0.9778 \quad \textbf{Ans.}$$

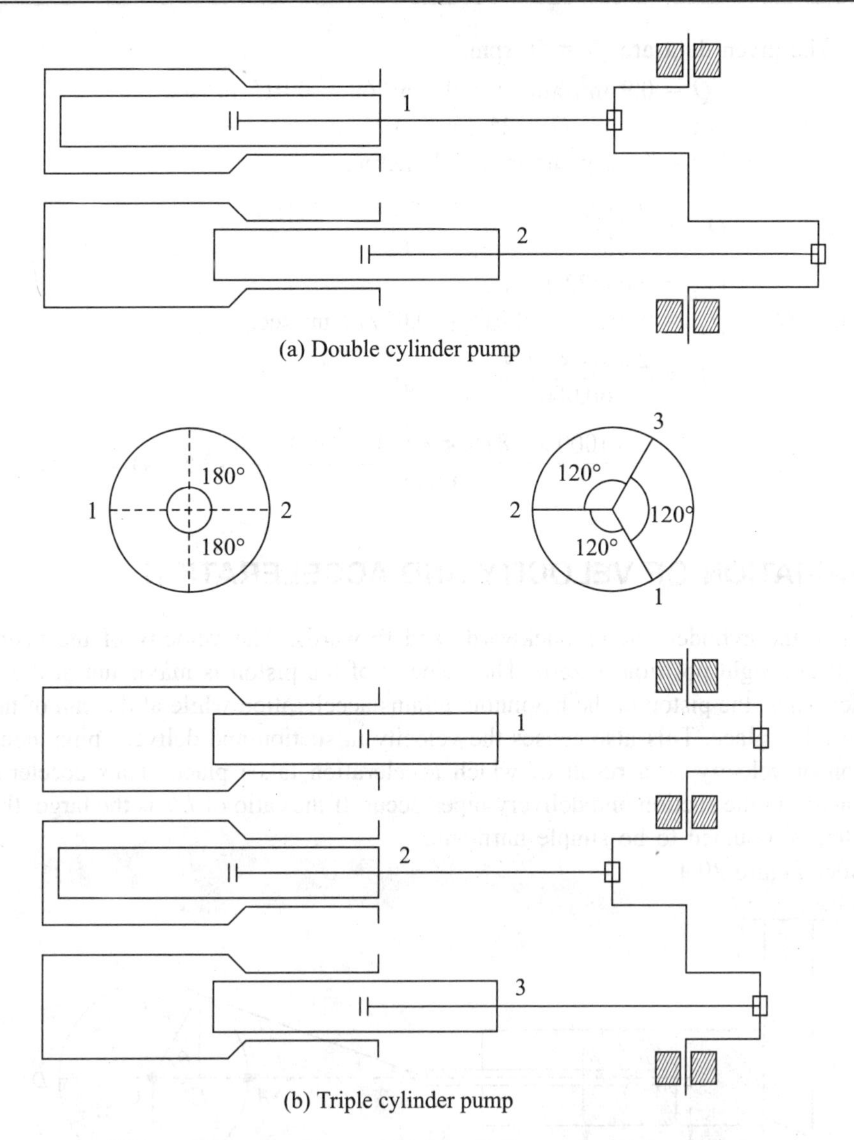

Figure 20.3 Double and triple cylinder reciprocating pump.

$$\% \text{ of slip} = \left(\frac{Q_{th} - Q_{ac}}{Q_{th}}\right) \times 100 = \left(\frac{0.01227 - 0.012}{0.01227}\right) \times 100$$

$$= 2.2\%$$ **Ans.**

EXAMPLE 20.2 A double-acting reciprocating pump running at 50 rpm is discharging 0.99 m^3 of water per minute. The stoke lengths 40 cm. The diameter of the piston is 25 cm, the delivery and suction heads are 25 m and 4 m, respectively. Find the slip of the pump and the power required to drive the pump.

Solution: The given data are: $N = 50$ rpm,

$$Q = 0.9 \text{ m}^3/\text{minute} = 0.9 \text{ m}^3/60 = 0.015 \text{ m}^3/\text{sec}$$

$$L = 40 \text{ cm} = 0.4 \text{ m}, D = 25 \text{ cm} = 0.25 \text{ m}$$

$$h_d = 25 \text{ m}, hs = 4 \text{ m, pump is double-acting}$$

$$\therefore \quad Q_{th} = \frac{2ALN}{60} = \frac{2 \times \pi/4(0.25)^2 \times 0.4 \times 50}{60}$$

$$Q_{th} = 0.03272 \text{ m}^3/\text{sec}$$

$$\text{Slip} = Q_{th} - Q_{ac} = 0.03272 - 0.015 = 0.01772 \text{ m}^3/\text{sec.}$$ **Ans.**

$$P = \frac{2 \times \rho g \times ALN}{60{,}000}(h_s + h_d)$$

$$= \frac{2 \times 1000 \times 9.81 \times \pi/4 \times 0.25^2 \times .4 \times 50}{60{,}000}(2.5 + 4) = 9.3099 \text{ kW}$$ **Ans.**

20.7 VARIATION OF VELOCITY AND ACCELERATION

The piston in the cylinder moves backwards and forwards. The velocity of the piston at the extreme left and right position is zero. The velocity of the piston is maximum at the centre of the cylinder. Thus, the piston at the beginning attains acceleration while at the end of the stroke, retardation takes place. This also causes the velocity in suction and delivery pipe, non-uniform or variation of velocity as a result of which acceleration takes place. Thus accelerating and retarding heads in the suction and delivery pipes occur. If the ratio of L/r is the large, the motion of the piston is assumed to be simple harmonic.

Consider Figure 20.4.

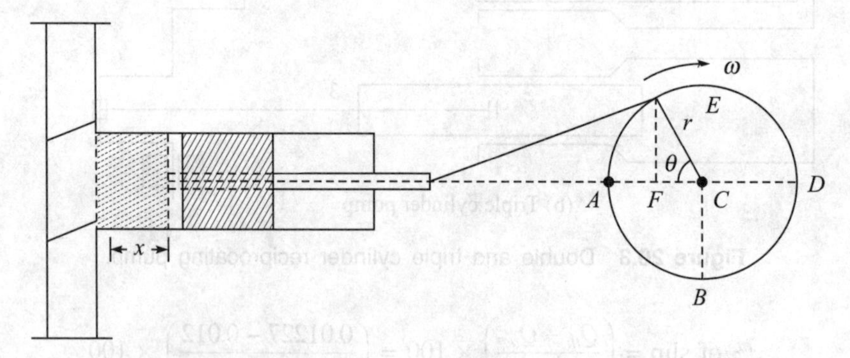

Figure 20.4 Velocity and acceleration of piston.

Let ω be the angular velocity of the crank in rad/sec

r is the radius of crank

A is the area of the cylinder

a is the area of suction and delivery pipe

l is length of suction and delivery pipe.

In the beginning, the crank is at the inner dead centre A and the piston in the cylinder is shown in the dotted line. Let in time t, the crank rotate through an angle θ with an angular velocity ω from A. The displacement of piston in time t is x.

$$\therefore \quad \theta = \omega t$$

$$\therefore \quad x = AF = r - r\cos\theta = r(1 - \cos\theta) = r(1 - \cos\omega t)$$

If V is the velocity of the piston,

$$V = \frac{dx}{dt} = \frac{d}{dx}[r - r\cos\omega t]$$

or

$$V = \frac{dr}{dt} - \frac{d(r\cos\omega t)}{dt}$$

$$V = 0 - r(-\sin\omega t) \times \omega \qquad \because r \text{ constant}, \frac{dr}{dt} = 0$$

$$V = r\omega\sin\omega t \tag{20.9}$$

The continuity equation gives,

Area of cylinder × velocity in the cylinder = Area of the pipe × Velocity in the pipe

$A \times V = av$, where v is the velocity in pipe

$Arw \sin\omega t = av$

$$\therefore \quad v = \frac{A}{a} r\omega \sin\omega t \tag{20.10}$$

Acceleration in the pipe is: $\dfrac{dv}{dt} = \dfrac{d}{dt}\left(\dfrac{A}{a} r\omega \sin\omega t\right)$

or

$$\frac{dv}{dt} = \frac{A}{a} r\omega \cos\omega t \times \omega$$

$$\therefore \quad \frac{dv}{dt} = \frac{A}{a} r\omega^2 \cos\omega t \tag{20.11}$$

Mass of water in pipe = ρ × Volume of water in pipe

$$= \rho(al)$$

Force required to accelerate this mass:

$$= (\rho al)\left(\frac{A}{a} r\omega^2 \cos\omega t\right)$$

Intensity of pressure due to acceleration:

$$= \frac{\text{Force}}{a} = \frac{(\rho al)\left[\left(\frac{A}{a}\right) r\omega^2 \cos\omega t\right]}{a}$$

$$= \rho l \frac{A}{a} r\omega^2 \cos\omega t$$

Pressure head (h_a) due to this acceleration:

$$h_a = \frac{\rho l \frac{A}{a} r\omega^2 \cos\omega t}{(\rho g)}$$

$$h_a = \rho l \frac{A}{a} r\omega^2 \cos(\theta/\rho g) \because \omega t = \theta$$

$$h_a = \frac{l}{g}\frac{A}{a} r\omega^2 \cos\theta \tag{20.12}$$

Thus pressure due to acceleration in suction and delivery pipe is obtained from Equation (20.12)

$$h_{a_s} = \frac{l_s}{g} \cdot \frac{A}{a_s} r\omega^2 \cos\theta \tag{20.13}$$

$$h_{a_d} = \frac{l_d}{g} \cdot \frac{A}{a_d} r\omega^2 \cos\theta \tag{20.14}$$

Since pressure head varies with θ,

When $\theta = 0$, $\quad h_a = \frac{l}{g} \cdot \frac{A}{a} r\omega^2 \quad \because \cos\theta = 1$

When $\theta = 90°$, $\quad h_a = 0 \quad \because \cos 90° = 0$

When $\theta = 180°$, $\quad h_a = -\frac{l}{g}\frac{A}{a} r\omega^2 \quad \because \cos 180° = -1$

$$\therefore \quad (h_{a\max}) = \frac{l}{g} \cdot \frac{A}{a} r\omega^2 \tag{20.15}$$

20.8 EFFECTS OF FRICTION

The velocity of water in suction and delivery pipe is already given by Equation (20.10), i.e.:

$$v = \frac{A}{a} r\omega \sin\omega t = \frac{A}{a} r\omega \sin\theta \tag{A}$$

Loss of head due to friction is $hf = \frac{flv^2}{2gd}$ where f is the friction coefficient

Substituting v in the head loss equation:

$$h_f = \frac{fl}{2gd}\left(\frac{A}{a} r\omega \sin\theta\right)^2 \tag{20.16}$$

Thus in the suction pipe:

$$h_{fs} = \frac{fl_s}{2gd_s}\left[\frac{A}{a_s} r\omega \sin\theta\right]^2 \tag{20.17}$$

and in the delivery pipe:

$$h_{fd} = \frac{fl_d}{2gd_d}\left[\frac{A}{a_d} r\omega \sin\theta\right]^2 \tag{20.18}$$

When $\theta = 0$, $h_f = 0 \because \sin 0° = 0$

When $\theta = 90$, $h_f = \frac{fl}{2gd}\left[\frac{A}{a} r\omega\right]^2 \because \sin 90° = 1$

When $\theta = 180°$, $h_f = 0 \because \sin 180° = 0$

Thus the maximum value of the head lost due to friction is:

$$h_{f\max} = \frac{fl}{2gd}\left[\frac{A}{a} r\omega\right]^2 \tag{20.19}$$

EXAMPLE 20.3 A double-acting reciprocating pump has a cylinder of diameter 15 cm and stroke length 30 cm. The centre of the pump is 4 m above the sump. The atmospheric pressure head is 10.3 m of water and pump N = 40 rpm. If l_d = 5 m, d_d = 10 cm, determine pressure head due to acceleration in the cylinder at the beginning of the suction stroke and at the middle of the suction stroke.

Solution: The given double data are: double-acting. D = 15 cm, = 0.15 m,

$L = 30 \text{ cm} = 0.3 \text{ m},\ h_s = 4 \text{ m}\ \frac{p_a}{w} = 10.3 \text{ m of water}$

$N = 40 \text{ rpm},\ l_s = 5 \text{ m},\ d_s = 10 \text{ cm} = 0.1 \text{ m}$

$$\text{Pressure head } h_g = \frac{\rho l(A/a)\, r\omega^2 \cos \omega t}{\rho g}$$

$$= \frac{l}{g} \cdot \frac{A}{a} r\omega^2 \cos \theta$$

Here l = 5 m

$$A = \frac{\pi}{4}(D^2) = \pi/4(0.5)^2 = 0.017671 \text{ m}$$

$$a = \pi/4(0.1)^2 = 0.007554 \text{ m}$$

$$r = \frac{L}{2} = \frac{0.3}{2} = 0.15 \text{ m}$$

$$\omega = \frac{2\pi N}{60} = \frac{2\pi \times 40}{60} = 4.188 \text{ rad/sec}$$

$$h_a = \frac{l}{g} \cdot \frac{A}{a} r\omega^2 \cos \theta$$

At the beginning $\theta = 0 \therefore \cos \theta = 1$

$$h_a = \frac{5}{9.81} \times \frac{0.017671}{.007854} \times .15 \times 4.188$$

$$h_a = 3.017 \text{ m}$$ **Ans.**

In the middle of the stroke, $\theta = 90°$, $\cos 90° = 0$

$$\therefore \quad h_a = 0$$ **Ans.**

EXAMPLE 20.4 A single-acting reciprocating pump has a cylinder diameter of 0.15 m, stroke length of 0.3 m, N = 50 rpm lifts water to height 25 m. ld = 22 m and 0.1 m in diameter. Determine Q_{th} and P_{th} required to run the pump. If $Q_{ac} = 0.0042 \text{ m}^3\text{/sec}$, find Pc of the slip. Also find h_a at the beginning and middle of the delivery stroke.

Solution: The given data are: $D = 0.15 \text{ m},\ L = 0.3 \text{ m},\ \therefore\ r = \frac{0.3}{2} = 0.15 \text{ m},\ N = 50 \text{ rpm}$

$$\therefore \quad \omega = \frac{2\pi N}{60} = 5.236 \text{ rad/sec},\ H = 25 \text{ m},\ l_d = 22 \text{ m},\ d_d = 0.7 \text{ m},$$

$$Q_{ac} = 0.0042 \text{ m}^3\text{/sec},\ A = \pi/4(0.15)^2 = 0.017671 \text{ m}^2$$

Now $Q_{th} = \frac{A \times L \times N}{60} = \frac{0.017671 \times 0.3 \times 50}{60} = 0.00441775 \text{ m}^3/\text{sec}$ **Ans.**

$$P_{th} = \frac{\rho g Q_{th} \times H}{1000}$$

$$= \frac{1000 \times 9.81 \times .00441775 \times 25}{1000} = 1.0345 \text{ kW}$$ **Ans.**

$$\text{P.C. of slip} = \left(\frac{Q_{th} - Q_{ac}}{Q_{th}}\right) \times 100 = \left(\frac{0.00441775 - 0.0042}{0.00441775}\right) \times 100 = 4.929\%$$

Ans.

Acceleration head in delivery pipe $= h_{ad} = \frac{l_d}{g} \cdot \frac{A}{a_d} r\omega^2 \cos\theta$

$$h_{ad} = \frac{22}{9.81} \times \frac{0.017671}{\pi/4(0.1)^2} \times .15 \times 5.236 \cos\theta$$

$$h_{ad} = 3.962 \cos\theta$$

At the beginning $\theta = 0$ $\therefore$ $\cos\theta = 1$

$$h_{ad} = 3.962 \text{ m}$$ **Ans.**

At the middle of the stroke $\theta = 90°$, $\cos 90° = 0$

$\therefore$ $$h_{ad} = 0$$ **Ans.**

20.9 INDICATOR DIAGRAM

An indicator diagram is a graph between the pressure head and the stroke length of the piston for one complete revolution. The pressure is taken as an ordinate and stroke length as an abscissa.

20.9.1 Ideal Indicator Diagram

Under ideal conditions, the graph between the head in the cylinder and the stroke length of the piston for one complete revolution of the crank is known as an ideal indication diagram. (Figure 20.5).

As shown in Figure 20.5, Hatm = 10.3 m of water = Atmospheric pressure head. Let L, h_s and h_d be suction stroke, suction head and delivery head, respectively. During the suction stroke, the pressure in the cylinder is constant and equal to the suction head, which is below atmospheric (Hatm) by a height h_s. The pressure head during the suction stroke is given by line AB which is below EF by a height h_s. Similarly, during delivery, the stoke pressure head in the cylinder above atmospheric head by a head h_d. It is represented by the line CD which is above EF by a height h_d. Thus for one complete revolution of the crank, pressure head in the cylinder is represented by the diagram $ABCD$. This diagram is a known ideal indicator diagram as the effect of acceleration is not considered.

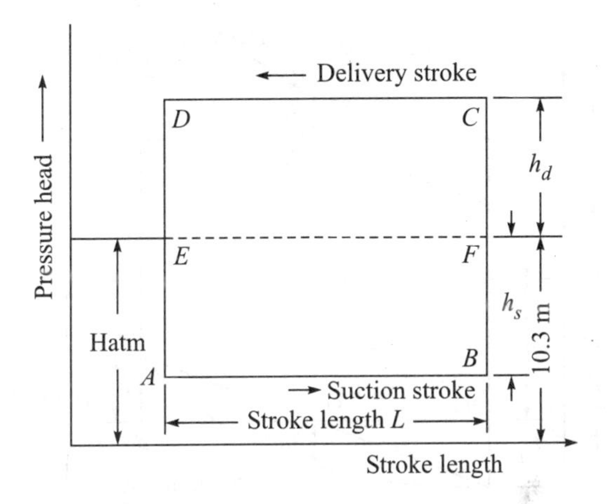

Figure 20.5 Ideal indicator.

Here work done by the pump per second

$$= \frac{\rho gALN}{60}(h_s + h_d)$$

$$= \left(\frac{\rho gAN}{60}\right)L(h_s + h_d)$$

$$= CL(h_s + h_d) \text{ where } C = \frac{\rho gAN}{60} = \text{constant}$$

$\therefore$ Work done $\propto L(h_s + h_d)$

But $L(h_s + h_d)$ = area of indicator diagram.

$\therefore$ Work done by the pump $\propto$ area of indicator diagram.

20.9.2 Effect of Acceleration on Indicator Diagram

Considering the effect of acceleration, pressure head h_a is given by Equation (20.12)

i.e.
$$h_a = \frac{l}{g} \cdot \frac{A}{a} r\omega^2 \cos\theta$$

Considering first the suction pressure head h_{a_s}. The pressure head at the beginning (when $\theta = 0$) in the cylinder will be $(h_s + h_{a_s})$ below atmospheric (shown in Figure 20.6). At middle $\theta = 90°$, $h_{a_s} = 0$, pressure head in the cylinder will be below atmospheric by h_s. At the end of suction stroke (i.e. $\theta = 180°$), h_{a_s} is negative and pressure head in the cylinder will be $(h_s - h_{a_s})$ below atmospheric.

In a similar process, the indicator diagram for a delivery stroke can be drawn represented by $C'D'$. At the beginning, pressure head within the cylinder is $(h_d + h_{a_d})$ above atmospheric. In the middle of the delivery stroke, the pressure head is only h_d above atmospheric as h_{a_d} is zero and at the end of the stroke, pressure is $(h_d - h_{a_d})$ above atmospheric. Thus the indicator diagram in the delivery stroke is represented by $C'HD'$.

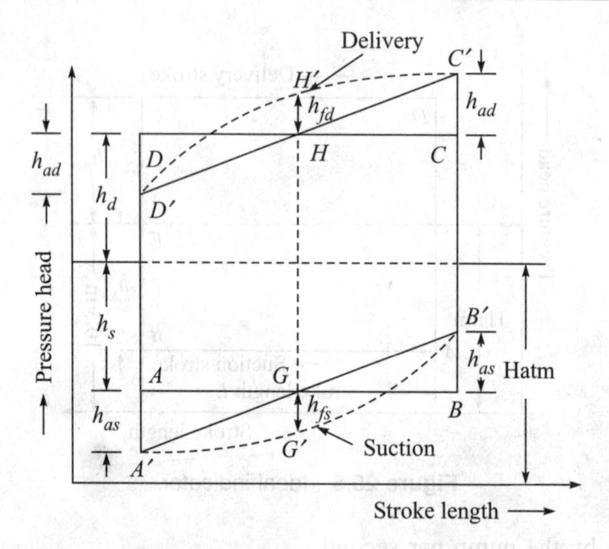

Figure 20.6 Indicator diagram with the effect of acceleration.

Thus the ideal indicator diagram is changed from $ABCD$ to $A'B'C'D'$ when there is the effect of acceleration, but

$$\text{Area of } ABCD = \text{area of } A'B'C'D'.$$

Work done by the pump ∝ area of indicator diagram. Thus work done remains the same even if the effect of acceleration is considered. When the effect of friction is considered, the indicator diagram will be $AG'B'C'H'D'$. Head loss due to friction for both suction pipe h_{fs} (20.17) and delivery pipe h_{fd} (i.e. Equation 20.18) have been considered.

EXAMPLE 20.5 A single-acting centrifugal pump has a plunger of diameter 0.2 m and a stroke length 0.4 m. The length of the suction pipe is 6 m and the diameter is 0.1 m. The centre of the pump is 3.5 m above the sump. Atmospheric pressure is 10.3 m of water and the pump is running at 35 rpm. Determine: (i) Pressure due to acceleration at the beginning of the suction stroke, (ii) Maximum pressure head due to acceleration, and (iii) Pressure head in the cylinder at the beginning and end of the stroke.

Solution: The given data are: $D = 0.2$ m, stroke length $L = 0.4$ m

$$\therefore \quad r = \frac{L}{2} = 0.2 \text{ m}, \; l_s = 6 \text{ m}, \; d_s = 0.1 \text{ m}$$

$$\therefore \quad a_s = \pi/4(.1)^2 = 0.007854 \text{ m}^2$$

$$A = \pi/4(.2)^2 = 0.0314159 \text{ m}^2$$

$$h_s = 3.5 \text{ m}, \text{ Hatm} = 10.3 \text{ m of water}$$

$$N = 35 \text{ rpm.}$$

$$\text{Angular speed} = \omega = \frac{2\pi N}{60} = \frac{2\pi \times 35}{60} = 3.6652 \text{ rad/sec}$$

$$\therefore \quad h_{a_s} = \frac{l_s}{g} \cdot \frac{A}{a_s} r\omega^2 \cos\theta$$

At the beginning $\theta = 0$, $h_{a_s} = \dfrac{l_s}{g} \cdot \dfrac{A}{a_s} r\omega^2$

$$= \frac{6}{9.81} \cdot \frac{0.0314159}{.007854} \times 0.2 \times 3.6652$$

$= 6.55$ m. **Ans.**

Maximum Pressure head $= \dfrac{l_s}{9.81} \times \dfrac{0.0314159}{0.007854} \times 0.2 \times 3.6652$

$= 6.55$ m **Ans.**

$\therefore$ Pressure head at beginning of suction stroke

$= h_s + 6.55 = 3.5 + 6.55 = 10.05$ m **Ans.**

Absolute pressure at the beginning of suction stroke

$= 10.3 - 10.05 = 0.25$ **Ans.**

EXAMPLE 20.6 A single-acting reciprocating pump has a piston diameter of 12.5 cm and stroke length 30 cm. The centre of the pump is 4.5 m above water level in the sump. The diameter and length of suction pipe are 7.5 cm and 6.8 m, respectively. The separation occurs if the absolute pressure head in the cylinder during suction stroke falls below 2.6 m. Calculate the maximum speed at which the pump can run without separation. Take atmospheric pressure head = 10.3 m of water.

Solution: The given data are: $D = 12.5$ cm $= 0.125$ m $\therefore A = \pi/4(0.125)^2 = 0.01227$ m^2

$L = 30$ cm $= 0.3$ m $\therefore r = 0.15$ m

Suction head $h_s = 4.5$ m

$d_s = 7.5$ cm $= 0.075$ m $\therefore a_s = \pi/4(0.075)^2 = 0.004818$ m^2

$ls = 6.8$ m, $h_{sep} = 2.6$ m, Hatm $= 10.3$ m.

The indicator diagram shows that absolute pressure head in suction stroke is minimum at the beginning of the stroke. Hence, separation may take place at the beginning of the stroke and that *pr.* head $= (h_s + h_{a_s})$ below atmospheric.

$= [\text{Hatm} - (h_s + h_{a_s})]$m of absolute.

$= 10.3 - (4.5 + h_{a_s})$

$\therefore \quad h_{sep} = 10.3 - (4.5 + h_{a_s})$

$2.6 = 10.3 - 4.5 - h_{a_s}$

$\therefore \quad h_{a_s} = 10.3 - 4.5 - 2.6 = 3.2$ m

But $\quad h_{a_s} = \dfrac{l_s}{g} \cdot \dfrac{A}{a_s} \omega^2 r \because \theta = 0, \cos\theta = 1$

$$3.2 = \frac{6.8}{9.81} \cdot \frac{0.01227}{0.004818} \times \omega^2 \times 0.15$$

$\omega^2 = 12.4177$

$\therefore \quad \omega = 3.5238 = \dfrac{2\pi N}{60}$

$\therefore \quad N = 33.65$ rpm **Ans.**

20.10 AIR VESSELS

The flow rate in the delivery pipe at any instant varies considerably in a single-acting reciprocating pump. The flow in delivery pipe during suction stroke is zero and rises to maximum in the delivery stroke. Thus flow or discharge is intermittent. To obtain a continuous uniform flow in both suction and delivery strokes, large air vessels are fitted in both suction and delivery pipe close to the pump as shown in Figure 20.7. Besides maintaining continuous uniform flow, air vessels save a considerable amount of work in overcoming frictional resistance in both suction and delivery pipes, and allow the pump to run with high speed without separation. An air vessel is a closed chamber having an opening at its base through which liquid flows into the vessel or out of the vessel. The top portion of the vessel contains compressed air which is further compressed when liquid flows in and it expands when the liquid flows out from the vessel.

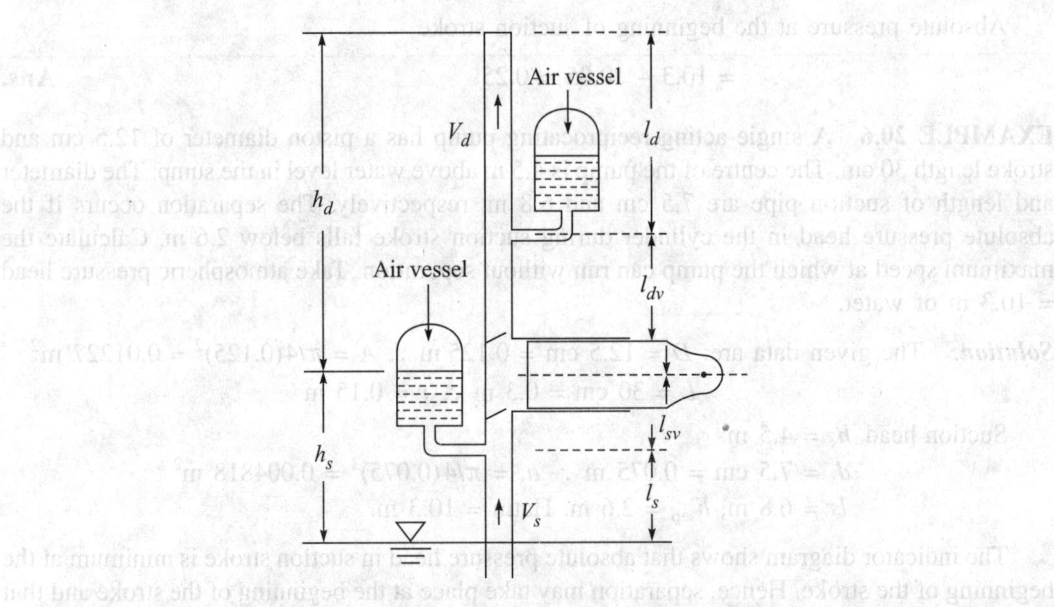

Figure 20.7 Air vessels in single acting reciprocating pump.

During the first half of the suction stroke, the piston moves with acceleration and hence liquid in suction pipe is to be accelerated, i.e. the liquid velocity in the suction pipe is more than the mean velocity. This demand of increased velocity and hence more flow to the cylinder at this stage is supplied from the air vessel. During the second half of the suction stroke, retardation of flow takes place. The velocity in the suction pipe is to be less than mean velocity of flow. Since the portion of pipe below the air vessel flows with mean velocity, the extra water for retardation to occur, flows into air vessel and thus the air in the air vessel is compressed. Only a small portion of the water enters the cylinder from the suction pipe. The liquid thus stored in the air vessel will be supplied during the first half of the next suction stroke and the same cycle will thus be repeated.

During the first half of the delivery stroke, the piston moves with an acceleration and forces the water into the delivery pipe with a velocity which is more than the mean velocity. The quantity in excess of mean discharge enters the air vessel which compresses the air in the vessel. During the second half of the delivery stroke, the piston moves with retardation and the velocity in the delivery pipe becomes less than the mean velocity of the flow. The water already enters the air vessel during the first half of the stroke, is under pressure of compressed air and starts flowing out to meet the demand for uniform velocity.

Thus the two air vessels in both the delivery and suction pipes help in maintaining velocity required in both the pipes.

20.10.1 Work Done with Air Vessels

Let l_d = length of delivery pipe beyond the air vessel

l_{dv} = length of delivery pipe between cylinder and air vessel

V_d = uniform velocity in l_d.

l_s = length of suction pipe below air vessel

l_{sv} = length of suction pipe between cylinder and air vessel.

The effect of acceleration is observed only on l_{dv} and l_{sv} and these two lengths may be very small by fitting the air vessels very close to the cylinder. Thus, the acceleration heads in these two portions and loss of heads due to friction may also be neglected due to very small length as compared to l_d and l_s. Thus work done against the total head may be equal to $(h_s + h_d + h_{fs} + h_{fd})$, where h_{fs} and h_{fd} are the loss of head due to friction in l_s and l_d respectively, and

$$\left.\begin{aligned} h_{fs} &= \frac{fls}{2gd}\left[\frac{A}{a} r\omega\right]^2 \\ h_{fd} &= \frac{fld}{2gd}\left[\frac{A}{a} r\omega\right]^2 \end{aligned}\right\}$$

Equation (20.19) gives the maximum friction loss in l_d and l_s, taking the diameter of l_d and l_s to be the same.

Work done = Weight of water per second × total head

$$= \left(\frac{\rho gALN}{60}\right)(h_s + h_d + h_{fs} + h_{fd}) \tag{20.20}$$

20.10.2 Work Saved by Fitting Air Vessels

The head lost due to friction in the suction and delivery pipes is saved by fitting the air vessels. This reduction of head loss saves some energy. This can be obtained by finding work done against friction without air vessels and with air vessels.

Work done against friction without air vessels is obtained from the indicator diagram (Figure 20.8)

$$\text{Head loss due to friction, } h_f = \frac{fl}{2gd}\left(\frac{A}{a} r\omega^2 \sin\theta\right)^2$$

Variation of θ with h_f is parabolic and therefore the indicator diagram for the loss of head due to friction in pipes is parabolic. Work done against friction per stroke is equal to the area of the indicator diagram due to friction.

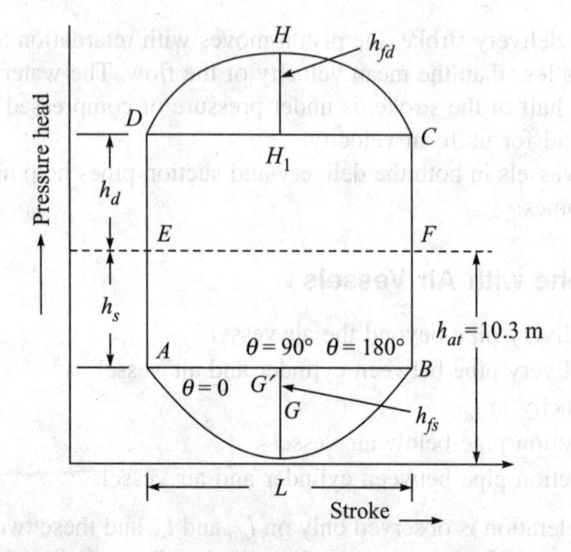

Figure 20.8 Indicator diagram to show the effect of friction.

Work done per stroke against friction without air vessel:

W_{fo} = area of the parabola

$$= \frac{2}{3} \times \text{base} \times \text{height}$$

$$= \frac{2}{3} \times L \times \left[\frac{fl}{2gd}\left(\frac{A}{a} r\omega\right)^2\right]$$

$$= \frac{2}{3} \times L \times \frac{fl}{2gd}\left(\frac{A}{a}\omega r\right)^2 \tag{20.21}$$

Work done per stroke against friction with air vessel:

W_{fAV} = Base × height

$$= L \times \frac{fl}{2gd}\left(\frac{A}{a}\cdot\frac{\omega r}{\pi}\right)^2$$

$$= \frac{1}{\pi^2} L \times \frac{fl}{2gd}\left(\frac{A}{a}\omega r\right)^2 \tag{20.22}$$

Work Saved = $W_{fo} - W_{fAV}$

$$= \frac{2}{3} L \left[\frac{fl}{2gd}\left(\frac{A}{a}\omega r\right)^2\right] - \frac{1}{\pi^2} L \frac{fl}{2gd}\left(\frac{A}{a}\omega r\right)^2$$

$$= L \times \frac{fl}{2gd}\left(\frac{A}{a}\omega r\right)^2 \left[\frac{2}{3} - \frac{1}{\pi^2}\right] \tag{20.23}$$

$$P.C. \text{ of work saved} = \frac{W_{fo} - W_{fAV}}{W_{fo}} = \left(\frac{\frac{2}{3} - \frac{1}{\pi^2}}{\frac{2}{3}}\right) \times 100$$

$$= 84.8\%$$

EXAMPLE 20.7 A single-acting reciprocating pump has its piston in simple harmonic motion. Show that the ratio of work done against friction when air vessels are fitted to that in the absence of air vessels is $\frac{3}{2\pi^2}$.

Solution: In the above, i.e. P.C. of work saved (84.8%),

Work done against friction without air vessel

$$= W_{fo} = \frac{2}{3}\left[\frac{fl}{2gd}\left(\frac{A}{a}\omega r\right)^2\right]$$

And work done against friction with air vessel

$$= W_{fAV} = \frac{1}{\pi^2} L \frac{fl}{2gd}\left(\frac{A}{a}\omega r\right)^2$$

$$\therefore \qquad \text{Ratio} = \frac{W_{fAV}}{W_{fV}} = \frac{\left[\dfrac{Lfl}{\pi^2 \cdot 2gd}(A/a\,\omega r^2)\right]}{\dfrac{2}{3}L\left[\dfrac{fl}{2gd}\left(\dfrac{A}{a}\omega r\right)^2\right]}$$

$$= \frac{3}{2\pi^2} \text{ shown.}$$

20.11 CONCLUSION

Reciprocating pump in present time is not very common in use. Yet for small discharge, it can be used. It is used in hydraulics laboratory for small discharge. In drip and trickle irrigation, watering in small farm, tea garden, water from small pond or well is lifted by this pump to a tank at a medium height and watering from that tank is done by pipe allowing water to flow to the crop area by gravity. It is easy to handle. For lifting of smaller discharge in the above different situations, knowledge of reciprocating pump has remained a continued interest. Therefore, chapter deals with details of its components, working principles, discharge equation, work done, classification, frictional effects, use of air vessels etc. Numerical problems as examples are solved along with the theory. Few problems are given for students to solve and list of reference books is enlisted for ambitious students.

PROBLEMS

20.1 A single-acting reciprocating pump has a piston diameter of 0.25 m and stroke length of 0.18 m. If the pump runs at 40 rpm, calculate average discharge of the pump.

(**Ans.** 0.00589 m^3/sec)

20.2 Find the slip of reciprocating pump in percentage when a single-acting pump cylinder diameter is 0.3 m, stroke length 0.25 m. The pump runs at 45 rpm and the actual discharge is 0.0128 m^3/sec.

(**Ans.** 3.4%)

20.3 A single-acting reciprocating pump has a piston diameter of 0.4 m and stroke length 0.3 m. The pump lifts water to a total height of 12 m. The speed of the pump is 50 rpm. Actual average discharge of the pump is 0.03 m^3/sec. Calculate slip, p.c. of slip and power required to drive the pump (**Ans.** 0.0014 m^3/sec, 4.46%, 3.5316 kW)

20.4 A single-acting reciprocating is 125 mm in diameter and 250 mm stroke. The speed of the pump is 40 rpm, it discharges water to a height of 15 m. The diameter and length of the pipe (delivery) are 100 mm and 30 m respectively. If a large air vessel is fitted in the delivery pipe at a distance of 1.5 m, from the centre of the pump, find the pressure head in the cylinder: (i) at the beginning of the delivery stroke, (ii) in the middle of the delivery stroke. Take $f = 0.04$ (**Ans.** 15.566 m, 15.07 m)

20.5 Show that the p.c. of work saved per stroke in a double-acting reciprocating pump by fitting an air vessel is equal to 39.2%.

20.6 A single-acting reciprocating pump has a piston of 18 cm diameter a stroke of 40 cm. The delivery pipe is 8 cm in diameter and 25 m long. Water is lifted to a height of 25 m above the centre of the pump. Find the maximum speed at which the pump can be run so that no separation takes place during the delivery stroke. Assume separation to occur at an absolute pressure head of 2.5 m of water. Take Hatmosphere = 10.3 m of water (**Ans.** 34 rpm)

REFERENCES

1. Subramanya, K., *Theory and Applications of Fluid Mechanic Including Hydraulic Machines*, Tata McGraw-Hill, New Delhi, 1992.
2. Baumeister, T., "Turbomachinery", Section 119 of *Handbook of Fluid Dynamics*, edited by V.L. Streeter, McGraw-Hill, Inc., 1961.
3. Daily, J.W., "Hydraulic Machinery," Chapter 13 of *Engineering Hydraulics*, edited by H. Rouse, John Wiley & Sons, Inc., New York, 1950.
4. Shepherd, D.G., *Principles of Turbomachinery*, Macmillan, New York, 1956.

Chapter 21

Miscellaneous Fluid Machines

21.1 INTRODUCTION

Several fluid or hydraulic machines are used for either storing the fluid energy and then transmitting it when required, or for magnifying this energy in the form of pressure several times and then transmitting the same. In general, these fluid machines are based on hydrostatics and hydrokinetics.

21.2 DIFFERENT MISCELLANEOUS FLUID MACHINES

The different miscellaneous fluid machines are:

(1) Hydraulic press
(2) Hydraulic ram
(3) Hydraulic lift
(4) Hydraulic crane
(5) Hydraulic intensifier
(6) Hydraulic accumulator
(7) Fluid or hydraulic couplings
(8) Fluid and hydraulic torque converter
(9) Air lift pump
(10) Gear wheel pump

21.3 HYDRAULIC PRESS

This is a device used for lifting heavy weights with the application of a much smaller weight or force. It is based on Pascal's law, which states that the intensity of pressure in a static fluid is transmitted equally in all directions. It was first built by Joseph Bramah in 1795 and is still in use since it was first developed. Its working principle is explained in Figure 21.1. It consists of a ram and a plunger operating in two cylinders of different diameters which are interconnected at the bottom through a chamber filled with liquid. If A is the area of ram and a is the area of the plunger, a force F is applied to the plunger and pressure intensity developed is $p = \frac{F}{a}$. According to Pascal's law, the same pressure intensity is transmitted throughout the liquid and therefore, the ram will also be subjected to the same pressure intensity. If W is the total load lifted by the ram, then $W = pA$.

$$\therefore \quad p = \frac{F}{a} = \frac{W}{A}$$

$$\therefore \quad W = \frac{A}{a} F \tag{21.1}$$

Thus the lifting force or load W is dependent on the ratio of $\frac{A}{a}$. If the area of the ram and plunger, are a small force may be multiplied many times.

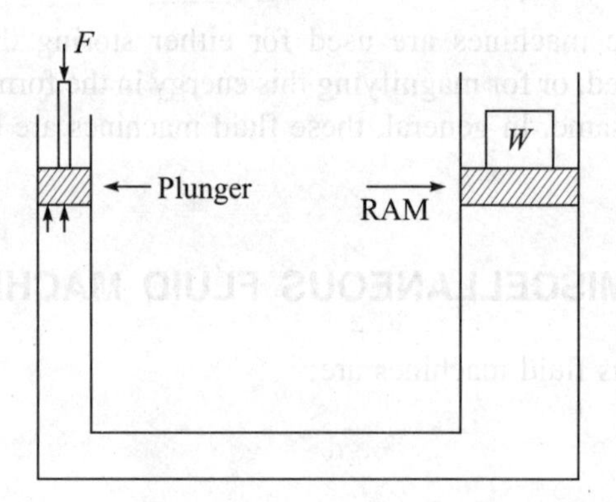

Figure 21.1 Hydraulic press.

21.4 HYDRAULIC RAM

Hydraulic ram is a pump in which the energy of large quantity of water falling through a small height is utilized to lift small quantity of water to greater height. For operation, no external power is required.

As shown in Figure 21.2, it consists of a valve chamber, supply tank, supply pipe, inlet valve, air vessel, delivery pipe, waste valve and a delivery valve.

When the inlet valve of the supply is opened, water starts flowing to the valve chamber having a waste valve and delivery valve. The delivery valve is fitted to an air vessel and the delivery pipe is connected to the air vessel as shown in Figure 21.2. When the valve chamber level begins to rise, the waste valve moves upward. A stage will come when the waste valve is suddenly closed which increases pressure in the valve chamber. The increase of pressure in the chamber forces the delivery valve to open. Water begins to enter the air vessel first which compresses the air above the water. The compressed air exerts pressure on the water in the vessel and a small quantity of water is raised to a greater height as shown in the figure.

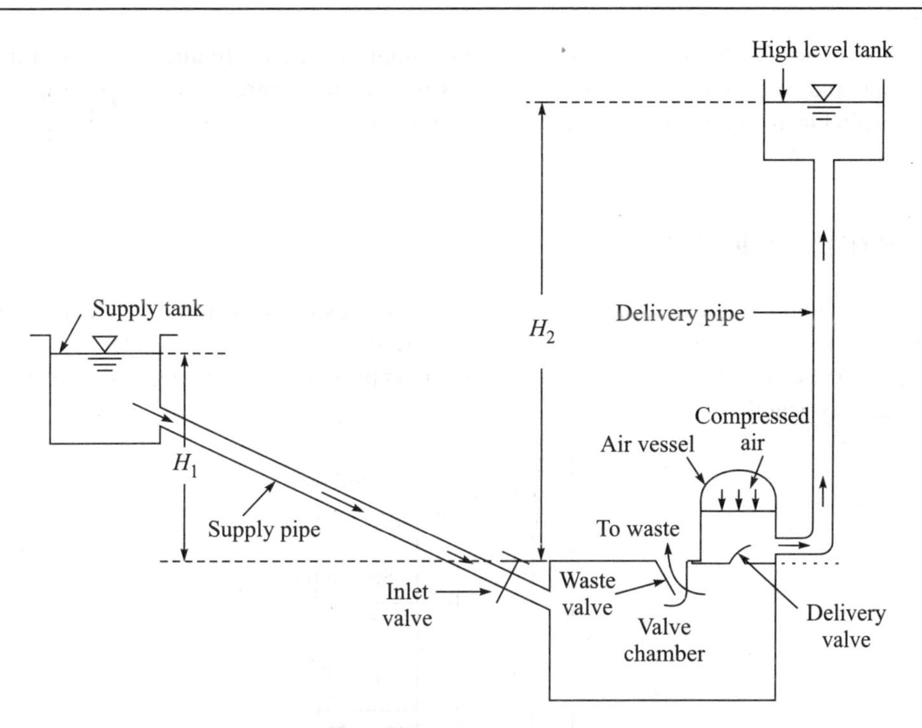

Figure 21.2 Hydraulic ram.

Let W and w be the weights of water flowing per second to the chamber and in the delivery valve respectively and H_1 and H_2 are the heights of the supply tank and the height raised above the level of the chamber, respectively.

Then energy supplied by the supply tank = $W \times H_1$

Energy delivered by the ram = $w \times H_2$

$$\text{Efficiency of the ram} = \frac{w \times H_2}{W \times H_1} \tag{21.2}$$

which is known as D' Aubuisson's efficiency.

According to Rankine, the weight of water W is raised to a height $(H_2 - H_1)$ and the energy delivered = $w(H_2 - H_1)$ and the energy supplied $(W - w)H_1$.

$$\therefore \qquad \text{Rankine efficiency} = \frac{w(H_2 - H_1)}{(W - w)H_1} \tag{21.3}$$

The above efficiencies in terms of discharges:

$$D' \text{ Aubuisson's } \eta = \frac{q \times H_2}{Q \times H_1} \tag{21.4}$$

$$\text{Rankine's } \eta = \frac{q(H_2 - H_1)}{(Q - q)H_1} \tag{21.5}$$

where q and Q are discharges in the delivery and supply pipes. Hydraulic ram is useful in the country-side and in remote areas for providing water supply where a source of large quantity of water available at some height and there is no power to use any other pump to raise the water to a greater height.

21.5 HYDRAULIC LIFT

This is a device which is used for carrying goods as well as persons from one floor to another. A simple form of hydraulic lift consists of a ram sliding in a cylinder as shown in Figure 21.3. At the top of the ram, a cage or platform is provided on which goods may be placed or persons may stand.

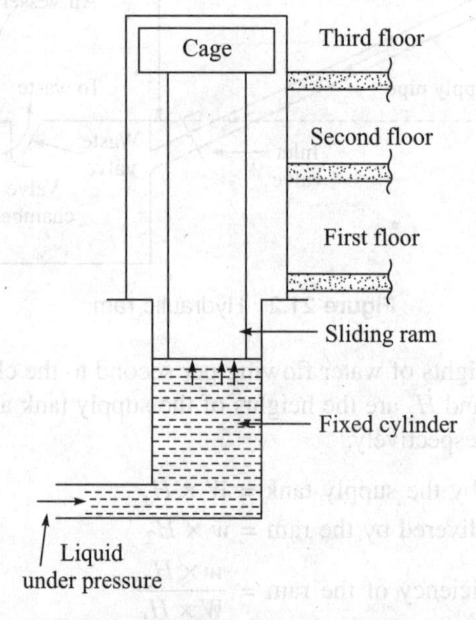

Figure 21.3 Hydraulic lift: Direct acting.

Liquid under pressure is allowed or admitted in the cylinder which pushes the ram vertically upwards. The platform on the ram is made to move downwards by removing the liquid from the cylinder. This type of lift is called a direct acting lift.

Another modified type of lift called suspended lift is shown in Figure 21.4. It consists of a cage, a jigger with fixed cylinder, sliding ram, a set of two pulley blocks of which one is movable, and the other fixed. The principle of working of a hydraulic lift is similar to that of a hydraulic crane explained later. The cage runs between guides suspended by four ropes. Water under pressure enters the fixed cylinder of the jigger, the sliding ram is forced to move towards the left of the distance between the two pulleys, the wire rope connected to the pulley is pulled and the cage is lifted.

This hydraulic lift is now-a-days replaced by an electric lift, which is most commonly used.

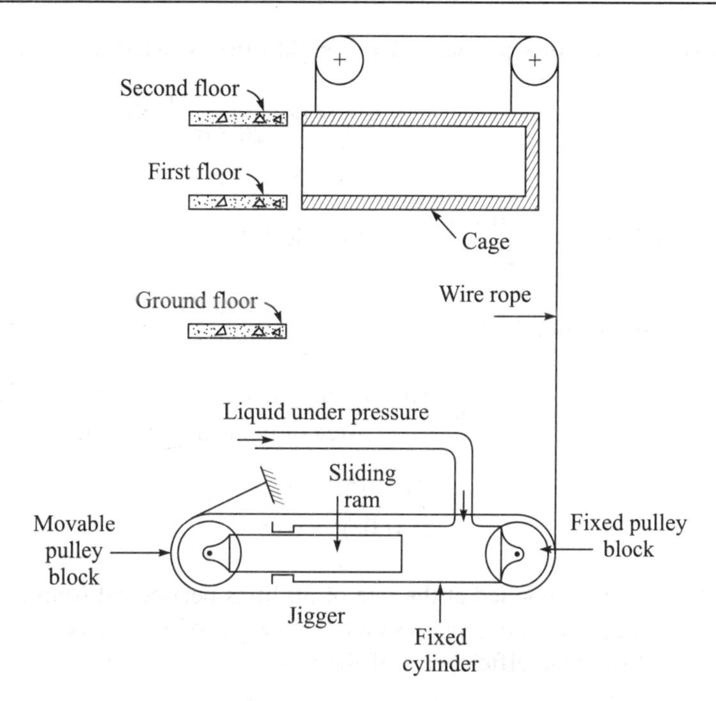

Figure 21.4 Hydraulic lift: Suspended type.

EXAMPLE 21.1 A hydraulic press has a ram of 30 cm diameter and a plunger of 5 cm diameter. Find the weight lifted by the hydraulic press when the force applied at the plunger is 50 N.

Solution: $A = (\pi/4)\ (0.3)^2\ \text{m}^2$

$A = (\pi/4)\ (0.05)^2\ \text{m}^2$

Applying the formula in Equation (21.1): $W = \dfrac{A}{a}F = \dfrac{\pi/4(.03)^2}{\pi/4(.05)^2} \times 50$

$W = 1800$ N **Ans.**

EXAMPLE 21.2 A hydraulic press has a ram of 15 cm diameter and a plunger of 3 cm. The stroke of the plunger is 25 cm and the weight lifted is 600 N. If the distance moved by the weight is 1.2 m in 20 minutes, determine:

(i) Force applied at the plunger, (ii) Power required to drive the plunger, and (iii) Number of strokes performed by the plunger.

Solution: $A = \dfrac{\pi}{4}(0.15)^2\ \text{m}^2,\ a = \pi/4(0.03)^2\ \text{m}^2$

Since $W = \dfrac{FA}{a}$

(i) $F = \dfrac{a}{A}W = 600\left(\dfrac{.03}{.15}\right)^2 = 24$ N **Ans.**

(ii) Work done by the press per second = Weight lifted × distance travelled/sec

$$= \left(600 \times \frac{1.2}{20 \times 60}\right) \text{ N-m/sec}$$

$$= 0.6 \text{ N-m/sec}$$

$\therefore$ Power required = $\left(\frac{0.6}{1000}\right)$ kW = 0.0006 kW **Ans.**

(iii) Number of strokes required = $\dfrac{\text{Total volume of liquid displaced}}{\text{Volume displaced per stroke}}$

$$= \frac{\text{Area of ram} \times \text{distance moved by the weight}}{\text{Area of plunger} \times \text{stroke length}}$$

$$= \frac{\pi/4(.15)^2 \times 1.2}{\pi/4(.03)^2 \times 0.25} = 120$$ **Ans.**

EXAMPLE 21.3 Water is supplied at the rate of 30 litres per second from a height of 4 m to a hydraulic ram which raises 3 litres per second to a height of 18 m from the ram. Determine the D' Aubuisson and Rankine efficiencies of the ram.

Solution: $W = 0.03 \text{ m}^3/\text{sec} \times \rho g = (1000 \times 9.81 \times .03)$ N

$w = 0.003 \text{ m}^3/\text{sec} \times \rho g = (1000 \times 9.81 \times .003)$ N

$H_2 = 4$ m

$H_1 = 18$ m

$\therefore$ D'Aubuisson's efficiency $= \dfrac{wH_2}{WH_1} = \dfrac{1000 \times 9.81 \times .003 \times 18}{1000 \times 9.81 \times 0.03 \times 4}$

$$= \frac{18}{10 \times 4}$$

$$= 0.45 = 45\%$$ **Ans.**

Rankine efficiency $= \dfrac{w(H_2 - H_1)}{(W - w)H_1}$

$$= \frac{q(H_2 - H_1)}{(Q - q)H_1} = \frac{.003 \times (18 - 4)}{(.03 - .003) \times 4}$$

$$= 0.3888$$

$$= 38.88\%$$ **Ans.**

EXAMPLE 21.4 A hydraulic lift is required to lift a load of 98.1 kN through a height of 12 m once in every 100 secs. The speed of the lift is 600 mm/sec. Determine:

(i) Power required to drive the belt
(ii) Working period of lift in seconds
(iii) Idle period of the lift in seconds.

Solution: $H = 12$ m, Time for one operation = 100 seconds, $W = 98.1$ kN = 98100 N

Speed of the lift = 600 mm/sec = 0.6 m/sec

Work done in lifting the weight in 100 secs = (98100 × 12) Nm/100 sec

= (981 × 12) Nm/sec

$\therefore$ Power required to drive the belt = $\left(\dfrac{981 \times 12}{1000}\right)$ kW

= 11.772 kW **Ans.**

$$\text{Working period of the lift} = \frac{\text{Height of lift}}{\text{Velocity of lift}}$$

$$= \frac{12}{0.6} = 20 \text{ seconds}$$

Idle period of lift = Total time – Working period

= (100 – 20) secs

= 80 seconds **Ans.**

21.6 HYDRAULIC CRANE

Hydraulic crane is a device used to lift heavy loads. It is widely used to lift and transfer heavy loads in workshops, warehouses, and docks. It consists of a vertical mast and a jib as shown in Figure 21.5. The mast has its own pedestal and foundation from which a jib or arm is suspended. The jib can be raised or lowered to reduce or increase the radius of action. The

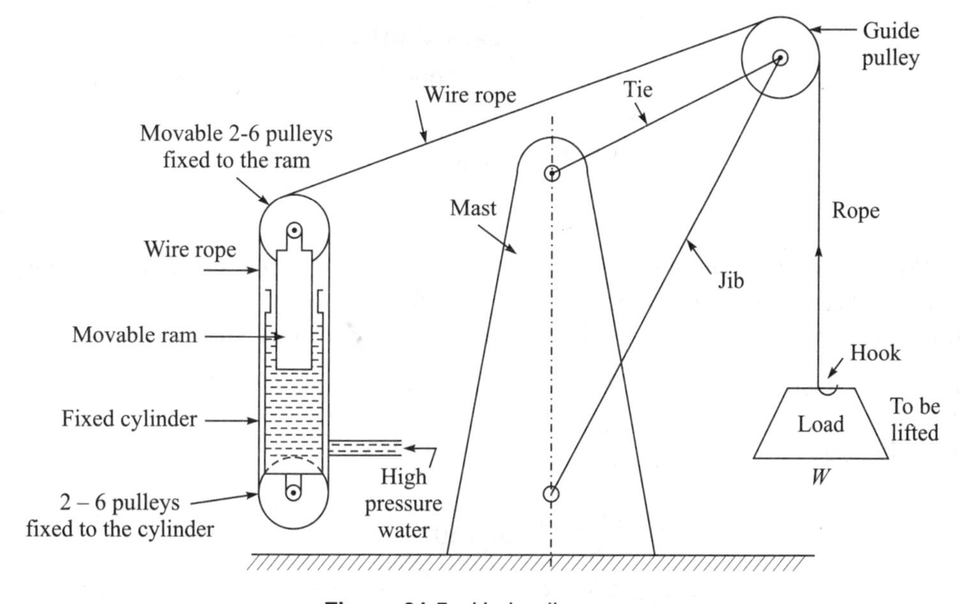

Figure 21.5 Hydraulic crane.

pedestal along with the mast can revolve about a vertical axis. Thus it consists mainly of two parts—crane and jigger. The jigger is attached to the mast of the crane. The jigger consists of a cylinder and a ram which slides in it. A set of 2 to 6 movable pulleys is provided at the top end of the ram. Another set of 2 to 6 pulleys is provided at the lower end of the fixed cylinder.

For lifting loads, water under pressure is allowed to enter the cylinder which pushes the ram in the upward direction. The fixed pulleys with the ram also move up, thereby increasing the distance between the two pulley blocks. Thus the rope wire, which passes over the guide pulley, is pulled by the jigger and the load attached to the rope with a hook is raised or transferred from one place to another.

EXAMPLE 21.5 Find the efficiency of a hydraulic crane which is supplied with 0.4 m^3 of water under pressure of 490.5 N/cm^2 for lifting a weight of 98.1 kN through a height of 10 m.

Solution: Water supplied = 0.4 m^3

$$\text{Pressure } p = 490.5 \text{ N/cm}^2 = 490.5 \times 10^4 \text{ N/m}^2$$

$$W = 98.1 \text{ kN}, \; H = 10 \text{ m}$$

$$\text{Output of the crane} = W \times H = (98.1 \times 10) = 981 \text{ kNm}$$

$$= 981000 \text{ Nm}.$$

Input of the crane = energy supplied by water

= Work done on the ram

= pressure × ram area × stroke of ram

$= pAL$

$= 490.5 \times 10^4 \times (AL)$

But (AL) = Volume of water

$= 0.4 \text{ m}^3$

$= 490.5 \times 10^4 \times .4$

$= (4905000 \times .4) \text{ Nm} = 1962000 \text{ Nm}$

$$\therefore \quad \text{Efficiency} = \frac{\text{Output}}{\text{Input}} = \frac{981000}{1962000} = 0.5$$

$= 50\%$ **Ans.**

EXAMPLE 21.6 The ram of a hydraulic crane is 0.2 m in diameter and its velocity ratio is 6. Water is supplied at pressure of 735.75 N/cm^2. The efficiency of the crane is 50%. Find the load lifted by the crane and quantity of water needed to lift the load by 10 m.

Solution: Pressure force = $735.75 \times [\pi/4 \times 20]^2$ N = 231142.6795 N

$$\text{Load lifted by the crane} = \frac{\text{Pressure force}}{\text{Velocity ratio}} \times \text{efficiency}$$

$$= \frac{231142.6795}{6} \times 0.50$$

$$= 19261.89 \text{ N}$$

Quantity of water required to use = Area of ram × displacement of ram

$$= \left[\frac{\pi}{4}(0.2)^2 \times \frac{10}{6}\right] \text{m}^3$$

$$= 0.05236 \text{ m}^3$$ **Ans.**

EXAMPLE 21.7 A hydraulic crane is supplied with water under a pressure intensity of 490.5 N/cm^2 to raise a load of 44.145 KN to a height of 10 m. The efficiency of the crane is 55%. The stroke of the ram is five times its diameter. Find the volume displaced by the ram and the diameter of the ram.

Solution: Work done by the crane = Load × displacement

$= (44.145 \times 10)$ KNm

$= 441.45$ KNm

$= 441450$ Nm

Pressure intensity = 490.5 N/cm^2 = 490.5×10^4 N/m^2

If d is the diameter of the ram in m,

Pressure force = Intensity of pressure × area of the ram section

$= 490.5 \times 10^4 \times (\pi/4)\ d^2 = 3852378\ d^2$ N

Energy supplied = Pressure force × stroke length

$= 3852378\ d^2 \times 5d$

$= 19261890\ d^3$ N-m.

$$\text{Now efficiency} = \frac{\text{Work done}}{\text{Energy supplied}}$$

$$0.55 = \frac{441450}{19261890 d^3}$$

$$\therefore \quad d^3 = \frac{441450}{0.55 \times 19261890} = 0.041669657$$

$$\therefore \quad d = 0.3466889 \text{ m} \simeq 0.346689 \text{ m}$$

$= 34.6689$ cm **Ans.**

Volume displaced by the ram = Volume in one stroke

$= \pi/4(.346689)^2 \times (5 \times 0.346689)$

$= 0.163636$ m^3 **Ans.**

EXAMPLE 21.8 The supply pipe in a hydraulic ram installation is 6 cm in diameter and 4 metres high. The waste valve is 13 cm in diameter having an effective lift of 6.5 mm and weight of 13.25 N and it makes 125 beats per minute. Find the discharge through the delivery pipe per minute against a delivery head of 8 m above water level in the supply reservoir. Neglect losses.

Solution: Base area of waste value = $\pi/4\ (13)^2$ = 132.7323 cm^2 upward pressure intensity needed to close the valve

$$= \frac{13.25}{132.7323} = 0.0998 \text{ N/cm}^2$$

Pressure head corresponds to the above pressure intensity

$$= \left(\frac{0.0998 \times 10^4}{9.81 \times 1000}\right) \text{ m of water}$$

$$= 0.1017 \text{ m of water}$$

$\therefore$ Velocity of flow past the waster valve $= \sqrt{2 \times 9.81 \times 0.1017}$ m/sec

$$= 1.413 \text{ m/sec.}$$

$$\text{Area past the waste valve} = \pi(13) \times 0.65 = 26.546 \text{ cm}^2$$

$$\text{Area of supply pipe} = \pi/4(6)^2 = 28.274 \text{ cm}^2$$

max^m velocity of water in supply pipe

$$V_{max} = \frac{28.546}{28.274} \times 1.413 = 1.3266 \text{ m/sec.}$$

$$\text{Retarding head} = \frac{l}{g} \cdot \frac{V}{t} = \frac{4}{9.81} \times \frac{1.413}{t}$$

$$\therefore \quad 8 = \frac{4}{9.81} \times \frac{1.3266}{t}$$

$$\therefore \quad t = 0.0676 \text{ sec}$$

Taking average velocity as half of max^m velocity:

$$V_{av.} = \frac{V_{max}}{2} = \frac{1.3266}{2} = .6633 \text{ m/sec}$$

$\therefore$ Quantity of water passing past the delivery valve per beat

$$= \text{Average velocity} \times \text{area of pipe} \times t$$

$$= [0.6633 \times 100) \times 28.274 \times 0.0676] \text{ cm}^3$$

$$= 126.778 \text{ cm}^3$$

$\therefore$ Discharge per minute $= 126.778 \times$ no. of beats/minute

$$= (126.778 \times 125) \text{ cm}^3\text{/min}$$

$$= 15847.25 \text{ cm}^3\text{/min}$$

$$= 15.84725 \text{ litres/min}$$

Ans.

21.7 HYDRAULIC INTENSIFIER

It is a device which is used for increasing the intensity of pressure of liquid, by utilising the energy of a larger quantity of liquid at low pressure. Such a device is needed when the pressure of water supplied to a machine is not requisitely high.

Figure 21.6 shows the hydraulic intensifier which consists of a fixed ram conveying the high pressure liquid to the machine. A hollow sliding ram is mounted over the fixed ram. The hollow sliding ram contains the high pressure water. The sliding ram is surrounded by a fixed cylinder containing the low pressure liquid from the main supply. The weight of the low pressure liquid

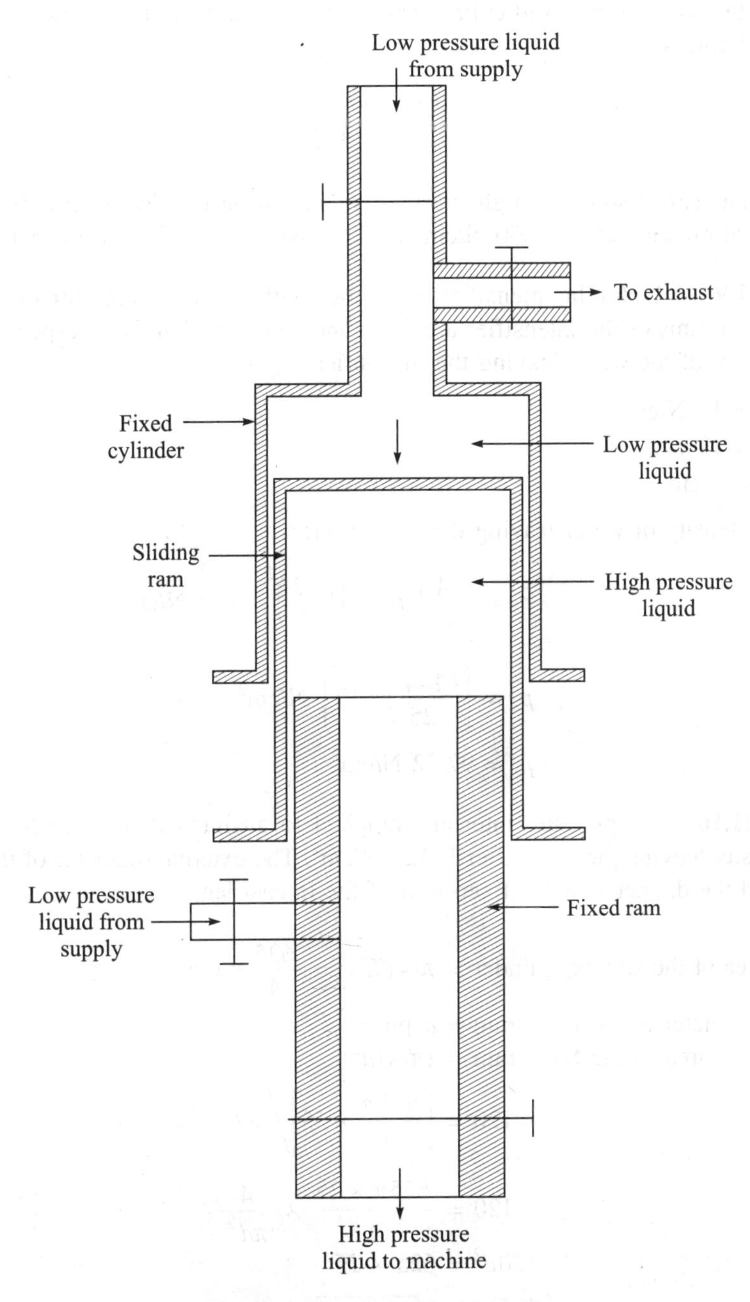

Figure 21.6 Hydraulic intensifier.

acts on the upper end of the sliding ram. In this process, water in the sliding ram is compressed and its pressure is thus increased. The pressure liquid is squeezed out of the sliding cylinder through the fixed ram to the machine.

If A and a are the areas of the upper end of the sliding ram and fixed ram, and p_f and p_s are the pressure intensities on the fixed cylinder and sliding ram, respectively, equating the upward and downward forces:

$$p_s a = p_f A$$

$$p_s = \frac{A}{a} p_f$$

Thus the intensifier supplies high pressure water during the downward stroke only. It is possible to raise pressure from 6800 N/cm^2 to 7800 N/cm^2 by using an intensifier.

EXAMPLE 21.9 A hydraulic intensifier is supplied with water at a pressure of 17 N/cm^2. The sliding and fixed rams of the intensifier are 5 cm and 12 cm in diameter, respectively. Find the pressure intensity of the water leaving the intensifier.

Solution: p = 17 N/cm^2

D = 12 cm.

d = 5 cm

Pressure intensity of water leaving the intensifier:

$$p_s = \left(\frac{A}{a}\right) p_f = \left[\left(\frac{12}{5}\right)^2 \times 17\right] \text{ N/cm}^2$$

$$p_s = \left[\left(\frac{144}{25}\right) \times 17\right] \text{ N/cm}^2$$

$$p_s = 97.92 \text{ N/cm}^2$$ **Ans.**

EXAMPLE 21.10 The pressure intensity supplied to an intensifier is 25 N/cm^2 while the pressure intensity leaving the intensifier is 120 N/cm^2. The external diameter of the sliding ram is 25 cm. Find the diameter of the fixed ram of the intensifier.

Solution: Area of the sliding cylinder = $\pi/4(25)^2 = \frac{625}{4}\pi$ cm^2

Let the diameter of the fixed ram = d pm.

area of the fixed ram = $(\pi/4)d^2$

$$\therefore \quad 120 = \left[\frac{625\pi}{4} \times 25\right] \Big/ (\pi/4d^2)$$

$$120 = \frac{625\pi \times 25}{4} \times \frac{4}{\pi d^2}$$

$$120d^2 = 625 \times 25$$

$$\therefore \quad d = \sqrt{\frac{625 \times 25}{120}} = 11.41 \text{ cm}$$ **Ans.**

21.8 HYDRAULIC ACCUMULATOR

It is a device that is used to store or accumulate the liquid under pressure supplied by the pump when the liquid is not required by the machine during its idle period. The demand of water under pressure by hydraulic machines like lift, crane, press etc. is not uniform whereas pump supplies water under pressure at a more or less uniform rate throughout its supply period. As such, if an accumulator is introduced between the pump and the machine, liquid under pressure can be stored in the accumulator during the idle period of the machine. This stored water in the accumulator can be used by the machine along with the uniform supply of the pump when the machine needs a large quantity of water during its working stroke. This arrangement of the accumulator between the pump and the machine helps in saving the amount of water supplied by the pump during the idle period of the machine. Besides, the pump capacity need not be as large as required by the machine as the stored water can be used along with the pump capacity.

As shown in Figure 21.7, a hydraulic accumulator consists of a fixed vertical cylinder containing a sliding plunger or ram which is loaded with weights. One side of the cylinder is connected to the pump and other side to the machine. At the beginning, the ram remains at its lowermost position. When the machine does not require water and the pump continues to deliver water under pressure to the cylinder, the loaded ram begins to move upwards till the cylinder is full of water which is under pressure and thus it accumulates the unused water during the idle period of the machine.

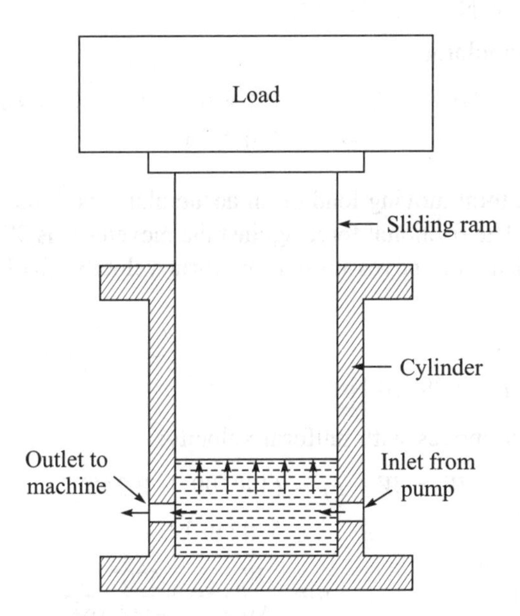

Figure 21.7 Hydraulic accumulator.

When the machine requires water after the idle period, water from the accumulator flows to the machine and the ram gradually moves down. This is the downward stroke of the ram during which liquid under pressure is delivered to the machine.

The maximum water under pressure or hydraulic energy that the accumulator can store is known as the capacity of the accumulator.

Let D be the diameter of the sliding ram
L be the stroke length
p be the pressure intensity at which the liquid is supplied by the pump
W be the total weight of the ram including the weight a load placed on it.

Then $$W = p \times \frac{\pi}{4} D^2$$

Work done by the accumulator in lifting the ram or the capacity of the accumulator = WL

$$= (p \times (\pi/4)D^2 \times L)$$

$$= p \times \left(\frac{\pi}{4} D^2 L\right)$$

$$= p \times \text{Volume of the accumulator}$$

EXAMPLE 21.11 An accumulator has a ram of 20 cm diameter and a lift of 6 m. Water is supplied at a pressure of 588.6 N/cm². Find the necessary load on the ram along with its own weight and capacity of the accumulator.

Solution: D = 20 cm = 0.2 m, L = 6 m, p = 588.6 N/cm² = 588.6 × 10⁴ N/m²

$\therefore$ $W(\text{Load}) = p \cdot A = (588.6 \times 10^4) \times \pi/4(0.2)^2$

$= 184914.14$ N = 184.914 kN **Ans.**

Capacity of the accumulator (i.e. work done)

$$(pA)L = WL = 184.914 \times 6 = 1109.484 \text{ kN-m}$$

$= 1109484$ N-m **Ans.**

EXAMPLE 21.12 The total moving load of an accumulator is 490500 N and the diameter of the sliding ram is 40 cm. The frictional force against the movement is 39240 N. Find the pressure of water when: (a) Loaded ram moves up with uniform velocity, (b) Loaded ram moves down with uniform velocity.

Solution: W = 490500 N

Frictional resistance F_R = 39240 N

(a) When loaded ram moves with uniform velocity:

$$W_u = W + F_R = (490500 + 39240) \text{ N}$$

$$= 529740 \text{ N}$$

$\therefore$ Pressure intensity $$p = \frac{W_u}{\text{Area}} = \frac{529740}{\pi/4(40)^2}$$

$= 421.5536$ N/cm² **Ans.**

(b) When loaded ram moves down with uniform velocity:

$$W_d = W - F_R = (490500 - 39240) \text{ N}$$

$$= 451260 \text{ N}$$

$$\therefore \qquad p = \frac{W_d}{\text{area}} = \frac{451280}{\pi/4\,(40)^2}$$

$$= 359.1 \text{ N/cm}^2 \qquad \textbf{Ans.}$$

EXAMPLE 21.13 An accumulator is loaded with a total load of 588600 N. The frictional resistance is 4% of the total load. The plunger diameter is 35 cm and effective stroke is 5 m. The time taken by the plunger to fall through a full stroke is 2.5 minutes with a steady movement and if the same pump is discharging 0.008 m^3/sec, find the power delivered to the machine by the accumulator.

Solution: Net Load $= \left(588600 - 588600 \times \frac{4}{100}\right)$

(while falling)

$= 565056$ N

$\therefore$ Pressure intensity $p = \frac{565056}{\pi/4\,(35)^2} = 587.3 \text{ N/cm}^2$

Corresponding head to this pressure intensity $= \frac{p}{\rho g}$

$$= \left(\frac{587.3 \times 10^4}{1000 \times 9.81}\right) \text{m}$$

$$= 598.67 \text{ m}$$

Work supplied by the pump/sec = (Weight of water by the pump/sec) × pressure head.

$= [(0.008) \times 100 \times 9.81 \times 598.67]$

$= 46983.621$ N-m

Work done by the accumulator per sec $= 565056 \times \left(\frac{5}{2.5 \times 60}\right)$

$= 18835.2$ N-m

$\therefore$ Total work done/power delivered per sec $= (46983.621 + 18835.2)$ N-m

$= 65818.821 \text{ N-m} = 65.8188 \text{ kW}$ **Ans.**

21.9 FLUID OR HYDRAULIC COUPLINGS

The fluid or hydraulic coupling shown in Figure 21.8 is a device used for transmitting power from driving shaft *A* to driven shaft *B* with the help of a fluid, preferably oil. There is no mechanical connection between the two shafts. The coupling consists of a radial pump impeller mounted on a driving shaft and a radial flow reaction turbine mounted on the driven shaft. Both the impeller and the runner are identical in shape and they together form a casing which is completely enclosed and filled with oil.

From rest, when the driving shaft is first rotated, the oil starts moving from the inner radius to the outer radius of the pump impeller. The pressure energy and kinetic energy of the oil increase at the outer radius of the pump impeller. This oil of increased energy enters the runner of the reaction turbine at the outer radius of the turbine runner, and flows inward. The oil transfers the energy to the blades and the runner rotates. The oil from the runner then flows back to the pump impeller and thus a continuous circulation is made. Power is transmitted from the driving shaft to the driven shaft whose speed is always less than the driving shaft by about 2%.

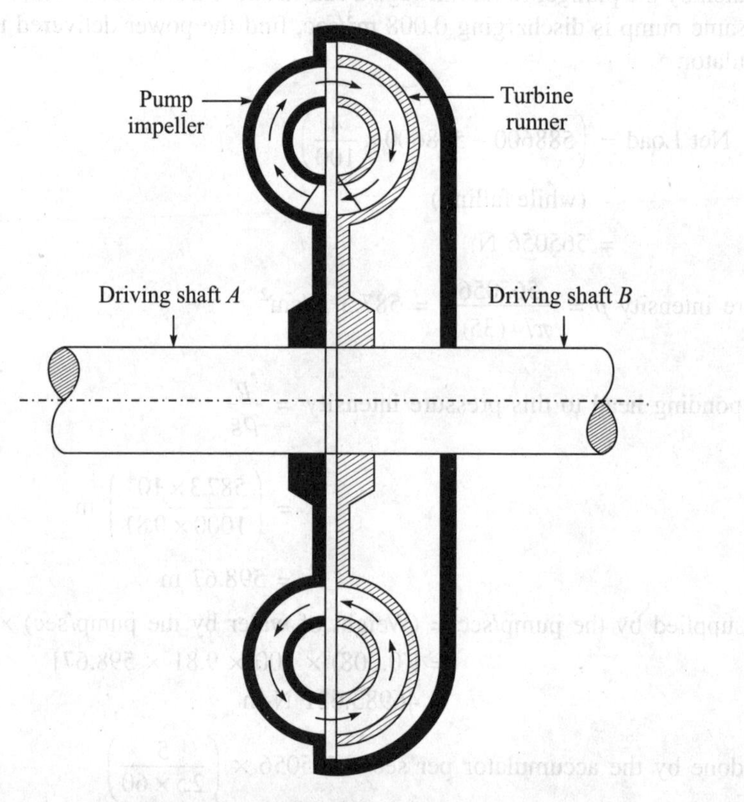

Figure 21.8 Fluid or hydraulic coupling.

Efficiency of coupling is obtained as follows:

$$\text{Efficiency of coupling } \eta = \frac{\text{Power output}}{\text{Power input}} = \frac{\text{Power transmitted to } B}{\text{Power available at shaft } A}$$

If N is the speed and T is the Torque:

$$P(\text{in any shaft}) \propto N \times T$$

$$\therefore \quad P_A \propto N_A T_A \quad \text{and} \quad P_B \propto N_B T_B$$

$$\therefore \quad \eta = \frac{N_B T_B}{N_A T_A}$$

But $T_A = T_B$, Torque transmitted is the same

$$\therefore \qquad \eta = \frac{N_B}{N_A}$$

$$\text{Slip of fluid coupling} = \frac{N_A - N_B}{N_A} = 1 - \frac{N_B}{N_A} = 1 - \eta.$$

EXAMPLE 21.14 In a hydraulic coupling, the speeds of the driving and driven shaft are 800 rpm and 780 rpm, respectively. Find the efficiency and slip.

Solution:

$$\eta = \frac{N_B}{N_A} = \frac{780}{800} = 0.975 = 97.5\% \qquad \textbf{Ans.}$$

$$\text{Slip} = 1 - \frac{N_B}{N_A} = 1 - 0.975$$

$$= 0.025$$

$$= 2.5\% \qquad \textbf{Ans.}$$

21.10 FLUID OR HYDRAULIC TORQUE CONVERTER

The hydraulic torque converter is similar to the fluid or hydraulic coupling except for a series of fixed guide vanes which are provided between the impeller and the turbine runner. In this case, the liquid exerts a torque on the driven unit through these stationary guide vanes as it flows from the pump to the turbine. As such, by suitably designing the stationary guide vanes, the torque transmitted to the driven unit can be either increased or decreased.

Thus the torque relationship is given as:

$$T_t = T_p + T_v$$

where T_t is the torque transmitted to the turbine runner,
T_p is the torque of the pump impeller, and
T_v is the variation in the torque caused by the fixed guide vanes.

21.11 AIR LIFT PUMP

An air lift pump is a device used for lifting water from a well or a sump by compressed air. As shown in Figure 21.9, compressed air is used in the inlet pipe which is fitted to the delivery pipe. This compressed air forms an air bubble mixed with water in the delivery pipe. The density of this water with the air bubble is much lower than water. Hence a small height of water column can balance the air–water mixture in the delivery pipe.

The length $(L - H)$ is called 'useful lift'. The result is found to be satisfactory if the useful lift $(L - H)$ is less than the submergence depth H.

This pump has its own advantage though the efficiency is not high. As there is no moving part like foot valve, etc., below the water level, so the chances of damaging the pump by suspended particles is zero. Besides, in this pump, limitation of suction height is not required.

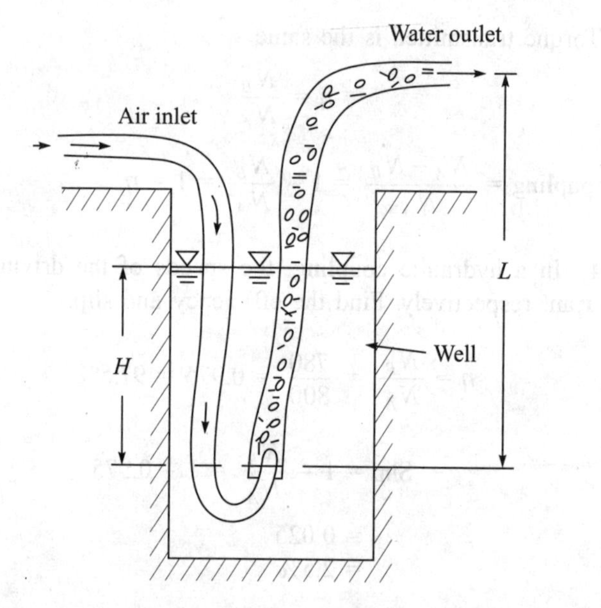

Figure 21.9 Air lift pump.

21.12 GEAR WHEEL PUMP

It is a rotary pump in which two identical intermeshing spur pinions work in a fine clearance inside a casing. One of the pinions is keyed to a driving shaft. The other pinion revolves idly. The space between the two teeth and the casing is filled with water. The oil is carried around between the gears from the suction pipe to the delivery pipe.

Let l = axial length of the teeth
a = area enclosed between two successive teeth and casing
n = total number of teeth in each pinion.
N = speed of the pump in rpm

$\therefore$ Volume of oil discharged in one revolution:

$$Q = 2 \times l \times a \times n$$

$$\text{Discharge per second} = (2 \times l \times a \times n) \times \frac{N}{60} \text{ m}^3$$

If q is the actual discharge:

$$\text{Volumetric efficiency} = \frac{q}{Q} \times 100\%$$

21.13 CONCLUSION

The miscellaneous fluid machines discussed in this chapter have wide applications in the practical field. Hydraulic press is used for lifting heavy weight by application of much smaller weight, hydraulic ram provides to raise water to a higher level without the use of electricity, hydraulic lift is a device to carry goods as well as persons from one floor to another, hydraulic crane is another device used to lift and transfer heavy loads in workshop, warehouse, docks etc., hydraulic intensifier is used for increasing intensity of pressure etc. Similarly other miscellaneous hydraulic machines have been used for some other important applications. Therefore, all types of these machines have been presented in a separate chapter. The basic principles, applications, numerical examples for all the different types are discussed and presented. Few problems are enclosed at the end for readers to solve.

PROBLEMS

21.1 A hydraulic accumulator is 45 cms in diameter and the total moving load is 588.6 KN. The frictional resistance against the movement of the ram is 49.05 KN. Calculate the pressure of water when the:

(a) Loaded ram is moving up with uniform velocity,
(b) Loaded ram is moving down with uniform velocity.

(**Ans.** 400.8366 N/cm^2, 339.23 N/cm^2)

21.2 A hydraulic crane utilises 0.375 m^3 of water at a pressure intensity of 490.5 N/cm^2 in order to lift a load of 106.93 KN through a height of 10 m. Calculate the efficiency.

(**Ans.** 58%)

21.3 A ram of a hydraulic crane is 25 cm in diameter, and the velocity ratio is 6. Water is supplied at a pressure of 748.8 N/cm^2. The efficiency of the crane is 50%. Find (i) Load lifted by the crane, (ii) Quantity of water needed to lift the load by 10 m.

(**Ans.** 32103.225 N, 81.8 litres)

21.4 Water is supplied at a pressure 15 N/cm^2 to an accumulator having a ram of 2 m diameter. If the total lift of the ram is 10 m, determine:

(i) Capacity of the accumulator,
(ii) Total weight placed on the ram including its self-weight.

(**Ans.** 4712.4 KNm, 471240 N)

21.5 The diameter of the fixed ram and fixed cylinder of an intensifier are 100 mm and 250 mm, respectively. If the pressure of water supplied to the fixed cylinder is 25 N/cm^2, find the pressure of water flowing through the fixed ram. (**Ans.** 156.25 N/cm^2)

21.6 Find the efficiency of a hydraulic crane which is supplied with 300 litres of water under a pressure of 60 N/cm^2 for lifting a weight of 12 KN through a height of 11 m.

(**Ans.** 73.33%)

21.7 A hydraulic lift is required to lift a load of 8 KN through a height of 10 m once in every 80 secs. The speed of the lift is 0.5 m/sec. Determine:

(i) Power required to drive the lift,
(ii) Working period of lift in seconds,
(iii) Idle period of the lift in seconds. (**Ans.** 1 kW, 20 sec, 60 secs).

21.8 Water is supplied at a rate of 0.02 m^3/sec from a height of 3 m to a hydraulic ram, which raises 0.002 m^3/sec to a height of 20 m from the ram. Determine D'Aubuisson's and Rankine's efficiencies of the ram. (**Ans.** 66.67%, 62.96%)

21.9 The pressure intensity of water supplied to an intensifier is 20 N/cm^2 while the pressure intensity of water leaving the intensifier is 100 N/cm^2. The external diameter of the sliding cylinder is 20 cm. Find the diameter of the fixed ram of the intensifier.

(**Ans.** 8.94 cm)

REFERENCES

1. Daily, J.W. "Hydraulic Machinery", Chapter 13 of *Engineering Hydraulics*, edited by H. Rouse, John Wiley and Sons. Inc., New York, 1950.
2. Bansal, R.K., *Fluid Mechanics and Hydraulic Machines*, Laxmi Publications, New Delhi, 1983.
3. Modi, P.N. and S.M. Seth, *Hydraulic and Fluid Mechanics including Hydraulic Machines*, Standard Book House, 3rd. ed., New Delhi, 1977.
4. Olson, R.M., *Engineering Fluid Mechanics*, International Text Book Company, Seranton Pennsylvania, 1967.

Chapter 22

Discharge Measurements: Principles, Techniques and Instruments

22.1 INTRODUCTION

A brief review of the subject of 'hydrometry', which is a science of water measurement, is presented. Twenty-four existing methods of discharge measurements, their principles, techniques, equations and figures are given.

22.2 HISTORICAL REVIEW OF HYDROMETRY

The subject of hydrometry is as old as human civilisation. The early history of hydrometry presented by Kolupaila[1] is quite comprehensive. The oldest hydrometric evidences are markings of flood stages of the river Nile cut off on steep rock faces. Stage-discharge relationship was conceived long back in the twelfth century in the irrigation system of Central Asia. The famous Pitot tube for measuring the velocity in river Seine was proposed by Henri de Pitot, a French engineer, in 1732. The Price current meter, which has been extremely widely used throughout the world to measure river velocity, was initially devised by T.G. Ellis in 1870, and subsequently re-designed by W.G. Price. The famous venturimeter was demonstrated by the Italian physicist, G.B. Venturi in 1797. The Parshall flume, which is extensively used in the USA to measure discharge, was developed by R.L. Parshall in 1920. The ultrasonic method of velocity measurement was first reported in Swengel in 1955 on the basis of different hydraulic and electric principles. Hydrometry is developed step by step. Many modern methods and new techniques have been developed in stream flow measurements. Conventional methods of flow measuring devices have been replaced by modern automatic instrument methods like bubble gauge, echo sounder, ultrasonic flow meter, dynamometer, inertia pressure methods, radio active tracers, oxygen polarography, hotwire anemometer, electronic flow meter and Laser Doppler Anemometer. Some of the automatic instruments are connected to high-speed digital computers to help obtain the result directly from the computer.

Thus, the subject of hydrometry has been developed extensively since the old days due to the contributions of scientists and engineers in the past and with the application of electronics in the present. Further works on different methods of discharge measurement have done by U.S.B.R.[2], Corbet[3], Chow[4,7] Herschy[5], Addison[6], Das and Goswami[8,30] Parshall[10,19] Rouse[11], Bour and Graf[12], Smith and Nayak[13], Allen and Taylor[14], Dalrymple and Benson[16], Schuster, J.C.[17], Ackers et at[20], Cone[21], Subramanya and Kumar[22], Herschy and Loosemor[23], Benson[24], Sitter[25], Langbein[26], DeMarchi[27], Inglis[28], Crump[29], Balloffet[31], Linford[32], Engel[33], Jameson[34,35] Boss[36], Frazier[37] and by others.

22.3 DIFFERENT METHODS OF MEASUREMENTS

Discharge Q is obtained when the area of flow A is multiplied by average velocity V. Therefore, in the different methods described below, either Q is obtained directly or the average velocity V or stage of flow H is measured. For a known cross-section A where the velocity is measured Q is obtained as $Q = A \cdot V$ and if the stage H is measured, the rating curve (Q vs H) is used to compute Q.

Following are the different methods of discharge measurement:

(1) Current meter method
(2) Slope area method
(3) Area-velocity method
(4) Pitot tube method
(5) Float method
(6) Variable area method: Rotameter
(7) Water stage recorder: Rating curve method
(8) Brink or end depth method
(9) Obstruction meter method: Venturimeter, nozzle meter, orifice meter, etc.
(10) Contraction meter method: Weir, notch, spillway dam, culvert, sluices, etc. i.e. channel control method
(11) Critical flow flumes: Channel transition method, Venturiflume, critical depth flume, Parshall flume, etc.
(12) Salt velocity method based on electrical conductivity.
(13) Dilution technique, i.e. tracer technique
(14) Moving boat method
(15) Echo sunder method
(16) Hydraulic model experimental method
(17) Inertia pressure method
(18) Ultrasonic flow meter method, i.e. use of transducers
(19) Laser Doppler Anemometer (LDA) method
(20) Hot wire anemometer or warm film anemometer
(21) Thrupple ripple method
(22) Bubble gauge: Automatic recorders
(23) Electromametic method
(24) Dynamometer method

22.4 CURRENT METER METHOD

It is the most popular and widely used instrument to measure the average velocity of river. As shown in Figure 22.1, it consists of cups or the propeller mounted on a shaft which is rotated by the flow velocity. An electric circuit powered by battery from above the free surface of water is used to determine the number of revolutions of the meter shaft from audible signals to a head phone used by the observer. It is suspended by a cable from the top of the water level to a depth of 0.2 time and 0.8 time of depth: Calibration curve for the meter is available from where velocity at depths $0.2D$ and $0.8D$ are obtained corresponding to revolutions or speed of the spindle. The average velocity V is then obtained:

$$V = \frac{V_{0.2D} + V_{0.8D}}{2} \tag{22.1}$$

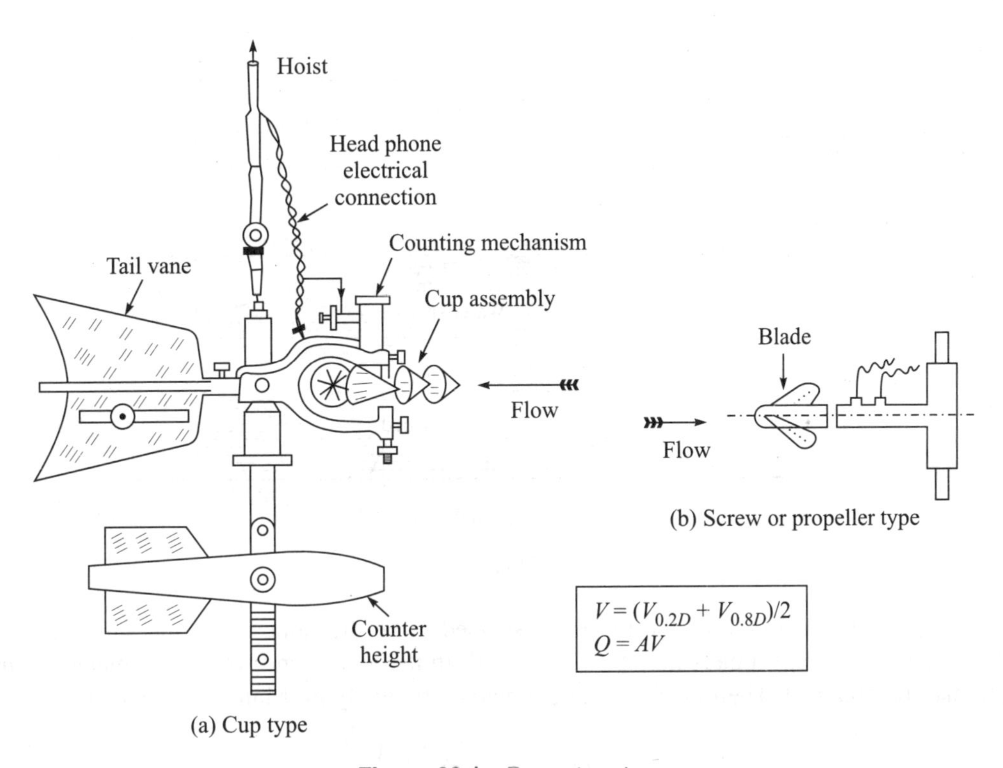

Figure 22.1 Current meter.

This in brief is the procedure to be followed for measurement of velocity. But there lie hydraulic principles, which, is why velocities are measured at $0.2D$ and $0.8D$, where D is the depth at the point of measurement. This principle can by shown be making an analytical solution starting from Reynolds stress equation and Prandtl's mixing length theory, and by applying the velocity defect law. This analytical solution implies that the average velocity is best obtained when the two velocities are measured at $0.16D$ and $0.84D$. For practical purposes, therefore, both velocities are measured as $0.2D$ and $0.8D$.

22.5 SLOPE AREA METHOD

If gauging stations is not possible, this method is used. A reach of the river of about 1 km is selected. The difference of water surface level (h_L) between upstream and downstream of reach L is obtained. Slope of the water surface is: (From Figure 22.2),

$$s = \frac{h_L}{L} \tag{22.2}$$

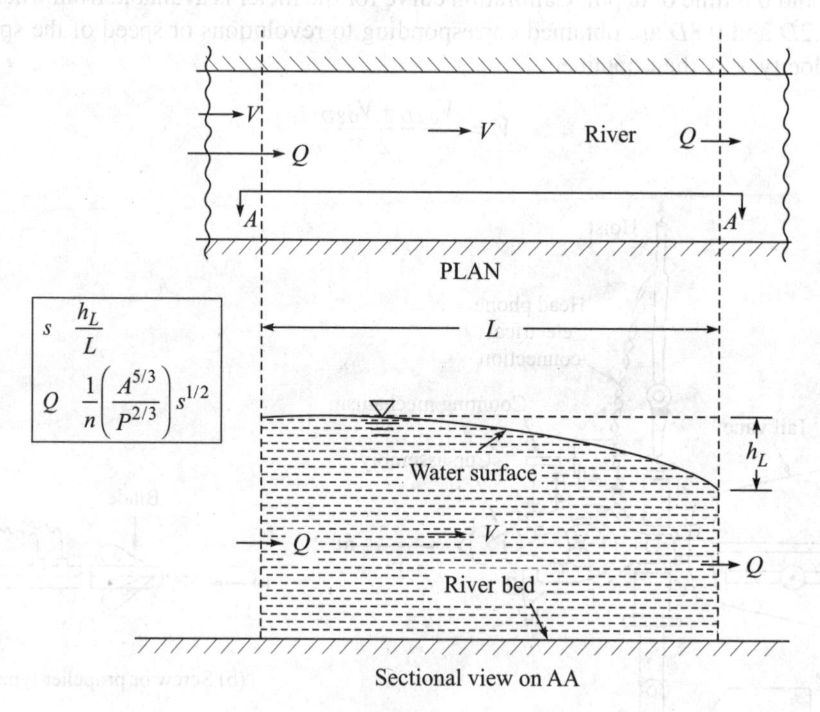

Figure 22.2 Slope area method.

The cross-sectional area of the reach is assumed to be the same or by measuring area at different points, average area A may be obtained. Both the area A and wetted perimeter against depth may be obtained. Then discharge Q is obtained by applying Manning's[9] equation, i.e.

$$Q = A \cdot V = A \frac{1}{n}\left(\frac{A}{P}\right)^{2/3} s^{1/2}$$

$$Q = \frac{1}{n}\left(\frac{A^{5/3}}{P^{2/3}}\right) s^{1/2} \tag{22.3}$$

where n is Manning's roughness coefficient, which is the bed roughness and may be obtained from Chow[7]. This method has been described in U.S. Geological Survey by Dalrymple and Benson[16].

22.6 AREA VELOCITY METHOD

It is based on the continuity equation $Q = A \cdot V$. The cross-sectional area of the stream at the point of measurement is known or measured. Velocity measurements at different depths like 0.2D, 0.4D, 0.6D, 0.8D and at the surface are measured by using float or Pitot tube or current meter. Velocity profile along the cross-section and vertical are shown in Figures 22.3(a) and 22.3(b). From the velocity profile along vertical, the average velocity V is determined. Then the discharge is obtained by

$$Q = A \cdot V \tag{22.4}$$

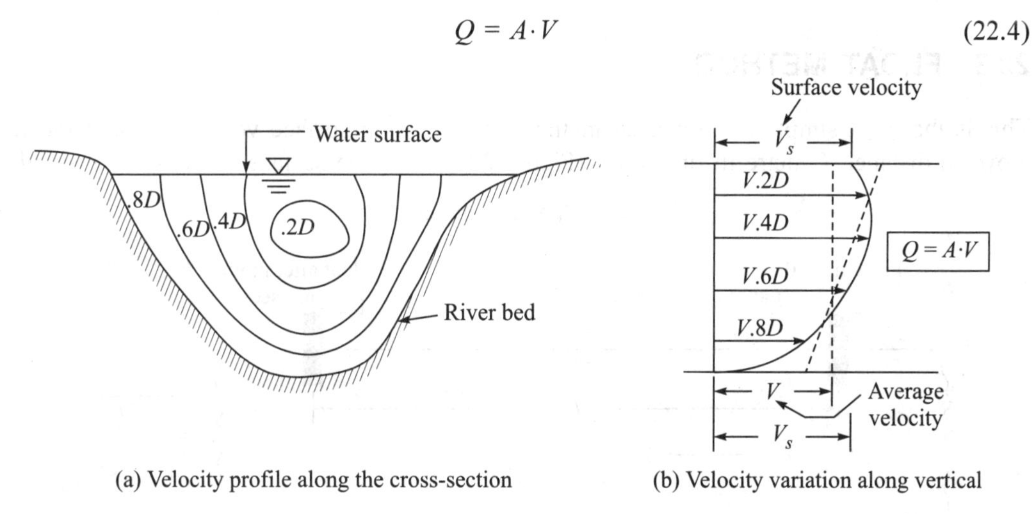

Figure 22.3 Velocity profiles of stream cross-section.

22.7 PITOT TUBE METHOD

This method is named in honour of its originator, Henri de Pitot, a French engineer, who adopted this principle for measuring velocity in the river. Seine. It is a very simple device. The basic principle used in this method is that the velocity at stagnation point S (Figure 22.4) is reduced to zero. Kinetic energy at that point is converted into pressure energy and the liquid therefore, rises in this bent and open tube as shown in the figure to a height of h above the liquid surface.

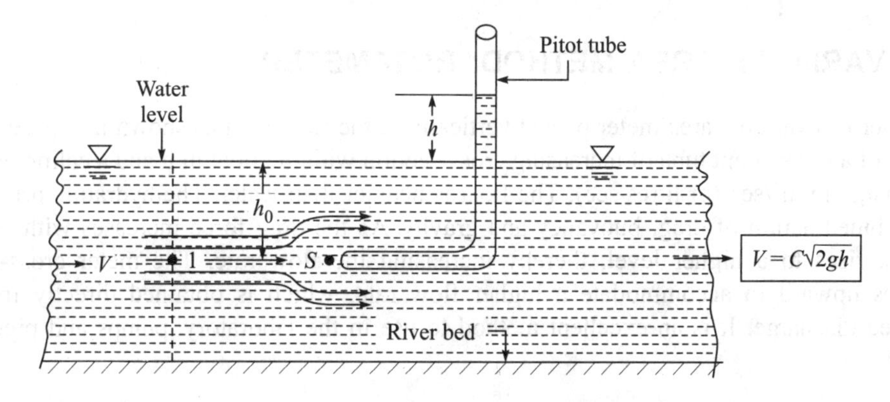

Figure 22.4 Pitot tube method.

Now if Bernoulli's Theorem is applied at u/s and at S, the average velocity becomes:

$$Z_1 + \frac{V^2}{2g} + h_0 = Z_1 + \frac{0^2}{2g} + h_0 + h$$

from which
$$V = C\sqrt{2gh} \tag{22.5}$$

where C is a tube constant, usually $\simeq$ 0.8, but the actual value is to be celebrated to obtain the correct results.

22.8 FLOAT METHOD

This is the most simple and quickest method of measuring surface velocity (V_s). If the float moves a distance L metre in time t sec (Figure 22.5) then the surface velocity is obtained as:

$$V_s = L/t \text{ m/sec} \tag{22.6}$$

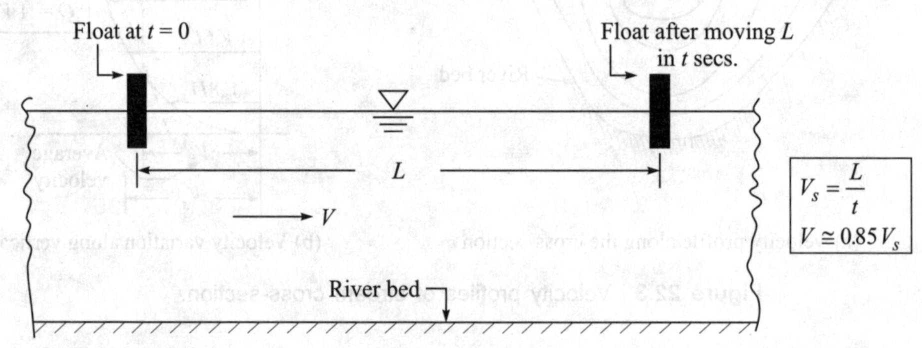

Figure 22.5 Float method to find surface velocity V_s.

From this surface velocity, the average velocity is obtained by multiplying a reduction factor. This factor varies from 0.79 to 0.92. The acceptable value is 0.85. The Mysore engineering research station has given the following average velocity equation:

$$V = 0.8529V_s + 0.0085 \tag{22.7}$$

22.9 VARIABLE AREA METHOD: ROTAMETER

Rotameter is a variable area meter placed vertically in the flow path as shown in Figure 22.6. It consists of a transparent tube of increasing cross-section with the float in it and a calibrated scale of discharge in m^3/sec (or litres/sec). The float is heavier than water. The hydraulic principle is the combined action of drag, buoyancy and gravity. As the drag force increases with velocity, the float rises to a higher level, thereby increasing the flow rate. The meter cross-section increases upward to accommodate a higher flow rate, which is obtained directly from the calibrated discharge. It is an excellent method to use in the laboratory, pumps and pipes, etc.

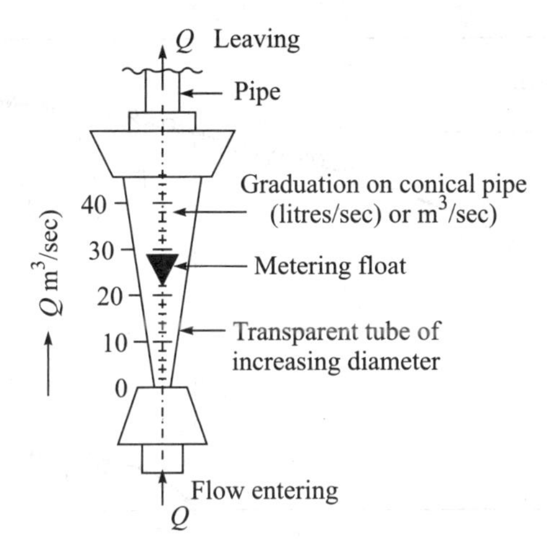

Figure 22.6 Rotameter method.

22.10 RATING CURVE METHOD: WATER STAGE RECORDER

This is the commonest and easiest method of recording stage or depth of water level in a river. Various water level recorders are the staff gauge shown in Figure 22.7 or the weight gauge or self-registering gauge. The gauges are fitted on a solid foundation shown in the enlarged section, or on abutments or piers if a bridge exists at the gauge site. Graduations on the gauge should be bold enough so as to enable the observer to read from a distance. From the recorded stage H on the gauge, the discharge is obtained from the previously calibrated rating curve at the gauge site (Figure 22.8). Figure 22.8 shows that for a gauge height of 200 m, the corresponding

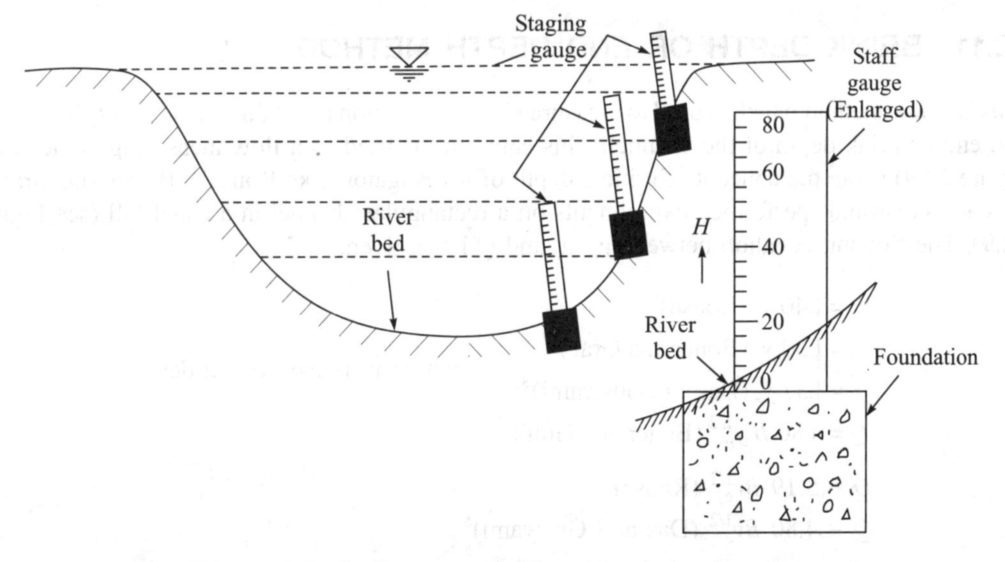

Figure 22.7 Staff gauges at different levels for deep river.

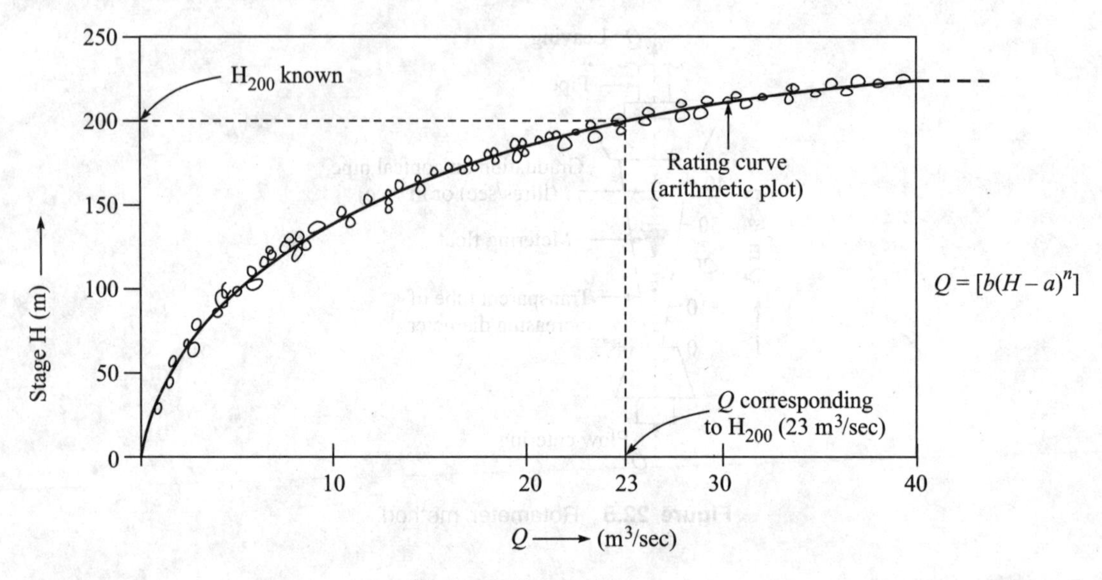

Figure 22.8 Rating curve.

discharge Q is 23 m^3/sec. This rating curve has lot of applications in river engineering. Statistical methods are employed to extrapolate the rating curve to measure high discharge. It is developed on the unique relationship between Q and H by the equation

$$Q = b(H - a)^n \tag{22.8}$$

where b, a and exponent n are calibrated from the previous plot of the rating curve at a particular gauge site.

22.11 BRINK DEPTH OR END DEPTH METHOD

This is a new method which is used to measure Q in an irrigation canal fall. Let y_c and y_e be critical and end or brink depth of the channel. This canal fall is used as a flow measuring device (see Figure 22.9) using the concept of critical depth of investigators like Rouse[11], Bouer and Graf[12], Das and Goswami[8] performed experiments on a rectangular channel in a canal fall (see Figure 22.9). The flowing equation between y_e, y_c and Q have given:

$$\left.\begin{aligned} y_c &= 1.40\, y_e \text{ (Rouse)}^{11} \\ y_c &= 1.30\, y_e \text{ (Bouer and Graf)}^{12} \\ y_c &= 1.37\, y_e \text{ (Das and Goswami)}^{8} \\ Q &= 4.80\, B\, y_n^{1/2} \text{ (Bouer and Graf)}^{12} \end{aligned}\right\} \text{ where } y_n \text{ is the normal depth}$$

$$Q = 5.19\; By_e^{1/2} \text{ (Rouse)}^{11}$$
$$Q = 4.80\; By_e^{1/2} \text{ (Das and Goswami)}^{8}$$

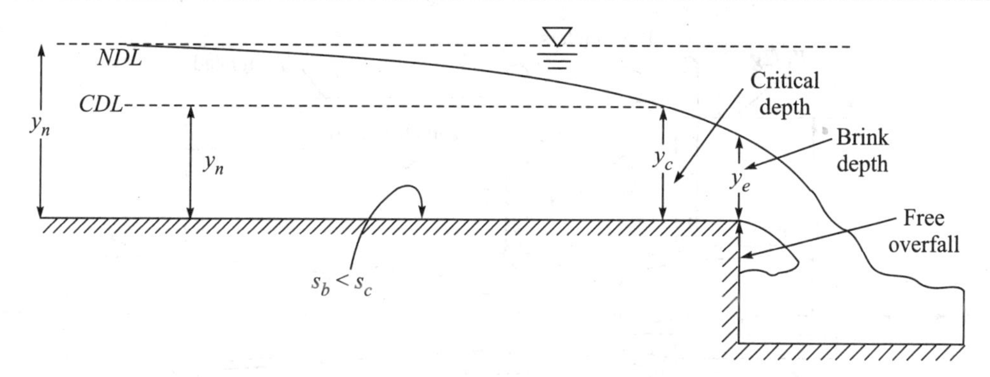

Figure 22.9 Brink or end depth method.

Thus directly measuring the end depth y_e, Q can be determined by using one of the formulae.

22.12 OBSTRUCTION METER METHOD: VENTURIMETER, ORIFICE METER AND NOZZLE METER

Conventional obstruction meters include the venturimeter, orifice meter and nozzle meter. These are based on applications of Bernoulli's equation. Due to obstruction the velocity increases, pressure fall as Bernoulli's equation. The difference of pressure at the inlet and the outlet is normally measured by the mercury manometer. Eventually applying Bernoulli's equation at inlet and throat

The equation discharge Q is:

$$Q = C\frac{A_1 A_2}{\sqrt{A_1^2 - A_2^2}}\sqrt{2gh} \tag{22.9}$$

where C, constant of meter, varies from 0.97 to 0.99.

A_1 and A_2 are the area at inlet and the throat, respectively, h is the difference of pressure head at the beta inlet and the throat as shown in Figure 22.10.

22.13 CONTRACTION METER METHOD BASED ON CHANNEL CONTROL: WEIRS, NOTCHES, SPILLWAY DAMS, CULVERTS, SLUICES

The use of above contraction meters to measure Q in a hydraulic laboratory is well known. They are also used in the field with some limitations. For example, notches are used to measure the discharge from a large tank for sprinkler and drip irrigation. Weirs are used to measure Q in canals used for irrigation in crop fields. In large natural streams, they have some limitations due to large head, debris, sediment load and back water effect. Other structures like spillway dams, culverts and sluices, whose primary functions are distinct from flow measurement, can conveniently be utilised in the field for discharge measurement.

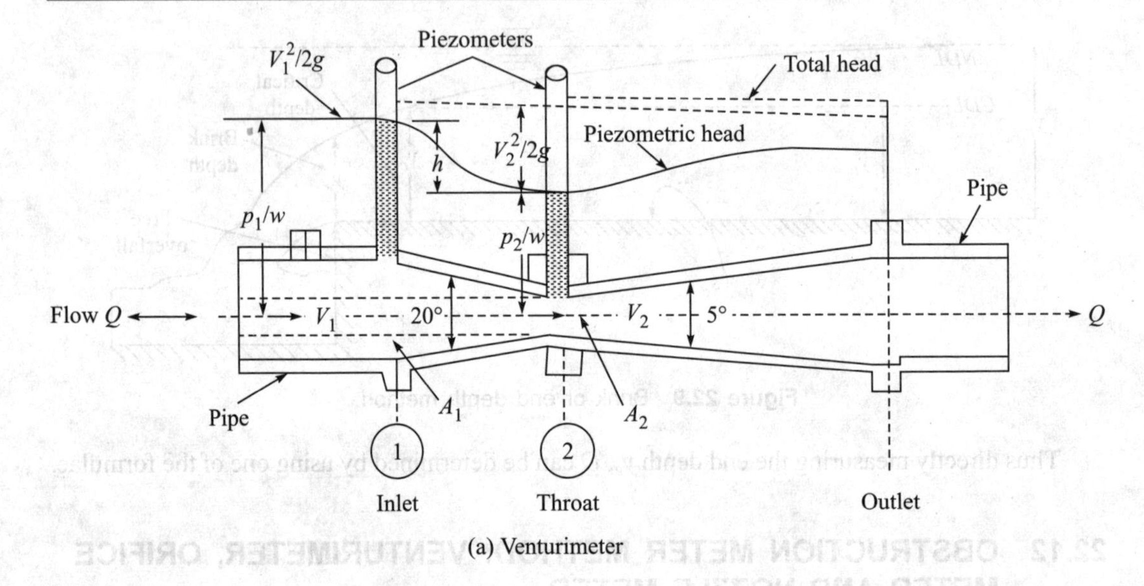

(a) Venturimeter

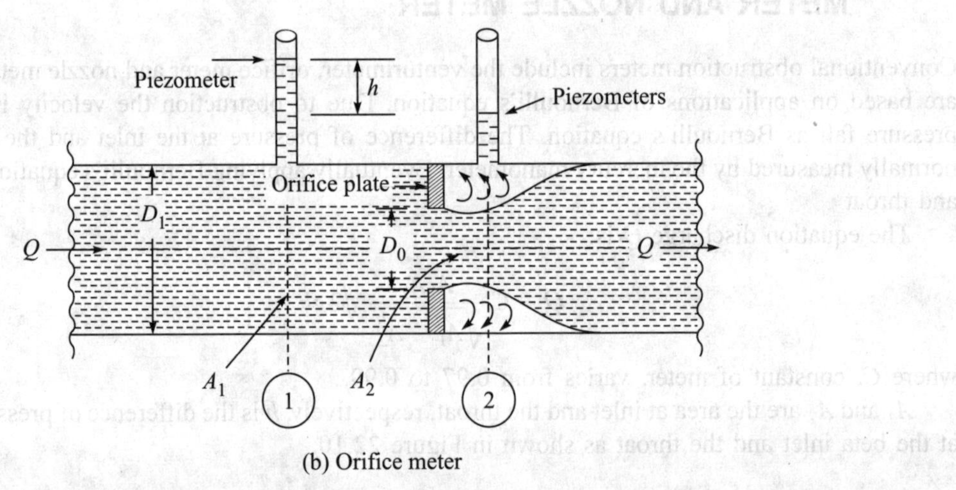

(b) Orifice meter

Figure 22.10 Two different obstruction meters.

Various equations used for above contraction meters are:

(a) Rectangular weir or notch, $Q = \frac{2}{3} C_d \sqrt{2g}\ H^{3/2}$ (22.10)

(b) Triangular weir or notch, $Q = \frac{8}{15} C_d \sqrt{2g}\ \tan \theta/2\ H^{5/2}$ (22.11)

(c) Trapezoidal notch or weir, $Q = C_d \sqrt{2g}\ H^{3/2} \left[\frac{2}{3} L + \frac{8}{15} H \tan \theta/2 \right]$ (22.12)

(d) Broad-crested weir, $Q = C_d L h \sqrt{2g(H - h)}$ (22.13)

(e) Spillway dam, $Q = C_d L\left(H + \frac{V^2}{2g}\right)^{3/2}$ (22.14)

(f) Sluices $Q = C_d La\sqrt{2g(H_1 - C_c a)}$ (22.15)

The various parameters of above equations are defined in Figures 22.11, 22.12, 22.13, 22.14, 22.15 and 22.16.

$$Q = \frac{2}{3} C_d \sqrt{2g}\ LH^{3/2}$$

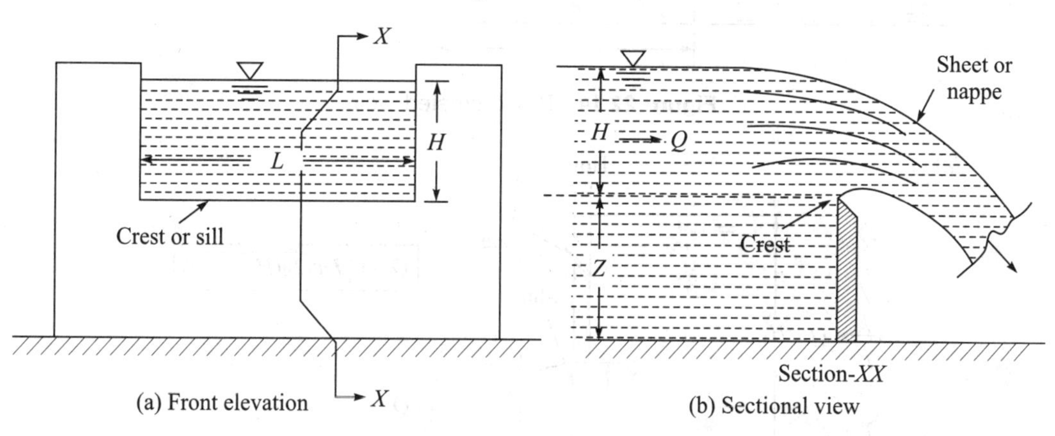

Figure 22.11 Thin plated rectangular weir or notch.

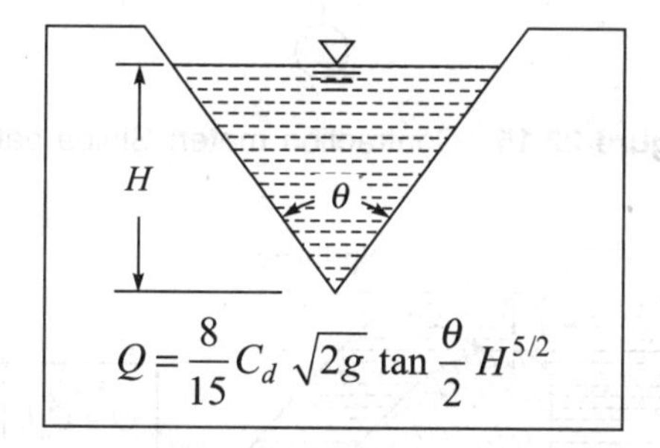

Figure 22.12 Triangular notch or weir.

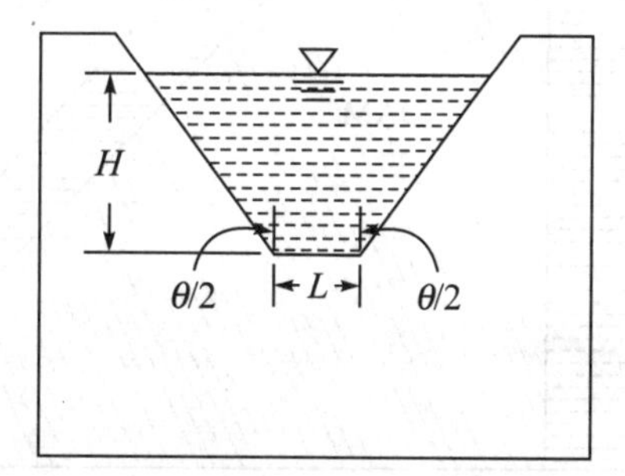

Figure 22.13 Trapezoidal notch or weir.

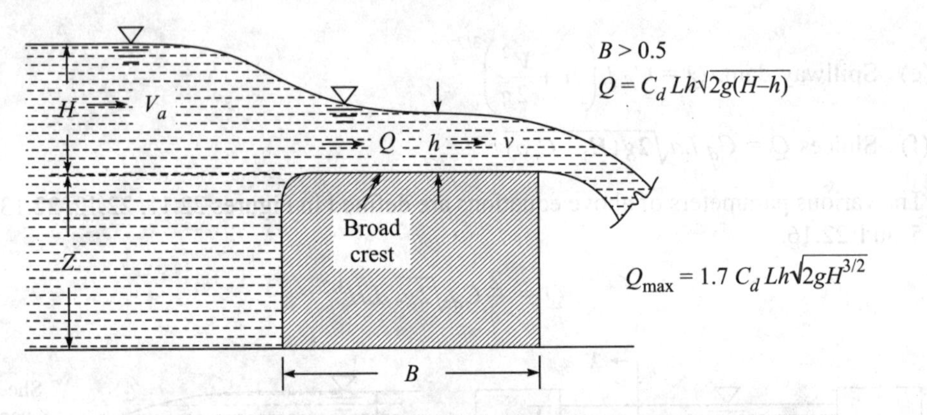

Figure 22.14 Broad crested weir.

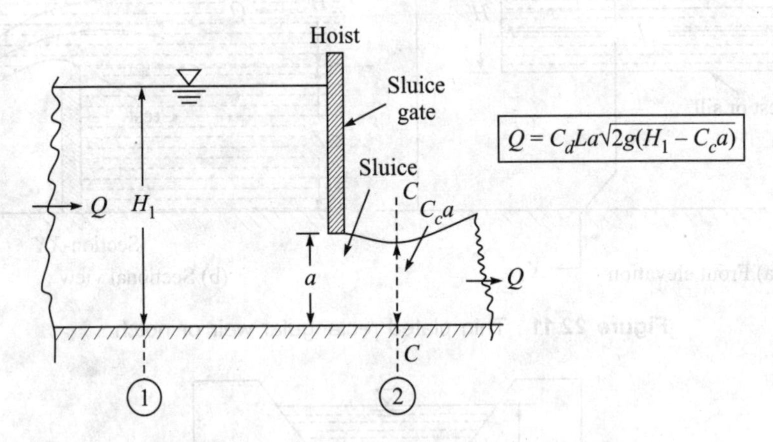

Figure 22.15 Contraction meter: Sluice gate.

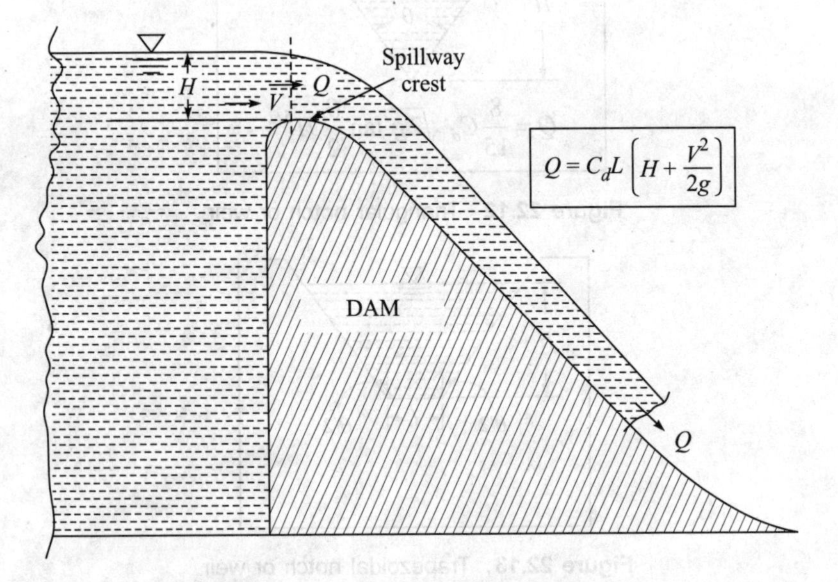

Figure 22.16 Contraction meter: Spillway dam.

22.14 CRITICAL FLOW FLUMES BASED ON CHANNEL TRANSITION

Critical flow flumes are developed by channel control creating critical flow conditions or sections. The hydraulic principle is used here besides the critical depth and minimum specific energy concepts. Although they are very good for a laboratory channel, they can satisfactorily be applied to artificial canals made for irrigation, water supply and power generation. Such flumes are shown in Figure 22.17(a), (b), and (c) with their parameters.

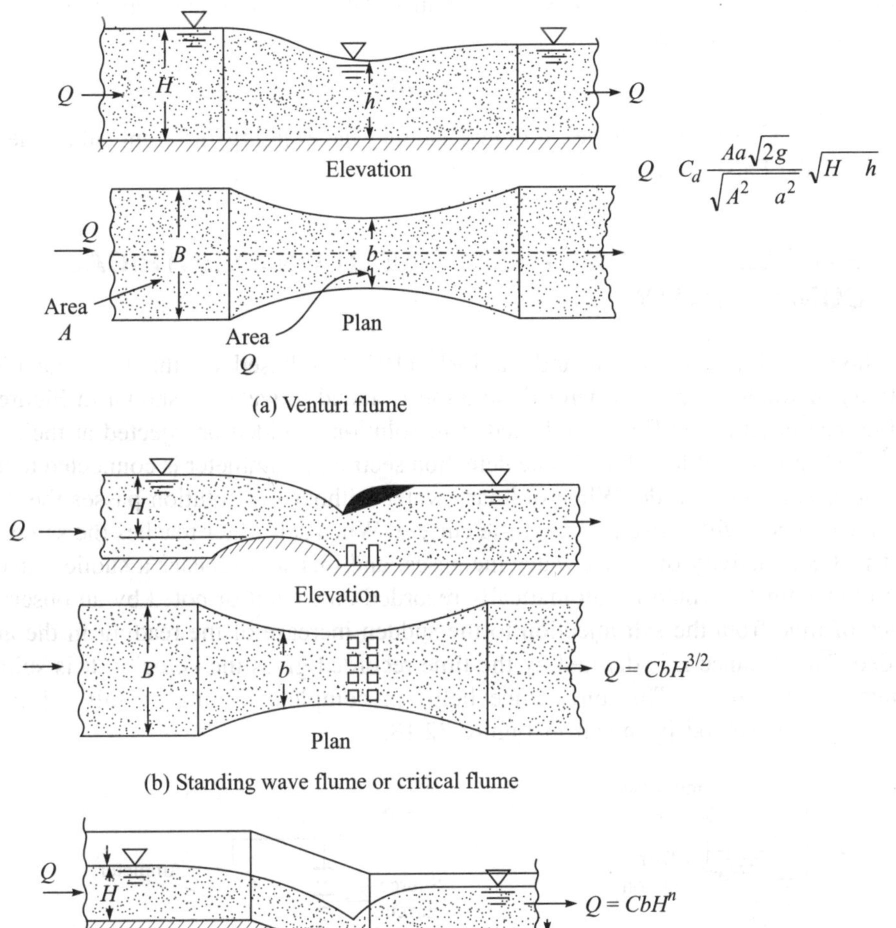

(a) Venturi flume

(b) Standing wave flume or critical flume

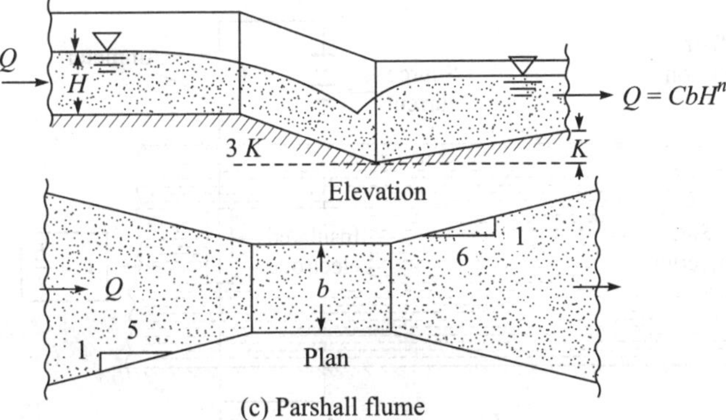

(c) Parshall flume

Figure 22.17 Critical flow flumes: Channel transition method.

General equations of discharges are essentially of the same pattern in all the flumes.

(a) Venturiflumes $Q = C_d \dfrac{A_a}{\sqrt{A^2 - a^2}} \sqrt{2g(H - h)}$ (22.16)

(b) Critical depth flumes

$$Q = C_d\, bH^{3/2} \tag{22.17}$$

(c) Parshal flume developed by R.L. Parshall in USA in 1920, and reported by Colarado Experimental Station[10]

$$Q = C_d\, bH^n \tag{22.18}$$

The values of C, b and n are to be calibrated and they have slightly different values, depending on the size of the Parshall flumes.

22.15 SALT VELOCITY METHOD: BASED ON ELECTRICAL CONDUCTIVITY

This method developed by Allen and Taylor[14] (1923) is based on the fact that electrical conductivity of water increases when salt solution is added to water as shown in Figure 22.18. A suitable length (L) of the flow is selected. Salt solution is added or injected at the source as shown at the beginning of length L. At the detection section, an ammeter is connected to a source current and insulated electrode. When ordinary water without salt solution passes the detection section, the ammeter will register a very little current. When salt water reaches the same section, the electrical conductivity of salted water increases, and this will register a sudden increase of current in the ammeter which is automatically recorded on a chart or noted by an observer. The difference of time from the salt injection to the sudden increase of the reading in the ammeter is recorded. The distance L is divided by the time recorded gives the velocity. It is suitable for a turbulent small stream. This turbulence helps in complete mixing of salt solution. The mechanism of the method is shown in Figure 22.18.

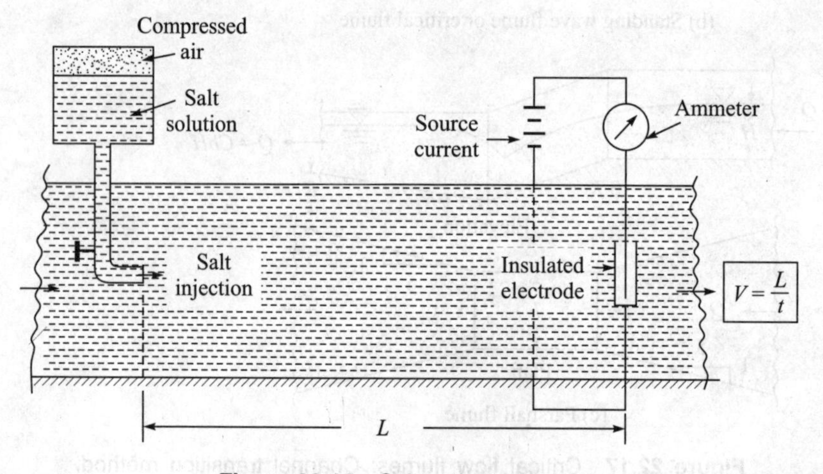

Figure 22.18 Salt velocity method.

22.16 DILUTION TECHNIQUE: TRACER TECHNIQUE

This method is known as the tracer technique or chemical method as it is based on the continuity principle applied to the tracer, which is allowed to mix completely with the flow. In this method, a tracer (common salt or sodium dichromate or radio active material) of known rate (qm^3/sec) and known concentration C_t is injected into the stream as shown in Figure 22.19. The solution will be diluted by the turbulent action of the stream. The concentration of tracer at *d/s* section is determined. If Q is the flow in the river, C_o is the concentration of the tracer initially present in the water and C_s is the concentration measured *d/s* after mixing then, and applying continuity,

$$QC_o + qC_t = (Q + q)C_s$$

Solving for Q,

$$Q = \left(\frac{C_s - C_b}{C_o - C_s}\right)q \tag{20.19}$$

This is based on steady flow assumption in the stream. If the flow is unsteady, there will be a change in storage volume in the reach under consideration.

However, this method has a major advantage of estimating discharges in a absolute way. It is particularly attractive for a small turbulent stream.

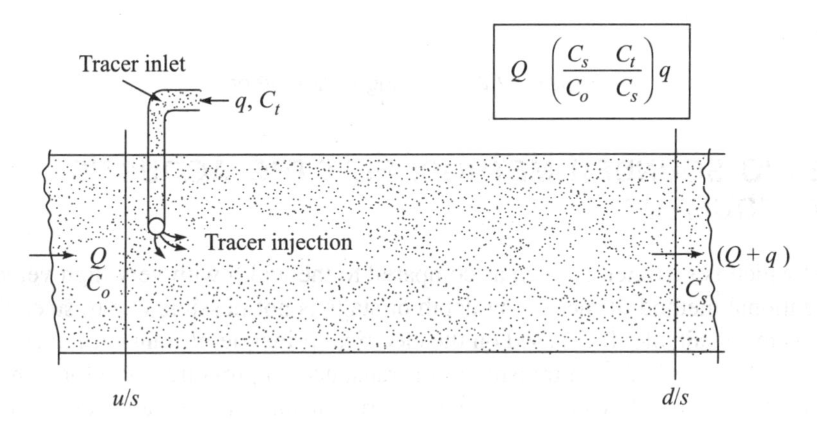

Figure 22.19 Dilution technique or tracer technique.

22.17 MOVING BOAT METHOD

This method has been developed by the U.S. Geological Survey[13]. It has some special advantages over the standard current meter method, when river is large and in spate. It is also suitable for use in estuaries. Let V_b be the velocity of the moving boat (Figure 22.20) with the current meter at right angle to the velocity V_f. Then the current meter will align itself in the direction of the resultant velocity V_R. Now from Figure 22.20, $V_b = V_R \cos \theta$ and $V_f = V_R \sin \theta$. If the time of transit between the two verticals placed at a distance W is Δt, V_b, $\Delta t = W$. If the flow between the two vertical depths y_1 and y_2 is ΔQ then

$$\Delta Q = [(y_1 + y_2)/2]WV_f$$

Putting the value of W above

$$\Delta Q = [(y_1 + y_2)/2] V_b \, V_f \, \Delta t$$

again substituting the values of V_f and V_b,

$$\Delta Q = [(y_1 + y_2)/2] \, V_R^2 \sin \theta \cdot \cos \theta \, \Delta t$$

$$\therefore \qquad \text{Total } Q = \Sigma \Delta Q = \sum_{i=1}^{n} \left(\frac{y_i + y_{i+1}}{2} \right) V_R^2 \sin \theta \cos \Delta t \qquad (22.20)$$

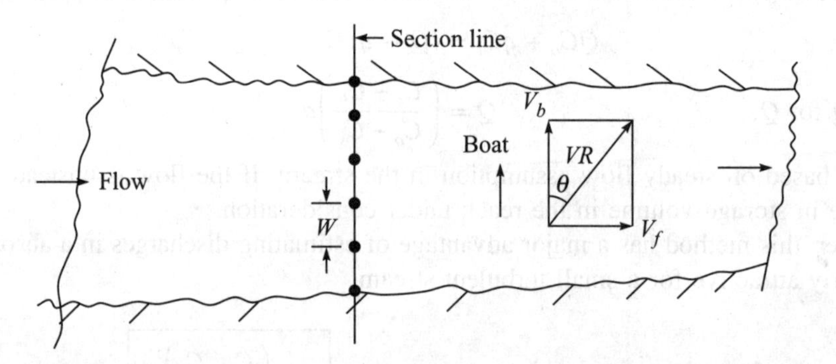

Figure 22.20 Moving boat method.

22.18 ECHO SOUNDER METHOD: ELECTRO ACOUSTIC INSTRUMENT

The channel which is sufficiently deep as compared to the width with very high velocity wherein the conventional method of measuring depth or stage is not possible due to steep topography, this method is most suitable. It is based on the electrical principle of transmitting from the water surface level to the river bed by a transmitter or transducer, a pressure wave for a short duration and receiving back the reflected wave in the form of an echo on a receiver. The time of transmission and reception are plotted automatically. From this time, if one knows the velocity of sound in water media ($\simeq$ 1470 m/sec), the depth of flow can be known.

22.19 HYDRAULIC MODEL METHOD

Hydraulic models are useful in the estimation of discharges in rivers particularly at high flow. Very high flows are difficult to measure. Low flood data are available if a sufficiently big size model (scale near to 1:50) is constructed. The stage-discharge relationship, i.e. the rating curve of the model is established by experiments and its calibration is checked by available low flood data of the river. If those data come very close to the model rating curve as shown in Figure 22.21, then extrapolating the model rating curve, high discharge of the river where gauging is extremely impossible, may be determined. It was successfully applied in Shirat Gorge in England.

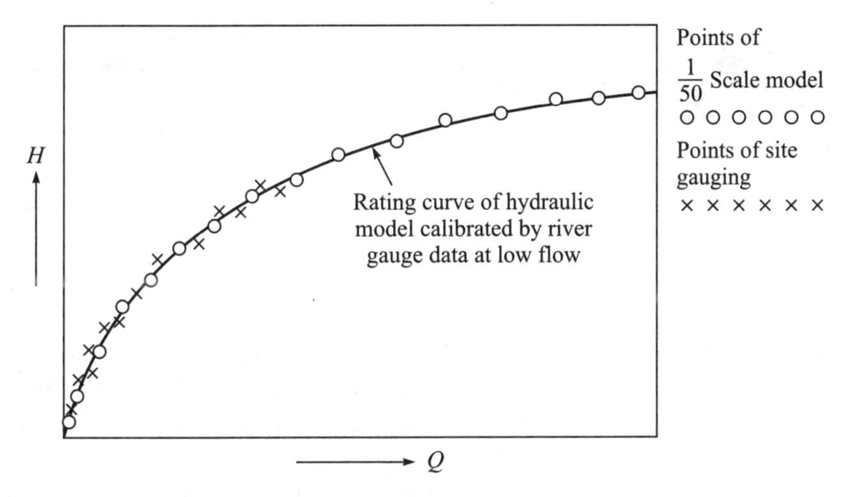

Figure 22.21 Rating Curve Hydraulic Model Method.

22.20 INERTIA PRESSURE METHOD

This method was devised by N.R. Gibson and was used to determine the average velocity of the liquid flowing in a long pipeline. This method involves in the case of rapidly closing valve at the *d/s* end of a pipe and recording the diagram of pressure p against time t for a point immediately *u/s* the valve.

Then
$$\int p dt = \frac{wLV}{g} \tag{22.21}$$

The value of the integral $\int p dt$ is determined graphically and the mean velocity of flow V may then be calculated from Equation (22.21).

22.21 ULTRASONIC FLOW METER METHOD: USE OF TRANSDUCERS

This is essentially an area velocity method wherein the average velocity is measured by using ultrasonic signals with the help of two transducers A, B in Figure (22.22). The method was first reported by Swengal (1955).

As shown in Figure 22.22, the transducers A and B are fixed at the same level h above the bed on either side of the channel. These transducers can receive as well as send ultrasonic signals. Let A send an ultrasonic signal to be received by B after an elapse of time t_1 sec. Similarly, B sends a signal to be received by A after an elapse of time t_2 sec. If C is the velocity of sound in water, then

$$t_1 = \frac{L}{(C + V_p)} \quad \text{where } L \text{ is the distance between } A \text{ and } B$$

and V_p is the component of velocity of sound path, i.e. $V_p = V \cos \theta$, where V is the average velocity of flow.

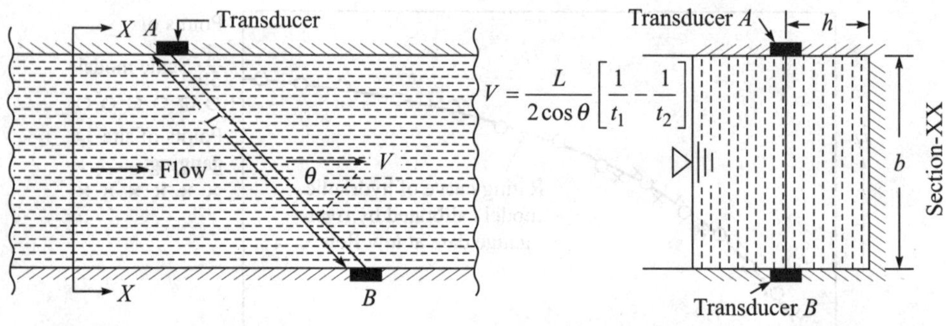

(a) Plan of the flow channel with the transducers

(b) Sectional view of the channel showing transducers in same level

Figure 22.22 Ultrasonic Flow Meter: Transducers.

and $\quad t_2 = L/(C - V_p)$, $\; V_p$ is negative as opposite to the direction of C.

Now $\quad \frac{1}{t_1} - \frac{1}{t_2} = \frac{2V_p}{L} = 2V \cos\theta/L$

$$\therefore \quad V = \frac{L}{2\cos\theta}\left[\frac{1}{t_1} - \frac{1}{t_2}\right] \tag{22.22}$$

which gives the velocity at height h from the bed. Similarly, velocity at different heights can be obtained by fixing transducers at different levels and thus average velocity V may be obtained to get $Q = A \cdot V$. Herschy and Loosemor[23] presented a paper on this ultrasonic method in the symposium of university of Reading (1974).

22.22 LASER DOPPLER ANEMOMETER (LDA) METHOD

It is very recent scientic method or device based on the use of a laser beam and the Doppler effect, and hence is called Laser Doppler Anemometer (LDA) Method (Figure 22.23).

In LDA, a low power laser beam is focused on a fluid in motion through a transparent duct. A portion of the beam is transmitted through the stream and some part of the beam is scattered by solid particles present in water. The component of the beam passing straight through the medium is colliminated by a lens outside the duct and reflected by a mirror to beam splitter. The scattered beam is condensed through another lens, which is also focused on the beam splitter. In other words, the two components of the original laser beam now combine at beam splitter. The reflecting mirror position is so adjusted that two components travel identical path lengths, whereas the direct transmitted component of the beam maintains the same frequency of the laser beam emanating from the source, the frequency of scattered beam is slightly different. The difference arises due to the well-known Doppler effect and is proportional to the velocity of scattering particles. In other words, the difference of frequency of the two beams combining at the splitter is a measure of the velocity of scattering particles. Since the particle moves with the same velocity as the water, the Doppler frequency shift is also a measure of liquid velocity. The combined beams are sensed by a photomultiplier, the output of which is connected to a frequency analyser for measuring the Doppler frequency. Thus the velocity of the flowing liquid in the duct is measured.

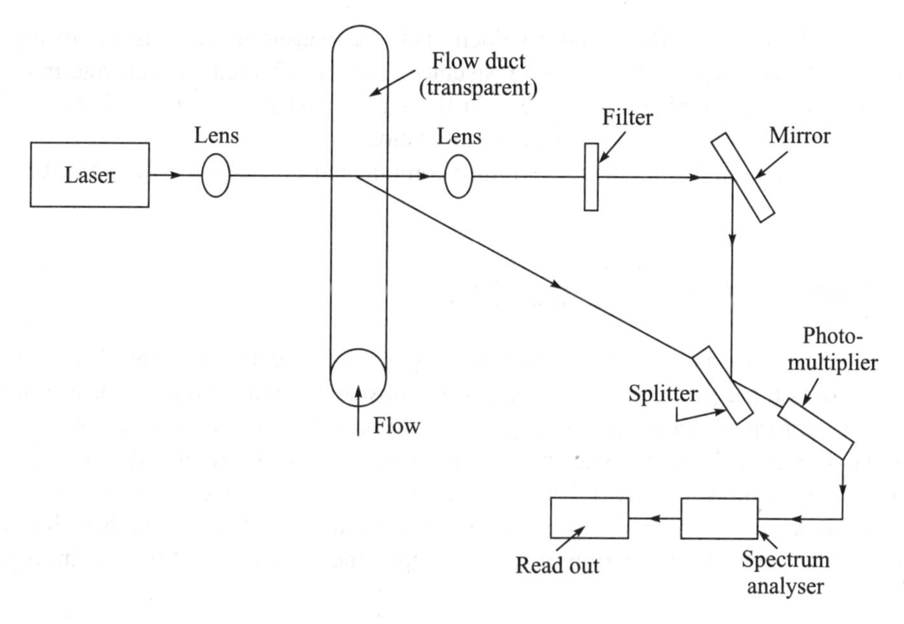

Figure 22.23 Laser Doppler Anemometer (LDA) layout.

22.23 HOT WIRE ANEMOMETER METHOD

This method is useful for compressible fluid flow measurement. Yet the use of this method in case of water to measure its velocity, especially in the laboratory, is quite fruitful. Iowa Institute of Hydraulic Research, USA, developed this anemometer to use it in the laboratory.

In an instrument (Figure 22.24) used to measure velocity. It consists of a platinum or tungsten wire of diameter 5×10^{-3} to 8×10^{-3} mm and about 6 mm long. The wire is mounted on prongs placed perpendicular to the flow. The temperature of the wire is raised relative to

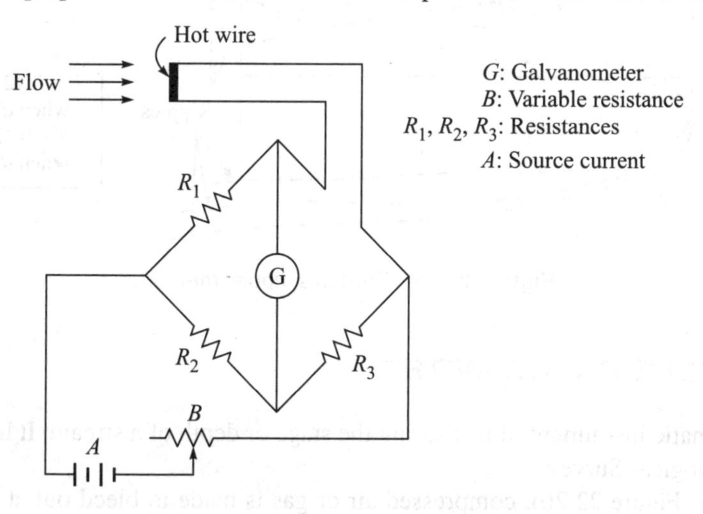

Figure 22.24 Hot wire anemometer.

temperature of the liquid or fluid whose velocity is to be measured. The passage of liquid flow over the wire, cools it which changes the resistance. This is reflected by a change in current or voltage across the wire which can be seen in the galvanometer shown in Figure 22.24. This change in current or voltage is calibrated against velocity.

The instrumentation of this device is rather complicated in use in the field but it is suitable for laboratory use.

22.24 THRUPP'S RIPPLE METHOD

It is an approximate but rapid method of determining surface velocity, given by E.C. Thrupp. This method is based on the principle that when a small obstacle is placed on the surface of flowing water ripples are formed when velocity exceeds 0.2 m/sec and as the velocity increases the angle between the diverging lines of ripples become more acute. In order to afford a simple means of divergence rate, Thrupp used two 7.5 cm wire nails of about 0.3 cm diameter at a definite distance apart. According to him, if l is the distance (Figure 22.25) from the base line between the nails and the point of intersection of the last ripple, the velocity V_s of the stream is given by him:

$$V_s = 0.12 + 0.025\, l \text{ for } d = 15 \text{ cm} \tag{22.23}$$
$$V_s = 0.12 + 0.034\, l \text{ for } d = 10 \text{ cm} \tag{22.24}$$

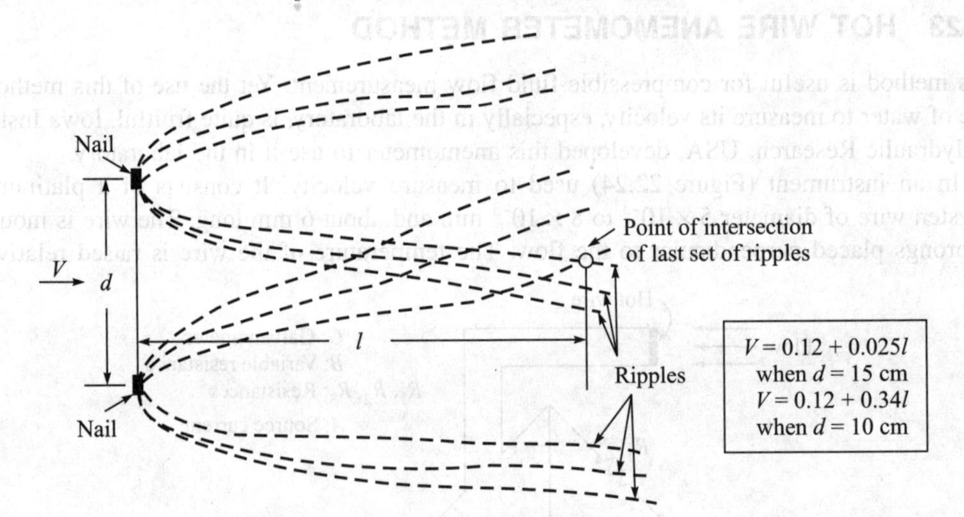

Figure 22.25 Thrupp's ripple method.

22.25 BUBBLE GAUGE METER

This is an automatic instrument of measuring the stage or depth of a stream. It has been modified by the US Geological Survey.

In this type (Figure 22.26), compressed air or gas is made to bleed out at a very small rate through an outlet placed at the bottom of the river as shown in the figure. A pressure gauge measures the gas pressure p, and depth H obtained by $H = \frac{p}{w}$. It consists of a gas bottle, servo manometer, transistor, central gas surge system and a recorder as shown in the figure.

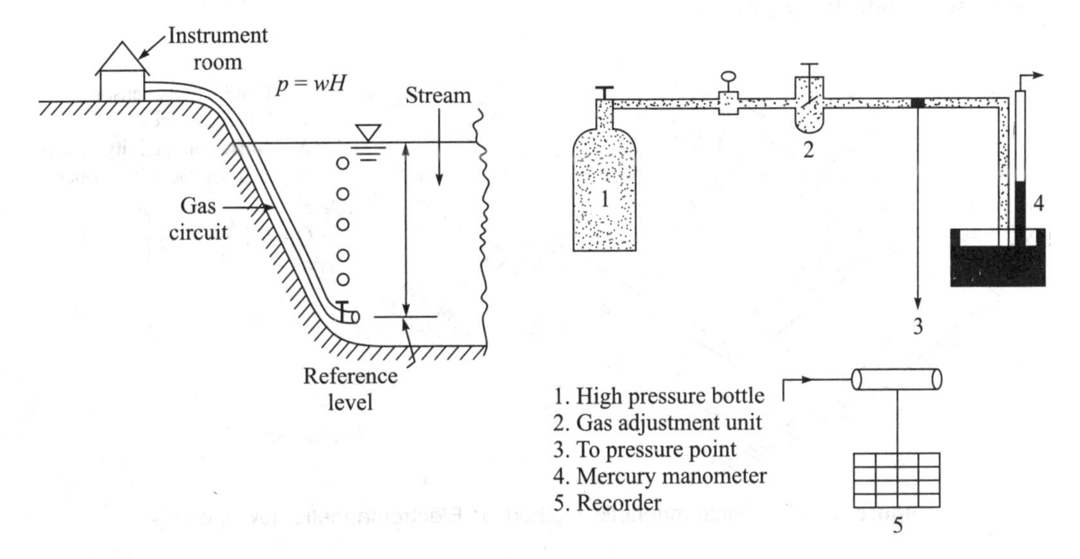

Figure 22.26 Bubble gauge method.

22.26 ELECTROMAGNETIC FLOW METER: ELECTROMAGNETIC METHOD

The main feature of this method is a large coil carrying a current which is buried below the river bed to generate a vertical magnetic field across the full river width.

It is based on Faraday's[15] principle that an emf is induced in the conductor (i.e. the water of the river in this case) when it cuts a normal magnetic field. Electrodes provided at the sides of the channel section measure the small voltage produced due to the flow of water in the channel. It has been found that the signal output E will be of the order of milli volts and is related to discharge Q as:

$$Q = K_1\left[\frac{Ed}{I} + K_2\right]^n \tag{22.25}$$

where d is the depth, Q is the flow, I is the current in the coil, and n, K_1, K_2 are the system constants.

This method involves sophisticated and expensive experimentation and has been successfully tried in a number of installations. It is suitable for rivers upto 100 m width. It gives a very accurate discharge with only upto ±3% error. Velocity can even be measured and detected upto 0.005 m/sec. The U.S. Geological Survey used one of these instruments, furnished by the US Navy, to conduct experiments at Jacksonville, Florida and eminently satisfactory results were obtained. Figure 22.27 represents a sketch of this electromagnetic flow meter.

C = Conductivity sensor
V = Voltage probes
N = Noise conductivity probes
B = Bed conductivity probes.

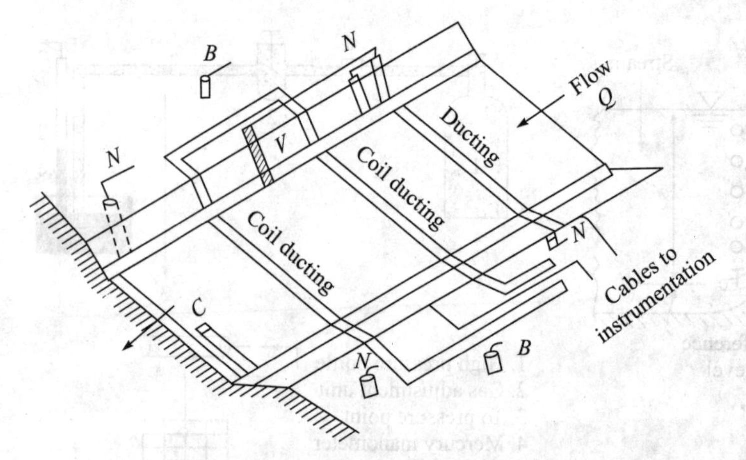

Figure 22.27 Electromagnetic method or Electromagnetic flow meter.

22.27 DYNAMOMETER

This is developed on the basis of translation of the momentum force of the stream velocity into either deflection or stress. This is measured and calibrated against velocity or discharge. The deflection principle has been incorporated into the Keeler meter and successfully applied to the measurement of outflow from a lake in New Hampshire. Based on the same principle, a hydrometric pendulum, developed at the Hydraulic Laboratory of Delft, has been utilised to measure velocity in the extremely turbulent rivers of Nigeria.

EXAMPLE 22.1 The following data in Table 22.1 were obtained for a river of 28 m wide at a gauging station. Current meter calibration equation is $V = 0.5\ N + 0.03$. Compute the discharge.

Table 22.1 Data of Example 22.1

Distance from one end of water surface (m)	*Depth in (m)*	*Current meter data at 0.6 times of depth*	
		Revolution	*Sec (t)*
4	1.5	10	25
8	3.5	20	40
12	5.0	40	50
16	6.5	45	60
20	6.0	42	50
24	3.5	30	50
28	1.5	18	36

Solution:

Diatance from one end of water surface (m)	*Depth* (m)	*Current meter data*		$N = R/t$ *(rps)*	$V = 0.5\,N + 0.03$ *(Average) m/sec*	*Discharge in Strip* $\Delta Q = AV$ (m^3/sec)
		Revolution R	*Time (t) sec*			
4	1.5	10	25	0.4	0.23	$\Delta Q = \frac{1}{2} \times 4 \times 1.5 \times .23 = 0.69$
8	3.5	20	40	0.5	0.28	$\Delta Q = 4 \times \left(\frac{1.5+3.5}{2}\right) \times 0.28 = 5.6$
12	5.0	40	50	0.8	0.43	$\Delta Q = 4\left(\frac{3.5+5}{2}\right) \times 0.43 = 7.31$
16	6.5	45	60	0.75	0.405	$\Delta Q = 4\left(\frac{5.0+6.5}{2}\right) \times 0.405 = 9.315$
20	6.0	42	50	0.82	0.44	$\Delta Q = 4\left(\frac{6.5+6.0}{2}\right) \times 0.44 = 11.0$
24	3.5	30	50	0.6	0.33	$\Delta Q = 4\left(\frac{6.0+3.5}{2}\right) \times 0.33 = 6.27$
28	1.5	18	36	0.5	0.28	$\Delta Q = \frac{1}{2}\ 4 \times 1.5 \times .28 = 0.84$

* Area at sides is triangular, at middle, all are trapezoidal.

$Q = 41.025$ m^3/sec

Ans. $Q = 41.025$ m^3/sec

EXAMPLE 22.2 A surface float takes 15 seconds to travel a straight distance of 30 m. What is the approximate mean velocity of flow?

Solution: $L = 30$ m

$T = 15$ seconds

$\therefore$ Surface velocity

$$V_s = \frac{L}{T}$$

$$= \frac{30}{15} \text{ m/sec}$$

$$= 2 \text{ m/sec}$$

To obtain mean or average velocity, a reduction factor is to be taken. The factor varies from 0.79 to 0.92. Taking the average reduction factor,

$$V = \left[\frac{(0.79 + 0.92)}{2} \times 2\right] \text{ m/sec}$$

or

$$V = (0.855 \times 2) = 1.71 \text{ m/sec.}$$ **Ans.**

Again using Equation (22.7)

$$V = 0.8529 \times V_s + 0.0085$$

$$= 0.8529 \times 2 + 0.0085$$

$$= 1.7143 \text{ m/sec}$$ **Ans.**

EXAMPLE 22.3 During a high flow water surface elevations of a river following data at section *A* and *B* are recorded. The distance between *A* and *B* is 10 km. Estimate the discharge.

Section	*Water surface elevation (m)*	*Area of section (m^2)*	*Hydraulic Radius (m)*	*Remarks*
A	50.8	73.3	2.75	*A* is upstream of *B*
B	50.5	93.4	3.1	Mannings $n = 0.02$

Solution: Slope of water surface $= \dfrac{(50.8 - 50.5)\,\text{m}}{(10 \times 1000)\,\text{m}}$

$$= 0.00003$$

$$\text{Average Area } A = \left(\frac{73.3 + 93.4}{2}\right) \text{m}^2$$

$$= 83.35 \text{ m}^2$$

$$\text{Applying equation and Average } R = \left(\frac{2.75 + 3.1}{2}\right) \text{m}$$

$$= 2.925 \text{ m}$$

$$Q = AV$$

or
$$Q = A\,\frac{1}{n}\left(\frac{A}{P}\right)^{2/3} s^{1/2}$$

or
$$Q = A\frac{1}{n}(R)^{2/3}s^{1/2}$$

or
$$Q = \left[83.35 \times \frac{1}{0.02}(2.925)^{2/3}(0.00003)^{1/2}\right]$$
$$= 46.686 \text{ m}^3/\text{sec}$$ **Ans.**

EXAMPLE 22.4 A Pitot tube is used to measure the discharge in river. When the tube was placed at a depth approximately 6 times of the depth of flow, water rises to height of 0.75 m above the water surface. If the average area of the river at that section is 40 m^2, estimate the discharge. Assume a suitable value of tube constant.

Solution: The height of water in the Pitot tube above the water level is h. h = 0.75, A = 40 m^2 From Equation 22.5, average velocity V is

$$V = C\sqrt{2gh}$$

Here $h = 0.75$ m

The reasonable value of $C = 0.8$

$\therefore$ $V = 0.8\sqrt{2 \times 9.81 \times 0.75}$ m/sec

$= 3.06881$ m/sec

$\therefore$ $Q = AV = (40 \times 3.06881)$ m^3/sec

$\therefore$ $Q = 122.752$ m^3/sec **Ans.**

22.28 CONCLUSION

This chapter briefly presents the different methods of discharge measurements along with their theoretical principles, techniques and instruments used from the old days to the modern age of electronic usage. The time, perhaps, is not very far when scientists will develop a package computer model connecting the computer to the flow system, and the flow or velocity will immediately be displayed in the monitor. It is hoped that hydrometry will very soon come under the purview of the modern computer.

PROBLEMS

22.1. A stream is assumed to be trapezoidal in cross-section with the base width of 12 m and side slope 2 horizontal: 1 vertical in a reach of 8 km. During flood time, high water levels recorded at both ends of the reach are as follows. (Table 22.2)

Table 22.2 Data of Problem 22.1

Section	*Elevation of bed (m)*	*Water surface*
Upstream	100.20	102.70
Downstream	98.60	101.30

If Manning's n = 0.03, estimate the discharge in the stream Table 22.2 Data of Problem 22.1. **Ans.** 30.18 m^3/sec

22.2. Draw the rating curve from the stage-discharge relationship from Table 22.3. Assume the value of zero discharge to be 20.50 m^3/sec. Determine the stage of the river corresponding to a discharge of 2600 m^3/sec.

Table 22.3 Data of Problem 22.2

Stage (m)	21.95	22.45	22.80	23.00	23.40	23.75	23.65	24.05	24.55	24.85	25.40	25.15	25.55	25.90
Discharge (m^3/sec)	100	220	290	400	490	500	640	780	1010	1220	1300	1420	1550	1760

Ans. 26.90 m^3/sec

22.3. In the moving boat method of discharge measurement the magnitude V_R and direction θ of the Velocity of the stream relative to the moving boat are measured. The depth of the stream also simultaneously measured and recorded in Table 22.4. Estimate the discharge in a river that gave the following moving boat data. Assume the mean Velocity is 0.95 times the surface Velocity measured by the instument.

Table 22.4 Data of Problem 22.3

Section	V_R *(m/sec)*	θ *(degrees)*	*Depth (m)*	*Remark*
0	–	–	–	Right bank
1	1.75	55	1.8	θ is the
2	1.84	57	2.5	angle
3	2.00	60	3.5	made by V_R
4	2.28	64	3.8	with boat
5	2.30	65	4.0	direction
6	2.20	63	3.8	
7	2.00	60	3.0	The various
8	1.84	57	2.5	sections are spaced at a
9	1.70	54	2.0	constant distance of 75 m apart
10	–	–	–	Left bank

22.4. In a stream gauging operation, observations recorded with a current are tabulated below (Table 22.5). Estimate discharge in the river if the current metre rating is $V = 0.8\ N + 0.05$, V is m/sec N is the revolution/sec.

Table 22.5 Data of Problem 22.4

Distance from the bank (m)	*Depth of flow D (m)*	*Depth of meter (m)*	*Revolutions*	*Time*	*Remarks: Current meter*
0.6	1.0	0.6	15	50	at 0.6 D
1.2	4.0	3.2	30	55	0.8 D
		0.8	48	53	0.2 D
2.0	5.5	4.4	40	46	0.8 D
		1.1	60	54	0.2 D
3.0	6.5	5.2	45	48	0.8 D
		1.3	67	52	0.2 D
3.8	4.5	3.6	33	54	0.8 D
		0.9	51	50	0.2 D
4.5	2.5	2.0	26	48	0.8 D
		0.5	44	55	0.2 D
5.0	1.0	0.6	20	47	0.6 D
5.6	0	–	–	–	Other bank

Ans. 14.97 m/sec

REFERENCES

1. Kolupiala, S., *Early History of Hydrometry in the United States* A.S.C.E., Hy. Div., Vol. 86, Hy.1, Jan. 1960.
2. U.S.B.R., *Water Measurement Manual*, 1953.
3. Corbet, D.M., et al., *Stream Gauging Proceeding*, U.S. Geological Survey, Water Supply Paper, 888, 1943.
4. *Handbook of Applied Hydrology*, Ven Tr Chow (Ed.), McGraw-Hill, New York, 1964.
5. Herschy, R.W., *Hydrometry*, Wiley Interscience, John Wiley, Chichester, 1978.
6. Addison, H., *Hydraulic Measurements*, Chapman & Hall, London, 1946.
7. Chow, Ven Te, *Open Channel Hydraulics*, McGraw-Hill, New York, 1959.
8. Das, M.M and M.D. Goswami, "Irrigation Measurement in Free Over Fall Field Channels", *Jour. IWRS*, Roorkee, Vol. 21, 2001.
9. Manning, R., *On the Flow of Water on Open Channel Flow and Pipes*, Trans, ICE, Ireland, Vol. 20, 1891.
10. Parshall, R.L., "The Parshall Measuring Flume", *Colorado Agri. Expt. Station Bull.*, No. 423, March, 1923.
11. Rouse, H., "Discharge Characteristics of Free Overfall", *Jour. Civil Engg.*, ASCE, Vol. 16, No. 4, pp. 257–60, April, 1936.
12. Bouer, S.W. and W.H. Graf, "Free Overfall as a Flow Measuring Device", *Jour. Irrig. Engg. and Drainage Div., Proc., ASCE*, Vol. 97, March, 1971.
13. Smooth, G.F. and C.E. Nayak, *Measurement of Discharge by Moving Boat Method*, US Geological Survey, Tech. Water Resource Invest., bk_3, A 11, 1969.
14. Allen, C.H. and E.A. Taylor, *The Salt Velocity Method of Water Measurement*, Trans., ASCE, Vol. 45, 1923.

15. Faraday, M., *Experimental Resources*, Phil. Trans. Roy. Society, 1832.
16. Dalrymple, T. and M.A. Benson, *Measurement of Peak Discharge by Slope Area Method*, US Geolo. Sur. Tech. Water Resource. Invest. bk$_3$, Chap. Az, 1967.
17. Schuster, J.C., *Measuring Water Velocity by Ultrasonic Flow Meter*, J. Hy. Div., ASCE, Vol. 101, pp. 1503–1517, 1975.
18. Parshall, R.L., *The Improved Venturiflume*, Trans. ASCE. Vol. 80, 1926.
19. Ackers, P., W.R. White, J.A. Perkins, and A.J.M. Harrison, *Weirs and Flumes in Flow Measurement*, John Wiley, New York, 1978.
20. Cone, V.M., "The Venturiflume", *Jour. Agri. Research*, Vol. 9, No. 4, 1917.
21. Subramanya, K. and N. Kumar, "End Depth in Horizontal Circular Free Overfall", *Jour. ICE (India)*, Vol. 73, 1993.
22. Herschy, R.H. and W.R. Loosemor, *Ultrasonic Method of River Flow Measurement*, Symp of River Gauging by Ultrasonic and Electromagnetic Methods, University of Reading, 1974.
23. Benson, M.A., *Measurement of Peak Discharge by Indirect Method*, W.H.O., Tech note 90, Geneva, 1968.
24. Sitter, W.T., *Extension of Rating Curve by Field Survey*, Jour. Hy. Div., ASCE, Vol., 89, March 1963.
25. Langbein, W.B., *Stream Gauging Networks*, Publ. 38, International Assoc. of Hydrology, Sci., General Assembly, 1954.
26. De Marchi, G., *New Experimental Researches on Standing Wave Flume Venturiflume*, Ministry of Agriculture, Paris, France, 1937.
27. Inglis, C.C., Notes on standing wave flumes and flume meter Waffle falls", Public Works Department, Govt. of Bombay, Tech. Paper No. 15, India, 1928.
28. Crump, E.S., *Modeling Irrigation Channels*, Punjab Irrigation Branch Publications, Paper No. 26 and 30 A, Lahore, India, 1922, 1923.
29. Das, M.M. and M.D. Goswami, "Brink Depth Technique for Free Overfall in Field Channel", *Jour, Irrig*. Vol. 37(1), pp. 55-60, Jan–March, 2000.
30. Balloffet, A., "Critical flow meter (Venturiflume)" Paper No. 743, Proc. ASCE, Vol. 81, p. 31, July, 1955.
31. Linford, A., "Venturiflume Flow Meter", *Civ. Engg. and Public Works Review*, London, Vol. 36, No. 44, 1941.
32. Engel, F.V. A.E. "The Venturiflume", *The Engineer*, Vol. 158, pp. 101–107, Aug. 1934.
33. Jameson, A.H., *The Venturiflume and Fffect of Contractions in Open Channels*, Trans., Instn. of Water Engrs., Vol. 30, pp. 19–24, 1925.
34. Jamson, A.H., *Development of Venturiflume*, Water Engg., London, Vol. 32, pp. 105–107, March, 1930.
35. Boss, M.G., *Discharge Measurement Structures*, "International Instn. for Land Reclamation and Improvement, Wageningen, Netherlands, Pub. No. 20, 1976.
36. Frazier, A.H., William Gurin Price and Price Current Meter" *U.S. Nat'l Mus. Bull*, Vol. 252, pp. 37–68, 1967.

Index